Mathematical Thinking

Problem-Solving and Proofs

Second Edition

John P. D'Angelo

Douglas B. West

University of Illinois — Urbana

PRENTICE HALL

Upper Saddle River, NJ 07458

Library of Congress Cataloging-in-Publication Data

D'Angelo, John P.
  Mathematical thinking: problem-solving and proofs / John P. D'Angelo, Douglas B. West.–2nd ed.
    p. cm.
  Includes bibliographical references and index.
  ISBN 0-13-014412-6
    1. Mathematics. 2. Problem solving. I. West, Douglas Brent. II. Title.

QA39.2 .D25 2000
510–dc21

99-050074

Acquisitions Editor: George Lobell
Assistant Vice President of Production and Manufacturing: David W. Riccardi
Executive Managing Editor: Kathleen Schiaparelli
Senior Managing Editor: Linda Mihatov Behrens
Production Editor: Betsy Williams
Manufacturing Buyer: Alan Fischer
Manufacturing Manager: Trudy Pisciotti
Marketing Manager: Melody Marcus
Marketing Assistant: Vince Jansen
Director of Marketing: John Tweeddale
Editorial Assistant/Supplements Editor: Gale Epps
Art Director: Jayne Conte

Printed in the United States of America

10

ISBN 0-13-014412-6

Prentice-Hall International, (UK) Limited, *London*
Prentice-Hall of Australia Pty Limited, *Sydney*
Prentice-Hall Canada Inc., *Toronto*
Prentice-Hall Hispanoamericana, (S.A.) *Mexico*
Prentice-Hall of India Provate Limited, *New Delhi*
Pearson Education Asia Pte. Ltd.
Editora Prentice-Hall do Vrasil, Ltda, *Rio d Janeiro*

To all who enjoy mathematical puzzles,
and to our loved ones,
who tolerate our enjoyment of them

Contents

Preface for the Instructor

This book arose from discussions about the undergraduate mathematics curriculum. We asked several questions. Why do students find it difficult to write proofs? What is the role of discrete mathematics? How can the curriculum better integrate diverse topics? Perhaps most important, why don't students enjoy and appreciate mathematics as much as we might hope?

Upperclass courses in mathematics expose serious gaps in the preparation of students; the difficulties are particularly evident in elementary real analysis courses. Such courses present two obstacles to students. First, the concepts of analysis are subtle; it took mathematicians centuries to understand limits. Second, proofs require both attention to exposition and a different intellectual attitude from computation. The combination of these difficulties defeats many students. Basic courses in linear or abstract algebra pose similar difficulties and can be overly formal.

Due to their specialized focus, upperclass courses cannot adequately address the issue of careful exposition. If students first learn techniques of proof and habits of careful exposition, then they will better appreciate more advanced mathematics when they encounter it.

The excitement of mathematics springs from engaging problems. Students have natural mathematical curiosity about problems such as those listed in the Preface for the Student. They then care about the techniques used to solve them; hence we use these problems as a focus of development. We hope that students and instructors will enjoy this approach as much as we have.

A course introducing techniques of proof should not specialize in one area of mathematics; later courses offer ample opportunities for specialization. This book considers diverse problems and demonstrates relationships among several areas of mathematics. One of the authors studies complex analysis in several variables, the other studies discrete mathematics. We explored the interactions between discrete and continuous mathematics to create a course on problem-solving and proofs.

When we began the revisions for the second edition, neither of us had any idea how substantial they would become. We are excited about the improvements. Our primary aim has been to make the book easier to use by making the treatment more accessible to students, more mathematically coherent, and better arranged for the design of courses. In the remainder of this preface we discuss the changes in more detail; here we provide a brief summary.

- We added almost 300 exercises; many are easy and/or check basic understanding of concepts in the text.
- We added sections called "How to Approach Problems" in Chapters 1–5 and 13–14 to help students get started on the exercises.
- We greatly expanded Appendix B: "Hints for Selected Exercises".
- Chapters 1–4 form the core of a coherent "Transition" course that can be completed in various ways using initial sections of other chapters.
- The real number system is the starting point. All discussion of the construction of $\mathbb{R}$ from $\mathbb{N}$ is in Appendix A.
- Induction comes earlier, immediately following the background material discussed in Chapters 1 and 2.
- Individual chapters have a sharper focus, and the development flows more smoothly from topic to topic.
- Terms being defined are in bold type, mostly in Definition items.
- The language is friendlier, the typography better, and the proofs a bit more patient, with more details.

Content and Organization

Our text presents elementary aspects of algebra, number theory, combinatorics, and analysis. We cover a broad spectrum of material that illustrates techniques of proof and emphasizes interactions among the topics.

Part I (Elementary Concepts) begins by deriving the quadratic formula and using it to motivate the axioms for the real numbers, which we agree to assume. We discuss inequalities, sets, logical statements, and functions, with careful attention to the use of language. Chapter 1 establishes the themes of mathematical discussion: numbers, sets, and functions. We added lively material on inequalities and level sets. The background terminology about functions moved to Chapter 1. The more abstract discussion of injections and surjections appears in Chapter 4, introduced by the base q representation of natural numbers. This allows induction to come early; the highlight of Part I is the use of induction to solve engaging problems. Part I ends with an optional treatment of the Schroeder-Bernstein Theorem.

Part II (Properties of Numbers) studies $\mathbb{N}$, $\mathbb{Z}$, and $\mathbb{Q}$. We explore elementary counting problems, binomial coefficients, permutations (as functions), prime factorization, and the Euclidean algorithm. Equivalence relations lead to the discussion of modular arithmetic. We emphasize geometric aspects of the rational numbers. Features include Fermat's Little Theorem (with several proofs), the Chinese Remainder Theorem, criteria for irrationality, and the description of Pythagorean triples.

Part III (Discrete Mathematics) explores more subtle combinatorial arguments. We consider conditional probability and discrete random variables, the pigeonhole principle, the inclusion-exclusion principle, graph theory, and recurrence relations. Highlights include Bertrand's Ballot Problem (Catalan numbers), Bayes' Theorem, Simpson's Paradox, Euler's totient function, Hall's Theorem on systems of distinct representatives, Platonic solids, and the Fibonacci numbers. With the focus on probability in Chapter 9, the optional discussion of generating functions has moved to the end of Chapter 12, where it is used to solve recurrences.

Part IV (Continuous Mathematics) begins with the Least Upper Bound Property for $\mathbb{R}$ and its relation to decimal expansions and uncountability of $\mathbb{R}$. We prove the Bolzano-Weierstrass Theorem and use it to prove that Cauchy sequences converge. We develop the theory of calculus: sequences, series, continuity, differentiation, uniform convergence, and the Riemann integral. We define the natural logarithm via integration and the exponential function via infinite series, and we prove their inverse relationship. Defining sine and cosine via infinite series, we use results on interchange of limiting operations to verify their properties (we do not rely on geometric intuition for technical statements). We include convex functions and van der Waarden's example of a continuous and nowhere differentiable function, but we omit many applications covered adequately in calculus courses, such as Taylor polynomials, analytic geometry, Kepler's laws, polar coordinates, and physical interpretations of derivatives and integrals. Finally, we develop the properties of complex numbers and prove the Fundamental Theorem of Algebra.

In Appendix A we develop the properties of arithmetic and construct the real number system using Cauchy sequences. There we begin with $\mathbb{N}$ and subsequently construct $\mathbb{Z}$, $\mathbb{Q}$, and $\mathbb{R}$. This foundational material establishes the properties of the real number system that we assume in the text. We leave this material to Appendix A because most students do not appreciate it until after they become familiar with writing proofs. Beginning instead by assuming the real numbers makes the theoretical development flow smoothly and keeps the interest of the students.

Chapters 1 and 2 provide the language for subsequent mathematical work. Formal discussion of mathematical language is problematic; students master techniques of proofs through examples of usage, not via memorization of terminology and symbolism of formal logic. Instead of formal manipulation of logical symbols, we emphasize the understanding of words. After the discussion in Chapter 2 that emphasizes the *use* of logic, familiarity with logical concepts is conveyed by repeated use throughout the book. Chapter 2 can be treated lightly in class; students can refer to it when they need help manipulating logical statements.

The rearrangement of material in Part I makes it more accessible to students and avoids using results before they can be proved. Students find induction easier and less abstract than bijections, and now it comes first. Placing the basic language about functions in Chapter 1 allows them to be used as a precise concept in Chapter 2, allows us to prove needed statements about them by induction in Chapter 3, and permits a sharper focus on the properties of injections and surjections in Chapter 4.

The material in Part II has been reorganized to give the chapters a clearer focus and to place the more fundamental material early in each chapter. Instead of combining cardinality and counting in Chapter 5, the material on cardinality has moved to Chapter 4 to better illuminate the properties of bijections. The discussion of binomial coefficients is in Chapter 5; in the first edition some of this was in Chapter 9. Chapter 5 also has new material on permutations that further explores aspects of functions. Because students have trouble producing combinatorial proofs, we provide additional examples here in the Approaches section.

We reorganized Chapter 6 to start with divisibility and factorization, allowing the Euclidean algorithm and diophantine equations to be skipped. We also added an optional section on algebraic properties of (the ring of) polynomials in one variable. In Chapter 7 we separated the discussion of general equivalence relations from the discussion of congruence. We reorganized Chapter 8 to remove the construction of $\mathbb{Q}$, beginning instead with geometric aspects of rational numbers. We moved the material on probability to Chapter 9, which now focuses completely on this topic. This clarifies the treatment of conditional probability and random variables. We moved the optional section on generating functions from Chapter 9 to Chapter 12, where it is applied.

In Part IV, we provided more details in proofs, plus friendlier language and typesetting. The treatment of decimal expansions in 13 is more natural and more precise. In Chapter 14, the material on Cauchy Sequences now appears after the material on limits of sequences.

Pedagogy and Special Features

Certain pedagogical issues require careful attention. In order to benefit from this course, students need a sense of intellectual progress. An axiomatic development of the real numbers is painfully slow and frustrates students. They have learned algebraic computational techniques throughout their schooling, and it is important to build on this foundation. This dictates our starting point.

In Chapter 1 we list the axioms for the real numbers and their elementary algebraic consequences, and we accept them for computation and reasoning. We defer the construction of the real numbers and verification of the field axioms to Appendix A, for later appreciation. In the second edition, we have made this pedagogically valuable approach more firmly consistent, obtaining $\mathbb{N}$ within $\mathbb{R}$ in Chapter 3 and moving the details of the rational number system from Chapter 8 to Appendix A. This simplifies the treatment of induction and eliminates most comments (and student uncertainty) about what we do and do not know at a given time. We exclude the use of calculus until it is developed in Part IV.

The exercises are among the strongest features of this book. Many are fun, some are routine, and some are difficult. Exercises designated by "$(-)$" are intended to check understanding of basic concepts; they require neither deep insight nor long solutions. The "$(+)$" problems are more difficult. Those designated by "$(!)$" are especially interesting or instructive. Most exercises emphasize thinking and writing rather than computation. The understanding and communication of mathematics through problem-solving should be the driving force of the course.

We have reorganized the exercises and added many, especially of the "$(-)$" type. We increased the number of exercises by 60% in Parts I–II and 40% overall; there are now well over 900 exercises. We have gathered the routine exercises at the beginning of the exercise sections. Usually a line of dots separates these from the other exercises to assist the instructor in selecting problems; after the dots the exercises are ordered roughly in parallel to the presentation of material in the text. Many of the exercise sets also have true/false questions, where students are asked to decide whether an assertion is true or false and then to provide a proof or a counterexample.

The purpose of the exercises is to encourage learning, not to frustrate students. Many of the exercises in the text carry hints; these represent what we feel will be helpful to most students. Appendix B contains more elementary hints for many problems; these are intended to give students a starting point for clearer thinking if they are completely stumped by a problem. We have expanded Appendix B so that now we give hints for more than half of the problems in the book.

We have also added sections called "How to Approach Problems" in Chapters 1–5 and 13–14. These are the chapters emphasized in courses with beginning students. In these sections, we summarize some thoughts from the chapters and provide advice to help students avoid typical pitfalls when starting to solve problems. The discussion here is informal.

The Preface to the Student lists many engaging problems. Some of these begin chapters as motivating "Problems"; others are left to the exercises. Solutions of such problems in the text are designated as "Solutions". Items designated as "Examples" are generally easier than those designated as "Solutions" or "Applications". "Examples" serve primarily to illustrate concepts, whereas "Solutions" or "Applications" employ the concepts being developed and involve additional reasoning.

Students have some difficulty recognizing what material is important. The book has two streams of material: the theoretical mathematical development and its illustrations or applications. "Definitions", "Propositions", "Lemmas", "Theorems", and "Corollaries" are set in an indented style. Students may use these results to solve problems and may want to learn them. Other items generally provide examples or commentary.

This book does not assume calculus and hence in principle can be used in a course taught to freshmen or to high school students. It does require motivation and commitment from the students, since problems can no longer be solved by mimicking memorized computations. The book is appropriate for students who have studied standard calculus and wonder why the computations work. It is ideal for beginning majors in mathematics and computer science. Readers outside mathematics who enjoy careful thinking and are curious about mathematics will also profit by it. High school teachers of mathematics may appreciate the interaction between problem-solving and theory.

The second author maintains a web site for this book with course materials, listing of errors or updates, etc. Please visit

$$\text{http://www.math.uiuc.edu/}{\sim}\text{west/mt}$$

Comments and corrections are welcome at west@math.uiuc.edu.

Design of Courses

We developed this book through numerous courses, beginning with a version we team-taught in 1991 at the University of Illinois. Various one-semester courses can be constructed from this material. The changes made for the second edition facilitate the design of courses.

Many schools have a one-semester "transition" course that introduces students to the notions of proof. Such a course should begin with Chapters 1–4 (omitting the Schroeder-Bernstein Theorem). Depending on the local curriculum and the students, good ways to complete such a course are with Chapters 5–8 or Chapters 13–14 (or both). The second edition makes these chapters more independent and places the more elementary material in each chapter near the beginning. This makes it easy to present just the fundamental material in each chapter. With good students, it is possible to present Chapters 1–10 and 13–15 in one semester, omitting the optional material.

A one-semester course on discrete mathematics that emphasizes proofs can cover Parts I–III, omitting most of Chapter 8 (rational numbers) and the more algebraic material from Chapters 6 and 7. Depending on the preparation of the students, Chapters 1–2 can be treated as background reading for a faster start. It should be noted that Part II maintains a more elementary atmosphere than Part III, and that the topics in Part III are more specialized.

A one-semester course in elementary analysis covers Chapters 3 and 4, perhaps some of Chapter 8 (many such courses discuss the rational numbers), and Chapters 13–17. Students should read Chapters 1 and 2 for background. This yields a thorough course in introductory analysis. The first author has twice taught successful elementary real analysis courses along these lines, covering chapters 13–17 completely after spending a few weeks on these earlier chapters.

The full text is suitable for a patient and thorough one-year course culminating in the Fundamental Theorem of Algebra.

Acknowledgments

Our preparation of the first edition was helped by comments from Art Benjamin, Dick Bishop, Kaddour Boukaabar, Peter Braunfeld, Tom Brown, Steve Chiappari, Everett Dade, Harold Diamond, Paul Drelles, Sue Goodman, Dan Grayson, Harvey Greenwald, Deanna Haunsperger, Felix Lazebnik, N. Tenney Peck, Steve Post, Sara Robinson, Craig Tovey, Steve Ullom, Josh Yulish, and other readers. The Mathematics Department of the University of Illinois gave us the opportunity to develop the course that inspired this book; we thank our students for struggling with preliminary versions of it. Our editor George Lobell provided the guidance and prodding needed to bring the book to its final form.

Additional comments for the preparation of the second edition were contributed by Charles Epstein, Dan Grayson, Corlis Johnson, Ward Henson, Ranjani Krishnan, Maria Muyot, Jeff Rabin, Mike Saks, Hector Sussmann, Steve Ullom, C.Q. Zhang. Many students who used the book spotted typographical errors or opaque passages and suggested additional improvements. The careful eye of our production editor Betsy Williams corrected many glitches and design problems.

The second edition was typeset using TEX, with illustrations created using the **gpic** program, a product of the Free Software Foundation. We thank Maria Muyot for assistance in the preparation of the index.

The authors thank their wives Annette and Ching, respectively, for their love, encouragement, and patience. The first author also thanks his children John, Lucie, and Paul for inspiration.

John P. D'Angelo, jpda@math.uiuc.edu
Douglas B. West, west@math.uiuc.edu
Urbana, Illinois

Preface for the Student

This book demands careful thinking; we hope that it also is enjoyable. We present interesting problems and develop the basic undergraduate mathematics needed to solve them. Below we list 37 such problems. We solve most of these in this book, while at the same time developing enough theory to prepare for upperclass math courses.

In Chapters 1–5 and 13–14 we have included sections called "How to Approach Problems". These provide advice on what to do in solving problems and warnings on what not to do. The "Approaches" evolved from using the book in the classroom; we have learned what difficulties students encountered and what errors occurred repeatedly. We have also provided, in Appendix B, hints to many exercises. These hints are intended to get students started in the right direction when they don't know how to approach a problem.

Many exercises are designated by "(−)", "(!)", or "(+)". The "(−)" exercises are intended to check understanding; a student who cannot do these is missing the basics. A student who can do an occasional "(+)" problem is showing some ability. The "(!)" problems are particularly instructive, important, or interesting; their difficulty varies. Many chapters contain true/false questions; here the student is asked to decide whether something is true and provide a proof or a counterexample.

This is a mathematics book that emphasizes writing and language skills. We do not ask that you memorize formulas, but rather that you learn to express yourself clearly and accurately. You will learn to solve mathematical puzzles as well as to write proofs of theorems from elementary algebra, discrete mathematics, and calculus. This will broaden your knowledge and improve the clarity of your thinking.

A proof is nothing but a complete explanation of why something is true. We will develop many techniques of proof. It may not be obvious what technique works in a given problem; we will sometimes give different proofs for a single result. Most students have difficulty when first asked to write proofs; they are unaccustomed to using language carefully and logically. Do not be discouraged; experience increases understanding and makes it easier to find proofs.

How can you improve your writing? Good writing requires practice. Writing out a proof can reveal hidden subtleties or cases that have been overlooked. It can also expose irrelevant thoughts. Producing a well-written solution often involves repeated revision. You must say what you mean and mean what you say. Mathematics encourages habits of writing precisely, because clear decisions can be made about whether sentences contain faulty reasoning. You will learn how to combine well chosen notation with clear explanation in sentences. This will enable you to communicate ideas concisely and accurately.

We invite you to consider some intriguing problems. We solve most of these in the text, and others appear as exercises.

1. Given several piles of pennies, we create a new collection by removing one coin from each old pile to make one new pile. Each original pile shrinks by one; 1, 1, 2, 5 becomes 1, 4, 4, for example. Which lists of sizes (order is unimportant) are unchanged under this operation?

2. Which natural numbers are sums of consecutive smaller natural numbers? For example, $30 = 9 + 10 + 11$ and $31 = 15 + 16$, but 32 has no such representation.

3. Including squares of sizes one-by-one through eight-by-eight, an ordinary eight-by-eight checkerboard has 204 squares. How many squares of all sizes arise using an n-by-n checkerboard? How many triangles of all sizes arise using a triangular grid with sides of length n?

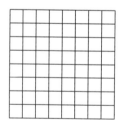 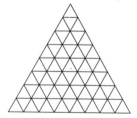

4. At a party with five married couples, no person shakes hands with his or her spouse. Of the nine people other than the host, no two shake hands with the same number of people. With how many people does the hostess shake hands?

5. We can tell whether two groups of weights have the same total weight by placing them on a balance scale. How many known weights are needed to balance each integer weight from 1 to 121? How should these weights be chosen? (Known weights can be placed on either side or omitted.)

6. Given a positive integer k, how can we obtain a formula for the sum $1^k + 2^k + \cdots + n^k$?

7. Is it possible to fill the large region below with non-overlapping copies of the small L-shape? Rotations and translations are allowed.

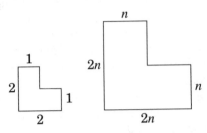

8. If each resident of New York City has 100 coins in a jar, is it possible that no two residents have the same number of coins of each type (pennies, nickels, dimes, quarters, half-dollars)?

9. How can we find the greatest common divisor of two large numbers without factoring them?

10. Why are there infinitely many prime numbers? Why are there arbitrarily long stretches of consecutive non-prime positive integers?

11. Consider a dart board having two regions, one worth a points and the other worth b points, where a and b are positive integers having no common factors greater than 1. What is the largest point total that cannot be obtained by throwing darts at the board?

12. A math professor cashes a check for x dollars and y cents, but the teller inadvertently pays y dollars and x cents. After the professor buys a newspaper for k cents, the remaining money is twice as much as the original value of the check. If $k = 50$, what was the value of the check? If $k = 75$, why is this situation impossible?

13. Must there be a Friday the 13th in every year?

14. When two digits in the base 10 representation of an integer are interchanged, the difference between the old number and the new number is divisible by nine. Why?

15. A positive integer is **palindromic** if reversing the digits of its base 10 representation does not change the number. Why is every palindromic integer with an even number of digits divisible by 11?

16. What are all the integer solutions to $42x + 63y = z$? To $x^2 + y^2 = z^2$?

17. Given a prime number L, for which positive integers K can we express the rational number K/L as the sum of the reciprocals of two positive integers?

18. Are there more rational numbers than integers? Are there more real numbers than rational numbers? What does "more" mean for these sets?

19. Can player A have a higher batting average than B in day games and in night games but a lower batting average than B over all games?

Player	Day	Night	Overall
A	.333	.250	.286
B	.300	.200	.290

20. Suppose A and B gamble as follows: On each play, each player shows 1 or 2 fingers, and one pays the other x dollars, where x is the total number of fingers showing. If x is odd, then A pays B; if x is even, then B pays A. Who has the advantage?

21. Suppose candidates A and B in an election receive a and b votes, respectively. If the votes are counted in a random order, what is the probability that candidate A never trails?

22. Can the numbers $0, \ldots, 100$ be written in some order so that no 11 positions contain numbers that successively increase or successively decrease? (An increasing or decreasing set need not occupy consecutive positions or use consecutive numbers.)

23. Suppose each dot in an n by n grid of dots is colored black or white. How large must n be to guarantee the existence of a rectangle whose corners have the same color?

24. How many positive integers less than 1,000,000 have no common factors (greater than 1) with 1,000,000?

25. Suppose n students take an exam, and the exam papers are handed back at random for peer grading. What is the probability that no student gets his or her own paper back? What happens to this probability as n goes to infinity?

26. There are n girls and n boys at a party, and each girl likes some of the boys. Under what conditions is it possible to pair the girls with boys so that each girl is paired with a boy that she likes?

27. A computer plotter must draw a figure on a page. What is the minimum number of times the pen must be lifted while drawing the figure?

28. Consider n points on a circle. How many regions are created by drawing all chords joining these points, assuming that no three chords have a common intersection?

29. A Platonic solid has congruent regular polygons as faces and has the same number of faces meeting at each vertex. Why are the tetrahedron, cube, octahedron, dodecahedron, and icosahedron the only ones?

30. Suppose n spaces are available for parking along the side of a street. We can fill the spaces using Rabbits, which take one space, and/or Cadillacs, which take two spaces. In how many ways can we fill the spaces? In other words, how many lists of 1's and 2's sum to n?

31. Repeatedly pushing the "x^2" button on a calculator generates a sequence tending to 0 if the initial positive value is less than 1 and tending to ∞ if it is greater than 1. What happens with other quadratic functions?

32. What numbers have more than one decimal representation?

33. Suppose that the points in a tennis game are independent and that the server wins each point with probability p. What is the probability that the server wins the game?

34. How is $\lim_{n\to\infty}(1 + x/n)^n$ relevant to compound interest?

35. One baseball player hits singles with probability p and otherwise strikes out. Another hits home runs with probability $p/4$ and otherwise strikes out. Assume that a single advances each runner by two bases. Compare a team composed of such home-run hitters with a team composed of such singles hitters. Which generates more runs per inning?

36. Let $T_1, T_2, \ldots$ be a sequence of triangles in the plane. If the sequence of triangles converges to a region T, can we then conclude that $\text{Area}(T) = \lim_{n\to\infty} \text{Area}(T_n)$?

37. Two jewel thieves steal a circular necklace with $2m$ gold beads and $2n$ silver beads arranged in some unknown order. Is it always true that there is a way to cut the necklace along some diameter so that each thief gets half the beads of each color? Does a heated circular wire always contains two diametrically opposite points where the temperature is the same? How are these questions related?

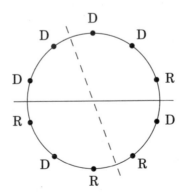

PART I

ELEMENTARY

CONCEPTS

Chapter 1

Numbers, Sets, and Functions

The ancient Babylonians considered the problem of finding two numbers when given their sum and product. They expressed the solution in words, not in formulas. We begin this book by deriving the quadratic formula and using it to solve this ancient problem. We then discuss the properties of the real numbers and the basic concepts of sets and functions that enable us to state and solve mathematical problems.

THE QUADRATIC FORMULA

Given two numbers s and p, the Babylonians wanted to find x and y such that $x + y = s$ and $xy = p$. To do so, we write $y = s - x$ and substitute to obtain $x(s - x) = p$, which we rewrite as $x^2 - sx + p = 0$. Every solution x to the problem of the Babylonians must satisfy this quadratic equation.

Solving this equation is equivalent to solving the general quadratic equation. We don't change the solutions if we multiply the equation by a nonzero constant a to obtain $ax^2 - asx + ap = 0$, and then we can name $b = -as$ and $c = ap$ to obtain $ax^2 + bx + c = 0$.

The familiar **quadratic formula** expresses the solution for x in terms of a, b, and c. First we rewrite the equation in a manner where the unknown value x appears only once:

$$0 = a(x^2 + \frac{b}{a}x) + c = a(x^2 + \frac{b}{a}x + \frac{b^2}{4a^2}) - \frac{b^2}{4a} + c$$

$$= a(x + \frac{b}{2a})^2 + c - \frac{b^2}{4a}.$$

Hence $(x + \frac{b}{2a})^2 = \frac{b^2 - 4ac}{4a^2}$. Solving for x yields the quadratic formula:

$$x = \frac{-b \pm \sqrt{b^2 - 4ac}}{2a}$$

This formula describes all the solutions to the general quadratic equation. When $b^2 - 4ac > 0$, it yields two values. When $b^2 - 4ac = 0$, these values are equal. When $b^2 - 4ac < 0$, there is no solution in real numbers. Rewriting the solution formula in terms of s and p yields the expressions

$$\frac{s + \sqrt{s^2 - 4p}}{2}, \quad \frac{s - \sqrt{s^2 - 4p}}{2} \tag{$*$}$$

to solve the Babylonian problem. When $s^2 - 4p < 0$, there is no solution.

The quadratic formula gave us $(*)$ as the solutions to the quadratic equation $x^2 - sx + p = 0$, when $s^2 - 4p \geq 0$. Note that the sum of the numbers in $(*)$ is s and their product is p. This checks our solution.

For any real numbers α and β, we can create a quadratic polynomial that is zero at α and β by letting $x - \alpha$ and $x - \beta$ be factors. Since $(x - \alpha)(x - \beta) = x^2 - (\alpha + \beta)x + \alpha\beta$, the product of the solutions is the constant term, and their sum is the negative of the coefficient of x.

What properties of numbers did we use in solving the Babylonian problem? First, we used basic rules about addition and multiplication. The result of adding several numbers does not depend on the order of writing them or on the order of performing pairwise additions. Multiplication has the same property. We also used the more subtle *distributive law*: $x(y + z) = xy + xz$.

We also used properties of subtraction and division. Every number u has an *additive inverse* $-u$, and subtracting u has the same effect as adding $-u$. Their sum is 0, and adding 0 causes no change. Similarly, every nonzero number u has a multiplicative inverse u^{-1}. Their product is 1, and multiplying by 1 causes no change. An important distinction is that we cannot divide by 0. The properties of inverses allow us to cancel equal terms or *nonzero* common factors from both sides of an equality.

These rules about arithmetic are *algebraic* properties. We also used properties of *inequality* and *order*. Because the product of two nonzero numbers with the same sign is positive, square roots exist only for non-negative numbers. Furthermore, if $u^2 = v$, then also $(-u)^2 = v$. Thus we write $\pm$ on the square root sign in the quadratic formula and say that there is no solution in real numbers when $b^2 - 4ac < 0$.

The Babylonians would not have accepted our solution, because their number system did not include negative numbers! In the real number system, the formula $(-b \pm \sqrt{b^2 - 4ac})/(2a)$ makes sense when $b^2 - 4ac \geq 0$. It remains to express the square root of $b^2 - 4ac$ in an acceptable form. Expressing square roots in decimal form generally requires infinite non-repeating decimal expansions. This requires a *completeness* property of the number system and is related to infinite processes and limits.

In developing mathematical ideas in this book, we take the real number system with its elementary properties as given; this allows us to focus on the logical structure of mathematical arguments. At the end of this chapter, we list the properties that characterize the real numbers, describe what we assume about them, and discuss ways to approach solving problems. Meanwhile, we discuss other background material.

ELEMENTARY INEQUALITIES

Manipulating inequalities requires care. Multiplying both sides of an equation by the same number preserves equality, but this fails for inequalities. If $a < b$, then $ac < bc$ if and only if $c > 0$.

In this section, we derive several inequalities about real numbers. They rely on two properties: positive real numbers have positive square roots, and the square of every real number is nonnegative. We prove first that taking squares or square roots of positive numbers preserves order.

1.1. Proposition. If $0 < a < b$, then $a^2 < ab < b^2$ and $0 < \sqrt{a} < \sqrt{b}$.

Proof: Multiplying an inequality by a positive number does not change whether the inequality is true. Thus we multiply $a < b$ by a to obtain $a^2 < ab$, and we multiply $a < b$ by b to obtain $ab < b^2$.

We also must have $\sqrt{a} < \sqrt{b}$; otherwise, applying the first statement to $\sqrt{b} \leq \sqrt{a}$ yields $b \leq a$, which violates the hypothesis $a < b$. ∎

We use bold type in this book for terms being defined.

1.2. Definition. The **absolute value** of a real number x, written as $|x|$, is defined by

$$|x| = \begin{cases} x & \text{if } x \geq 0, \\ -x & \text{if } x \leq 0. \end{cases}$$

We think of $|x|$ as the distance from x to 0; this motivates our next proof (see Example 1.50 for another approach). Note that always $x \leq |x|$ and $|xy| = |x| \, |y|$.

1.3. Proposition. (Triangle Inequality) If x and y are real numbers, then $|x + y| \leq |x| + |y|$.

Proof: We start with the inequality $2xy \leq 2 |x| \, |y|$. By adding $x^2 + y^2$ to both sides and using $z^2 = |z|^2$, we obtain

$$x^2 + 2xy + y^2 \leq x^2 + 2 |x| \, |y| + y^2 = |x|^2 + 2 |x| \, |y| + |y|^2.$$

By Proposition 1.1, we may take the positive square root of both sides and preserve the inequality. Thus $|x + y| \leq |x| + |y|$, as desired. ∎

In order to prove a statement, we derive it from known facts. Before we find a proof, we may not know which known facts to use. To discover a proof, it may be helpful to ask what is needed to make the conclusion true. In this approach, we try to "reduce" the desired conclusion to a statement known to be true. The written proof must be a rigorous justification of the conclusion from known facts.

The next proposition illustrates this. Manipulating the desired inequality leads to a known inequality, but the proof starts with the known inequality and derives the desired one from it. The **arithmetic mean** (or "average") of x and y is $(x + y)/2$. The **geometric mean** of nonnegative numbers x and y is $\sqrt{xy}$. The term **AGM Inequality** stands for *Arithmetic Mean–Geometric Mean Inequality*; it states that the arithmetic mean of two nonnegative numbers is always at least their geometric mean.

1.4. Proposition. (AGM Inequality) If x and y are real numbers, then $2xy \le x^2 + y^2$ and $xy \le (\frac{x+y}{2})^2$. If x and y are also nonnegative, then $\sqrt{xy} \le (x + y)/2$. Equality holds in each only when $x = y$.

Proof: We begin with $0 \le (x - y)^2 = x^2 - 2xy + y^2$ and observe that equality holds only when $x = y$. Adding $2xy$ yields $2xy \le x^2 + y^2$. Adding another $2xy$ yields $4xy \le x^2 + 2xy + y^2 = (x + y)^2$, which we divide by 4 to obtain $xy \le (\frac{x+y}{2})^2$.

If $x \ge 0$ and $y \ge 0$, then also $xy \ge 0$, and we can take positive square roots in $xy \le (\frac{x+y}{2})^2$. Proposition 1.1 yields $\sqrt{xy} \le (x + y)/2$. ∎

1.5. Corollary. If $x, y > 0$, then $\frac{2xy}{x+y} \le \sqrt{xy} \le \frac{x+y}{2}$. Equality holds in each inequality only when $x = y$.

Proof: Proposition 1.4 yields $\sqrt{xy} \le \frac{x+y}{2}$. We obtain the other inequality from this by multiplying both sides by the positive number $\frac{2\sqrt{xy}}{x+y}$. ∎

1.6. Application. The expression $\frac{2xy}{x+y}$ is the **harmonic mean** of x and y. It arises in the study of average rates. When we travel a distance d at rate r in time t, we have $d = rt$, in appropriate units.

If we travel a distance d at a rate r_1 in time t_1 and make the return trip at rate r_2 in time t_2, then $r_1 t_1 = d = r_2 t_2$. What is the average rate r for the full trip? The computation is $2d = r(t_1 + t_2)$, and hence

$$r = \frac{2d}{t_1 + t_2} = \frac{2d}{\frac{d}{r_1} + \frac{d}{r_2}} = \frac{2r_1 r_2}{r_1 + r_2}.$$

Thus the average rate for the full trip is the harmonic mean of the rates in the two directions. By Corollary 1.5, the rate for the full trip is less than the average of the two rates one-way when those rates differ.

For example, if the rate one way on a plane trip is 380 mph, and the return rate over the same distance is 420 mph, then the average rate is

$$\frac{2(380)(420)}{380+420} = \frac{800(19)(21)}{800} = (20-1)(20+1) = 399 \text{ mph}$$

This is less than 400 because more time is spent at the slower rate. ∎

In this book, we reserve the label **Example** for direct illustrations of mathematical concepts. We use **Solution** and **Application** to designate examples that incorporate additional reasoning. Results that can be used to solve problems here and elsewhere have the labels **Definition**, **Proposition**, **Lemma**, **Theorem**, and **Corollary**.

SETS

We begin our formal development with basic notions of set theory. Our most primitive notion is that of a **set**. This notion is so fundamental that we do not attempt to give a precise definition. We think of a set as a collection of distinct objects with a precise description that provides a way of deciding (in principle) whether a given object is in it.

1.7. Definition. The objects in a set are its **elements** or **members**. When x is an element of A, we write $x \in A$ and say "x **belongs to** A". When x is not in A, we write $x \notin A$. If every element of A belongs to B, then A is a **subset** of B, and B **contains** A; we write $A \subseteq B$ or $B \supseteq A$.

When we list the elements of a set explicitly, we put braces around the list; "$A = \{-1, 1\}$" specifies the set A consisting of the elements -1 and 1. Writing the elements in a different order does not change a set. We write $x, y \in S$ to mean that both x and y are elements of S.

1.8. Example. By convention, we use the special characters $\mathbb{N}, \mathbb{Z}, \mathbb{Q}, \mathbb{R}$ to name the sets of **natural numbers, integers, rational numbers**, and **real numbers**, respectively. Each set in this list is contained in the next, so we write $\mathbb{N} \subseteq \mathbb{Z} \subseteq \mathbb{Q} \subseteq \mathbb{R}$.

We take these sets as familiar. We use the convention that 0 is not a natural number; $\mathbb{N} = \{1, 2, 3, \ldots\}$. The set of integers is $\mathbb{Z} = \{\ldots, -2, -1, 0, 1, 2, \ldots\}$. The set $\mathbb{Q}$ of rational numbers is the set of real numbers that can be written as $\frac{a}{b}$ with $a, b \in \mathbb{Z}$ and $b \neq 0$. ∎

1.9. Definition. Sets A and B are **equal**, written $A = B$, if they have the same elements. The **empty set**, written $\emptyset$, is the unique set with no elements. A **proper subset** of a set A is a subset of A that is not A itself. The **power set** of a set A is the set of all subsets of A.

Note that the empty set is a subset of every set.

1.10. Example. Let S be the set {Kansas, Kentucky}. Let T be the set of states in the United States whose names begin with "K". The sets S and T are equal. The set S has four subsets: $\emptyset$, {Kansas}, {Kentucky}, and {Kansas, Kentucky}. These four are the elements of the power set of S. ∎

1.11. Remark. *Specifying a set.* In Example 1.10, we specified a set both by listing its elements and by describing it as a subset of a larger set. In order to specify a set S consisting of the elements in a set A that satisfy a given condition, we write "$\{x \in A : \text{condition}(x)\}$". We read this as "the set of x in A such that x satisfies 'condition' ". For example, the expression $S = \{x \in \mathbb{R} : ax^2 + bx + c = 0\}$ specifies S as the set of real numbers satisfying the equation $ax^2 + bx + c = 0$, where a, b, c are known constants. We may omit specifying the universe A when the context makes it clear. ∎

1.12. Remark. What must be done to determine the solutions to a mathematical problem? In order to prove that the set of solutions is T, we must prove that every solution belongs to T, and we must prove that every member of T is a solution.

Letting S denote the set of solutions, our goal is to prove that $S = T$, where T is a list or has a simple description. The statement "$S = T$" conveys two pieces of information: "$S \subseteq T$ and $T \subseteq S$". The first containment states that every solution belongs to T, and the second states that every member of T is a solution. ∎

1.13. Example. *Equality of sets.*

1) *The inequality $x^2 < x$.* Let $S = \{x \in \mathbb{R}: x^2 < x\}$, and let $T = \{x \in \mathbb{R}: 0 < x < 1\}$. We claim that $S = T$. To prove this, we show that $T \subseteq S$ and that $S \subseteq T$. First consider $x \in T$. Since $x > 0$, we can multiply the known inequality $x < 1$ by x to obtain $x^2 < x$, so $x \in S$. Conversely, consider $x \in S$. Since $x^2 < x$, we have $0 > x^2 - x = x(x - 1)$. This requires that x and $x - 1$ are nonzero and have opposite signs, which yields $x \in T$.

2) *The quadratic equation $ax^2 + bx + c = 0$.* Let S be the set of solutions, and let $T = \{\frac{-b+\sqrt{b^2-4ac}}{2a}, \frac{-b-\sqrt{b^2-4ac}}{2a}\}$. When we proved $S = T$, we could have proved both $S \subseteq T$ and $T \subseteq S$. The latter involves plugging each purported solution into the original equation and checking that it works. Our reasoning was more efficient; we operated on the equation in ways that preserved the set of solutions. This produced a string of equalities of sets, starting with S and ending with T. Note that plugging the members of T into the equation enables us to check that we have not made an error in manipulating the equation. ∎

Our next example again illustrates the process of describing a solution set by proving two containments. We are proving that an object is a solution *if and only if* it belongs to the desired set.

1.14. Application. *The Penny Problem.* Given piles of pennies, we re-move one coin from each pile to make one new pile. Each original pile shrinks by one, so each pile of size one disappears: 1,1,2,5 becomes 1,4,4, for example. We consider different orderings of the same list of sizes to be equivalent, so we restrict our attention to lists of positive integers in nondecreasing order. Let S be the set of lists that do not change.

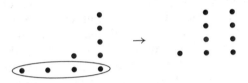

Let a be a list with n piles, and let b be the resulting new list. If $a \in S$, meaning that a and b are the same, then b also has n piles. Since we introduce one new pile, exactly one pile must disappear. Thus a has exactly one pile of size 1. Thus b also has exactly one pile of size 1. This forces a to have exactly one pile of size 2.

We continue this reasoning for i from 1 to $n - 1$. From a having one pile of size i, we conclude that b has one pile of size i, and therefore that a has one pile of size $i + 1$. This gives us one pile of each size 1 through n.

Let T be the set of lists consisting of one pile of each size from 1 through some natural number n. We have shown that every unchanged configuration has this form, so $S \subseteq T$. To complete the solution, we also check that all elements of T remain unchanged.

Consider the element of T with piles of sizes $1, 2, \ldots, n$. For each i from 2 to n, the pile of size i becomes a pile of size $i - 1$. The pile of size 1 disappears, and the n piles each contribute one coin to form a new pile of size n. The result is the original list. Now we have proved that $S \subseteq T$ and $T \subseteq S$, so $S = T$. We have described all the unchanged lists. ■

The next three definitions introduce notation and terminology for special sets that we will use throughout this book.

1.15. Definition. *Sets of integers.* When $a, b \in \mathbb{Z}$ with $a \le b$, we use $\{a, \ldots, b\}$ to denote $\{i \in \mathbb{Z} : a \le i \le b\}$. When $n \in \mathbb{N}$, we write $[n]$ for $\{1, \ldots, n\}$. The set of **even numbers** is $\{2k : k \in \mathbb{Z}\}$. The set of **odd numbers** is $\{2k + 1 : k \in \mathbb{Z}\}$.

Note that 0 is an even number. Every integer is even or odd, and no integer is both. The **parity** of an integer states whether it is even or odd. We say "even" and "odd" for numbers *only* when discussing integers. Similarly, when we say that a number is positive without specify the number system containing it, we mean that it is a positive real number. Thus, "consider $x > 0$" means "let x be a positive real number".

1.16. Definition. *Intervals.* When $a, b \in \mathbb{R}$ with $a \leq b$, the **closed interval** $[a, b]$ is the set $\{x \in \mathbb{R}: a \leq x \leq b\}$. The **open interval** (a, b) is the set $\{x \in \mathbb{R}: a < x < b\}$.

Consider $S \subseteq \mathbb{R}$. If an element x belonging to S is at least as large as every element of S, then x is a **maximum** of S. A set can only have one maximum. The concept of **minimum** is defined analogously. The open interval (a, b) has no maximum and no minimum.

There are several natural ways to obtain new sets from old sets.

1.17. Definition. *k-tuples and Cartesian product.* A **list** with entries in A consists of elements of A in a specified order, with repetition allowed. A k-**tuple** is a list with k entries. We write A^k for the set of k-tuples with entries in A.

An **ordered pair** is a list with two entries. The **Cartesian product** of sets S and T, written $S \times T$, is the set $\{(x, y): x \in S, y \in T\}$.

Note that $A^2 = A \times A$ and $A^k = \{(x_1, \ldots, x_k): x_i \in A\}$. We read "$x_i$" as "$x$ sub i". Since we use the notation (a, b) for ordered pairs, we often write "the interval (a, b)" to avoid confusion when specifying an open interval. When $S = T = \mathbb{R}$, the Cartesian product $S \times T$ or $\mathbb{R}^2$ can be viewed as the set of all points in the plane, designated by horizontal and vertical coordinates, called the **Cartesian coordinates** of the point. The concept of Cartesian product is named for René Descartes (1596–1650). The Cartesian product of two intervals in $\mathbb{R}$ is a rectangle in the plane.

1.18. Definition. *Set operations.* Let A and B be sets. Their **union**, written $A \cup B$, consists of all elements in A or in B. Their **intersection**, written $A \cap B$, consists of all elements in both A and B. Their **difference**, written $A - B$, consists of the elements of A that are not in B. Two sets are **disjoint** if their intersection is the empty set $\emptyset$. If a set A is contained in some universe U under discussion, then the **complement** A^c of A is the set of elements of U *not* in A.

1.19. Example. Let E and O denote the sets of even numbers and odd numbers. We have $E \cap O = \emptyset$ and $E \cup O = \mathbb{Z}$. Within $\mathbb{Z}$, we have $E^c = O$. ∎

Pictures give life to mathematical concepts and illuminate essential ideas. We encourage the reader to draw pictures to clarify concepts. We do this for the operations in Definition 1.18. Diagrams illustrating sets and their relationships are named for John Venn (1834–1923), though he was not the first person to use them.

1.20. Remark. *Venn diagrams.* In a **Venn diagram**, an outer box represents the universe under consideration, and regions within the box correspond to sets. Non-overlapping regions correspond to disjoint sets. The four regions in the Venn diagram for two sets A and B represent $A \cap B$, $(A \cup B)^c$, $A - B$, and $B - A$.

Since $A - B$ consists of the elements in A and not in B, we have $A - B = A \cap B^c$. Similarly, the diagram suggests that B^c is the union of $A - B$ and $(A \cup B)^c$, which are disjoint. Also $A - B$ and $B - A$ are disjoint.

Slightly more subtle is $(A - B) \cup (B - A) = (A \cup B) - (A \cap B)$. A rigorous proof shows that an element belongs to one set if and only if it belongs to the other. Exercise 41 lists other elementary relationships. ∎

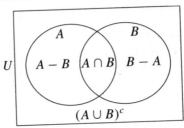

FUNCTIONS

"Function" is the name we use for a mathematical machine with inputs and outputs. The inputs are the elements from one set; the outputs are elements of a (possibly) different set. Familiar ways to specify a function include an algebraic formula, a list of the outputs associated with the inputs, a description in words of how an input determines its output, and various graphical representations.

1.21. Definition. A **function** f *from* a set A *to* a set B assigns to each $a \in A$ a single element $f(a)$ in B, called the **image** of a under f. For a function f from A to B (written $f: A \to B$), the set A is the **domain** and the set B is the **target**. The **image** of a function f with domain A is $\{f(a): a \in A\}$.

1.22. Remark. *Schematic representation.* A function $f: A \to B$ is **defined on** A and **maps** A into B. To visualize a function $f: A \to B$, we draw a region representing A and a region representing B, and from each $x \in A$ we draw an arrow to $f(x)$ in B.

The image of a function is contained in its target. Thus we draw the region for the image inside the region for the target. ∎

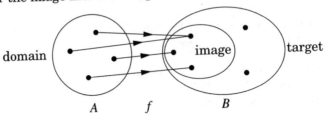

There are many ways to describe a function; for each $a \in A$, we must specify $f(a)$. We can list the pairs $(a, f(a))$, provide a formula for computing $f(a)$ from a, or describe the rule for obtaining $f(a)$ from a in words. Note that $f(a)$ denotes an element of the target of f and does not denote the function f. Thus x^2 is a number (when we know x); it should be distinguished from the function $f: \mathbb{R} \to \mathbb{R}$ defined by $f(x) = x^2$.

1.23. Example. *Descriptions of functions.*

Functions given by formulas. The "squaring" function $S: \mathbb{R} \to \mathbb{R}$ is defined by $S(x) = x \cdot x = x^2$. The "addition" and "multiplication" functions are defined from $\mathbb{R} \times \mathbb{R}$ to $\mathbb{R}$ by $A(x, y) = x + y$ and $M(x, y) = xy$.

A function given by listing its values. Define $g: [7] \to \mathbb{N}$ by listing $g(1) = 6$, $g(2) = 6$, $g(3) = 7$, $g(4) = 9$, $g(5) = 8$, $g(6) = 6$, $g(7) = 8$.

A function given by words. Define $h: [7] \to \mathbb{N}$ by letting $h(n)$ be the number of letters in the English word for the nth day of the week, starting with Sunday. The function h is the same as g defined above. ∎

1.24. Remark. *The meaning of "well-defined".* A function $f: A \to B$ may be specified by different rules on different subsets of A. The statement "f is well-defined" means that the rules assign to each element of A exactly one element, belonging to B. When different rules apply to an element of A, we must check that they give the same element of B (see Exercise 45).

For example, the absolute value of x (Definition 1.2) is defined using two rules, both applying when $x = 0$. Since $0 = -0$, the rules agree at 0, and thus absolute value is well-defined. ∎

1.25. Definition. A function f is **real-valued** if its image is a subset of $\mathbb{R}$; in this case $f(x)$ is a number. For real-valued functions f and g with domain A, the **sum** $f + g$ and **product** fg are real-valued functions on A defined by $(f + g)(x) = f(x) + g(x)$ and $(fg)(x) = f(x)g(x)$.

1.26. Definition. A (real) **polynomial** in one variable is a function $f: \mathbb{R} \to \mathbb{R}$ defined by $f(x) = c_0 + c_1 x^1 + \ldots + c_k x^k$, where k is a nonnegative integer and $c_0, \ldots, c_k$ are real numbers called the **coefficients** of f. The **degree** of f is the largest d such that $c_d \neq 0$; the polynomial with all coefficients 0 has no degree. Polynomials of degrees $0,1,2,3$, are **constant**, **linear**, **quadratic**, **cubic**, respectively.

We can study polynomials in more variables. A **monomial** in variables $x_1, \ldots, x_n$ is an expression $c x_1^{a_1} \cdots x_n^{a_n}$, where c is a real number and each a_j is a nonnegative integer. A polynomial in n variables is a finite sum of monomials in n variables. For example, the function f defined by

$$f(x, y, z) = x^2 + y^2 + z^2 + 2xy + 2xz + 2yz$$

is a polynomial in three variables. It is a polynomial in the single variable x when y and z are held constant.

We can also describe functions using geometric ideas.

1.27. Definition. The **graph** of a function $f: A \to B$ is the subset of $A \times B$ consisting of the ordered pairs $\{(x, f(x)): x \in A\}$.

1.28. Remark. *Pictures of graphs.* Let f be a real-valued function defined on a set $A \subseteq \mathbb{R}$. We draw two copies of $\mathbb{R}$ as horizontal and vertical axes, associating the horizontal axis with the domain. The graph of f is then a set of points in the plane. A set S of points in the plane is the graph of a function if and only if it contains at most one element (x, y) for each real number x; in other words, each vertical line intersects S at most once. ∎

1.29. Example. *Alternative representations.* We describe a particular function $f: [4] \to [4]$ using each method we have discussed. Define f by $f(n) = 5 - n$ to give a formula. Define f by $f(1) = 4$, $f(2) = 3$, $f(3) = 2$, $f(4) = 1$, listing values. Define f by saying that f interchanges 1 and 4 and interchanges 2 and 3. The graph of f is $\{(1, 4), (2, 3), (3, 2), (4, 1)\}$. ∎

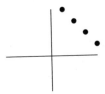

For a function defined by a formula, the image may not be obvious.

1.30. Example. *For the function $f: \mathbb{R} \to \mathbb{R}$ defined by $f(x) = x/(1 + x^2)$, the image is the interval $[-1/2, 1/2]$.* To prove that this is the image, we show first that $|f(x)| \leq 1/2$ for $x \in \mathbb{R}$. This claim is equivalent to $|x| \leq (1 + x^2)/2$, which follows from $(1 - |x|)^2 \geq 0$.

We have proved that the interval contains the image; we must also prove that the image contains the interval. For $y \in [-1/2, 1/2]$, we prove that there exists $x \in \mathbb{R}$ such that $f(x) = y$. Note that $f(0) = 0$. For $y \neq 0$ and $y \in [-1/2, 1/2]$, we set $y = x/(1 + x^2)$ and solve for x in terms of y. Applying the quadratic formula to $yx^2 - x + y = 0$ yields $x = (1 \pm \sqrt{1 - 4y^2})/2y$. Since $|y| \leq 1/2$, we now have $x \in \mathbb{R}$ such that $f(x) = y$. ∎

1.31. Definition. A set $S \subseteq \mathbb{R}$ is **bounded** if there exists $M \in \mathbb{R}$ such that $|x| \leq M$ for all $x \in S$. A set is **unbounded** if no such M exists. A **bounded function** is a real-valued function whose image is bounded; that is, a real-valued function f for which there is some M in $\mathbb{R}$ such that $|f(x| \leq M$ for all x in the domain.

1.32. Definition. Let $f: \mathbb{R} \to \mathbb{R}$, and let A be a set of real numbers. We say that f is **increasing** (on A) if $f(x) < f(x')$ whenever $x < x'$ and $x, x' \in A$. It is **nondecreasing** (on A) if $f(x) \leq f(x')$ whenever $x < x'$ and $x, x' \in A$. Changing $<$ to $>$ and $\leq$ to $\geq$ yields definitions for **decreasing** and **nonincreasing**. A function is **monotone** on A if it is nondecreasing on A or if it is nonincreasing on A.

The properties "increasing" and "nondecreasing" are also called **strictly increasing** and **weakly increasing**, respectively. Similarly, a function is **strictly monotone** on A if it is increasing on A or if it is decreasing on A. We use the word "monotone" to avoid repetition; many results apply in both cases. A function that is increasing on one interval and decreasing on another is *not* monotone. The function of Example 1.30 is bounded but not monotone.

unbounded, increasing

bounded, not monotone

bounded, increasing

1.33. Remark. *Geometric interpretations.* A function from $\mathbb{R}$ to $\mathbb{R}$ is increasing if and only if for every horizontal line intersecting its graph, the graph is above that line to the right of the intersection and below it to the left. The function is bounded if and only if every point in the graph lies in the band between some pair of horizontal lines. ∎

The use of "if" in Definitions 1.31–1.32 has the same meaning as the use of "if and only if" in Remark 1.33. In *defining* X, we often say that X occurs "if" some property holds, yet we mean that the new concept and the condition are equivalent. This is a convention; in some sense the concept does not exist until it is defined, so the implication can only hold in one direction. In this book, the definition usage of "if" is recognizable by the use of bold type for the concept being defined.

1.34. Definition. The **identity** function on a set S is the function $f: S \to S$ defined by $f(x) = x$ for all $x \in S$. A **fixed point** of a function $f: S \to S$ is an element $x \in S$ such that $f(x) = x$.

Every element of S is a fixed point for the identity function on S. In the Penny Problem (Application 1.14), we studied a function from the set of nondecreasing lists of natural numbers to itself; our aim was to find all fixed points. A function f from $\mathbb{R}$ to $\mathbb{R}$ has a fixed point if and only if the line $\{(x, x)\}$ through the origin intersects the graph of f.

The identity functions on $\mathbb{N}$, $\mathbb{Z}$, and $\mathbb{R}$ are graphed below. The graphs show that these are different functions; a function cannot be specified by a formula alone. Two functions are **equal** if they have the same domain, have the same target, and agree in value at each element of the domain.

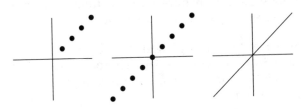

INVERSE IMAGE AND LEVEL SETS

We can interpret solution sets for equations using the language of functions. For any function f and value y in its target, we consider the set of solutions to $f(x) = y$.

1.35. Definition. Given $f\colon A \to B$ and $y \in B$, the **inverse image** of y under f, written $I_f(y)$, is the set $\{x \in A\colon f(x) = y\}$.

If $f(p)$ is the temperature at the point p, then $I_f(32)$ is the set of all points where the temperature is 32. The inverse image is called an *isotherm*; sketches of isotherms appear on most weather maps.

The inverse image of y under a function $f\colon A \to B$ is a subset of the domain A. Generally speaking, inverse image is not a function from B to A, because it may associate many elements of A with an element of B.

Real-valued functions often arise as measurements. Consider for example the function h that assigns to a point in the United States the height of this point above sea level. A topographical map shows points with the same height above sea level connected by a *level curve* (the curve may have many pieces). Each level curve for h is $I_h(c)$ for some c; the number c gives the height above sea level.

1.36. Definition. For $h\colon \mathbb{R} \times \mathbb{R} \to \mathbb{R}$, the **level set** of h with value c is $I_h(c)$.

1.37. Example. Let $A(x, y) = x + y$. For each c, $I_A(c)$ is a line in $\mathbb{R}^2$. The level sets are parallel lines whose union is all of $\mathbb{R}^2$.

Let $M(x, y) = xy$. The level set $I_M(0)$ consists of the two coordinate axes. For $c \neq 0$, the level sets are hyperbolas; each has two branches.

Let $D(x, y) = x^2 + y^2$. The level set $I_D(c)$ is empty when $c < 0$, consists of one point when $c = 0$, and is a circle of radius $\sqrt{c}$ when $c > 0$. The figure shows these level sets when $c \in \{-2, -1, 0, 1, 2\}$. ■

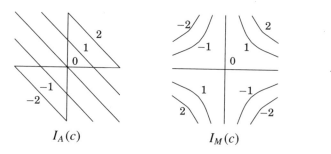

$I_A(c)$ $I_M(c)$ $I_D(c)$

1.38. Application. Given two real numbers whose sum is between -8 and 8 and whose product is between -20 and 20, what is the largest that one of these numbers can be?

We use level sets to solve this problem. We are given $|x + y| \le 8$ and $|xy| \le 20$. In the graph of the solution set the boundary is determined by the level sets $x + y = 8$, $x + y = -8$, $xy = 20$, and $xy = -20$. The level sets for intermediate values lie between them.

By plotting the level sets, we see that the largest value x can have (when both inequalities hold) occurs when $xy = -20$ and $x + y = 8$. Solving these equations as in the discussion of the Babylonian problem of Chapter 1 yields $x = 10$ and $y = -2$. Thus the maximum value is 10. ∎

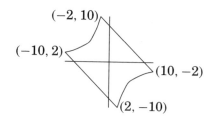

THE REAL NUMBER SYSTEM

The real numbers satisfy a short list of properties called **axioms** from which all the other properties can be derived. In this section we state these properties and some of their consequences. Our purpose here is *not* to study these in detail, but rather to state our starting point and clarify what the student may assume when solving exercises.

A structure satisfying Definitions 1.39–1.41 below is a **complete ordered field**. In Appendix A, we prove that all such structures are essentially equivalent. Furthermore, we build such a structure and verify that it satisfies the axioms. The construction begins with $\mathbb{N}$ (satisfying appropriate axioms) and successively builds $\mathbb{Z}$, $\mathbb{Q}$, and finally what we call $\mathbb{R}$, each time defining the new objects in terms of the previous objects.

These constructions are somewhat formal and dull. In the text, we instead begin with the real numbers and their properties and emphasize techniques of reasoning. *We assume that the real number system $\mathbb{R}$ exists and satisfies the properties in Definitions 1.39–1.41.* These imply all other properties of real numbers, such as those in Propositions 1.43–1.46. For now we treat $\mathbb{N}$ informally; in Chapter 3 we give a formal definition of $\mathbb{N}$ as a subset of $\mathbb{R}$. Whether we begin with $\mathbb{N}$ and define $\mathbb{R}$ as in Appendix A, or begin with $\mathbb{R}$ as in the text, the same results hold. In each case, the real number system satisfies Definitions 1.39–1.41.

1.39. Definition. *Field Axioms.* A set S with operations $+$ and $\cdot$ and distinguished elements 0 and 1 with $0 \neq 1$ is a **field** if the following properties hold for all $x, y, z \in S$.

A0: $x + y \in S$	M0: $x \cdot y \in S$	Closure
A1: $(x+y)+z = x+(y+z)$	M1: $(x \cdot y) \cdot z = x \cdot (y \cdot z)$	Associativity
A2: $x + y = y + x$	M2: $x \cdot y = y \cdot x$	Commutativity
A3: $x + 0 = x$	M3: $x \cdot 1 = x$	Identity
A4: given x, there is a $w \in S$	M4: for $x \neq 0$, there is a $w \in S$	Inverse
such that $x + w = 0$	such that $x \cdot w = 1$	
	DL: $x \cdot (y + z) = x \cdot y + x \cdot z$	Distributive Law

The operations $+$ and $\cdot$ are called **addition** and **multiplication**. The elements 0 and 1 are the **additive identity element** and the **multiplicative identity element**.

It follows from these axioms that the additive inverse and multiplicative inverse (of a nonzero x) are unique. The additive inverse of x is the **negative** of x, written as $-x$. To define **subtraction** of y from x, we let $x - y = x + (-y)$. The multiplicative inverse of x is the **reciprocal** of x, written as x^{-1}. The element 0 has no reciprocal. To define **division** of x by y when $y \neq 0$, we let $x/y = x \cdot (y^{-1})$. We write $x \cdot y$ as xy and $x \cdot x$ as x^2. We use parentheses where helpful to clarify the order of operations.

1.40. Definition. *Order Axioms.* A **positive set** in a field F is a set $P \subseteq F$ such that for $x, y \in F$,

P1: $x, y \in P$ implies $x + y \in P$ Closure under Addition
P2: $x, y \in P$ implies $xy \in P$ Closure under Multiplication
P3: $x \in F$ implies exactly one of   Trichotomy
    $x = 0$, $x \in P$, $-x \in P$

An **ordered field** is a field with a positive set P. In an ordered field, we define $x < y$ to mean $y - x \in P$. The relations $\leq$, $<$, and $\geq$ have analogous definitions in terms of P.

Note that $P = \{x \in F : x > 0\}$. Another phrasing of trichotomy is that each ordered pair (x, y) satisfies exactly one of $x < y$, $x = y$, $x > y$.

If $S \subseteq F$, then $\beta \in F$ is an **upper bound** for S if $x \leq \beta$ for all $x \in S$.

1.41. Definition. *Completeness Axiom.* An ordered field F is **complete** if every nonempty subset of F that has an upper bound in F has a least upper bound in F.

Until Part IV, we do not need the Completeness Axiom for $\mathbb{R}$, except to be aware that it ensures the existence of square roots of positive real numbers. The axioms in Definitions 1.39–1.40 imply that arithmetic has its familiar properties. We list some of these below. We agree to assume all these properties of numbers. Note that $\mathbb{Q}$ also is an ordered field, and thus the properties listed below also hold for arithmetic in $\mathbb{Q}$. The set $\mathbb{Z}$ of integers satisfies all the field and order axioms except the existence of multiplicative inverses.

1.42. Proposition. *Arithmetic in* $\mathbb{N}, \mathbb{Z}, \mathbb{Q}$. Each of $\mathbb{N}, \mathbb{Z}, \mathbb{Q}$ is closed under addition and multiplication, $\mathbb{Z}$ and $\mathbb{Q}$ are closed under subtraction, and the set of nonzero numbers in $\mathbb{Q}$ is closed under division.

The next four propositions state properties of an ordered field F. All statements apply for each choice of $x, y, z, u, v \in F$.

1.43. Proposition. *Elementary consequences of the field axioms.*
   a) $x + z = y + z$ implies $x = y$ e) $(-x)(-y) = xy$
   b) $x \cdot 0 = 0$ f) $xz = yz$ and $z \neq 0$ imply $x = y$
   c) $(-x)y = -(xy)$ g) $xy = 0$ implies $x = 0$ or $y = 0$
   d) $-x = (-1)x$

1.44. Proposition. *Properties of an ordered field.*
   O1: $x \leq x$ — Reflexive Property
   O2: $x \leq y$ and $y \leq x$ imply $x = y$ — Antisymmetric Property
   O3: $x \leq y$ and $y \leq z$ imply $x \leq z$ — Transitive Property
   O4: at least one of $x \leq y$ and $y \leq x$ holds — Total Ordering Property

1.45. Proposition. *More properties of an ordered field.*
   F1: $x \leq y$ implies $x + z \leq y + z$ — Additive Order Law
   F2: $x \leq y$ and $0 \leq z$ imply $xz \leq yz$ — Multiplicative Order Law
   F3: $x \leq y$ and $u \leq v$ imply $x + u \leq y + v$ — Addition of Inequalities
   F4: $0 \leq x \leq y$ and $0 \leq u \leq v$ imply $xu \leq yv$ — Multiplication of Inequalities

1.46. Proposition. *Still more properties of an ordered field.*
   a) $x \leq y$ implies $-y \leq -x$ e) $0 < 1$
   b) $x \leq y$ and $z \leq 0$ imply $yz \leq xz$ f) $0 < x$ implies $0 < x^{-1}$
   c) $0 \leq x$ and $0 \leq y$ imply $0 \leq xy$ g) $0 < x < y$ implies $0 < y^{-1} < x^{-1}$
   d) $0 \leq x^2$

Properties (a) and (b) of Proposition 1.46 tell us that multiplying an inequality by a negative number requires reversing the inequality.

There are other equivalent formulations of the axioms. Hence it is not important to remember which are axioms and which are consequences in our list; we take the entire list as our starting point.

HOW TO APPROACH PROBLEMS

In this chapter, we discussed mathematical objects; in Chapter 2, we will discuss mathematical statements. As a warmup, the exercises here begin with translations between mathematics and English. Most problems in this chapter demand precise understanding of language but require little calculation. Computations become just one part of a mathematical tool box. We will develop more tools in later chapters.

We mention some simple strategies to help students get started on unfamiliar problems. Although they seem self-evident, these strategies will be helpful throughout the book; keep them in mind.

1) Understand the problem and approach it logically.

2) Substitutions allow us to simplify expressions or to introduce useful new expressions.

3) When there are only a few possibilities, analysis by cases may help eliminate all possibilities except the desired conclusion.

4) Check whether answers are reasonable.

Understanding problems.

Exercises 1–26 provide practice in translating words into mathematical concepts. One must also understand the definitions of mathematical concepts used (see Exercise 18 and beyond).

To gain an understanding of a problem, one sometimes analyzes a special case. For example, one could analyze the Penny Problem for small values of n to discover a pattern and then extend the argument that works for special values to prove the desired result in general.

Distinguish what is given or known from what is to be shown. Understand what is needed to obtain the desired conclusion from the known information. Break a complicated problem into simpler steps.

Substitution.

As change for a dollar we might receive four quarters, or we might exchange four quarters for a dollar. Mathematical equations also can be read two ways. Substitution is the process of replacing a mathematical expression by a more convenient expression with the same value. Substitution has many facets; we substitute when we apply a general formula in a special case, when we wish to simplify, or when we eliminate variables.

1.47. Example. Since $x^2 - y^2 = (x+y)(x-y)$ holds for all x and y, we may replace one side of this equality by the other. For example, to multiply 598 by 602 mentally, think

$$(600 - 2)(600 + 2) = 600^2 - 2^2 = 360000 - 4 = 359996.$$

Here it was convenient to replace $(x + y)(x - y)$ with $x^2 - y^2$. On the other hand, to find the roots to an equation we might replace $x^2 - y^2$ with its factored form $(x + y)(x - y)$. ∎

Substitution can sometimes be used to eliminate an irrelevant variable. In Application 1.6, we wrote $r = 2d/(t_1 + t_2)$, but we wanted to express r in terms of r_1 and r_2. We substituted expressions for t_1 and t_2 in terms of the desired variables, and the dependence on d canceled out.

1.48. Example. In Exercise 31, the hint for part (a) suggests using the inequality $2tu \leq t^2 + u^2$ from Proposition 1.4. In fact, we use six instances of this inequality, substituting various quantities for t and u, to obtain the inequality $4xyzw \leq x^4 + y^4 + z^4 + w^4$.

To obtain $3abc \leq a^3 + b^3 + c^3$ from this, we again use substitution. We want to reduce the expression from four variables to three variables with symmetric roles; letting $w = (xyz)^{1/3}$ accomplishes this in a useful way. After this, substituting a, b, c for appropriate expressions in x, y, z yields the desired identity.

The substitutions in the last step are natural, but finding the substitution $w = (xyz)^{1/3}$ is more difficult. Experience, intelligent guessing, and trial and error all help decide what substitutions might be useful. ∎

Analysis by cases.

The form of an answer may depend on the values of the variables; the cases in Exercise 37 arise in this way. Alternatively, deductions we want to make might be valid only for restricted choices of the variables.

1.49. Example. We seek all integer solutions to $a^2 b > 2a$. In other words, we seek an explicit description of the set $\{(a, b) \in \mathbb{Z}^2 : a^2 b > 2a\}$. We rewrite the inequality as $a(ab - 2) > 0$. This inequality holds if and only if both factors have the same sign. Thus we are led to the two cases below.

  1) $a > 0$ and $ab > 2$
  2) $a < 0$ and $ab < 2$

The first case contains all integer pairs in the first quadrant except $(1, 1)$, $(1, 2)$, and $(2, 1)$. The second case contains all integer pairs in the second quadrant and also $(-1, -1)$ from the third quadrant. The answer is the union of the sets of solutions in the two cases. ∎

1.50. Example. Analysis by cases may arise when studying the absolute value function. For real numbers x, y, we can prove the Triangle Inequality $|x + y| \leq |x| + |y|$ (Proposition 1.3) in this way.

When x, y are both nonnegative, both sides equal $x + y$. When x, y are both nonpositive, both sides equal $-x - y$. When x, y have opposite signs, we may assume that $x > 0 > y$. The inequality then holds because

$$|x + y| = \max\{x + y, -x - y\} < x - y = |x| + |y|. \qquad \blacksquare$$

Finding a way to avoid analysis by cases can lead to deeper understanding of a problem or method. Many questions related to absolute value and distance are best understood by studying squared distance.

The use of sets facilitates analysis by cases. The word "or" corresponds to union of sets, and the word "and" corresponds to intersection.

Checking answers.

Checking answers can expose errors in reasoning. When finding a general answer, one should check it in special cases. When a formula describes areas or lengths (as in Exercise 19), the resulting values must be nonnegative. We recommend checking answers for reasonableness; how to do this depends on the problem.

EXERCISES

Words like "determine", "show", "obtain", or "construct" include a request for justification; these are very similar to "prove". Answers to problems in this book should be given full explanations. Explanations include *sentences*; reasoning cannot be explained without words.

Easier problems are indicated by "(−)", harder problems by "(+)". Those designated "(!)" are particularly interesting or instructive.

1.1. (−) We have many tables and many chairs. Let t be the number of tables, and let c be the number of chairs. Write down an inequality that means "We have at least four times as many chairs as tables."

1.2. (−) Fill in the blanks. The equation $x^2 + bx + c = 0$ has exactly one solution when _____, and it has no solutions when _____.

1.3. (−) Given that $x + y = 100$, what is the maximum value of xy?

1.4. (−) Explain why the square has the largest area among all rectangles with a given perimeter.

1.5. (−) Consider the Celsius (C) and Fahrenheit (F) temperature scales.

C	0	5	10	15	20	25	30
F	32	41	50	59	68	77	86

Express the sentence "The temperature was 10° C and increased by 20° C" using the Fahrenheit scale.

1.6. (−) At a given moment, let f and c be the values of the temperature on the Fahrenheit and Celsius scales, respectively. These values are related by

$f = (9/5)c + 32$. At what temperatures do the following events occur?
a) The Fahrenheit and Celsius values of the temperature are equal.
b) The Fahrenheit value is the negative of the Celsius value.
c) The Fahrenheit value is twice the Celsius value.

1.7. (−) The statement below is not always true for $x, y \in \mathbb{R}$. Give an example where it is false, and add a hypothesis on y that makes it a true statement.
"If x and y are nonzero real numbers and $x > y$, then $(-1/x) > (-1/y)$."

1.8. (!) In the morning section of a calculus course, 2 of the 9 women and 2 of the 10 men receive the grade of A. In the afternoon section, 6 of the 9 women and 9 of the 14 men receive A. Verify that, in each section, a higher proportion of women than of men receive A, but that, in the combined course, a lower proportion of women than of men receive A. Explain! (See Exercises 9.19–9.20 for related exercises and Example 9.20 for a real-world example.)

1.9. (−) If a stock declines 20% in one year and rises 23% in the next, is there a net profit? What if it goes up 20% in the first year and down 18% in the next?

1.10. (−) On July 4, 1995, the *New York Times* reported that the nation's universities were awarding 25% more Ph.D. degrees than the economy could absorb. The headline concluded that there was a 1 in 4 chance of underemployment. Here "underemployment" means having no job or having a job not requiring the Ph.D. degree. What should the correct statement of the odds have been?

1.11. (−) A store offers a 15% promotional discount for its grand opening. The clerk believes that the law requires the discount to be applied first and then the tax computed on the resulting amount. A customer argues that the discount should be applied to the total after the 5% sales tax is added, expecting to save more money that way. Does it matter? Explain.

1.12. (−) A store offers an "installment plan" option, with no interest to be paid. There are 13 monthly payments, with the first being a "down payment" that is half the size of the others, so payment is completed one year after purchase. If a customer buys a $1000 stereo, what are the payments under this plan?

1.13. (−) Let A be the set of integers expressible as $2k - 1$ for some $k \in \mathbb{Z}$. Let B be the set of integers expressible as $2k + 1$ for some $k \in \mathbb{Z}$. Prove that $A = B$.

1.14. (−) Let a, b, c, d be real numbers with $a < b < c < d$. Express the set $[a, b] \cup [c, d]$ as the difference of two sets.

1.15. (−) For what conditions on sets A and B does $A - B = B - A$ hold?

1.16. (−) Starting with a single pile of 5 pennies, determine what happens when the operation of Application 1.14 is applied repeatedly. Determine what happens when the initial configuration is a single pile of 6 pennies.

1.17. (−) What are the domain and the image of the absolute value function?

1.18. (−) Determine which real numbers exceed their reciprocals by exactly 1.

•   •   •   •   •

1.19. What are the dimensions of a rectangular carpet with perimeter 48 feet and area 108 square feet? Given positive numbers p and a, under what conditions does there exist a rectangular carpet with perimeter p and area a?

1.20. Suppose that r and s are distinct real solutions of the equation $ax^2+bx+c = 0$. In terms of a, b, c, obtain formulas for $r + s$ and rs.

1.21. Let a, b, c be real numbers with $a \neq 0$. Find the flaw in the following "proof" that $-b/2a$ is a solution to $ax^2 + bx + c = 0$.

Let x and y be solutions to the equation. Subtracting $ay^2 + by + c = 0$ from $ax^2 + bx + c = 0$ yields $a(x^2 - y^2) + b(x - y) = 0$, which we rewrite as $a(x + y)(x - y) + b(x - y) = 0$. Hence $a(x + y) + b = 0$, and thus $x + y = -b/a$. Since x and y can be any solutions, we can apply this computation letting y have the same value as x. With $y = x$, we obtain $2x = -b/a$, or $x = -b/(2a)$.

1.22. We have two identical glasses. Glass 1 contains x ounces of wine; glass 2 contains x ounces of water ($x \geq 1$). We remove 1 ounce of wine from glass 1 and add it to glass 2. The wine and water in glass 2 mix uniformly. We now remove 1 ounce of liquid from glass 2 and add it to glass 1. Prove that the amount of water in glass 1 is now the same as the amount of wine in glass 2.

1.23. A digital 12-hour clock is defective: the reading for hours is always correct, but the reading for minutes always equals the reading for hours. Determine the minimum number of minutes between possible correct readings of the clock.

1.24. Three people register for a hotel room; the desk clerk charges them $30. The manager returns and says this was an overcharge, instructing the clerk to return $5. The clerk takes five $1 bills, but pockets $2 as a tip and returns only $1 to each guest. Of the original $30 payment, each guest actually paid $9, and $2 went to the attendant. What happened to the "missing" dollar?

1.25. A census taker interviews a woman in a house. "Who lives here?" he asks. "My husband and I and my three daughters," she replies. "What are the ages of your daughters?" "The product of their ages is 36 and the sum of their ages is the house number." The census taker looks at the house number, thinks, and says, "You haven't given me enough information to determine the ages." "Oh, you're right," she replies, "Let me also say that my eldest daughter is asleep upstairs." "Ah! Thank you very much!" What are the ages of the daughters? (The problem requires "reasonable" mathematical interpretations of its words.)

1.26. (+) Two mail carriers meet on their routes and have a conversation. A: "I know you have three sons. How old are they?" B: "If you take their ages, expressed in years, and multiply those numbers, the result will equal your age." A: "But that's not enough to tell me the answer!" B: "The sum of these three numbers equals the number of windows in that building." A: "Hmm [pause]. But it's still not enough!" B: "My middle son is red-haired." A: "Ah, now it's clear!" How old are the sons? (Hint: The ambiguity at the earlier stages is needed to determine the solution for the full conversation.) (G. P. Klimov)

1.27. Determine the set of real solutions to $|x/(x + 1)| \leq 1$.

1.28. (!) *Application of the AGM Inequality.*
a) Use Proposition 1.4 to prove that $x(c - x)$ is maximized when $x = c/2$.
b) For $a > 0$, use part (a) to find the value of y maximizing $y(c - ay)$.

1.29. Let x, y, z be nonnegative real numbers such that $y + z \geq 2$. Prove that $(x + y + z)^2 \geq 4x + 4yz$. Determine when equality holds.

1.30. (!) Let x, y, u, v be real numbers.
    a) Prove that $(xu + yv)^2 \le (x^2 + y^2)(u^2 + v^2)$.
    b) Determine precisely when equality holds in part (a).

1.31. (+) *Extensions of the AGM Inequality.*
    a) Prove that $4xyzw \le x^4 + y^4 + z^4 + w^4$ for real numbers x, y, z, w. (Hint: Use the inequality $2tu \le t^2 + u^2$ repeatedly.)
    b) Prove that $3abc \le a^3 + b^3 + c^3$ for nonnegative a, b, c. (Hint: In the inequality of part (a), set w equal to the cube root of xyz.)

1.32. (!) Assuming only arithmetic (not the quadratic formula or calculus), prove that $\{x \in \mathbb{R}: x^2 - 2x - 3 < 0\} = \{x \in \mathbb{R}: -1 < x < 3\}$.

1.33. Let $S = \{(x, y) \in \mathbb{N}^2 : (2 - x)(2 + y) > 2(y - x)\}$. Prove that $S = T$, where $T = \{(1, 1), (1, 2), (1, 3), (2, 1), (3, 1)\}$.

1.34. Let $S = \{(x, y) \in \mathbb{R}^2 : (1 - x)(1 - y) \ge 1 - x - y\}$. Give a simple description of S involving the signs of x and y.

1.35. (!) Determine the set of ordered pairs (x, y) of nonzero real numbers such that $x/y + y/x \ge 2$.

1.36. Let $S = [3] \times [3]$ (the Cartesian product of $\{1, 2, 3\}$ with itself). Let T be the set of ordered pairs $(x, y) \in \mathbb{Z} \times \mathbb{Z}$ such that $0 \le 3x + y - 4 \le 8$. Prove that $S \subseteq T$. Does equality hold?

1.37. Determine the set of solutions to the general quadratic inequality $ax^2 + bx + c \le 0$. Express the answer using linear inequalities or intervals. (Use the quadratic formula; the complete solution involves many cases.)

1.38. Let $S = \{x \in \mathbb{R}: x(x - 1)(x - 2)(x - 3) < 0\}$. Let T be the interval $(0, 1)$, and let U be the interval $(2, 3)$. Obtain a simple set equality relating S, T, U.

1.39. (!) Given $n \in \mathbb{N}$, let $a_1, a_2, \ldots, a_n$ be real numbers such that $a_1 < a_2 < \cdots < a_n$. Express $\{x \in \mathbb{R}: (x - a_1)(x - a_2) \cdots (x - a_n) < 0\}$ using the notation for intervals. (For convenience, use $(-\infty, a)$ to denote $\{x \in \mathbb{R}: x < a\}$.)

1.40. Let A and B be sets. Explain why the two sets $(A - B) \cup (B - A)$ and $(A \cup B) - (A \cap B)$ must be equal. Check this when A is the set of states in the United States whose names begin with a vowel and B is the set of states in the United States whose names have at most six letters.

1.41. (−) Let A, B, C be sets. Explain the relationships below. Use the definitions of set operations and containment, with Venn diagrams to guide the argument.
    a) $A \subseteq A \cup B$, and $A \cap B \subseteq A$.   d) $A \subseteq B$ and $B \subseteq C$ imply $A \subseteq C$.
    b) $A - B \subseteq A$.   e) $A \cap (B \cap C) = (A \cap B) \cap C$.
    c) $A \cap B = B \cap A$, and $A \cup B = B \cup A$.   f) $A \cup (B \cup C) = (A \cup B) \cup C$.

1.42. Let $A = \{$January, February, $\ldots$, December$\}$. Given $x \in A$, let $f(x)$ be the number of days in x. Does f define a function from A to $\mathbb{N}$?

1.43. (−) Let $S = \{(x, y) \in \mathbb{R}^2: 2x + 5y \le 10\}$. Graph S. Explain how the answer changes when the constraint is $2x + 5y < 10$.

1.44. (!) Let $S = \{(x, y) \in \mathbb{R}^2: x^2 + y^2 \le 100\}$. Let $T = \{(x, y) \in \mathbb{R}^2: x + y \le 14\}$.
    a) Graph $S \cap T$.
    b) Count the points in $S \cap T$ whose coordinates are both integers.

1.45. (−) Determine whether the rules below define functions from $\mathbb{R}$ to $\mathbb{R}$.
  a) $f(x) = |x - 1|$ if $x < 4$ and $f(x) = |x| - 1$ if $x > 2$.
  b) $f(x) = |x - 1|$ if $x < 2$ and $f(x) = |x| - 1$ if $x > -1$.
  c) $f(x) = ((x + 3)^2 - 9)/x$ if $x \neq 0$ and $f(x) = 6$ if $x = 0$.
  d) $f(x) = ((x + 3)^2 - 9)/x$ if $x > 0$ and $f(x) = x + 6$ if $x < 7$.
  e) $f(x) = \sqrt{x^2}$ if $x \geq 2$, $f(x) = x$ if $0 \leq x \leq 4$, and $f(x) = -x$ for $x < 0$.

1.46. Determine the images of the functions $f \colon \mathbb{R} \to \mathbb{R}$ defined as follows:
  a) $f(x) = x^2/(1 + x^2)$.   b) $f(x) = x/(1 + |x|)$.

1.47. Let $f \colon \mathbb{N} \times \mathbb{N} \to \mathbb{R}$ be defined by $f(a, b) = (a + 1)(a + 2b)/2$.
  a) Show that the image of f is contained in $\mathbb{N}$.
  b) (+) Determine exactly which natural numbers are in the image of f. (Hint: Formulate a hypothesis by trying values.)

1.48. Give several descriptions of the function $f \colon [0, 1] \to [0, 1]$ defined by $f(x) = 1 - x$. Compare with Example 1.29.

1.49. (!) Let f and g be functions from $\mathbb{R}$ to $\mathbb{R}$. For the sum and product of f and g (see Definition 1.25), determine which statements below are true. If true, provide a proof; if false, provide a counterexample.
  a) If f and g are bounded, then $f + g$ is bounded.
  b) If f and g are bounded, then fg is bounded.
  c) If $f + g$ is bounded, then f and g are bounded.
  d) If fg is bounded, then f and g are bounded.
  e) If both $f + g$ and fg are bounded, then f and g are bounded.

1.50. (!) For S in the domain of a function f, let $f(S) = \{f(x) \colon x \in S\}$. Let C and D be subsets of the domain of f.
  a) Prove that $f(C \cap D) \subseteq f(C) \cap f(D)$.
  b) Give an example where equality does not hold in part (a).

1.51. When $f \colon A \to B$ and $S \subseteq B$, we define $I_f(S) = \{x \in A \colon f(x) \in S\}$. Let X and Y be subsets of B.
  a) Determine whether $I_f(X \cup Y)$ must equal $I_f(X) \cup I_f(Y)$.
  b) Determine whether $I_f(X \cap Y)$ must equal $I_f(X) \cap I_f(Y)$.
(Hint: Explore this using the schematic representation described in Remark 1.22.)

1.52. Let M and N be nonnegative real numbers. Suppose that $|x + y| \leq M$ and $|xy| \leq N$. Determine the maximum possible value of x as a function of M and N.

1.53. Solve Application 1.38 by using inequalities rather than graphs.

1.54. (!) Let $S = \{(x, y) \in \mathbb{R}^2 \colon y \leq x \text{ and } x + 3y \geq 8 \text{ and } x \leq 8\}$.
  a) Graph the set S.
  b) Find the minimum value of $x + y$ such that $(x, y) \in S$. (Hint: On the graph from part (a), sketch the level sets of the function f defined by $f(x, y) = x + y$.)

1.55. (+) Let $\mathbf{F}$ be a field consisting of exactly three elements $0, 1, x$. Prove that $x + x = 1$ and that $x \cdot x = 1$. Obtain the addition and multiplication tables for $\mathbf{F}$.

1.56. (+) Is there a field with exactly four elements? Is there a field with exactly six elements?

Chapter 2

Language and Proofs

Understanding mathematical reasoning requires familiarity with the precise meaning of words like "every", "some", "not", "and", "or", etc.; these arise often in analyzing mathematical problems. Relevant aspects of language include word order, quantifiers, logical statements, and logical symbols. With these, we can discuss elementary techniques of proof.

TWO THEOREMS ABOUT EQUATIONS

We begin with two problems that illustrate both the need for careful use of language and the variety of techniques in proofs.

2.1. Definition. A **linear equation** in two variables x and y is an equation $ax + by = r$, where the coefficients a, b and the constant r are real numbers. A **line** in $\mathbb{R}^2$ is the set of pairs (x, y) satisfying a linear equation whose coefficients a and b are not both 0.

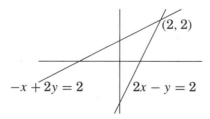

Geometric intuition suggests three possibilities for a pair of linear equations in two variables. If each equation describes a line, then the lines may intersect in one point, may be parallel, or may be identical. The equations then have one, none, or infinitely many common solutions, respectively. We can analyze this without relying on geometric intuition, because we have defined "line" using only arithmetic of real numbers.

25

2.2. Theorem. Let $ax + by = r$ and $cx + dy = s$ be linear equations in two variables x and y. If $ad - bc \neq 0$, then there is a unique common solution. If $ad - bc = 0$, then there is no common solution or there are infinitely many, depending on the values of r and s.

Proof: If all four coefficients are zero, then there is no solution unless $r = s = 0$, in which case all pairs (x, y) are solutions. Otherwise, at least one coefficient is nonzero. By interchanging the equations and/or interchanging the roles of x and y, we may assume that $d \neq 0$. We can now solve the second equation for y, obtaining $y = (s - cx)/d$. By substituting this expression for y into the first equation and simplifying, we obtain $(a - \frac{bc}{d})x + \frac{bs}{d} = r$. Multiplying by d yields $(ad - bc)x + bs = rd$.

When $ad - bc \neq 0$, we may divide by $ad - bc$ to obtain $x = \frac{rd - bs}{ad - bc}$. Substituting this into the equation for y yields the unique solution

$$(x, y) = \left(\frac{rd - bs}{ad - bc}, \frac{as - rc}{ad - bc} \right).$$

When $ad - bc = 0$, the equation for x becomes $bs = rd$. If $bs \neq rd$, then there is no solution. If $bs = rd$, then for each x we obtain the solution $(x, y) = (x, (s - cx)/d)$; here there are infinitely many solutions. ∎

When $ad - bc \neq 0$, the equations define lines with one common point. When $ad - bc = 0$ and both equations describe lines, there may be no solution (parallel lines) or infinitely many solutions (the lines coincide). An equation does not describe a line if both its coefficients are 0; here there is no solution unless the equation is $0x + 0y = 0$, in which case the common solutions are the solutions to the other equation in the pair.

In the proof, avoiding division by 0 leads us to consider cases. No single solution formula holds for all pairs of linear equations; the form of the solution changes when $ad - bc = 0$. The solution statement itself requires careful attention to language.

Our next argument uses the fundamental method of *proof by contradiction*; we suppose that the desired conclusion is false and then derive a contradiction from this hypothesis. The method is particularly useful for proving statements of nonexistence. Here we combine the method of proof by contradiction with an understanding of rational numbers and several elementary observations about odd and even numbers.

2.3. Theorem. If a, b, c are odd integers, then $ax^2 + bx + c = 0$ has no solution in the set of rational numbers.

Proof: Suppose that there *is* a rational solution x. We write this as p/q for integers p, q. We may assume that p/q expresses x "in lowest terms", meaning that p and q have no common integer factor larger than 1. From $ax^2 + bx + c = 0$ we obtain $ap^2 + bpq + cq^2 = 0$ after multiplying by q^2.

We obtain a contradiction by showing that $ap^2 + bpq + cq^2$ cannot equal 0. We do this by proving the stronger statement that it is odd.

Because we expressed x as a rational number in lowest terms, p and q cannot both be even. If both are odd, then the three terms in the sum are all odd, since the product of odd numbers is odd. Since the sum of three odd numbers is odd, we have the desired contradiction in this case. If p is odd and q is even (or vice versa), then we have the sum of two even numbers and an odd number, which again is odd. In each case, the assumption of a rational solution leads to a contradiction.   ∎

QUANTIFIERS AND LOGICAL STATEMENTS

Understanding a subject and writing clearly about it go together. We next discuss the use of well-chosen words and symbols to express mathematical ideas precisely. The language of mathematical statements will become familiar as we use it in later chapters to solve problems.

Using proof by contradiction requires understanding what it means for a statement to be false. Consider the sentence "Every classroom has a chair that is not broken". Without using words of negation, can we write a sentence with the opposite meaning? This will be easy once we learn how logical operations are expressed in English.

2.4. Example. *Negation of simple sentences.* What is the negation of "All students are male"? Some would say, incorrectly, "All students are not male". The correct negation is "At least one student is not male". Similarly, the negation of "all integers are odd" is *not* "all integers are not odd"; the correct negation is "at least one integer is even".   ∎

Common English permits ambiguities; the listener can obtain the intended meaning from context. Mathematics must avoid ambiguities.

2.5. Example. *Word order and context.* Consider the sentence "There is a real number y such that $x = y^3$ for every real number x". This seems to say that some number y is the cube root of all numbers, which is false. To say that every number has a cube root, we write "For every real number x, there is a real number y such that $x = y^3$".

In both English and mathematics, meaning depends on word order. Compare "Mary made Jane eat the food", "Eat, Mary; Jane made the food", and "Eat the food Mary Jane made". Meaning can also depend on context, as in "The bartender served two aces". This may have different meanings, depending on whether we are watching tennis or relaxing in a bar on an airbase. Mathematics can present similar difficulties; words such as "square" and "cycle" have several mathematical meanings.   ∎

The fundamental issue in mathematics is whether mathematical statements are true or false. Before discussing proofs, we must agree on what to accept as mathematical statements. We first require correct grammar for both words and mathematical symbols. Grammar eliminates both "food Mary Jane" and "1+ =".

The sentences "$1 + 1 = 3$" and "$1 + 1 < 3$" are mathematical statements, even though the first is false. Similarly, "$(1 + 1)^{4.3}$ is 96 more than 4000" is acceptable. We accept grammtically correct assertions where performing the indicated computations determines truth or falsity. This computational criterion extends to more complicated operations and to objects defined using sets and numbers.

We also consider general assertions about many numbers or objects, such as "the square of each odd integer is one more than a multiple of 8". This statement is closely related to the list of statements "$1^2 = 1 + 0 \cdot 8$", "$3^2 = 1 + 1 \cdot 8$", "$5^2 = 1 + 3 \cdot 8$", $\cdots$. We can describe many related mathematical statements by introducing a **variable**. If $P(x)$ is a mathematical statement when the variable x takes a specific value in a set S, then we accept as mathematical statements the sentences below. They have different meanings when S has more than one element.

"For all x in S, the assertion $P(x)$ is true."
"There exists an x in S such that the assertion $P(x)$ is true."

2.6. Example. The sentence "$x^2 - 1 = 0$" by itself is not a mathematical statement, but it becomes one when we specify a value for x. Consider

"For all $x \in \{1, -1\}$, $x^2 - 1 = 0$."
"For all $x \in \{1, 0\}$, $x^2 - 1 = 0$."
"There exists $x \in \{1, 0\}$ such that $x^2 - 1 = 0$."

All three are mathematical statements. The first is true; there are two values of x to check, and each satisfies the conclusion. The second statement is false, and the third is true. ∎

If it is not possible to assign "True" or "False" to an assertion, then it is not a mathematical statement. Consider the sentence "This statement is false"; call it P. If the words "this statement" in P refer to another sentence Q, then P has a truth value. If "this sentence" refers to P itself, then P must be false if it is true, and true if it is false! In this case, P has no truth value and is not a mathematical statement.

2.7. Definition. We use uppercase "$P, Q, R \cdots$" to denote mathematical statements. The truth or falsity of a statement is its **truth value**. Negating a statement reverses its truth value. We use $\neg$ to indicate **negation**, so "$\neg P$" means "not P". If P is false, then $\neg P$ is true.

In the statement "For all x in S, $P(x)$ is true", the variable x is **universally quantified**. We write this as $(\forall x \in S)P(x)$ and say that $\forall$ is a **universal quantifier**. In "There exists an x in S such that

$P(x)$ is true", the variable x is **existentially quantified**. We write this as $(\exists x \in S)P(x)$ and say that $\exists$ is an **existential quantifier**. The set of allowed values for a variable is its **universe**. ∎

2.8. Remark. *English words that express quantification.* Typically, "every" and "for all" represent universal quantifiers, while "some" and "there is" represent existential quantifiers. We can also express universal quantification by referring to an arbitrary element of the universe, as in "Let x be an integer," or "A student failing the exam will fail the course". Below we list common indicators of quantification.

Universal (∀)	(helpers)	Existential (∃)	(helpers)
for [all], for every		for some	
if	then	there exists	such that
whenever, for, given		at least one	for which
every, any	satisfies	some	satisfies
a, arbitrary	must, is	has a	such that
let	be		

The "helpers" may be absent. Consider "The square of a real number is nonnegative." This means $x^2 \geq 0$ for *every* $x \in \mathbb{R}$; it is not a statement about one real number and cannot be verified by an example. ∎

In conversation, a quantifier may appear after the expression it quantifies. "I drink whenever I eat" differs from "Whenever I eat, I drink" only in what is emphasized. Similarly, we easily understand "The AGM Inequality states that $(a + b)/2 \geq \sqrt{ab}$ for every pair a, b of positive real numbers" and "The value of $x^2 - 1$ is 0 for some x between 0 and 2". These quantifiers appear at the end for smoother reading. Error is unlikely in sentences with only one quantifier, but the order of quantification matters when there is more than one.

2.9. Remark. *Order of quantifiers.* We adopt a convention to avoid ambiguity. Consider "If n is even, then n is the sum of two odd numbers". Letting E and O be the sets of even and odd integers, and letting $P(n, x, y)$ be "$n = x + y$", the sentence becomes

$$(\forall n \in E)(\exists x, y \in O)P(n, x, y).$$

In this format, the value chosen for a quantified variable remains unchanged for later expressions but can be chosen in terms of variables quantified earlier. When we reach $(\exists x, y \in O)P(n, x, y)$, we treat "$n$" as a constant, already chosen. We use the same convention when writing mathematics in English: quantifiers appear in order at the beginning of the sentence so that the value of each variable is chosen independently of subsequently quantified variables. ∎

2.10. Example. *Parameters and implicit quantifiers.* Consider the exercise "Let a and b be real numbers. Prove that the equation $ax^2 + bx = a$ has a real solution." Using quantifiers, this becomes $(\forall a, b \in \mathbb{R})(\exists x \in \mathbb{R})(ax^2 + bx = a)$. In solving the problem, we treat a and b as *parameters*. Although these are variables and we must find a solution for each choice of these variables, the scope of the quantification is that we treat a and b as constants when we study x.

We find a suitable x in terms of a and b. When $a = 0$, $x = 0$ works for all b. When $a \neq 0$, the quadratic formula tells us that $x = (-b + \sqrt{b^2 + 4a^2})/2a$ works. This is real (since positive real numbers have square roots), and it satisfies the equation.

The negative square root also yields a solution. We do not need it, because the statement asked only for the existence of a solution. ∎

2.11. Example. *Order of quantifiers.* Compare the statements below.

$$(\forall x \in A)(\exists y \in B)P(x, y) \qquad (\exists y \in B)(\forall x \in A)P(x, y)$$

Regardless of the meanings of A, B, P, the second statement always implies the first. The first statement is true if for each x we can pick a y that "works". For the second statement to be true, there must be a single y that will always work, no matter which x is chosen.

Simple examples clarify the distinction. Let A be the set of children, let B be the set of parents, and let $P(x, y)$ be "y is the parent of x". The first statement is true, but the second statement is too strong and is not true. Another example occurred in Example 2.5, with $A = B = \mathbb{R}$, and with $P(x, y)$ being "$x = y^3$". Consider also the statement in Remark 2.9.

Sometimes both statements are true. For example, let $A = B = \mathbb{R}$, and let $P(x, y)$ be "$xy = 0$". ∎

2.12. Remark. *Negation of quantified statements.* After placing a statement involving quantifiers in the conventional order, negating the statement is easy. If it is false that $P(x)$ is true for every value of x, then there must be some value of x such that $P(x)$ is false, and vice versa. Similarly, if it is false that $P(x)$ is true for some value of x, then $P(x)$ is false for every value of x. Thus in notation,

$\neg[(\forall x)P(x)]$ has the same meaning as $(\exists x)(\neg P(x))$.
$\neg[(\exists x)P(x)]$ has the same meaning as $(\forall x)(\neg P(x))$.

Note that when using logical symbols, we may add matched parentheses or brackets to clarify grouping. ∎

Understanding negation of quantified statements by passing the negation through the quantifier and changing the type of quantifier is imperative for understanding the mathematics in this book.

When negating quantified statements with specified universes, one must not change the universe of potential values. Also, when negating $(\forall x)P(x)$ or $(\exists x)P(x)$, it may be that $P(x)$ itself is a quantified statement.

2.13. Example. *Negation involving universes.* The negation of "Every Good Boy Does Fine" (a mnemonic for reading music) is "some good boy does not do fine"; it says nothing about bad boys. The negation of "Every chair in this room is broken" is "Some chair in this room is not broken"; it says nothing about chairs outside this room.

Similarly, the negation of the statement $(\forall n \in \mathbb{N})(\exists x \in A)(nx < 1)$ is $(\exists n \in \mathbb{N})(\forall x \in A)(nx \geq 1)$. The negated sentence means that the set A has a lower bound that is the reciprocal of an integer. It does not mention values of n outside $\mathbb{N}$ or values of x outside A. ∎

2.14. Example. Let us rephrase "It is false that every classroom has a chair that is not broken". The quantifiers make it improper to cancel the "double negative"; the sentence "every classroom has a chair that is broken" has a different meaning.

The original statement has a universal quantifier ("every") and an existential quantifier ("has a"). By successively negating these quantifiers, we obtain first "There is a classroom that has no chair that is not broken" and then "There is a classroom in which every chair is broken".

We can also express this manipulation symbolically. Let R denote the set of classrooms. Given a room r, let $C(r)$ denote the set of chairs in r. For a chair c, let $B(c)$ be the statement that c is broken. The successive statements (all having the same meaning) now become

$$\neg\big[(\forall r \in R)(\exists c \in C(r))(\neg B(c))\big]$$
$$(\exists r \in R)\big(\neg\big[(\exists c \in C(r))(\neg B(c))\big]\big)$$
$$(\exists r \in R)(\forall c \in C(r))B(c).$$ ∎

2.15. Example. In Definition 1.31, we defined bounded function. We negate this to obtain "f is *unbounded* if for every real number M, some real number x satisfies $|f(x)| > M$." In notation, the two conditions are

$$\text{bounded}: \quad (\exists M \in \mathbb{R})(\forall x \in \mathbb{R})(|f(x)| \leq M)$$
$$\text{unbounded}: \quad (\forall M \in \mathbb{R})(\exists x \in \mathbb{R})(|f(x)| > M).$$

Thus unboundedness implies $(\forall n \in \mathbb{N})(\exists x_n \in \mathbb{R})(|f(x_n)| > n)$. ∎

COMPOUND STATEMENTS

The negation of a logical statement is another logical statement. We can also use the connectives "and", "or", "if and only if", and "implies" to build compound statements. For each choice of truth values for the component statements, the compound statement has a specified truth value; this constitutes the defininition of the connective.

2.16. Definition. *Logical connectives.* In the following table, we define the operations named in the first column by the truth values specified in the last column.

Name	Symbol	Meaning	Condition for truth
Negation	$\neg P$	not P	P false
Conjunction	$P \wedge Q$	P and Q	both true
Disjunction	$P \vee Q$	P or Q	at least one true
Biconditional	$P \Leftrightarrow Q$	P if & only if Q	same truth value
Conditional	$P \Rightarrow Q$	P implies Q	Q true whenever P true

2.17. Remark. *Disjunctions.* The meaning of "or" in mathematics differs from its common usage in English. In response to "Are you going home or not?", the answer "Yes" causes annoyance despite being logically correct; in common English the word "or" means "one or the other but not both". In mathematics, this usage is **exclusive-or**; we reserve **or** for disjunction.

Disjunction is more common in mathematics than exclusive-or because *and* and *or* act as quantifiers. A conjunction is true if *all* of its component statements are true; thus *and* is a universal quantifier. A disjunction is true if *at least one* of its component statements is true; thus *or* is an existential quantifier. ∎

In the conditional statement $P \Rightarrow Q$, we call P the **hypothesis** and Q the **conclusion**. The statement $Q \Rightarrow P$ is the **converse** of $P \Rightarrow Q$.

2.18. Remark. *Conditionals.* Conditional statements are the only type in Definition 2.16 whose meaning changes when P and Q are interchanged. There is no general relationship between the truth values of $P \Rightarrow Q$ and $Q \Rightarrow P$. Consider three statements about a real number x: P is "$x > 0$", Q is "$x^2 > 0$", and R is "$x + 1 > 1$". Here $P \Rightarrow Q$ is true but $Q \Rightarrow P$ is false. On the other hand, both $P \Rightarrow R$ and $R \Rightarrow P$ are true.

Note that here x is a variable. We have dropped x from the notation for the statements because the context is clear. Technically, when we write $P \Rightarrow Q$ here, we mean $(\forall x \in \mathbb{R})(P(x) \Rightarrow Q(x))$.

A conditional statement is false when and only when the hypothesis is true and the conclusion is false. When the hypothesis is false, the conditional statement will be true regardless of what the conclusion says and whether it is true. For example, if S is "This book was published in the year 73", then $S \Rightarrow P$ is true, no matter what P is.

It may be helpful to read the conditional as "if-then" instead of "implies". Below we list several ways to say $P \Rightarrow Q$ in English. ∎

If P (is true), then Q (is true). P is true only if Q is true.
Q is true whenever P is true. P is a sufficient condition for Q.
Q is true if P is true. Q is a necessary condition for P.

When a logical statement is built from elementary statements using connectives, we treat the elementary statements as variables in the universe {True, False}. Given their values, Definition 2.16 yields the truth value of the full expression. A listing of these computations for each choice of truth values of the elementary statements is a **truth table.**

2.19. Example. We give one example of a truth table to emphasize again the meaning of conditional statements. We want to know whether the expression R given by $(P \Rightarrow Q) \Leftrightarrow ((\neg P) \vee Q)$ is always true, no matter what P and Q represent. Such an expression is called a **tautology**. Each of P and Q may be true or false; we consider all cases.   ∎

P	Q	$P \Rightarrow Q$	$\neg P$	$(\neg P) \vee Q$	R
T	T	T	F	T	T
T	F	F	F	F	T
F	T	T	T	T	T
F	F	T	T	T	T

Two logical expressions X, Y are **logically equivalent** if they have the same truth value for each assignment of truth values to the variables. Equivalences allow us to rephrase statements in more convenient ways.

2.20. Remark. *Elementary logical equivalences.* We may substitute P for $\neg(\neg P)$ whenever we wish, and vice versa. Similarly, $P \vee Q$ is equivalent to $Q \vee P$, and $P \wedge Q$ is equivalent to $Q \wedge P$. Whenever P and Q are statements, we may substitute the expression in the right column below for the corresponding expression in the left column (or vice versa); they always have the same truth value. We could verify these equivalences by manipulating symbols in truth tables, but it is more productive to understand them using the English meanings of the connectives.

a)	$\neg(P \wedge Q)$	$(\neg P) \vee (\neg Q)$
b)	$\neg(P \vee Q)$	$(\neg P) \wedge (\neg Q)$
c)	$\neg(P \Rightarrow Q)$	$P \wedge (\neg Q)$
d)	$P \Leftrightarrow Q$	$(P \Rightarrow Q) \wedge (Q \Rightarrow P)$
e)	$P \vee Q$	$(\neg P) \Rightarrow Q$
f)	$P \Rightarrow Q$	$(\neg Q) \Rightarrow (\neg P)$

Equivalences (a) and (b) present our understanding of "and" and "or" as universal and existential quantifiers, respectively, over their component statements (see Remark 2.17). These two equivalences are called **de Morgan's laws** in honor of the logician Augustus de Morgan (1806–1871).

Equivalences (c) and (d) restate the definitions of the conditional and biconditional. A conditional statement is false precisely when the hypothesis is true and the conclusion is false. The biconditional is true precisely when the conditional and its converse are both true.

Each side of (e) is false precisely when P fails and Q fails. Each side of (f) fails precisely when P is true and Q is false.   ∎

2.21. Remark. *Logical connectives and membership in sets.* Let $P(x)$ and $Q(x)$ be statements about an element x from a universe U. Often we write a conditional statement $(\forall x \in U)(P(x) \Rightarrow Q(x))$ as $P(x) \Rightarrow Q(x)$ or simply $P \Rightarrow Q$ with an implicit universal quantifier.

The hypothesis $P(x)$ can be interpreted as a universal quantifier in another way. With $A = \{x \in U : P(x) \text{ is true}\}$, the statement $P(x) \Rightarrow Q(x)$ can be written as $(\forall x \in A)Q(x)$.

Another interpretation of $P(x) \Rightarrow Q(x)$ uses set inclusion. With $B = \{x \in U : Q(x) \text{ is true}\}$, the conditional statement has the same meaning as the statement $A \subseteq B$. The converse statement $Q(x) \Rightarrow P(x)$ is equivalent to $B \subseteq A$; thus the biconditional $P \Leftrightarrow Q$ is equivalent to $A = B$.

We can alternatively interpret operations with sets using logical connectives and membership statements. When P is the statement of membership in A and Q is the statement of membership in B, the statement $A = B$ has the same meaning as $P \Leftrightarrow Q$. Below we list the correspondence for other set operations.   ∎

$$
\begin{array}{ccccc}
x \in A^c & \Leftrightarrow & \text{not } (x \in A) & \Leftrightarrow & \neg(x \in A) \\
x \in A \cup B & \Leftrightarrow & (x \in A) \text{ or } (x \in B) & \Leftrightarrow & (x \in A) \vee (x \in B) \\
x \in A \cap B & \Leftrightarrow & (x \in A) \text{ and } (x \in B) & \Leftrightarrow & (x \in A) \wedge (x \in B) \\
A \subseteq B & \Leftrightarrow & (\forall x \in A)(x \in B) & \Leftrightarrow & (x \in A) \Rightarrow (x \in B)
\end{array}
$$

The understanding of union and intersection in terms of quantifiers allows us to extend the definitions of union and intersection to apply to more than two sets. The **intersection** of a collection of sets consists of all elements that belong to all of the sets. The **union** of a collection of sets consists of all elements that belong to at least one of the sets.

2.22. Remark. The correspondence between $P \Leftrightarrow Q$ and $A = B$ in Remark 2.21 highlights an important phenomenon. Expressions that represent "being the same" can be interpreted as two instances of comparison. When x and y are numbers, the statement $x = y$ includes two pieces of information, $x \le y$ and $y \le x$. When A and B are sets, the equality $A = B$ includes two pieces of information, $A \subseteq B$ and $B \subseteq A$. For logical statements P and Q, similarly, $P \Leftrightarrow Q$ means both $P \Rightarrow Q$ and $Q \Rightarrow P$.

In some contexts, we prove equality by proving both comparisons. In other contexts, we can prove equality directly, by using manipulations that preserve the value, set, or meaning while transforming the first description into the second.   ∎

2.23. Example. *de Morgan's laws for sets*. In the language of sets, de Morgan's laws (Remark 2.20a,b) become (1) $(A \cap B)^c = A^c \cup B^c$, and (2) $(A \cup B)^c = A^c \cap B^c$. We verify (1) by translation into a logical equivalence about membership, leaving (2) to Exercise 50. Given an element x, let P be the property $x \in A$, and let Q be the property $x \in B$. Remarks 2.20–2.21 imply that

$$x \in (A \cap B)^c \iff \neg(P \wedge Q) \iff (\neg P) \vee (\neg Q) \iff (x \notin A) \vee (x \notin B)$$

Alternatively, a Venn diagram makes the reasoning clear. ∎

Although relationships between sets correspond to logical statements about membership, the two expressions tell the same story in different languages. One must not mix them. For example, $A \cap B$ is a set, not a statement; it has no truth value. The notation "$(A \cap B)^c \iff A^c \cup B^c$" has no meaning, but $(A \cap B)^c = A^c \cup B^c$ is true whenever A and B are sets.

ELEMENTARY PROOF TECHNIQUES

The business of mathematics is deriving consequences from hypotheses—that is, proving conditional statements. Although we prove some biconditionals by chains of equivalences, as in Example 2.23, usually we prove a biconditional by proving a conditional and its converse, as suggested by Remark 2.20d. Also, we can prove the universally quantified statement "$(\forall x \in A) Q(x)$" by proving the conditional statement "If $x \in A$, then $Q(x)$"; the two have the same meaning. (For example, consider the two sentences when A is the set of even numbers and $Q(x)$ is "x^2 is even".)

2.24. Remark. *Elementary methods of proving* $P \Rightarrow Q$. The *direct method* of proving $P \Rightarrow Q$ is to assume that P is true and then to apply mathematical reasoning to deduce that Q is true. When P is "$x \in A$" and Q is "$Q(x)$", the direct method considers an *arbitrary* $x \in A$ and deduces $Q(x)$. This must not be confused with the invalid "proof by example". The proof must apply to every member of A as a possible instance of x, because "$(x \in A) \Rightarrow Q(x)$" is a universally quantified statement.

Remark 2.20f suggests another method. The **contrapositive** of $P \Rightarrow Q$ is $\neg Q \Rightarrow \neg P$. The equivalence between a conditional and its contrapositive allows us to prove $P \Rightarrow Q$ by proving $\neg Q \Rightarrow \neg P$. This is the *contrapositive method*.

Remark 2.20c suggests another method. Negating both sides $(P \Rightarrow Q) \iff \neg[P \wedge (\neg Q)]$. Hence we can prove $P \Rightarrow Q$ by proving that P and $\neg Q$ cannot both be true. We do this by obtaining a contradiction after assuming both P and $\neg Q$. This is the *method of contradiction* or **indirect proof**. We summarize these methods below:

Direct Proof: Assume P, follow logical deductions, conclude Q.
Contrapositive: Assume $\neg Q$, follow deductions, conclude $\neg P$.
Method of Contradiction: Assume P and $\neg Q$, follow deductions,
  obtain a contradiction.   ∎

We begin with easy examples of the direct method, including statements used in proving Theorem 2.3.

2.25. Example. *If integers x and y are both odd, then $x + y$ is even.* Suppose that x and y are odd. By the definition of "odd", there exist integers
k, l such that $x = 2k + 1$ and $y = 2l + 1$. By the properties of addition and
the distributive law, $x + y = 2k + 2l + 2 = 2(k + l + 1)$. This is twice an
integer, so $x + y$ is even.
  The converse is false. When x, y are integers, it is possible that $x + y$
is even but x, y are not both odd. Compare this with the next example. ∎

2.26. Example. *An integer is even if and only if it is the sum of two odd
integers.* First we clarify what must be proved. Formally, the statement
is $(\forall x \in \mathbb{Z})[(\exists k \in \mathbb{Z})(x = 2k) \Leftrightarrow (\exists y, z \in O)(x = y + z)]$, where O is the set
of odd numbers. If $x = 2k$ is even, then $x = (2k - 1) + 1$, which expresses
x as the sum of two odd integers. Conversely, let y and z be odd. By
the definition of "odd", there exist integers k, l such that $y = 2k + 1$ and
$z = 2l + 1$. Then $y + z = 2k + 1 + 2l + 1 = 2(k + l + 1)$, which is even. ∎

2.27. Example. *If x and y are odd, then xy is odd.* If x and y are odd,
then there are integers k, l such that $x = 2k + 1$ and $y = 2l + 1$. Now
$xy = 4kl + 2k + 2l + 1 = 2(2kl + k + l) + 1$. Since this is one more than
twice an integer, xy also is odd.   ∎

A special case of Example 2.27 is "x odd $\Rightarrow x^2$ odd". Here the conclusion is "There is an integer m such that $x^2 = 2m + 1$". We can prove
an existential conclusion by providing an *example*: in this case a value
m (constructed in terms of x) such that the statement is true. The direct
method often succeeds when the conclusion is existentially quantified.

2.28. Example. *An integer is even if and only if its square is even.* If n is
even, then we can write $n = 2k$, where k is an integer. Now $n^2 = 4k^2 =
2(2k^2)$, proving that "$n$ even" implies "n^2 even" by the direct method. For
the converse, we want to prove "n^2 even implies n even", but this we have
already done! Since integers are even or odd, the desired implication is
the contrapositive of "n odd implies n^2 odd".   ∎

2.29. Remark. *Converse versus contrapositive.* Proving the biconditional
statement $P \Leftrightarrow Q$ requires proving one statement from each column below. Each statement is the converse of the other in its row. Each statement

is the contrapositive of the other in its column. Every conditional is equivalent to its contrapositive, so proving the two statements in one column would be proving the same fact twice.

$$P \Rightarrow Q \qquad Q \Rightarrow P$$
$$\neg Q \Rightarrow \neg P \qquad \neg P \Rightarrow \neg Q$$

For example, consider "the product of two nonzero real numbers is positive if and only if they have the same sign". The axioms for real numbers imply that if x and y have the same sign, then xy is positive. We might then argue, "Now suppose that xy is negative. This implies that x and y have opposite signs." This accomplishes nothing; we have proved the contrapositive of the first conditional, not its converse. Instead, we must prove "If xy is positive, then x and y have the same sign" or "If x and y have opposite signs, then xy is negative".

We can interpret the first line of the display above as the direct method and the second line as the contrapositive method. To include the method of contradiction, we could add the line below:

$$\neg(P \wedge \neg Q) \qquad \neg(Q \wedge \neg P). \qquad \blacksquare$$

The next example uses the contrapositive and illustrates that care must be taken to avoid unjustified assumptions.

2.30. Example. Consider the statement *"If $f(x) = mx + b$ and $x \neq y$, then $f(x) \neq f(y)$."* The direct method considers $x < y$ and $x > y$ separately and obtains $f(x) < f(y)$ or $f(x) > f(y)$. This unsatisfying analysis by cases results from "not equals" being a messier condition than "equals".

We can use the contrapositive to retain the language of equalities and reduce analysis by cases. When $f(x) = f(y)$, we obtain $mx + b = my + b$ and then $mx = my$. If $m \neq 0$, then we obtain $x = y$.

If $m = 0$, then we cannot divide by m, and actually the statement is false. The difficulty is that m is a variable in the statement we want to prove, and we cannot determine its truth without quantifying m. The statement is true if and only if $m \neq 0$. $\blacksquare$

A universally quantified statement like "$(\forall x \in U)[P(x) \Rightarrow Q(x)]$" can be disproved by finding an element x in U such that $P(x)$ is true and $Q(x)$ is false. Such an element x is a **counterexample**. In Example 2.30, $m = 0$ is a counterexample to a claim that the implication holds for all m.

We continue with another example of proof by contrapositive.

2.31. Example. *If a is less than or equal to every real number greater than b, then $a \leq b$.* The direct method goes nowhere, but when we say "suppose not", the light begins to dawn. If $a > b$, then $a > \frac{a+b}{2} > b$. Thus a is *not* less than or equal to every number greater than b. We have proved the contrapositive of the desired statement. $\blacksquare$

When the hypothesis of $P \Rightarrow Q$ is universally quantified, its negation is existentially quantified. This can make the contrapositive easy; given $\neg Q$, we need only construct a counterexample to P. This is the scenario in Example 2.31; having assumed $a > b$, we need only construct a counterexample to "a is less than every real x that is greater than b".

The method of contradiction proves $P \Rightarrow Q$ by proving that P and $\neg Q$ cannot both hold, thereby proving that $P \Rightarrow Q$ cannot be false.

2.32. Example. *Among the numbers $y_1, \ldots, y_n$, some number is as large as the average.* Let $Y = y_1 + \cdots + y_n$. The **average** z is Y/n.

An indirect proof of the claim begins, "suppose that the conclusion is false". Thus $y_i < z$ for all y_i in the list. If we sum these inequalities, we obtain $Y < nz$, but this contradicts the definition of z, which yields $Y = nz$. Hence the assumption that each element is too small must be false.

A direct proof constructs the desired number. Let y^* be the largest number in the set. We prove that this candidate *is* as large as the average. Since $y_i \leq y^*$ for all i, we sum the inequalties to obtain $Y \leq ny^*$ and then divide by n to obtain $z \leq y^*$.   ∎

In Example 2.32, we did not derive the negation of the hypothesis; we obtained a different contradiction. This is the method of contradiction. Like the contrapositive method, it begins by assuming $\neg Q$ when proving $P \Rightarrow Q$. We need not decide in advance whether to deduce $\neg P$ or to use both P and $\neg Q$ to obtain some other contradiction.

2.33. Example. *There is no largest real number.* If there is a largest real number z, then for all $x \in \mathbb{R}$, we have $z \geq x$. When x is the real number $z + 1$, this yields $z \geq z + 1$. Subtracting z from both sides yields $0 \geq 1$. This is a contradiction, and thus there is no largest real number.   ∎

The method of contradiction works well when the conclusion is a statement of non-existence or impossibility, because negating the conclusion provides an *example* to use, like p/q in the proof of Theorem 2.3 or z in Example 2.33. In one sense the method of contradiction ("indirect proof") has more power than the contrapositive, since we start with more information (P *and* $\neg Q$), but in another sense it is less satisfying, because we start with a situation that (we hope) cannot be true.

2.34. Remark. *The consequences of false statements.* Recall that a conditional statement is false only if the hypothesis is true and the conclusion is false. When the hypothesis cannot be true, we say the conditional follows *vacuously*. Similarly, every statement universally quantified over an empty set is true; when there are no dogs in the class, the statement "Every dog in the class has three heads" is true. In contrast, every statement existentially quantified over an empty set is false; when there are no dogs in the class, the statement "Some dog in the class has four legs" is false!

Returning to the conditional, we have argued that $P \Rightarrow Q$ is true whenever P is false. This explains why a proof containing a single error in reasoning cannot be considered "nearly correct"; we can derive any conclusion from a single false statement (see Exercise 44a). Bertrand Russell (1872–1970) once stated this in a public lecture and was challenged to start with the assumption that $1 = 2$ and prove that he was God. He replied, "Consider the set {Russell, God}. If $1 = 2$, then the two elements of the set are one element, and therefore Russell = God." ∎

Students sometimes wonder about the meanings of the words "theorem", "lemma", "corollary", etc. The usage of these words is part of mathematical convention, like the notation $f: A \to B$ for functions and the designations $\mathbb{N}, \mathbb{Z}, \mathbb{Q}, \mathbb{R}$ for the number systems. (By the way, $\mathbb{Q}$ stands for "quotient" and $\mathbb{Z}$ stands for "Zahlen", the German word for numbers.)

In Greek, *lemma* means "premise" and *theorema* means "thesis to be proved". Thus a theorem is a major result whose proof may require considerable effort. A lemma is a lesser statement, usually proved in order to help prove other statements. A proposition is something "proposed" to be proved; typically this is a less important statement or requires less effort than a theorem. The word *corollary* comes from Latin, as a modification of a word meaning "gift"; a corollary follows easily from a theorem or proposition, without much additional work.

Theorems, Propositions, Corollaries, and Lemmas may all be used to prove other results. In this book, these embody the central mathematical development, while Examples, Solutions, Applications, and Remarks are particular uses of or commentary on the mathematics. These two streams are interwoven but can be distinguished by the titles of the items. The first stream comprises the mathematical results that students might want to remember for later application, while the second illuminates the first and provides additional examples of problem-solving.

HOW TO APPROACH PROBLEMS

In this chapter we have discussed the language of mathematics and elementary techniques of proof. We review some of these issues and discuss several additional ones that arise when solving problems.

Methods of proof.

The first step is making sure that one understands exactly what the problem is asking. Definitions may provide a road map for what needs to be verified. Sometimes, the desired statement follows from a theorem already proved, and then one needs to verify that its hypotheses hold.

Most problems request proofs of conditional statements. These state that given circumstances produce certain results. Such sentences are often written using "if" and "then", but implication can be expressed with universal quantifiers and in many other ways (see Remark 2.8, Remark 2.18, and Exercise 10). Examples cannot provide proofs of such statements. Implications need to be proved in Exercises 34–42.

The elementary techniques for proving implications are direct proof, proof by contradiction, and proof of the contrapositive. The latter two methods are called "indirect" proofs. When seeking a direct proof, one can work from both ends. List statements that follow from the hypothesis. List statements that suffice to imply the conclusion. When some statement appears in both lists, the problem is solved.

When unsuccesful with the direct method, consider what would happen if the conclusion were false. If this leads to impossibility of some consequence of the hypothesis (or of other known facts), then again the problem is solved, using the method of contradiction. If the negation of the hypothesis is obtained, then the contrapositive has been proved.

Students often wonder when to use indirect proof. The form of the conclusion can provide a clue; when its negation provides something useful to work with, indirect proof may be appropriate. This can happen with obvious-sounding statements like Example 2.31. Often indirect proof is appropriate for statements of nonexistence, as in Theorem 2.3, Example 2.33, and Exercise 40. The negation of the conclusion provides an example, an object with specified properties. (In contrast, one can often prove that something *does* exist by constructing an example and proving that it has the desired properties; this is the direct method.)

Be aware of hypotheses and quantifiers.

An implication is true when the truth of its hypotheses guarantees the truth of its conclusion. The sentence "if we add two even integers, then the result is even" is true and easily proved, but the sentence "if we add two integers, then the result is even" is false. The second sentence is obviously missing a hypothesis (that the integers are even) that is needed to make the conclusion true.

In more subtle statements, the same principles apply. Carefully distinguish the hypotheses and the desired conclusions. Remember that hypotheses can be expressed as universal quantifiers: "for all $x \in A$" means the same as "if $x \in A$". In writing a solution, check where the hypotheses are used. If a hypothesis is not used, then either it is unnecessary (and the proof yields a stronger statement) or an error has been made.

Solving a problem may require determining whether a statement with many quantified variables is true or false. One must be able to identify the universal and existential quantifiers, put them in proper order (see items 2.9–2.11), and negate a quantified statement (see items 2.12–2.15).

More about cases.

A universally quantified statement must be proved for all instances of the variables. This includes statements phrased in the singular, like "The square of an even number is even." Writing $(-4)^2 = 16 = 2 \cdot 8$ does not prove this, because here "an" means *each individual*. The sentence means "If x is an even number, then x^2 is an even number." Similarly, "Let x be a positive real number" and "For $x > 0$" are universal quantifications; the claim to be proved must be proved for every positive real number x.

Analysis by cases can arise when an argument is valid for some instances but not for all. Consider showing that $x(x + 1)/2$ is an integer whenever x is an integer. When x is even, we write $x = 2k$ and compute $2k(2k+1)/2 = k(2k+1)$, where k is an integer. For odd x, we need a different computation. We can avoid cases by observing that one of $\{x, x + 1\}$ is even and is divisible by 2. Combining cases via a unified argument leads to a concise solution that captures the essence of the proof.

When several cases are treated in the same way, it may be possible to reduce to a single case by using symmetry. We did this in proving Theorem 2.2. Having disposed of the case where all four coefficients are zero (which uses a different argument), we may assume that some coefficient is nonzero. We would use the same arguments no matter which it is. By writing the equations in the opposite order and/or switching the names of the variables, we can arrange that the coefficient d is nonzero. We say that symmetry allows us to reduce to the case where d is nonzero.

Similarly, when proving a statement about distinct real numbers x, y, it may be helpful to assume by symmetry that $x > y$. The same argument with the roles of x and y switched would apply when $y > x$, and we use the symmetry in the problem to avoid writing out the argument twice.

On the other hand, sometimes a problem becomes simpler when we introduce an additional hypothesis. This leads to two cases: when the assumption is true and when it is false. Consider Exercise 33. The first child knows that her hat is black or red. She considers these two cases to seek a contradiction that will eliminate one. Perhaps further assumptions will be needed, leading to subcases. Exercise 32 is similar; we consider various assumptions. Assuming that Person A tells the truth yields an immediate contradiction; knowing that A lies leads to further conclusions.

This method is known informally as "process of elimination". If a particular assumption seems to lead nowhere, try another! Remember that eventually all possibilities must be considered. For example, when the roles of variables x and y are not interchangeable in a problem, we cannot use symmetry to reduce to $x \leq y$, but considering the cases $x < y$, $x = y$, and $x > y$ separately might lead to different arguments that work.

Finally, beware of overlooking cases that result from introducing unwanted hypotheses. In particular, be aware of the conditions under which symbolic manipulations are valid. Since we cannot divide by zero, the equation $y = mx$ can be solved for x only when $m \neq 0$. For all real y and all nonzero real m, there is a unique x with $y = mx$. The case $m = 0$ has not been considered and must be treated in some other way.

Taking square roots also requires care. For example, Exercise 1.19 has no solution for some choices of the perimeter p and area a, because the algebraic solution involves a square root. Square roots exists only for nonnegative numbers; this constrains the values p and a.

Equations and algebraic manipulations.

Consider the equation $x^2 - 10x + 5 = -20$. Manipulating the equation yields $(x - 5)^2 = 0$, which implies $x = 5$. This can be interpreted as the conditional statement "If the equation holds, then $x = 5$". Checking the answer shows the converse assertion "If $x = 5$, then the equation holds". Together, the two steps yield the statement "The set of solutions to the equation $x^2 - 10x + 5 = -20$ is $\{5\}$".

Consider also the equation $x^2 = 5x$. Dividing both sides by x yields $x = 5$. Checking 5 in the equation yields "if $x = 5$, then the equation holds". The statement "If the equation holds, then $x = 5$" is false. The correct assertion is "If the equation holds, then $x = 0$ or $x = 5$". The problem is that the division was valid only under the hypothesis that $x \neq 0$. The solution in the remaining case was lost.

Algebraic manipulations can also introduce extraneous solutions. Consider the equation $x = 4$. If we next write $x^2 = 4x$, then we obtain $x^2 - 4x = 0$, with solutions $x = 4$ and $x = 0$. Multiplying by x introduced the extraneous solution $x = 0$; it changed the solution set. Substituting the results of invalid manipulations into the original equation may or may not expose the error.

Multiplying both sides of an equation in x by an expression $f(x)$ introduces all the zeroes of f as solutions; some may be extraneous. Dividing by $f(x)$ is invalid when $f(x)$ can be zero; in this case solutions may be lost. When manipulating an equation to seek equivalent statements, one must check that the set of solutions never changes or analyze separately the cases where it may change.

2.35. Example. The following argument alleges to prove that $2 = 1$; it must be wrong! What is the flaw?

Let x, y be real numbers, and suppose that $x = y$. This yields $x^2 = xy$, which implies $x^2 - y^2 = xy - y^2$ by subtracting y^2 from both sides. Factoring yields $(x + y)(x - y) = y(x - y)$, and thus $x + y = y$. In the special case $x = y = 1$, we obtain $2 = 1$. ∎

Sets and membership.

Various exercises in this chapter involve identities involving unions, intersections, and differences of sets. These can be understood using Venn

diagrams. Equality of expressions involving sets can be proved by showing that an element belongs to the set given by one expression if and only
if it belongs to the set given by the other.

Reasoning about sets and subsets is parallel to reasoning about conditional statements. The set-theoretic statement $S \subseteq T$ can be interpreted
as "If $x \in S$, then $x \in T$". Thus the logical statement $P \Leftrightarrow Q$ is parallel to
the set-theoretic equality $S = T$ (see Remarks 2.21–2.22).

Identities involving operations on sets (Exercises 50–53) and equivalences involving logical connectives and statements (Exercises 43–46) are
universally quantified, with variables representing sets or statements.
Thus the proof must be valid for all instances.

In several of the exercises, two sets of real numbers are specified by
numerical constraints, and the problem is to show that the two sets are
the same. One can prove that each set is contained in the other, or one can
manipulate the constraints in ways that do not change the set of solutions.
In either approach, words should be used to explain the arguments.

Communicating mathematics.

Solutions to problems should be written using sentences that explain
the argument. Notation introduced to represent concepts in the discussion should be clearly defined, and a symbol should not be used with
different meanings in a single discussion.

A convincing proof cannot depend on asking the reader to guess what
the writer intended. A well-written argument may begin with an overview
or with an indication of the method of proof. Such an indication is particularly helpful when using the contrapositive or the method of contradiction.

When the writer gives no explanation of the method of proof and
merely lists some formulas, the reader can only assume that a direct
proof is being given, with each line derived from the previous line. This
gets students into trouble when they reduce a desired statement to a
known statement. In attempting to prove the AGM Inequality for all
nonnegative real numbers x, y, some students will write

$$\sqrt{xy} \le (x + y)/2$$
$$xy \le (x + y)^2/4$$
$$4xy \le x^2 + 2xy + y^2$$
$$0 \le x^2 - 2xy + y^2$$
$$0 \le (x - y)^2, \quad \text{which is true.}$$

Here the student has derived a true statement from the desired statement; this does not prove the desired statement. Within the set of pairs of
nonnegative real numbers, these manipulations of the inequality have not
changed the set of solutions, so the steps are reversible to obtain the desired inequality. Without words to indicate that this is what is intended,

the proof is wrong. Note that the "proof" never used the hypothesis that $x, y \geq 0$, and when $x = y = -1$ the claimed inequality fails.

One must always distinguish a statement from its converse. Deriving a true statement Q from the desired statement P does not prove P! Let P be the assertion "$x + 1 = x + 2$". When we multiply both sides of P by 0, we obtain the true statement "$0 = 0$"; call this Q. Although Q is true for all x and we have proved $P \Rightarrow Q$, the statement P is true for no x.

EXERCISES

2.1. Find the flaw in Example 2.35.

2.2. Show that the following statement is false: "If a and b are integers, then there are integers m, n such that $a = m + n$ and $b = m - n$." What can be added to the hypothesis of the statement to make it true?

2.3. Consider the following sentence: "If a is a real number, then $ax = 0$ implies $x = 0$". Write this sentence using quantifiers, letting $P(a, x)$ be the assertion "$ax = 0$" and $Q(x)$ be the assertion "$x = 0$". Show that the implication is false, and find a small change in the quantifiers to make it true.

2.4. Let A and B be sets of real numbers, let f be a function from $\mathbb{R}$ to $\mathbb{R}$, and let P be the set of positive real numbers. Without using words of negation, for each statement below write a sentence that expresses its negation.

    a) For all $x \in A$, there is a $b \in B$ such that $b > x$.

    b) There is an $x \in A$ such that, for all $b \in B$, $b > x$.

    c) For all $x, y \in \mathbb{R}$, $f(x) = f(y) \Rightarrow x = y$.

    d) For all $b \in \mathbb{R}$, there is an $x \in \mathbb{R}$ such that $f(x) = b$.

    e) For all $x, y \in \mathbb{R}$ and all $\epsilon \in P$, there is a $\delta \in P$ such that $|x - y| < \delta$ implies $|f(x) - f(y)| < \epsilon$.

    f) For all $\epsilon \in P$, there is a $\delta \in P$ such that, for all $x, y \in \mathbb{R}$, $|x - y| < \delta$ implies $|f(x) - f(y)| < \epsilon$.

2.5. (−) Prove the following statements.

    a) For all real numbers y, b, m with $m \neq 0$, there is a unique real number x such that $y = mx + b$.

    b) For all real numbers y, m, there exist $b, x \in \mathbb{R}$ such that $y = mx + b$.

2.6. (−) *Usage of language.*

    a) The following sentence appeared on a restaurant menu: "Please note that every alternative may not be available at this time". Describe the unintended meaning. Rewrite the sentence to state the intended meaning clearly.

    b) Give an example of an English sentence that has different meanings depending on inflection, pronunciation, or context.

2.7. (−) Describe how the notion of an *alibi* in a criminal trial fits into our discussion of conditional statements.

2.8. From outside mathematics, give an example of statements A, B, C such that A and B together imply C, but such that neither A nor B alone implies C.

2.9. (−) The negation of the statement "No slow learners attend this school" is:[†]
a) All slow learners attend this school.
b) All slow learners do not attend this school.
c) Some slow learners attend this school.
d) Some slow learners do not attend this school.
e) No slow learners attend this school.

2.10. Express each of the following statements as a conditional statement in "if-then" form or as a universally quantified statement. Also write the negation (without phrases like "it is false that").
a) Every odd number is prime.
b) The sum of the angles of a triangle is 180 degrees.
c) Passing the test requires solving all the problems.
d) Being first in line guarantees getting a good seat.
e) Lockers must be turned in by the last day of class.
f) Haste makes waste.
g) I get mad whenever you do that.
h) I won't say that unless I mean it.

2.11. (!) Suppose I have a penny, a dime, and a dollar, and I say, "If you make a true statement, I will give you one of the coins. If you make a false statement, I will give you nothing." What should you say to obtain the best coin?

•   •   •   •   •

2.12. A telephone bill y (in cents) is determined by $y = mx + b$, where x is the number of calls during the month, and b is a constant monthly charge. Suppose that the bill is \$5.48 when 8 calls are made and is \$5.72 when 12 calls are made. Determine what the bill will be when 20 calls are made.

2.13. "In one year, my wife will be one-third as old as my house. In nine years, I will be half as old as my house. I am ten years older than my wife. How old are my house, my wife, and I?" Answer the question, stating the needed equations.

2.14. A **circle** is the set of ordered pairs $(x, y) \in \mathbb{R}^2$ such that x and y satisfy an equation of the form $x^2 + y^2 + ax + by = c$, where $c > -(a^2 + b^2)/4$. The circle is specified by the parameters a, b, c.
a) Using this definition, give examples of two circles such that
  i) the circles do not intersect.
  ii) the circles intersect in exactly one common element.
  iii) the circles intersect in two common elements.
b) Explain why the parameter c is restricted as given.

2.15. *The quadratic formula, revisited.* We derive the quadratic formula by solving a system of linear equations for the two unknown solutions. The equation $ax^2 + bx + c = 0$ with $a \neq 0$ has real solutions r, s if and only if $ax^2 + bx + c = a(x - r)(x - s)$ (see Exercise 1.20). The calculation below shows that the factorization exists if and only if $b^2 - 4ac \geq 0$ and expresses r, s in terms of a, b, c.
a) By equating coefficients of corresponding powers of x, obtain the equations

---

[†]From the 1955 High School Mathematics Exam (C. T. Salkind, *Annual High School Mathematics Examinations 1950–1960*, Math. Assoc. Amer. 1961, p. 37.)

$r + s = -b/a$ and $rs = c/a$. Use these to prove that $(r - s)^2 = (b^2 - 4ac)/a^2$.

b) From (a), obtain $r + s = -b/a$ and $r - s = \sqrt{b^2 - 4ac}/a$. Solve this system for r, s in terms of a, b, c.

c) What happens if the second equation in (b) is $r - s = -\sqrt{b^2 - 4ac}/a$?

2.16. (!) Let f be a function from $\mathbb{R}$ to $\mathbb{R}$.

a) Prove that f can be expressed in a unique way as the sum of two functions g and h such that $g(-x) = g(x)$ for all $x \in \mathbb{R}$ and $h(-x) = -h(x)$ for all $x \in \mathbb{R}$. (Hint: Find a system of linear equations for the unknowns $g(x)$ and $h(x)$ in terms of the known values $f(x)$ and $f(-x)$.)

b) When f is a polynomial, express g and h in terms of the coefficients of f.

2.17. Given $f: \mathbb{R} \to \mathbb{R}$, let $g(x) = \frac{x}{2} + \frac{x}{f(x)-1}$ for all x such that $f(x) \neq 1$. Suppose $g(x) = g(-x)$ for all such x. Prove that $f(x)f(-x) = 1$ for all such x.

2.18. (!) Given a polynomial p, let A be the sum of the coefficients of the even powers, and let B be the sum of the coefficients of the odd powers. Prove that $A^2 - B^2 = p(1)p(-1)$.

2.19. Abraham Lincoln said, "You can fool all of the people some of the time, and you can fool some of the people all of the time, but you can't fool all of the people all of the time." Write this sentence in logical notation, negate the symbolic sentence, and state the negation in English. Which statement seems to be true?

2.20. Using quantifiers, explain what it would mean for the first player to have a "winning strategy" in Tic-Tac-Toe. (Don't consider whether the statement is true.)

2.21. Consider the sentence "For every integer $n > 0$ there is some real number $x > 0$ such that $x < 1/n$". Without using words of negation, write a complete sentence that means the same as "It is false that for every integer $n > 0$ there is some real number $x > 0$ such that $x < 1/n$". Which sentence is true?

2.22. Let f be a function from $\mathbb{R}$ to $\mathbb{R}$. Without using words of negation, write the meaning of "f is not an increasing function".

2.23. Consider $f: \mathbb{R} \to \mathbb{R}$. Let S be the set of functions defined by putting $g \in S$ if there exist positive constants $c, a \in \mathbb{R}$ such that $|g(x)| \leq c|f(x)|$ for all $x > a$. Without words of negation, state the meaning of "$g \notin S$". (Comment: The set S (written as $O(f)$) is used to compare the "order of growth" of functions.)

2.24. In simpler language, describe the meaning of the following two statements and their negations. Which one implies the other, and why?

a) There is a number M such that, for every x in the set S, $|x| \leq M$.

b) For every x in the set S, there is a number M such that $|x| \leq M$.

2.25. For $a \in \mathbb{R}$ and $f: \mathbb{R} \to \mathbb{R}$, show that (a) and (b) have different meanings.

a) $(\forall \epsilon > 0)(\exists \delta > 0)[(|x - a| < \delta) \Rightarrow (|f(x) - f(a)| < \epsilon)]$

b) $(\exists \delta > 0)(\forall \epsilon > 0)[(|x - a| < \delta) \Rightarrow (|f(x) - f(a)| < \epsilon)]$

2.26. For $f: \mathbb{R} \to \mathbb{R}$, which of the statements below implies the other? Does there exist a function for which both statements are true?

a) For every $\epsilon > 0$ and every real number a, there is a $\delta > 0$ such that $|f(x) - f(a)| < \epsilon$ whenever $|x - a| < \delta$.

b) For every $\epsilon > 0$, there is a $\delta > 0$ such that $|f(x) - f(a)| < \epsilon$ whenever $|x - a| < \delta$ and a is a real number.

2.27. (+) For $c \in \mathbb{R}$ and $f: \mathbb{R} \rightarrow \mathbb{R}$, interpret each statement below.

a) For all $x \in \mathbb{R}$ and all $\delta > 0$, there exists $\epsilon > 0$ such that $|x| < \delta$ implies $|f(x) - c| < \epsilon$.

b) For all $x \in \mathbb{R}$, there exists $\delta > 0$ such that, for all $\epsilon > 0$, we have $|x| < \delta$ implies $|f(x) - c| < \epsilon$.

2.28. (!) Consider the equation $x^4 y + ay + x = 0$.

a) Show that the following statement is false. "For all $a, x \in \mathbb{R}$, there is a unique y such that $x^4 y + ay + x = 0$."

b) Find the set of real numbers a such that the following statement is true. "For all $x \in \mathbb{R}$, there is a unique y such that $x^4 y + ay + x = 0$."

2.29. (!) *Extremal problems.*

a) Let f be a real-valued function on S. In order to prove that the minimum value in the image of f is β, two statements must be proved. Express each of these statements using quantifiers.

b) Let T be the set of ordered pairs of positive real numbers. Define $f: T \rightarrow \mathbb{R}$ by $f(x, y) = \max\{x, y, \frac{1}{x} + \frac{1}{y}\}$. Determine the minimum value in the image of f. (Hint: What must a pair achieving the minimum satisfy?)

2.30. (!) Consider tokens that have some letter written on one side and some integer written on the other, in unknown combinations. The tokens are laid out, some with letter side up, some with number side up. Explain which tokens must be turned over to determine whether these statements are true:

a) Whenever the letter side is a vowel, the number side is odd.

b) The letter side is a vowel if and only if the number side is odd.

2.31. Which of these statements are believable? (Hint: Consider Remark 2.88.)

a) "All of my 5-legged dogs can fly."

b) "I have no 5-legged dog that cannot fly."

c) "Some of my 5-legged dogs cannot fly."

d) "I have a 5-legged dog that cannot fly."

2.32. A fraternity has a rule for new members: each must always tell the truth or always lie. They know who does which. If I meet three of them on the street and they make the statements below, which ones (if any) should I believe?

A says: "All three of us are liars."

B says: "Exactly two of us are liars."

C says: "The other two are liars."

2.33. Three children are in line. From a collection of two red hats and three black hats, the teacher places a hat on each child's head. The third child sees the hats on the first two, the second child sees the hat on the first, and the first child sees no hats. The children, who reason carefully, are told to speak out as soon as they can determine the color of the hat they are wearing. After 30 seconds, the front child correctly names the color of her hat. Which color is it, and why?

2.34. (!) For each statement below about natural numbers, decide whether it is true or false, and prove your claim using only properties of the natural numbers.

a) If $n \in \mathbb{N}$ and $n^2 + (n + 1)^2 = (n + 2)^2$, then $n = 3$.

b) For all $n \in \mathbb{N}$, it is false that $(n - 1)^3 + n^3 = (n + 1)^3$.

2.35. Prove that if x and y are distinct real numbers, then $(x + 1)^2 = (y + 1)^2$ if and only if $x + y = -2$. How does the conclusion change if we allow $x = y$?

2.36. Let x be a real number. Prove that if $|x - 1| < 1$, then $\left|x^2 - 4x + 3\right| < 3$.

2.37. Given a real number x, let A be the statement "$\frac{1}{2} < x < \frac{5}{2}$", let B be the statement "$x \in \mathbb{Z}$", let C be the statement $x^2 = 1$, and let D be the statement "$x = 2$". Which statements below are true for all $x \in \mathbb{R}$?

    a) $A \Rightarrow C$.   e) $C \Rightarrow (A \wedge B)$.

    b) $B \Rightarrow C$.   f) $D \Rightarrow [A \wedge B \wedge (\neg C)]$.

    c) $(A \wedge B) \Rightarrow C$.   g) $(A \vee C) \Rightarrow B$.

    d) $(A \wedge B) \Rightarrow (C \vee D)$.

2.38. Let x, y be integers. Determine the truth value of each statement below.

    a) xy is odd if and only if x and y are odd.

    b) xy is even if and only if x and y are even.

2.39. (!) A particle starts at the point $(0, 0) \in \mathbb{R}^2$ on day 0. On each day, it moves one unit in a horizontal or vertical direction. For $a, b \in \mathbb{Z}$ and $k \in \mathbb{N}$, prove that it is possible for the particle to reach (a, b) on day k if and only if (1) $|a| + |b| \leq k$, and (2) $a + b$ has the same parity as k.

2.40. (!) *Checkerboard problems.* (Hint: Use the method of contradiction.)

    a) Two opposite corner squares are deleted from an eight by eight checkerboard. Prove that the remaining squares cannot be covered exactly by dominoes (rectangles consisting of two adjacent squares).

    b) Two squares from each of two opposite corners are deleted as shown on the right below. Prove that the remaining squares cannot be covered exactly by copies of the "T-shape" and its rotations.

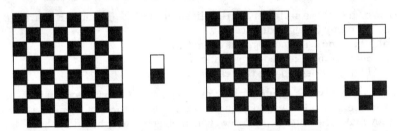

2.41. A clerk returns n hats to n people who have checked them, but not necessarily in the right order. For which k is it possible that exactly k people get a wrong hat? Phrase your conclusion as a biconditional statement.

2.42. A closet contains n different pairs of shoes. Determine the minimum t such that every choice of t shoes from the closet includes at least one matching pair of shoes. For $n > 1$, what is the minimum t to guarantee that two matching pairs of shoes are obtained?

2.43. Using the equivalences discussed in Remark 2.20, write a chain of symbolic equivalences to prove that $P \Leftrightarrow Q$ is logically equivalent to $Q \Leftrightarrow P$.

2.44. Let P and Q be statements. Prove that the following are true.

    a) $(Q \wedge \neg Q) \Rightarrow P$.   b) $P \wedge Q \Rightarrow P$.   c) $P \Rightarrow P \vee Q$.

2.45. Prove that the statements $P \Rightarrow Q$ and $Q \Rightarrow R$ imply $P \Rightarrow R$, and that the statements $P \Leftrightarrow Q$ and $Q \Leftrightarrow R$ imply $P \Leftrightarrow R$. (Comment: This is the justification for using a chain of equivalences to prove an equivalence.)

2.46. Prove that the logical expression S is equivalent to the logical expression $\neg S \Rightarrow (R \wedge \neg R)$, and explain the relationship between this equivalence and the method of proof by contradiction.

2.47. Let $P(x)$ be the assertion "x is odd", and let $Q(x)$ be the assertion "$x^2 - 1$ is divisible by 8". Determine whether the following statements are true:
  a) $(\forall x \in \mathbb{Z})[P(x) \Rightarrow Q(x)]$.
  b) $(\forall x \in \mathbb{Z})[Q(x) \Rightarrow P(x)]$.

2.48. Let $P(x)$ be the assertion "x is odd", and let $Q(x)$ be the assertion "x is twice an integer". Determine whether the following statements are true:
  a) $(\forall x \in \mathbb{Z})(P(x) \Rightarrow Q(x))$.
  b) $(\forall x \in \mathbb{Z})(P(x)) \Rightarrow (\forall x \in \mathbb{Z})(Q(x))$.

2.49. Let $S = \{x \in \mathbb{R}: x^2 > x + 6\}$. Let $T = \{x \in \mathbb{R}: x > 3\}$. Determine whether the following statements are true, and interpret these results in words:
  a) $T \subseteq S$.
  b) $S \subseteq T$.

2.50. Prove the following identities involving complementation of sets.
  a) $(A \cup B)^c = A^c \cap B^c$. (This is de Morgan's second law.)
  b) $A \cap [(A \cap B)^c] = A - B$.
  c) $A \cap [(A \cap B^c)^c] = A \cap B$.
  d) $(A \cup B) \cap A^c = B - A$.

2.51. *Distributive laws for set operations.* Using statements about membership, prove the statements below, where A, B, C are any sets. Use Venn diagrams to illustrate the results and guide the proofs.
  a) $A \cup (B \cap C) = (A \cup B) \cap (A \cup C)$.
  b) $A \cap (B \cup C) = (A \cap B) \cup (A \cap C)$.

2.52. Let A, B, C be sets. Prove that $A \cap (B - C) = (A \cap B) - (A \cap C)$.

2.53. (!) Let A, B, C be sets. Prove that $(A \cup B) - C$ must be a subset of $[A - (B \cup C)] \cup [B - (A \cap C)]$, but that equality need not hold.

2.54. (+) Consider three circles in the plane as shown below. Each bounded region contains a token that is white on one side and black on the other. At each step, we can either (a) flip all four tokens inside one circle, or (b) flip the tokens showing white inside one circle to make all four tokens in that circle show black. From the starting configuration with all tokens showing black, can we reach the indicated configuration with all showing black except the token in the central region? (Hint: Consider parity conditions and work backward from the desired configuration.)

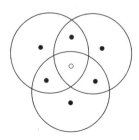

Chapter 3

Induction

Many mathematical problems involve only integers; computers perform operations in terms of integer arithmetic. The natural numbers enable us to solve problems by working one step at a time. After giving a definition of the natural numbers as a subset of the real numbers, we study the principle of mathematical induction. We use this fundamental technique of proof to solve problems such as the following.

3.1. Problem. *The Checkerboard Problem.* Counting squares of sizes one-by-one through eight-by-eight, an ordinary eight-by-eight checkerboard has 204 squares. How can we obtain a formula for the number of squares of all sizes on an *n*-by-*n* checkerboard? ∎

3.2. Problem. *The Handshake Problem.* Consider n married couples at a party. Suppose that no person shakes hands with his or her spouse, and the $2n - 1$ people other than the host shake hands with different numbers of people. With how many people does the hostess shake hands? ∎

3.3. Problem. *Sums of Consecutive Integers.* Which natural numbers are sums of consecutive smaller natural numbers? For example, $30 = 9 + 10 + 11$ and $31 = 15 + 16$, but 32 has no such representation. ∎

3.4. Problem. *The Coin-Removal Problem.* Suppose that n coins are arranged in a row. We remove heads-up coins, one by one. Each time we remove a coin we must flip the coins still present in the (at most) two positions surrounding it. For which arrangements of heads and tails can we remove all the coins? For example, $THTHT$ fails, but $THHHT$ succeeds. Using dots to denote gaps due to removed coins, we remove $THHHT$ via $THHHT$, $H.THT$, $..THT$, $..H.H$, $....H$, $......$ ∎

50

THE PRINCIPLE OF INDUCTION

In Chapter 1 we described the natural numbers $\mathbb{N}$ in an informal fashion as the set $\{1, 2, 3, \ldots\}$. We need a more precise definition in order to prove statements about $\mathbb{N}$. The idea is simple.

To generate $\mathbb{N}$ as a subset of $\mathbb{R}$, we begin with the number 1, which is defined to be the multiplicative identity in $\mathbb{R}$. We define 2 to equal $1 + 1$, and then we define 3 to equal $2 + 1$. We do not include 0 as a natural number; some authors do. This does not change what can be proved, but the statements or proofs may need to be modified.

We want $\mathbb{N}$ to be the subset of $\mathbb{R}$ obtained by beginning with 1 and successively adding 1. This motivates our formal definition of $\mathbb{N}$. We seldom use this definition directly; instead we use the principle of induction, which follows easily from it.

3.5. Definition. The set $\mathbb{N}$ of **natural numbers** is the intersection of all sets $S \subseteq \mathbb{R}$ that have the following two properties:
a) $1 \in S$.
b) If $x \in S$, then $x + 1 \in S$.

By definition, the intersection of a family of sets consists of the elements belonging to all of them. Since there is a set ($\mathbb{R}$ itself) satisfying properties (a) and (b), the family has at least one member. Note also that the intersection of all sets satisfying (a) and (b) also satisfies (a) and (b); thus $\mathbb{N}$ satisfies properties (a) and (b) of Definition 3.5. Thus $\mathbb{N}$ is contained in every set of real numbers satisfying (a) and (b).

Definition 3.5 yields the principle of induction, which is a method for proving that a set S of natural numbers is all of $\mathbb{N}$. It suffices to prove that S satisfies properties (a) and (b) of Definition 3.5. This observation underlies the principle of induction.

3.6. Theorem. (Principle of Induction) For each natural number n, let $P(n)$ be a mathematical statement. If properties (a) and (b) below hold, then for each $n \in \mathbb{N}$ the statement $P(n)$ is true.
a) $P(1)$ is true.
b) For $k \in \mathbb{N}$, if $P(k)$ is true, then $P(k + 1)$ is true.

Proof: Let $S = \{n \in \mathbb{N} : P(n) \text{ is true}\}$. By definition, $S \subseteq \mathbb{N}$. On the other hand, (a) and (b) here imply that S satisfies (a) and (b) of Definition 3.5. Since $\mathbb{N}$ is the smallest such set, $\mathbb{N} \subseteq S$. Therefore $S = \mathbb{N}$, and $P(n)$ is true for each $n \in \mathbb{N}$. ∎

Induction justifies all the elementary properties of arithmetic for natural numbers. Since addition and multiplication are defined on $\mathbb{R}$, and $\mathbb{N}$ is a subset of $\mathbb{R}$, the sum and product of two natural numbers are real numbers. As we expect, these are in fact *natural* numbers. Given a natural number n, let $S_n = \{m \in \mathbb{N}: n + m \in \mathbb{N}\}$. Exercise 25 uses induction to show for each n that $S_n = \mathbb{N}$. A similar proof then works for multiplication.

We also observe that natural numbers are positive. The order axioms for $\mathbb{R}$ imply that 1 is positive and state that sums of positive numbers are positive. Induction then implies that all natural numbers are positive. (Knowing $\mathbb{N}$ enables us to define $\mathbb{Z}$ precisely. A real number x is an integer if $x = 0$, $x \in \mathbb{N}$, or $-x \in \mathbb{N}$. In Appendix A we show that the arithmetic operations on $\mathbb{R}$ agree with those on $\mathbb{Z}$.)

A proof by induction involves two steps. Proving property (a) of Theorem 3.6 is the **basis step**, and proving property (b) is the **induction step**. Given statements $P(1), P(2), P(3), \ldots$ indexed by $\mathbb{N}$, we often seek a proof by induction when we can find a simple relationship between $P(k)$ and $P(k + 1)$, since the induction step requires us to prove for each k that $P(k)$ implies $P(k + 1)$. Our first application is a summation formula.

3.7. Proposition. For $n \in \mathbb{N}$, the formula $1 + 2 + \cdots + n = \frac{n(n+1)}{2}$ holds.

Proof: For $n \in \mathbb{N}$, let s_n be the sum of the integers 1 through n, and let $P(n)$ be the statement "$s_n = n(n + 1)/2$".

Basis step: Since $1 = 1 \cdot 2/2$, the statement $P(1)$ is true.

Induction step: The quantity s_{k+1} is obtained from s_k by adding $k + 1$. The hypothesis that $P(k)$ is true specifies the value of s_k and yields

$$s_{k+1} = s_k + (k + 1) = \frac{k(k+1)}{2} + (k + 1) = (k + 1)\left(\frac{k}{2} + 1\right) = \frac{(k+1)(k+2)}{2}.$$

Hence $P(k)$ implies $P(k + 1)$.

By the principle of induction, the formula holds for every $n \in \mathbb{N}$. ∎

The numbers $s_1, s_2, s_3, \ldots$ in Proposition 3.7 form a list indexed by $\mathbb{N}$. Every such list, including a list of statements to be proved, can be viewed as a function defined on $\mathbb{N}$. We introduce a term for such functions.

3.8. Definition. A **sequence** is a function whose domain is $\mathbb{N}$.

We usually think of a sequence as the infinite list of its values in order. When $f: \mathbb{N} \to S$, we say that $f(1), f(2), f(3), \ldots$ is a sequence of elements of S or a sequence in S. When $S = \mathbb{R}$, we speak of a sequence of real numbers. We often write the sequence as $a_1, a_2, a_3, \ldots$, where $a_n = f(n)$, and we call a_n the n**th term** of the sequence. We refer to the entire list using angled brackets; thus $\langle a \rangle$ is the sequence whose nth term is a_n.

Writing the sequence as $\{a_n\}$ is common but imprecise. In this form, n is unquantified. Also, interpreting this as $\{a_n: n \in \mathbb{N}\}$ names only the set of values. For example, if $a_n = (-1)^n$ for all n, then $\{a_n: n \in \mathbb{N}\} = \{1, -1\}$. If $b_n = (-1)^{n+1}$ for all n, then $\{a_n\} = \{b_n\}$, but $\langle a \rangle$ and $\langle b \rangle$ are different sequences. Hence we write $\langle a \rangle$ or $a_1, a_2, \ldots$.

Proposition 3.7 uses induction to prove a formula for the nth term of a sequence in terms of n; here $s_n = n(n+1)/2$. The sequence $\langle s \rangle$ is defined using summation; we next introduce concise notation for summation.

3.9. Remark. *Notation for summation and product.* We express summation using $\sum$, a large uppercase Greek sigma. When a, b are integers, the value of $\sum_{i=a}^{b} f(i)$ is the sum of the numbers $f(i)$ over the integers i satisfying $a \le i \le b$. Here i is the **index of summation**, and the formula $f(i)$ is the **summand**. In Proposition 3.7, the summand is i; in summation notation, the result is $\sum_{i=1}^{n} i = n(n+1)/2$ for $n \in \mathbb{N}$.

We write $\sum_{j \in S} f(j)$ to sum a real-valued function f over the elements of a set S in its domain. When no subset is specified, as in $\sum_j x_j$, we sum over the entire domain. When the summand has only one symbol that can vary, we may omit the subscript on the summation symbol, as in $\sum x_i$.

Similar comments apply to indexed products using the uppercase Greek pi. An important example is $\prod_{i=1}^{n} i = 1 \times 2 \times \cdots \times n$, commonly written as $n!$. ∎

Induction works particularly well for Proposition 3.7 because the summation for $n = k+1$ is obtained from the summation for $n = k$ by adding one additional term. Once the hypothesis of the induction step is invoked, we obtain the desired formula by algebraic manipulations.

3.10. Remark. *Alternative argument.* When still in grade school, Karl Friedrich Gauss (1777–1855) gave a direct proof that $\sum_{i=1}^{n} i = n(n+1)/2$. We list two copies of the sum, one above the other:

$$
\begin{array}{ccccccccc}
1 & + & 2 & + & 3 & + & \cdots & + & n \\
n & + & n-1 & + & n-2 & + & \cdots & + & 1
\end{array}
$$

For each i, the sum of the ith column is $i + (n+1-i) = n+1$. There are n columns, so $2\sum_{i=1}^{n} i = n(n+1)$.

This argument has a "geometric" interpretation. Consider $n(n+1)$ points in a grid with n columns of size $n+1$. Counted by columns, there are $n(n+1)$ points. We can also group the points into two disjoint subsets with columns of sizes 1 through n. In Chapter 5 we will return to this technique of "counting two ways". ∎

3.11. Remark. *Renaming the index.* An index of summation has meaning only within the summand; the value of $\sum_{i=1}^{n} f(i)$ cannot depend on i. Note that $\sum_{i=1}^{n} f(i) = \sum_{j=1}^{n} f(j)$. Both equal $f(1) + f(2) + \cdots + f(n)$, where no index appears. The result must be a function of n.

This allows us to rename the index of summation when convenient. We can also substitute to reverse the order of the summands. Below we repeat Gauss' argument using summations. We twice rename the index on the second copy of the sum, replacing i by j and then j by $n + 1 - i$.

$$2\sum_{i=1}^{n} i = \sum_{i=1}^{n} i + \sum_{j=1}^{n} j = \sum_{i=1}^{n} i + \sum_{i=1}^{n}(n+1-i) = \sum_{i=1}^{n}(n+1) = n(n+1) \quad \blacksquare$$

We write $\sum_{i=1}^{n}(n+1) = n(n+1)$ because summing n terms that each equal k is the same as multiplying k by n. A precise verification of this uses induction (Exercise 14). Although Remark 3.10 thus indirectly uses induction, it illustrates that alternatives may exist to proof by induction.

In this book we give careful proofs of some "obvious" statements. Studying the proof of a seemingly obvious statement gives us confidence in applying it. It also helps us understand the technique of proof. We may then be able to prove less obvious statements by the same technique. Our next proof is a model for using induction to extend statements about two objects to analogous statements about n objects.

Understanding a proof may also reveal the limitations of the argument and the difficulties that arise in generalizing it. The "obvious" statements (a) and (b) in the next proposition extend to infinite series; proving this requires the Completeness Axiom. In contrast, the "obvious" statement that the sum of n numbers is independent of the order of summation (Exercise 42) fails for infinite series! (See Exercises 14.53–14.54.)

3.12. Proposition. Suppose that $\langle a \rangle$ and $\langle b \rangle$ are sequences of real numbers and that $n \in \mathbb{N}$.
   a) If $c \in \mathbb{R}$, then $\sum_{i=1}^{n} ca_i = c\sum_{i=1}^{n} a_i$.
   b) If $a_i \le b_i$ for all $i \in \mathbb{N}$, then $\sum_{i=1}^{n} a_i \le \sum_{i=1}^{n} b_i$.
   c) If $0 \le a_i \le b_i$ for all $i \in \mathbb{N}$, then $\prod_{i=1}^{n} a_i \le \prod_{i=1}^{n} b_i$.

Proof: We leave the proof of (c) as Exercise 18. For $n \in \mathbb{N}$, let $P(n)$ and $Q(n)$ denote the conclusions in (a) and (b), respectively. We prove each claim by induction.

a) The distributive law (Definition 1.39DL) states that multiplication by a real number distributes over a sum of *two* real numbers: $x(y + z) = xy + xz$. We use induction to prove $P(n)$ for all $n \in \mathbb{N}$.

Basis step: The statement $P(1)$ is "$ca_1 = ca_1$", which is true.

Induction step: We use both the hypothesis that $P(k)$ is true and the distributive law to compute

$$\sum_{i=1}^{k+1} ca_i = ca_{k+1} + \sum_{i=1}^{k} ca_i = ca_{k+1} + c\sum_{i=1}^{k} a_i = c\left(a_{k+1} + \sum_{i=1}^{k} a_i\right) = c\sum_{i=1}^{k+1} a_i.$$

This computation shows that $P(k)$ implies $P(k+1)$.

By the principle of induction, $P(n)$ is true for all n.

b) In Chapter 1, we observed that $a < b$ and $c < d$ implies $a+c < b+d$. This leads to a proof by induction that $Q(n)$ holds for all $n \in \mathbb{N}$.

Basis step: $Q(1)$ states "$a_1 \le b_1$"; this is true by hypothesis.

Induction step: If $Q(k)$ is true, then $\sum_{i=1}^{k} a_i \le \sum_{i=1}^{k} b_i$, and we can use this and our ability to sum two inequalities to compute

$$\sum_{i=1}^{k+1} a_i = \left(\sum_{i=1}^{k} a_i\right) + a_{k+1} \le \left(\sum_{i=1}^{k} b_i\right) + b_{k+1} = \sum_{i=1}^{k+1} b_i.$$

This computation shows that $Q(k)$ implies $Q(k+1)$.

By the principle of induction, $Q(n)$ is true for all n. ∎

Proposition 3.12a allows us to extend the useful factorization $x^2 - y^2 = (x-y)(x+y)$ to nth powers.

3.13. Lemma. If $x, y \in \mathbb{R}$ and $n \in \mathbb{N}$, then

$$x^n - y^n = (x-y)(x^{n-1} + x^{n-2}y + \cdots + xy^{n-2} + y^{n-1}).$$

Proof: Using the distributive law (Proposition 3.12a), we multiply out the product on the right. Below we write the terms using the factor x on the first line and the terms using the factor $-y$ on the second line. Canceling like terms in the "columns" yields $x^n - y^n$, which completes the proof.

$$
\begin{array}{ccccccccc}
x^n & + & x^{n-1}y & + & \cdots & + & x^2 y^{n-2} & + & xy^{n-1} \\
& - & x^{n-1}y & - & \cdots & - & x^2 y^{n-2} & - & xy^{n-1} & - & y^n
\end{array}
$$
∎

Exercise 20 asks for this proof in summation notation. Note that terms arising from the factor $-y$ are shifted to combine with desired terms arising from x. This corresponds to rewriting $\sum_{j=1}^{n} f(j)$ as $\sum_{j=0}^{n-1} f(j+1)$. No summands change, and this is merely a special case of renaming the index called **shifting the index** of summation.

Exercise 35 requests a proof of the next statement by induction.

3.14. Corollary. (The Geometric Sum) If $q \in \mathbb{R}$, $q \ne 1$, and n is a nonnegative integer, then $\sum_{i=0}^{n-1} q^i = \frac{q^n - 1}{q - 1}$.

Proof: In the formula of Lemma 3.13, we set $x = q$ and $y = 1$. We obtain $q^n - 1 = (q-1)(q^{n-1} + q^{n-2} + \cdots + 1)$. Since $q \ne 1$, we can divide both sides by $q - 1$ to obtain the desired formula. ∎

3.15. Example. *A knockout tournament.* The NCAA basketball tournament starts with 64 teams. How many games are played to produce a winner? In the first round, there are 32 games. The 32 winners play 16 games in the second round. The subsequent rounds have 8, 4, 2, 1 games, respectively. By the Geometric Sum, the total number of games is $1 + 2 + 4 + 8 + 16 + 32 = \sum_{i=0}^{5} 2^i = 2^6 - 1 = 63$. We can also obtain this result by observing that every team other than the winner must lose exactly one game, and hence there must be exactly 63 games. The two sides of the equality give different methods for counting the games. ∎

When we use induction to prove a claim involving a parameter $n \in \mathbb{N}$, we say that the proof is "by induction on n" and call n the **induction parameter**. The induction step proves the conditional statement "$P(k)$ is true implies $P(k+1)$ is true". The hypothesis of this conditional ("$P(k)$ is true") is the **induction hypothesis**. Somewhere in the proof of the induction step, we say "by the induction hypothesis". If we have not used the induction hypothesis, then we have not written a proof by induction.

With the next example, we begin to relax the formal template for induction proofs to show the flexibility of the technique.

3.16. Proposition. If $n \in \mathbb{N}$ and $q \geq 2$, then $n < q^n$.

Proof: We use induction on n. For the basis step, we have $1 < q$ by the hypothesis on q, so the claim holds when $n = 1$. For the induction step, suppose that the claim holds when $n = k$, meaning that $k < q^k$. Using the induction hypothesis for the step of strict inequality, we compute

$$k + 1 \leq k + k = 2k \leq qk < q \cdot q^k = q^{k+1}.$$

Hence the claim also holds when $n = k + 1$, which completes the proof of the induction step. ∎

In the induction step, we show that the truth of one instance of the desired statement implies the truth of the next instance. We have one such proof for each natural number; starting from the basis, each provides another value of n for which $P(n)$ is known. Proving that $P(k+1)$ follows from $P(k)$ for a general natural number k writes all the proofs at once.

To visualize the process of induction, consider a sequence of upright dominoes, one for each natural number. If any domino falls, then it knocks over the next; this is the "induction step". The principle of induction says that if also the first domino falls (the "basis step"), then all the dominoes fall. The proof for the first domino cannot be omitted.

3.17. Example. $n = n + 1$ (!?). The induction step $P(k) \Rightarrow P(k+1)$ is a conditional statement. A conditional statement fails only if the hypothesis is true and the conclusion is false. From the hypothesis that $k = k + 1$, we can easily derive $k + 1 = k + 2$. This proof of the induction step is valid, but we have not proved that $n = n + 1$ for all $n \in \mathbb{N}$, because the basis step is false: $1 \neq 2$. The basis step cannot be omitted. ∎

Next we illustrate a similar error. The *proof* of the induction step must be valid for each value of the induction parameter where it is needed.

3.18. Example. *All people have the same sex (!?).* We try to prove by induction on n that the people in every set of n people have the same sex. Certainly all the people in a set consisting of one person have the same sex, so the claim holds for $n = 1$. Now suppose the claim holds for $n = k$, and let $S = \{a_1, \ldots, a_{k+1}\}$ be a set of $k+1$ people. Deleting a_1 yields a set T of k people. Deleting a_2 yields another set T' of k people. By the induction hypothesis, the people in T have the same sex, and the people in T' have the same sex. Now all people in S have the same sex as a_{k+1}.

The error is that the proof of the induction step is invalid when going from $n = 1$ to $n = 2$. In this case, the sets T and T' have no common element, so we cannot use the induction hypothesis to conclude that a_1 and a_2 have the same sex. ∎

In our next result, the statement has a natural number as a parameter and clearly is true when the parameter is 1. Also, the statement for $n = k + 1$ involves quantities used in the statement for $n = k$. These properties suggest trying induction as a technique of proof.

3.19. Proposition. If $x_1, \ldots, x_n$ are numbers in the interval $[0, 1]$, then

$$\prod_{i=1}^{n}(1 - x_i) \geq 1 - \sum_{i=1}^{n} x_i.$$

Proof: We use induction on n.

Basis step: For $n = 1$, the inequality is $1 - x_1 \geq 1 - x_1$, which is true.

Induction step: Assume the claim when $n = k$. Given numbers $x_1, \ldots, x_{k+1}$, applying the induction hypothesis to the first k of them yields $\prod_{i=1}^{k}(1-x_i) \geq 1-\sum_{i=1}^{k} x_i$. Since $x_{k+1} \leq 1$, multiplying by $1-x_{k+1}$ preserves the inequality. This explains the first line below.

$$\prod_{i=1}^{k+1}(1 - x_i) = (1 - x_{k+1})\prod_{i=1}^{k}(1 - x_i) \geq (1 - x_{k+1})\left(1 - \sum_{i=1}^{k} x_i\right)$$

$$= 1 - x_{k+1} - \sum_{i=1}^{k} x_i + \left(x_{k+1}\sum_{i=1}^{k} x_i\right) \geq 1 - \sum_{i=1}^{k+1} x_i$$

The next step expands the product. The term $x_{k+1}\sum_{i=1}^{k} x_i$ is nonnegative, since $x_i \geq 0$ for all i; thus dropping it does not increase the value. The remaining terms become the desired right side. Thus $\prod_{i=1}^{k+1}(1 - x_i) \geq 1 - \sum_{i=1}^{k} x_i$, which completes the induction step. ∎

3.20. Corollary. If $0 \leq a \leq 1$ and $n \in \mathbb{N}$, then $(1 - a)^n \geq 1 - na$.

Proof: Set $x_1 = \cdots = x_n = a$ in Proposition 3.19. ∎

APPLICATIONS

Applications of induction occur throughout mathematics. We solve the Checkerboard and Handshake Problems, characterize when polynomials are equal, and solve a problem about cutting regions into pieces.

To solve the Checkerboard Problem, we need to sum the squares of the first n natural numbers. Since the sum must be an integer, we obtain as a corollary that $n(n + 1)(2n + 1)/6$ is an integer for each $n \in \mathbb{N}$.

3.21. Proposition. For all $n \in \mathbb{N}$, $\sum_{i=1}^{n} i^2 = n(n + 1)(2n + 1)/6$.

Proof: We use induction on n.

Basis step: For $n = 1$, the sum is 1, and the right side is $1 \cdot 2 \cdot 3/6 = 1$.

Induction step: Suppose that the formula is valid when $n = k$. By the induction hypothesis, we have $\sum_{i=1}^{k} i^2 = k(k + 1)(2k + 1)/6$, and hence

$$\sum_{i=1}^{k+1} i^2 = \frac{k(k + 1)(2k + 1)}{6} + (k + 1)^2 = (k + 1)\left[\frac{2k^2 + k}{6} + (k + 1)\right]$$

$$= (k + 1)\frac{2k^2 + 7k + 6}{6} = \frac{(k + 1)(k + 2)(2k + 3)}{6}.$$

The last expression is the formula when $n = k + 1$, proving the induction step. By the principle of induction, the formula holds for all n. ∎

Since the formula for $n = k + 1$ involves the factor $k + 1$, we factored out $k + 1$ when it appeared instead of multiplying out the numerator. Keeping the goal in mind often saves time in computations.

3.22. Solution. *The Checkerboard Problem.* In the n-by-n checkerboard, there is one n-by-n square, and there are n^2 one-by-one squares. In general, there are $k \cdot k = k^2$ squares with sides of length $n - k + 1$, for $1 \le k \le n$. Hence the total number is $\sum_{k=1}^{n} k^2 = n(n + 1)(2n + 1)/6$, by Proposition 3.21. For $n = 8$, the value is $8 \cdot 9 \cdot 17/6 = 204$. ∎

Using induction to prove a summation formula requires knowing the formula in advance. If we can guess the formula from the first few values, then induction may provide an easy proof, but induction will not help us find the formula (see Exercises 28–29). In Chapter 9, we will develop summation methods that do not require knowing the formula in advance.

Some students wonder whether the technique of proof by induction uses what it is trying to prove. We want to prove, "for all n, $P(n)$ is true". In the induction step we prove, "for all n, $P(n)$ is true implies $P(n + 1)$ is true". The statements mean different things; the second is a *conditional* statement for each n. In our first few proofs by induction we used different letters (n and k) to avoid confusion about this difference, but doing so is not necessary. Henceforth we often use the same letter. When convenient, we phrase the induction step as "$P(n-1) \Rightarrow P(n)$ whenever $n > 1$" instead of "$P(k) \Rightarrow P(k+1)$ whenever $k \in \mathbb{N}$.

We can use induction to prove that $P(n)$ holds for $\{n \in \mathbb{Z}: n \geq n_0\}$ by replacing $P(1)$ by $P(n_0)$ in the basis step. This is equivalent to treating $n - n_0 + 1$ as the induction parameter. Any integer-valued function of the variables in our problem can serve as an induction parameter.

We use these remarks in determining when polynomials are equal. The **zeros** of a function f are the solutions to the equation $f(x) = 0$. Recall that the zero polynomial does not have a degree.

3.23. Lemma. If f is a polynomial of degree d, then a is a zero of f if and only if $f(x) = (x - a)h(x)$ for some polynomial h of degree $d - 1$.

Proof: By the definition of polynomial, we have $f(x) = \sum_{i=0}^{d} c_i x^i$, with d a nonnegative integer and $c_d \neq 0$. If the condition $f(x) = (x - a)h(x)$ holds, then $f(a) = 0$.

It remains only to assume that $f(a) = 0$ and obtain the factorization. Since $f(x) = c_0 + \sum_{i=1}^{d} c_i x^i$ and $f(a) = 0$,

$$f(x) = f(x) - f(a) = c_0 - c_0 + \sum_{i=1}^{d} c_i(x^i - a^i) = \sum_{i=1}^{d} c_i(x^i - a^i).$$

By Lemma 3.13, for $i \geq 1$ we have $x^i - a^i = (x - a)h_i(x)$, where $h_i(x) = \sum_{j=1}^{i} x^{i-j}a^{j-1}$. Using Proposition 3.12a, factoring $(x-a)$ from each term in $f(x) = \sum_{i=1}^{d} c_i(x^i - a^i)$ yields $f(x) = (x - a)h(x)$, where $h(x) = \sum_{i=1}^{d} h_i(x)$. Each h_i has degree $i - 1$, and thus h has degree $d - 1$.   ∎

3.24. Theorem. Every polynomial of degree d has at most d zeros.

Proof: We use induction on d. Let f be a polynomial of degree d.

Basis step: If $d = 0$, then $f(x) = c_0 \neq 0$ for all x, and f has no zeros.

Induction step: Consider $d \geq 1$. If f has no zero, then we are done, so let a be a zero of f. By Lemma 3.23, we have $f(x) = (x - a)h(x)$, where h is a polynomial of degree $d - 1$.

Since nonzero numbers have nonzero product, the only zeros of f are the zeros of $x - a$ and the zeros of h. Since $x - a = 0$ implies $x = a$, the first factor has exactly one zero. Since h has degree $d - 1$, the induction hypothesis implies that h has at most $d - 1$ zeros. We conclude that f has at most d zeros.   ∎

3.25. Corollary. Two real polynomials are equal if and only if their corresponding coefficients are equal.

Proof: Let f and g be polynomials. If their corresponding coefficients are equal, then $f(x) = g(x)$ for all $x \in \mathbb{R}$, because they have the same formula. Thus they are the same function.

Conversely, suppose that $f(x) = g(x)$ for all $x \in \mathbb{R}$. Let $h = f - g$. The difference of two polynomials is a polynomial (see Exercise 13). Since $h(x) = 0$ for all $x \in \mathbb{R}$, Theorem 3.24 implies that h cannot have degree d for any $d \geq 0$ and thus must be the zero polynomial. Therefore, for each i the coefficients of x^i in f and in g are equal.　　　　　　　■

The proof of Corollary 3.25 actually implies a stronger statement: If polynomials f and g are equal at more than d places, where d is the maximum of the degrees of f and g, then they are the same polynomial.

In the induction step of a proof by induction, we consider an arbitrary instance of the hypotheses for one value of the parameter, and we find an instance for a smaller value of the parameter in order to apply the induction hypothesis. In the proof of Theorem 3.24, we factored out $x - a$ to obtain a polynomial of smaller degree to which we could apply the induction hypothesis. Finding an appropriate smaller problem may take some effort. In the next example, the smaller instance emerges by stripping away pieces of the larger instance in an interesting way.

3.26. Solution. *The Handshake Problem.* Let a **handshake party** be a party with n married couples where no spouses shake hands with each other and the $2n - 1$ people other than the host shake hands with different numbers of people. We use induction on n to prove that in every handshake party, the hostess shakes hands with exactly $n - 1$ people.

If no one shakes with his or her spouse, then each person shakes with between 0 and $2n - 2$ people. Since the $2n - 1$ numbers are distinct, they must be 0 through $2n - 2$. The figure below illustrates for $n \in \{1, 2, 3\}$ the situation that is forced; each pair of encircled points indicates a married couple (host and hostess are rightmost), and two points are connected by a curve if and only if those two people shook hands.

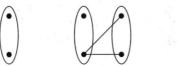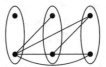

Basis step: If $n = 1$, then the hostess shakes with 0 (which equals $n - 1$), because the host and hostess don't shake.

Induction step: Suppose that $n > 1$. The induction hypothesis states that the claim holds for a handshake party with $n - 1$ couples. Let P_i denote the person (other than the host) who shakes with i people. Since P_{2n-2} shakes with all but one person, P_0 must be the person omitted. Hence P_0 is the spouse of P_{2n-2}. Furthermore, this married couple is not the host and hostess, since $S = \{P_0, P_{2n-2}\}$ does not include the host. Everyone not in S shakes with exactly one person in S, namely P_{2n-2}. If we delete S to obtain a smaller party, then we have $n - 1$ couples remaining (including the host and hostess), no person shakes with a spouse, and each person shakes with one fewer person than in the full party. Hence in the smaller party the people other than the host shake hands with different numbers of people.

By deleting the set S, we obtain a handshake party with $n-1$ couples (deleting the leftmost couple in the picture for $n=3$ yields the picture for $n=2$). We can thus apply the induction hypothesis to conclude that, outside of the couple S, the hostess shakes with $n-2$ people. Since she also shakes with P_{2n-2}, in the full party she shakes with $n-1$ people. ∎

In this proof, we cannot discard just any married couple to obtain the smaller party. We must find a couple S such that everyone outside S shakes with exactly one person in S. Only then can we apply the induction hypothesis to the smaller party, because only then will we know that it satisfies the hypothesis about distinct numbers of handshakes.

A similar problem arises if we start with a handshake party of n couples where the hostess shakes with $n-1$ people and add a couple S in which one person shakes with everyone else and the other person shakes with no one. This produces a handshake party of $n+1$ couples where the hostess shakes with n people. Unfortunately, it does not prove the induction step, because we have not shown that every handshake party of $n+1$ couples arises in this way.

We avoid this difficulty by proving that in any party of the larger size, P_0 must be the spouse of the person shaking the most. The induction step must consider every instance for the larger value of the induction parameter (see discussion after Application 11.46).

Sometimes we must verify more than $P(1)$ in the basis step. If we need both $P(n-1)$ and $P(n)$ to prove $P(n+1)$, then we must verify both $P(1)$ and $P(2)$ to get started. The reason is that since there is no $P(0)$, the proof of the induction step does not apply to prove $P(2)$. This occurs in the next example and in many proofs using recurrence relations (see Exercises 55–57 and Chapter 12).

3.27. Solution. *The L-Tiling Problem.* A child has a large number of L-shaped tiles as illustrated on the left below. Is it possible to form the large similar region on the right with non-overlapping copies of this tile?

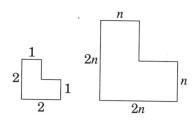

Let the large region be R_n and the small region be L. A partition of a region into copies of L is an **L-tiling**. We want to prove for $n \in \mathbb{N}$ that R_n has an L-tiling. Since R_1 is a copy of L itself, R_1 has an L-tiling.

In seeking a proof by induction, we can find a copy of R_{n-1} inside R_n by removing a strip of unit width along the top, left, bottom, and right edges. This does not help. The induction hypothesis would tell us that R_{n-1} can be tiled, but when $n \geq 3$ we cannot extend this to a tiling of R_n because the outer strip has no L-tiling.

To fix this flaw, we use an outer strip of width 2 and obtain an L-tiling of R_n from an L-tiling of R_{n-2}. Since R_1 has an L-tiling, this completes the proof for odd n, but to handle the even cases we also must treat R_2 in the basis. Below we explicitly tile R_2 and illustrate that every 2 by $3k$ rectangle has an L-tiling. (The decomposition of R_2 suggests a simple inductive proof that R_n has an L-tiling whenever n is a power of 2—Exercise 58.)

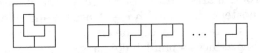

For the induction step, consider R_n, where $n \geq 3$. It suffices to cut R_n into regions that we already know have L-tilings. The induction hypothesis provides an L-tiling of the inner region R_{n-2}. To complete the proof, it suffices to tile the outer strip. For this we use copies of R_2 and copies of 2 by $3k$ rectangles, which we have already shown have L-tilings.

We tile the outer band in one of three ways, depending on whether n, $n-1$, or $n-2$ is a multiple of 3. To prove that the decomposition works in each case, we need only verify that the length of the long side of each rectangle used is a multiple of 3. For clarity in the pictures, we list only these lengths; the short sides all equal 2. The three cases occur when $n \geq 3$, $n \geq 4$, and $n \geq 5$, respectively, so the lengths of the rectangles are nonnegative multiples of 3. Verifying this completes the induction step. ∎

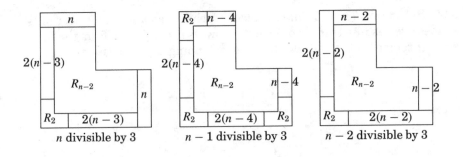

n divisible by 3   n − 1 divisible by 3   n − 2 divisible by 3

STRONG INDUCTION

In this section we present several variations of induction. They are useful alternative phrasings of the same idea.

Sometimes a proof of $P(k)$ in the induction step needs the hypothesis that $P(i)$ is true for every value of i before k. By assuming more in the induction hypothesis, we make the condition statement in the induction step weaker. Nevertheless, this weaker implication suffices to complete the proof, and so we call the method *strong induction*.

3.28. Theorem. (Strong Induction Principle) Let $\{P(n): n \in \mathbb{N}\}$ be a sequence of mathematical statements. If properties (a) and (b) below hold, then for every $n \in \mathbb{N}$, $P(n)$ is true.
a) $P(1)$ is true.
b) For $k \geq 2$, if $P(i)$ is true for all $i < k$, then $P(k)$ is true.

Proof: For $n \in \mathbb{N}$, let $Q(n)$ be the statement that $P(i)$ is true for all i with $1 \leq i \leq n$. We prove by induction on n that all $Q(n)$ are true, which implies that all $P(n)$ are true.

Basis step: By (a), $Q(1)$ is true.

Induction step: For $k > 1$, the hypothesis that $Q(k-1)$ is true is the statement that $P(i)$ is true for $i < k$. Thus (b) implies that $P(k)$ is true. With the truth of $Q(k-1)$, this yields $Q(k)$. The principle of induction now implies that all $Q(n)$ are true. ∎

Proving $Q(n)$ by ordinary induction on n is equivalent to proving $P(n)$ by strong induction on n. Again proving (a) is the **basis step** and proving (b) is the **induction step**. We can use strong induction to prove a statement for all *nonnegative* integers by starting with $P(0)$ in the basis step.

3.29. Solution. *The Coin-removal Problem.* Let a *string* be a row of coins without gaps and without other coins beyond the ends. We write a string as a list of Hs and Ts. When we remove an H, we leave a gap (marked by a dot), and we flip all of the (at most two) coins next to it that remain. Thus HHT becomes $T \cdot H$ when we remove the H in the middle, and then we get $T \cdot \cdot$ when we remove the new H. Removing a coin from a string leaves two strings except when we remove the end.

We begin with a string of length n. Examination of examples suggests that we can empty a string (remove all its coins) if and only if it has an odd number of Hs. We prove this by strong induction on n.

Basis step: We can empty a string of length 1 if and only if it is H.

Induction step: Consider a string S of length k, for $k > 1$. The induction hypothesis tells us that shorter strings can be emptied if and only if they have odd weight, where the *weight* of a string is the number of Hs.

Suppose first that S has odd weight. Let X be its leftmost H. Remove X and flip its neighbor(s). The portion before X now is empty or has one H (at its right end). The portion after X has even weight, but if it is nonempty we flip its first element to obtain odd weight. Thus each remaining string is shorter than S and has odd weight. By the induction hypothesis, each remaining string can be emptied, so S can be emptied.

If S has even weight, we show that removing any H leaves a shorter nonempty string with even weight. For each H in S, the number of other Hs is odd. Thus there is an even number of them to one side and an odd number to the other side. The side with an odd number is nonempty, and flipping its member next to the H being removed gives it even weight. Thus for each H we might remove, we leave a shorter nonempty string of even weight. By the induction hypothesis, that string cannot be emptied, so S cannot be emptied. ∎

Good	Bad
$TTTHHTTH$	$TTHHTHTH$
$TTH.TTTH$	$TTHHH.HH$

In Solution 3.29, removing a coin can yield much shorter strings. We need the hypothesis for all smaller lengths, so we use strong induction.

3.30. Proposition. (Well-Ordering Property) Every nonempty subset of $\mathbb{N}$ has a least element.

Proof: For $n \in \mathbb{N}$, let $P(n)$ be the statement that every subset of $\mathbb{N}$ containing n has a least element. Proving these statements proves the claim, because every nonempty $S \subseteq \mathbb{N}$ contains some n, and then $P(n)$ implies that S has a least element. We prove $P(n)$ by strong induction on n.

Basis step: Since 1 is the least natural number, every subset containing 1 has a least element, and $P(1)$ is true.

Induction step: Suppose that $P(i)$ is true for all $i < k$. Let S be a subset of $\mathbb{N}$ containing k. If S has no member less than k, then k is its least element. Otherwise, S contains an element i less than k, and $P(i)$ implies that S has a least element. Thus $P(k)$ is true. ∎

Exercise 64 requests a proof of the ordinary principle of induction from the well-ordering property, thus verifying that our three versions of induction are equivalent. We next describe yet another.

Suppose that $S \subset \mathbb{N}$, but that $S \neq \mathbb{N}$. By the well-ordering property, S^c has a least element. Thus when $P(n)$ fails for some $n \in \mathbb{N}$, there is a *least* n where it fails. This yields another approach to induction, called the **method of descent**. We can prove $P(n)$ for all $n \in \mathbb{N}$ by proving that there is no least n where $P(n)$ fails. To do this, we suppose that $P(n)$ fails for some n and show that $P(k)$ must fail for some k less than n. The existence of k implies that $n > 1$, and thus we have proved the contrapositive of property (b) from Theorem 3.28.

We give two proofs to illustrate the method of descent.

3.31. Theorem. $\sqrt{2}$ is irrational.

Proof: If $\sqrt{2}$ is rational, then we may write $\sqrt{2} = m/n$ for some $m, n \in \mathbb{N}$. We obtain another fraction equal to $\sqrt{2}$ with smaller positive denominator; the method of descent then implies that $\sqrt{2}$ has no representation as a quotient of natural numbers.

Since $1 < \sqrt{2} < 2$, we have $n < m < 2n$. Thus $0 < m - n < n$. Using also $2n^2 = m^2$, the computation below shows that $(2n - m)/(m - n)$ works.

$$\frac{2n - m}{m - n} = \frac{n(2n - m)}{n(m - n)} = \frac{2n^2 - mn}{n(m - n)} = \frac{m^2 - mn}{n(m - n)} = \frac{m(m - n)}{n(m - n)} = \frac{m}{n} \qquad \blacksquare$$

3.32. Proposition. Every natural number n can be expressed in exactly one way as the product of an odd number and a power of 2.

Proof: If the claim fails, then some least n does not have a unique such expression. If n is odd, then n is not divisible by 2, so $1 \cdot n$ is such an expression and the only one. If n is even, then we consider $n/2$. Each such expression for $n/2$ produces one for n by adjusting the power of 2, and vice versa. Thus if n is a counterexample, then $n/2$ is also a counterexample. We have proved that there is no least counterexample. $\qquad \blacksquare$

Alternatively, we can obtain Proposition 3.32 from Proposition 3.16. Since $n < 2^n$, there is a largest power of 2 that divides n, call it 2^l. Dividing n by a smaller power of 2 leaves an even number. Thus the only expression for n as a power of 2 times an odd number is $2^l \cdot (n/2^l)$.

Using Proposition 3.32, we determine which natural numbers can be written as a sum of consecutive smaller natural numbers.

3.33. Example. *Sums of Consecutive Positive Integers.* For $r \geq 1$, we can write the odd number $n = 2r + 1$ as $r + (r + 1)$. When n is twice an odd number $2r + 1$, we can then write $n = (r - 1) + (r) + (r + 1) + (r + 2)$. This works whenever $r - 1 \geq 1$ and fails when $n = 2$. When $n = 6$, we have $r - 1 = 0$ and can drop this 0 to obtain $6 = 1 + 2 + 3$.

When $n = 4(2r + 1)$, we can write

$$n = (r - 3) + (r - 2) + (r - 1) + (r) + (r + 1) + (r + 2) + (r + 3) + (r + 4).$$

This works whenever $r - 3 \geq 1$. When $r - 3 = -1$, we can drop the first three terms $(-1) + (0) + (1)$ to write n as a sum of five consecutive integers. When $r - 3 = 0$, we drop the 0.

This suggests the general procedure used in the proof below. We illustrate it for 11, 22, 44, 88, which are powers of 2 times the odd number $11 = 2 \cdot 5 + 1$. For $2^l 11$, we use 2^l pairs of numbers summing to 11. Thus we write $11 = 5 + 6$, $22 = 4 + 5 + 6 + 7$, $44 = 2 + 3 + \cdots + 9$, and

$88 = -2 + (-1) + 0 + 1 + 2 + \cdots + 13$. We cancel the first five terms in the last expression to obtain $88 = 3 + 4 + \cdots + 13$. ∎

3.34. Solution. *Sums of Consecutive Positive Integers.* We prove that a natural number n is a sum of consecutive smaller natural numbers if and only if n is not a power of 2.

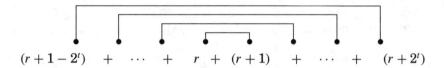

$$(r + 1 - 2^t) \quad + \quad \cdots \quad + \quad r \quad + \quad (r+1) \quad + \quad \cdots \quad + \quad (r + 2^t)$$

If n is not a power of 2, then $n = 2^t(2r+1)$ for some integers $t \geq 0$ and $r > 0$, by Proposition 3.32. Consider the 2^t numbers ending at r and the 2^t numbers starting at $r + 1$. Grouping these symmetrically around the middle yields 2^t pairs, each with sum $2r+1$, so the overall sum is n. When $r + 1 - 2^t \leq 0$, the numbers are not all positive. The two numbers in the middle are positive, so the number of positive terms exceeds the number of nonpositive terms by at least 2. In this case, the numbers $r + 1 - 2^t$ through $-(r + 1 - 2^t)$ have sum 0, and we delete them to express n as the sum of (at least two) consecutive natural numbers.

To prove the converse, suppose that n is the sum of p consecutive natural numbers starting with m. We will write n as an integer times an odd number larger than 1. When $p = 2$, n is odd. Otherwise, we use Proposition 3.7 to evaluate $\sum_{i=0}^{p-1} i$ and obtain

$$n = \sum_{i=0}^{p-1}(m + i) = mp + \sum_{i=0}^{p-1} i = mp + \frac{p(p-1)}{2} = \frac{p(2m + p - 1)}{2}.$$

Whether p is odd or even, exactly one of $\{p, 2m + p - 1\}$ is even and both exceed 2. Hence $n = \frac{p(2m+p-1)}{2}$ expresses n as the product of two integers, at least one of which is odd and larger than 1. We conclude that n is not a power of 2. (This expression for n shows that here we have the same answer as in Exercise 1.47.) ∎

HOW TO APPROACH PROBLEMS

In approaching the problems in this chapter, we consider how and when to use induction and how to deal with difficulties that may arise.

1) As in Chapter 1, it helps to express the desired conclusion in terms of known facts, especially when proving an induction step.

2) Not all statements involving natural numbers require induction.

3) Induction has many variations. When the proof of the induction step uses r earlier instances of the problem, r instances are needed in the basis. When the induction step needs arbitrary earlier instances, use strong induction.

Using induction.

Induction proofs of statements like the summation formulas in Exercises 14–17 should become routine. To prove such a formula by induction, verify that it holds for the first instance, and then for the induction step verify that each instance implies the next. This often amounts to little more than grouping the sum as the previous instance plus the new term, applying the induction hypothesis to the previous instance, and manipulating the resulting expression to obtain the desired formula in the new instance. Proposition 3.7 and Proposition 3.21 exemplify this.

Induction also applies to many statements other than summation formulas. Often one must explore a few small values of the parameter to find a pattern. Eventually a uniform way of using the statement for one value of the parameter to prove the next statement emerges. Describing and explaining that process for the general case becomes the proof of the induction step. Consider Solution 3.26 in this light.

What happens when we try to give an inductive proof of a statement that isn't always true? If the argument is valid only when the parameter is sufficiently large, then we may be able to prove that the statement holds for large n by finding an appropriate basis step.

3.35. Example. Suppose we try to prove that $n^3 + 20 > n^2 + 15n$ for all $n \in \mathbb{N}$. Setting $n = 1$ yields $21 > 16$, so the inequality holds when $n = 1$. Suppose it holds when $n = k$; we want $(k+1)^3 + 20 > (k+1)^2 + 15(k+1)$. Using the hypothesis $k^3 + 20 > k^2 + 15k$, we compute

$$
\begin{aligned}
(k+1)^3 + 20 &= (k^3 + 3k^2 + 3k + 1) + 20 = (3k^2 + 3k + 1) + (k^3 + 20) \\
&> (3k^2 + 3k + 1) + (k^2 + 15k) = (k^2 + 2k + 1) + (15k + 15) + 2k^2 + k - 15.
\end{aligned}
$$

To prove that the final expression is at least $(k+1)^2 + 15(k+1)$, we need only verify that $k(2k+1) \geq 15$. Unfortunately, this requires $k \geq 4$.

We can salvage something. When $n = 4$, we have $n^3 + 20 = 84 > 76 = n^2 + 15n$, so we can use $n = 4$ as a basis step. Now we only need the induction step to be valid when $k \geq 4$. If $k \geq 4$, then $k(2k+1) \geq 4 \cdot 9 > 15$, and we have proved the inequality for natural numbers at least 4.

Our argument for the induction step is not valid when $k = 1$, so we cannot use it to go from $n = 1$ to $n = 2$. In fact, the inequality $n^3 + 20 > n^2 + 15n$ fails when n is 2 or 3. ∎

3.36. Example. For $n \in \mathbb{N}$, when does $3^n > n^4$ hold? The statement is true for $n = 1$, but it fails for $n = 2$ and $n = 3$ by inspection. Nevertheless, we ask when $3^n > n^4$ implies $3^{n+1} > (n+1)^4$. The hypothesis yields $3^{n+1} = 3 \cdot 3^n > 3n^4$, and thus we want $3n^4 > (n+1)^4$. Solving for n yields $n > \frac{1}{3^{1/4}-1} = 3.16$, so the implication holds whenever $n \geq 4$.

Since $3^8 = 9^4$, we have $3^8 > 8^4$. Our proof of the induction step is valid when $n \geq 4$, so using $n = 8$ as the basis step yields a proof by induction that $3^n > n^4$ when $n \geq 8$. The proof of the induction step must break down for small n, because the inequality fails there. Checking the small values completes the answer; the inequality holds for $n = 1$ and $n \geq 8$. ∎

n:	1	2	3	4	5	6	7	8
3^n	3	9	27	81	243	729	2187	6561
n^4	1	16	81	256	625	1296	2401	4096

To induct or not to induct.

Some formulas follow easily from known formulas.

3.37. Example. For $x \neq 1$ and $n \in \mathbb{N}$, we evaluate the sum $b = \sum_{i=2}^{n} x^i$. The key idea is to recognize that the sum is much like one we already know: $a = \sum_{i=0}^{n} x^i$. The summation for b lacks the first two terms. Writing what we want to know in terms of what we already know yields

$$b = a - 1 - x = \frac{x^{n+1} - 1}{x - 1} - 1 - x. \qquad ∎$$

Some of the exercises need proofs by induction; others apply statements that we have proved by induction, such as $\sum_{i=1}^{n} i = n(n+1)/2$. Most can be done in several ways. For example, calculus can be used to analyze numerical inequalities like $3^n > n^4$ in Example 3.36. In Remark 3.10, we quickly evaluated the sum $\sum_{i=1}^{n} i$ by grouping two copies of the summands in a clever way. See also Exercise 39.

In the next example, we give two proofs. The inductive proof illustrates the technique of manipulating a desired inequality (without changing its validity) to reduce the inequality to a known true statement. The second proof uses substitution.

3.38. Example. If $n \in \mathbb{N}$ and $x, y \geq 0$, then $(\frac{x+y}{2})^n \leq \frac{x^n + y^n}{2}$.

Our first proof uses induction on n. When $n = 1$, the claim is an equality. For the induction step, we assume that the inequality holds for all x, y when $n = k$. The desired expression has $(\frac{x+y}{2})^{k+1}$ on the left side. The induction hypothesis tells us something about $(\frac{x+y}{2})^k$. Writing what we have in terms of what we know yields

$$\left(\frac{x+y}{2}\right)^{k+1} = \left(\frac{x+y}{2}\right)^k \frac{x+y}{2} \leq \left(\frac{x^k + y^k}{2}\right)\frac{x+y}{2} = \frac{x^{k+1} + y^{k+1} + x^k y + xy^k}{4}.$$

To complete the proof by this approach, we need to know that the last expression is at most $\frac{x^{k+1} + y^{k+1}}{2}$. Multiplying by 4 and collecting terms on one side reduces this desired inequality to the equivalent inequality $0 \le x^{k+1} - x^k y + y^{k+1} - y^k x$. Here we can factor the right side to reduce the needed statement to $0 \le (x - y)(x^k - y^k)$.

It is easy to prove this last inequality. The function that maps x to x^k is increasing. Thus the factors $(x - y)$ and $(x^k - y^k)$ always have the same sign, and the inequality holds. Since the steps of our reduction are reversible, this completes the proof.

Alternative proof (substitution). We prove $(\frac{x+y}{2})^n \le \frac{x^n+y^n}{2}$ directly. To simplify the left side, we give $\frac{x+y}{2}$ a new name a; this suggests also giving $\frac{x-y}{2}$ a new name b. We may assume that $x \ge y$; this keeps a and b nonnegative. The substitution yields the new desired inequality

$$a^n \le \frac{(a + b)^n + (a - b)^n}{2}.$$

When we multiply out the numerator, we obtain a polynomial in a and b. The terms with negative coefficients are canceled by corresponding terms with positive coefficents. The right side of the inequality is left with $a^n/2 + a^n/2$ plus only terms with positive coefficients. Since a and b are nonnegative, the desired inequality thus holds. ∎

When do we seek a proof by induction? This is a hard question. Although induction is a possible strategy for proving that a formula involving n holds for every natural number n, sometimes it fails. There may be no nice way to write the formula for $n + 1$ in terms of the formula for n, and a proof by induction may be impossible or require difficult calculations. When this happens, try another approach.

For example, consider the sum $\sum_{i=0}^{n} n^i$. The parameter n appears both in the summand and as an index of notation. Replacing n by $n + 1$ in the formula leads to a mess. One needs another approach to verify that the sum is $\frac{n^{n+1}-1}{n-1}$ (see Exercise 37).

Those who have studied calculus may also consider the integral $\int_0^{2\pi} \cos^{2n}(\theta)\, d\theta$. One could evaluate this explicitly for small values of n, try to guess a pattern for the formula, and then seek a proof by induction. Guessing the answer might be hard; it is $2\pi \frac{(2n)!}{(2^n n!)^2}$. Even after guessing the correct formula, it is not clear how to obtain one integral from the previous and apply the induction hypothesis to prove the induction step. Ideas from Chapters 4 and 18 permit a simultaneous evaluation of this integral for all n by the same calculation (Exercise 18.15), without induction.

Strong Induction.

Strong induction applies when we consider arbitrarily large steps.

3.39. Example. *The game of Nim* (special case). Two players move alternately in a game that starts with two equal-sized piles of coins. A move consists of removing some positive number of coins from one pile. The winner is the player who removes the last coin.

Using strong induction, we can show easily that Player 2 wins this game. Certainly this holds when both piles have one coin, as Player 1 takes one and Player 2 takes the other. This proves the basis step.

For the induction step, suppose we start with two piles of size n. If the Player 1 takes a complete pile, then Player 2 takes the other pile and wins. Otherwise, Player 1 takes j coins from one pile, for some j. Player 2 responds by taking j coins from the other pile. The remainder of the game is equivalent to a game with $n - j$ coins in each pile and Player 1 moving first. The induction hypothesis tells us that Player 2 wins this game. This completes the proof.   ∎

Example 3.39 used strong induction in that j may be any number from 1 to $n - 1$, leaving an arbitrarily smaller game after the first round.

Strong and ordinary induction can be closely related. When a statement about n involves 2^k, it may be possible to prove it by strong induction on n or by ordinary induction on k. Exercise 60 considers an example.

In Solution 3.27, the proof of the statement $P(n)$ in the induction step uses the statement $P(n-2)$. When we need the previous r instances of the statement to prove the next statement, we must verify r instances in the basis step to get started. This is slightly different from strong induction.

3.40. Example. Let $\langle a \rangle$ be a sequence satisfying $a_1 = 2$, $a_2 = 8$, and $a_n = 4(a_{n-1} - a_{n-2})$ for $n \geq 3$. We seek a formula for a_n.

Given no formula to prove, we may try to guess one. The definition of $\langle a \rangle$ tells us that $a_3 = 24$, $a_4 = 64$, $a_5 = 160$. Note that $a_n = n2^n$ fits all the data so far. Having guessed this as a possible formula for a_n, we can try to use induction to prove it.

When $n = 1$, we have $a_1 = 2 = 1 \cdot 2^1$. When $n = 2$, we have $a_2 = 8 = 2 \cdot 2^2$. In both cases, the formula is correct.

In the induction step, we want to show that the desired formula is correct when $n \geq 3$. The induction hypothesis is the hypothesis that the formula is correct for the instances $n - 1$ and $n - 2$. Using our expression for a_n in terms of earlier terms, we thus have

$$a_n = 4(a_{n-1} - a_{n-2}) = 4[(n - 1)2^{n-1} - (n - 2)2^{n-2}] = (2n - 2)2^n - (n - 2)2^n = n2^n.$$

The validity of the formula for a_n follows from its validity for a_{n-1} and a_{n-2}, which completes the proof.   ∎

In this proof, we must verify the formula for $n = 1$ and $n = 2$ in the basis step, because otherwise the induction step would prove nothing for $n = 3$. Also, although we generated additional values to guess the formula, they do not appear in the proof. This also applies to extra small instances we may explore to understand the inductive argument.

Induction works well to prove statements about such sequences (see Exercises 55–57), but it will not discover a formula for the terms. Methods for obtaining such formulas when not given appear in Chapter 12.

The method of descent is particularly useful when the statement to be proved is a statement of nonexistence. A smallest counterexample is an actual solution to the problem we are trying to show has no solutions, and then we work with this solution to obtain a smaller solution (see Theorem 3.31). Fermat named the method of descent and used it to prove that the equation $x^4 + y^4 = z^4$ has no solution for positive integers x, y, z.

Communicating mathematics.

Finally, we comment again on the proper presentation of an argument. In a proof by induction, the induction step requires particular care.

When proving $P(n)$ for all $n \in \mathbb{N}$ by induction, it can help to start by stating what is being proved. The induction step proves for all $k \in \mathbb{N}$ that $P(k)$ implies $P(k + 1)$. This can be phrased as "Assume that $P(n)$ is true when $n = k$. We prove that then also $P(n)$ is true when $n = k + 1$." Some students write "We prove that $n = k$ implies $n = k + 1$"; this is nonsense.

In the induction step as phrased above, $P(k)$ is known and $P(k + 1)$ is to be proved. One must not write a proof deriving $P(k)$ from $P(k + 1)$. When $P(n)$ is a formula involving n, this error occurs if the proof begins with the formula $P(k + 1)$ and manipulates it without words until $P(k)$ is obtained. When the manipulations are reversible, the proof can be corrected by arguing that the steps are reversible and that therefore the desired formula $P(k + 1)$ has been reduced to the hypothesis $P(k)$.

It may be more efficient to manipulate the assumed formula $P(k)$ to obtain $P(k + 1)$, but the reduction method may help discover the proof. A good compromise is to start with one side of the formula $P(k + 1)$ and manipulate it, invoking the truth of $P(k)$ at an appropriate point, to reach the other side (see Propositions 3.7, 3.12, 3.19, and 3.21).

EXERCISES

Words like "determine", "obtain", "construct", or "show" request proof.

3.1. $(-)$ Give a sentence $P(n)$ depending on a natural number n, such that $P(1), P(2), \ldots, P(99)$ are all true but $P(100)$ is false. Make your sentence as simple as possible.

3.2. $(-)$ Let $P(n)$ be a mathematical statement depending on a natural number n. Suppose that $P(1)$ is false. Suppose also that whenever $P(n)$ is false, also $P(n+1)$ is false. Show that $P(k)$ is false for all $k \in \mathbb{N}$. (There is a one-line proof!)

3.3. $(-)$ Let $P(n)$ be a mathematical statement depending on an integer n. Suppose that $P(0)$ is true. Suppose also that whenever $P(n)$ is true, also both $P(n+1)$ and $P(n-1)$ are true. Show that $P(k)$ is true for all $k \in \mathbb{Z}$.

3.4. $(-)$ Let $P(n)$ be a mathematical statement depending on an integer n. Suppose that $P(0)$ is true. Suppose also that whenever $P(n)$ is true, at least one of $P(n+1)$ and $P(n-1)$ is true. For which $n \in \mathbb{Z}$ must $P(n)$ be true?

• • • • •

In Exercises 5–9, determine whether the statement is true or false. If true, provide a proof. If false, provide a counterexample.

3.5. For $n \in \mathbb{N}$, $\sum_{k=1}^{n}(2k+1) = n^2 + 2n$.

3.6. If $P(2n)$ is true for all $n \in \mathbb{N}$, and $P(n) \Rightarrow P(n+1)$ for all $n \in \mathbb{N}$, then $P(n)$ is true for all $n \in \mathbb{N}$.

3.7. For $n \in \mathbb{N}$, $2n - 8 < n^2 - 8n + 17$.

3.8. For $n \in \mathbb{N}$, $2n - 18 < n^2 - 8n + 8$.

3.9. For $n \in \mathbb{N}$, $\frac{2n-18}{n^2-8n+8} < 1$.

• • • • •

3.10. $(-)$ Suppose that $n \in \mathbb{N}$ and that $x_1, \ldots, x_{2n+1}$ are odd integers. Prove that $\sum_{i=1}^{2n+1} x_i$ is odd and that $\prod_{i=1}^{2n+1} x_i$ is odd.

3.11. $(-)$ Use induction on n to prove that a set of n elements has 2^n subsets.

3.12. $(-)$ Given $x \in \mathbb{R}$ and $n \in \mathbb{N}$, use induction to prove that $\sum_{i=1}^{n} x = nx$.

3.13. $(-)$ Explain why the sum and difference of two polynomials are polynomials.

3.14. $(-)$ For each sum below, write it in summation notation and find and prove a formula in terms of n.
  a) $3 + 7 + 11 + \cdots + (4n - 1)$.
  b) $1 + 5 + 9 + \cdots + (4n + 1)$.
  c) $-1 + 2 - 3 + 4 - \cdots - (2n - 1) + 2n$.
  d) $1 - 3 + 5 - 7 + \cdots + (4n - 3) - (4n - 1)$.

3.15. For $n \in \mathbb{N}$, prove that $\sum_{i=1}^{n}(-1)^i i^2 = (-1)^n \frac{n(n+1)}{2}$.

3.16. For $n \in \mathbb{N}$, prove that $\sum_{i=1}^{n} i^3 = (\frac{n(n+1)}{2})^2$.

3.17. For $n \in \mathbb{N}$, prove that $\sum_{i=1}^{n} i(i+1) = \frac{n(n+1)(n+2)}{3}$.

3.18. Given $0 \le a_i \le b_i$ for all $i \in \mathbb{N}$, prove that $\prod_{i=1}^{n} a_i \le \prod_{i=1}^{n} b_i$.

3.19. For $k \in \mathbb{N}$, prove that $x < y$ implies $x^{2k-1} < y^{2k-1}$.

3.20. Write out the proof of Lemma 3.13 using summation notation.

3.21. Multiply out $\left(\sum_{i=1}^{n} x_i\right)^2$, writing the result in summation notation.

3.22. (!) For $n \in \mathbb{N}$, prove that $\left|\sum_{i=1}^{n} a_i\right| \le \sum_{i=1}^{n} |a_i|$.

3.23. Let a be a nonzero real number. Find the flaw in the following "proof" that $a^n = 1$ for every nonnegative integer n.
  "Basis step: $a^0 = 1$. Induction step: $a^{n+1} = a^n \cdot a^n / a^{n-1} = 1 \cdot 1/1 = 1$."

3.24. Let m be a natural number. Find the flaw in the statement below. Explain why the statement is not valid, and change one symbol to correct it.

"If T is a set of natural numbers such that 1) $m \in T$ and 2) $n \in T$ implies $n + 1 \in T$, then $T = \{n \in \mathbb{N}: n \geq m\}$."

3.25. Prove that the sum and product of natural numbers are natural numbers. (Hint: See the discussion after Theorem 3.6.)

3.26. Let $\langle a \rangle$ be a sequence such that $a_1 = 1$ and $a_{n+1} = a_n + 3n(n + 1)$ for $n \in \mathbb{N}$. Prove that $a_n = n^3 - n + 1$ for $n \in \mathbb{N}$.

3.27. For $n \in \mathbb{N}$, prove that $\sum_{i=1}^{n} \frac{1}{(3i-2)(3i+1)} = \frac{n}{3n+1}$.

3.28. For $n \in \mathbb{N}$, find and prove a formula for $\sum_{i=1}^{n} \frac{1}{i(i+1)}$.

3.29. For $n \in \mathbb{N}$, find and prove a formula for $\sum_{i=1}^{n} (2i - 1)$.

3.30. For $n \in \mathbb{N}$, prove that $\sum_{i=1}^{n} (2i - 1)^2 = \frac{n(2n-1)(2n+1)}{3}$.

3.31. For $n \in \mathbb{N}$ and $n \geq 2$, find and prove a formula for $\prod_{i=2}^{n} (1 - \frac{1}{i^2})$.

3.32. For $n \in \mathbb{N}$ and $n \geq 2$, find and prove a formula for $\prod_{i=2}^{n} (1 - \frac{(-1)^i}{i})$.

3.33. Obtain a simple formula for the number of closed intervals with integer endpoints contained in the interval $[1, n]$ (including one-point intervals).

3.34. Consider a set of 20 boxes, each containing 20 balls. Suppose every ball weighs one pound, except that the balls in one box are all one ounce too heavy or all one ounce too light. A precise scale is available that can weigh to the nearest ounce (not a balance scale). By selecting some balls to place on the scale, explain how to determine in one weighing which is the defective box and whether its balls are too heavy or too light.

3.35. Let q be a real number other than 1. Use induction on n to prove that $\sum_{i=0}^{n-1} q^i = (q^n - 1)/(q - 1)$.

3.36. Obtain a polynomial f such that $\sum_{i=2}^{n} x^i = f(x)/(x - 1)$.

3.37. For $n \in \mathbb{N}$, obtain a formula for $\sum_{i=1}^{n} n^i$. (Hint: Do not use induction.)

3.38. Starting with 0, two players alternately add 1, 2, or 3 to a single running total. The player who first brings the total to at least 1000 wins. Prove that the second player has a strategy to win against any strategy for the first player. (Hint: Use induction to prove a more general statement.)

3.39. (!) Let S_n be the hexagonal arrangement consisting of n rings of dots, as illustrated below for $n \in \{1, 2, 3\}$. Let a_n be the number of dots in S_n. Find formulas for a_n and $\sum_{k=1}^{n} a_k$ (simplify all sums).

3.40. Consider a cube of size n formed by assembling n^3 cubes of size 1. Prove that the number of cubes of all positive integer sizes in a cube of size n is $\frac{1}{4}n^2(n+1)^2$.

3.41. (!) Let $f\colon \mathbb{R} \to \mathbb{R}$ be a function such that $f(x+y) = f(x)+f(y)$ for $x, y \in \mathbb{R}$.
 a) Prove that $f(0) = 0$.
 b) Prove that $f(n) = nf(1)$ for all $n \in \mathbb{N}$.

3.42. Addition is defined as a function from $\mathbb{R} \times \mathbb{R}$ to $\mathbb{R}$; it sums pairs of numbers. Use induction on n to prove that the sum of n numbers is independent of the order in which the numbers are added into the total. This justifies the use of summation notation for a sum of n numbers.

3.43. (!) Suppose $f\colon \mathbb{R} \to \mathbb{R}$ satisfies $f(xy) = xf(y)+yf(x)$ for all $x, y \in \mathbb{R}$. Prove that $f(1) = 0$ and that $f(u^n) = nu^{n-1}f(u)$ for all $n \in \mathbb{N}$ and $u \in \mathbb{R}$.

3.44. (!) Determine the set of natural numbers that can be expressed as the sum of some nonnegative number of 3s and some nonnegative number of 10s.

3.45. (!) Determine the set of natural numbers n such that every sum of n consecutive natural numbers is divisible by n.

3.46. (−) Let $f(n) = n^2 - 8n + 18$. For which $n \in \mathbb{N}$ is $f(n) > f(n-1)$ true?

3.47. Prove that $5^n + 5 < 5^{n+1}$ for all $n \in \mathbb{N}$.

3.48. (!) Determine the set of positive real numbers x such that the inequality $x^n + x < x^{n+1}$ holds for all $n \in \mathbb{N}$.

3.49. For each of the following inequalities, determine the set of natural numbers n for which it holds.
 a) $3^n \geq 2^{n+1}$.
 b) $2^n \geq (n+1)^2$.
 c) $3^{n+1} > n^4$.
 d) $n^3 + (n+1)^3 > (n+2)^3$.

3.50. Let f be a function mapping $\mathbb{Z}$ into the set of positive real numbers. Suppose that $f(1) = 1$ and that f satisfies $f(x - y) = f(x)/f(y)$ for $x, y \in \mathbb{Z}$. Find $f(n)$ for $n \in \mathbb{N}$ and prove your formula by induction. Repeat for $f(1) = c$.

3.51. Construct a cubic polynomial such that the set of natural numbers where its value is at least 3 is $\{1\} \cup \{n \in \mathbb{N}\colon n \geq 5\}$.

3.52. *Partial fraction expansion.* Use Corollary 3.90 to obtain constants A, B, r, s such that $\frac{1}{x^2+x-6} = \frac{A}{x-r} + \frac{B}{x-s}$ for all $x \in \mathbb{R} - \{r, s\}$.

3.53. (!) Suppose that $f(x)$ is a polynomial of degree n and that the values $f(0), f(1), ..., f(n)$ are known. Describe a procedure for determining f, and justify that it works. (Hint: For $n \geq 1$, recall from the proof of Theorem 3.89 that $f(x) - f(n) = (x - n)h(x)$, where h is a polynomial of degree $n - 1$.)

3.54. (!) Suppose that F is defined by $f(x) = \sum_{i=0}^{n} c_i x^i$ and has zeros $\alpha_1, \ldots, \alpha_n$ such that $\alpha_i \neq 0$ for all i. Derive a formula for $\sum_{i=1}^{n}(1/\alpha_i)$ in terms of $c_0, \ldots, c_n$. (Hint: First show that $f(x) = c\prod(x - \alpha_i)$; see Lemma 3.88. Comment: A more general result is proved in Exercise 17.40.)

3.55. Let $\langle a \rangle$ be a sequence satisfying $a_1 = 1$, $a_2 = 8$, and $a_n = a_{n-1} + 2a_{n-2}$ for $n \geq 3$. Prove that $a_n = 3 \cdot 2^{n-1} + 2(-1)^n$ for $n \in \mathbb{N}$.

3.56. Let $\langle a \rangle$ be a sequence satisfying $a_n = 2a_{n-1} + 3a_{n-2}$ for $n \geq 3$.
   a) Given that a_1, a_2 are odd, prove that a_n is odd for $n \in \mathbb{N}$.
   b) Given that $a_1 = a_2 = 1$, prove that $a_n = \frac{1}{2}(3^{n-1} - (-1)^n)$ for $n \in \mathbb{N}$.

3.57. Let $\langle a \rangle$ be a sequence satisfying $a_1 = a_2 = 1$ and $a_n = \frac{1}{2}(a_{n-1} + 2/a_{n-2})$ for $n \geq 2$. Prove that $1 \leq a_n \leq 2$ for $n \in \mathbb{N}$.

3.58. (!) *L-tilings.* Prove that R has an L-tiling in the following situations.
   a) R is a 2^k by 2^k chessboard with one corner square removed.
   b) R is a 2^k by 2^k chessboard with *any* single square removed.

3.59. (+) Determine which rectangles have L-tilings.

3.60. (!) Consider a row of n boxes, each containing a number, such that the number in the ith box is the ith smallest number. Given a number x, one would like to know whether x appears in one of the boxes. Iteratively, one can look at the number in a box and then decide what box to look in next.
   a) Prove that when $n < 2^k$, there is a strategy that always determines whether x is present by looking in at most k boxes, no matter what x is or what numbers are in the boxes.
   b) Prove that when $n \geq 2^k$, there is no strategy that will always answer the question by looking in at most k boxes.

3.61. ($-$) Using the rules of Problem 3.69, remove the coins in $HTHTHHTHHH$. How many steps does it take?

3.62. (!) *The December 31 Game.* Two players alternately name dates. On each move, a player can increase the month or the day of the month but not both. The starting position is January 1, and the player who names December 31 wins. According to the rules, the first player can start by naming some day in January after the first or the first of some month after January. For example, (Jan. 5, Mar. 5, Mar. 15, Apr. 15, Apr. 25, Nov. 25, Nov. 30, Dec. 30, Dec. 31) is an instance of the game won by the first player. Derive a winning strategy for the first player. (Hint: Use strong induction to describe the "winning dates".)

3.63. Beginning at the origin, two players alternately move a token in the plane. When the token is at (x, y), the player chooses a natural number n and moves either to $(x + n, y)$ or to $(x, y + 5n)$. Show that the second player can arrange to always return to the line $y = 5x$.

3.64. Derive the principle of induction from the Well-Ordering Property for $\mathbb{N}$.

3.65. (!) In the village of Perfect Reasoning, each employer has an apprentice. At least one apprentice is a thief. To remedy this without embarrassment, the mayor proclaims the following true statements: "At least one apprentice in this town is a thief. Every thief is known to be a thief by everyone except his or her employer, and all employers reason perfectly. If n days from now you have concluded that your apprentice is a thief, you will come to the village square at noon that day to denounce your apprentice." The villagers gather at noon every day thereafter. If in fact $k \geq 1$ of the apprentices are thieves, when will they be denounced, and how do their employers reason? (Hint: Study small values of k, and use induction to prove the pattern for all k.)

Chapter 4

Bijections and Cardinality

We begin this chapter by discussing how to represent natural numbers. Our first main result is the analogue of decimal representation using any base q. The base q representation provides a unique name for each natural number and thus introduces the notion of a one-to-one correspondence. We study such correspondences via properties of functions and use this to develop the notion of cardinality of sets.

4.1. Problem. *The Weights Problem.* A balance scale has left and right pans; we can place objects in each pan and test whether the total weight is the same on each side. Suppose that five objects of known integer weight can be selected. How can we choose the weights to guarantee being able to check all integer weights from 1 through 121? Given an object believed to have weight $n \in [121]$, how should we place the known weights to check it? Is it possible to choose five values to check more weights? ∎

4.2. Problem. Is there a one-to-one correspondence between the set of points in the open interval $(0, 1)$ and the set of real numbers? ∎

REPRESENTATION OF NATURAL NUMBERS

The most naive way to represent the number "one hundred" is by a collection of one hundred dots; it is hard even to count them. We can arrange the dots in a ten-by-ten square, but no geometric method gives convenient representations of large natural numbers.

Roman numerals permit a reasonably concise description of large natural numbers, but they make arithmetic computations difficult. In Roman numerals, the symbols I, V, X, L, C, D, M represent 1, 5, 10, 50, 100, 500, 1000. Other numbers are represented by strings of these symbols using complicated rules involving addition and subtraction of adjacent symbols.

For example, 2 is written as II, 44 as XLIV, 88 as LXXXVIII, and 90 as XC; addition of 2 and multiplication by 2 are awkward operations!

The familiar decimal representation facilitates arithmetic computations and represents fairly large numbers concisely. The decimal (base 10) representation of a natural number is a string of symbols from $\{0, 1, \ldots, 9\}$, encoding the number as a sum of multiples of powers of 10. Chemists, physicists, and astronomers often need very large numbers and express them in "scientific notation", a variant of decimal representation where only the significant digits and the order of magnitude are recorded (6.02×10^{23} is scientific notation for 602,000,000,000,000,000,000,000). Computer scientists use binary (base 2), octal (base 8), and hexadecimal (base 16) representations, where the string representing the number encodes its expression as a sum of multiples of powers of the base.

The appropriate method of representation depends on the problem being solved. In base q, there are q elementary symbols, representing the numbers 0 through $q - 1$. Computers use binary digits ("bits") because there are two alternatives for a switch: "on" or "off". In solving the Weights Problem (Problem 4.1), we will apply base 3 representation.

4.3. Definition. Let q be a natural number greater than 1. A q-**ary** or **base q representation** of n is a list $a_m, \ldots, a_0$ of integers, each in $\{0, 1, \ldots, q - 1\}$, such that $a_m > 0$ and $n = \sum_{i=0}^{m} a_i q^i$. For clarity, we may use a subscript (q) to indicate that the base is q. We call representations in base 2, 3, or 10 **binary**, **ternary**, or **decimal**, respectively.

In Theorem 4.7, we will prove that every natural number has a unique base q representation. This allows us to write "*the* base q representation" instead of "*a* base q representation".

4.4. Example. The ternary representations for the first ten natural numbers in order are 1, 2, 10, 11, 12, 20, 21, 22, 100, 101. The corresponding representations in base 4 are 1, 2, 3, 10, 11, 12, 13, 20, 21, 22. ■

4.5. Example. Base 10 is the familiar base for representing numbers. For the natural number $354 = 3 \cdot 10^2 + 5 \cdot 10^1 + 4 \cdot 10^0$, the elements of the base 10 representation are $a_2 = 3$, $a_1 = 5$, $a_0 = 4$. We can also write $354 = 2 \cdot 5^3 + 4 \cdot 5^2 + 0 \cdot 5^1 + 4 \cdot 5^0$, expressed concisely as $2404_{(5)}$. Note that the coefficient of the highest power of q appears at the left.

There are several ways to find a representation of a number n in base q. One way is first to determine the largest nonzero index m by finding the largest power of q that is at most n. The coefficient a_m is the largest multiple of q^m that can be subtracted from n without making it negative. We then repeat the procedure with what remains, which is smaller than q^m. For example, 5^4 is larger than 354, but 5^3 is not, so the base 5 representation of 354 starts with a_3. Since $5^3 = 125$ can be subtracted twice from 354, the representation begins with $a_3 = 2$, followed by the representation of 104. By this procedure, we obtain $354_{(10)} = 2404_{(5)}$. In other bases, we have $354_{(10)} = 11202_{(4)} = 111010_{(3)} = 101100010_{(2)}$.   ∎

The procedure in Example 4.5 generates a base q representation for each natural number (Exercise 14). This procedure is much faster than the simpler one used in Theorem 4.7 to generate base q representations.

4.6. Theorem. Let q be a natural number greater than 1. Every natural number has a unique base q representation with no leading zeros.

Proof: We first use induction on n to construct a base q representation of n. For $n = 1$, we have the representation with $a_0 = 1$. For $n > 1$, we assume that $n - 1$ has a base q representation. Let $n - 1 = \sum_{j=0}^{m} a_j q^j$ with $a_m \neq 0$ be such a representation.

If $a_m = \cdots = a_0 = q - 1$, then we represent n by $a_{m+1} = 1$ and $a_j = 0$ for $j \leq m$. This works, since the geometric sum (Corollary 3.14) yields

$$n - 1 = \sum_{j=0}^{m}(q - 1)q^j = (q - 1)\sum_{j=0}^{m} q^j = (q - 1)\frac{q^{m+1}-1}{q-1} = q^{m+1} - 1,$$

and thus $n = q^{m+1}$.

Otherwise, some coefficient in the expansion is less than $q - 1$, and we let t be the smallest index such that $a_t < q - 1$. Define $b_0, \ldots, b_m$ by $b_j = a_j$ for $j > t$, $b_t = a_t + 1$, and $b_j = 0$ for $j < t$. Since $a_j = q - 1$ for all j (if any) with $j < t$, the geometric sum as above yields $\sum_{j=0}^{m} b_j q^j = 1 + \sum_{j=0}^{m} a_j q^j$, and thus $b_m, \ldots, b_0$ is a base q representation of n.

We prove uniqueness by the method of descent. Suppose that $a_r, \ldots, a_0$ and $b_s, \ldots, b_0$ are distinct base q representations of some $n \in \mathbb{N}$. If $r \neq s$, then by symmetry we may assume that $r > s$. Now the number represented by $a_r, \ldots, a_0$ is at least q^r, but the number represented by $b_s, \ldots, b_0$ is at most $\sum_{j=0}^{r-1}(q - 1)q^j = q^r - 1$, so this case never occurs.

Thus we have $r = s$. Both a_r and b_r are nonzero, and we can subtract 1 from each to obtain distinct base q representations of the smaller number $n - q^r$. Hence there is no smallest failure of uniqueness.   ∎

Base q provides both convenient representations of natural numbers and a system of computation. We can compute directly in base q as we do in base 10, but we "carry" or "borrow" q rather than 10. For example, $42_{(5)} + 14_{(5)} = 111_{(5)}$. The proof of Theorem 4.6 uses the notion of carrying in base q. It also illustrates the familiar statement that m exceeds n if and only if, in the highest-order position where their decimal representations differ, the digit for m is larger than the digit for n.

Ternary representations lead to a solution of Problem 4.1. For an approach using induction, see Exercise 15.

4.7. Example. To understand the Weights Problem, we first consider the analogous question with fewer weights. With two weights, we do best by choosing $\{1, 3\}$; we can then test 2 by putting 1 and 3 on opposite sides and test 4 by putting them on the same side, along with testing 1 or 3 by using that weight alone.

Exploring a mathematical problem may involve both experimentation and insightful thinking. Here we may experiment to find that choosing $\{1, 3, 9\}$ for three known weights allows us to balance all unknown weights up to 13, and no other choice goes farther. This pattern suggests that choosing powers of 3 as the unknown weights may be a good idea. Insightful thinking may suggest this choice directly, since for each weight we have the three options of "left pan", "right pan", and "omit". Using powers of 3 allows us to exploit these three options fully.

Now consider five weights. Using $\{1, 3, 9, 27, 81\}$, we can balance an unknown object A of weight 49 (for example) by using $\{9, 27\}$ and A on one side and using $\{1, 3, 81\}$ on the other side. The picture below shows how this example is solved by the general method in Solution 4.8. The largest weight that can be balanced is $1 + 3 + 9 + 27 + 81 = 121$. With these weights, we can balance each integer weight up to 121. We will also see that this is the best choice.   ∎

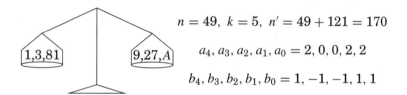

$$n = 49, \ k = 5, \ n' = 49 + 121 = 170$$

$$a_4, a_3, a_2, a_1, a_0 = 2, 0, 0, 2, 2$$

$$b_4, b_3, b_2, b_1, b_0 = 1, -1, -1, 1, 1$$

4.8. Solution. *The Weights Problem.* On a balance scale, we prove that the set $S_k = \{1, 3, \ldots, 3^{k-1}\}$ of k known weights permits the checking of all integer weights from 1 through $(3^k - 1)/2$, and that no other choice of k known weights permits more values to be checked.

Let $f(k) = (3^k - 1)/2$. We prove first that for $1 \le n \le f(k)$, the set S_k allows us to balance an object A of weight n. We need to use known weights from S_k so that the difference between the total weight on the side opposite A and the total weight on the side with A is n. Weights on the opposite side count positively, and those on the side with A count negatively. Thus it suffices to express n as $\sum_{i=0}^{k-1} b_i 3^i$, where each $b_i \in \{-1, 0, 1\}$. Interpreting the values $-1, 0, 1$ for b_i to mean "same side as A", "off the balance", and "side opposite to A" yields an explicit configuration that balances A.

We find $b_0, \ldots, b_{k-1}$ using the ternary representation of the number $n' = n + f(k)$. The equation $n = \sum_{i=0}^{k-1} b_i 3^i$ holds if and only if the equation $n' = \sum_{i=0}^{k-1} (b_i + 1) 3^i$ holds, because the geometric sum yields $(3^k - 1)/2 = \sum_{i=0}^{k-1} 3^i$. Since $n \le f(k)$, we have $n' \le 2f(k) = 3^k - 1$. Theorem 4.6 now guarantees a (unique) expression of n as $n = \sum_{i=0}^{k-1} a_i 3^i$ with each $a_i \in \{0, 1, 2\}$. Setting $b_i = a_i - 1$ yields the desired weighing of n.

We also must prove that no other set of weights can balance more values. We count the possible configurations: each weight can be placed on the left, on the right, or omitted, generating 3^k possible configurations. The configuration that omits all weights balances no nonzero weight. Of the remaining $3^k - 1$ configurations, each balances the same weight as the configuration obtained by switching the left pan and right pan. Hence at most $(3^k - 1)/2$ distinct values can be weighed.   ∎

BIJECTIONS

Given q, Theorem 4.6 provides a unique name for each natural number. This name is a sequence from $\{0, \ldots, q - 1\}$ that is 0 after some last nonzero term. Let S be the set of these sequences. The theorem establishes a **one-to-one correspondence** between S and $\mathbb{N}$. The function $f: S \to \mathbb{N}$ defined by $f(\langle a \rangle) = \sum a_j q^j$ assigns to each $\langle a \rangle \in S$ a natural number. Furthermore, each natural number is assigned to exactly one $\langle a \rangle \in S$. Thus we have matched up the elements of S with those of $\mathbb{N}$. The sequence matched with $n \in \mathbb{N}$ is its base q representation.

One-to-one correspondences have many applications. For example, we can interpret solving equations in this way. Given $f: A \to B$, the equation $f(x) = b$ has a unique solution whenever $b \in B$ if and only if f establishes a one-to-one correspondence between A and B.

We illustrate the subtlety of this notion by constructing a one-to-one correspondence between $\mathbb{N}$ and $\mathbb{Z}$, even though $\mathbb{N}$ is a proper subset of $\mathbb{Z}$!

4.9. Example. *One-to-one correspondence between $\mathbb{N}$ and $\mathbb{Z}$.* We define a function from $\mathbb{N}$ to $\mathbb{Z}$ by letting $f(n) = -(n-1)/2$ if n is odd and $f(n) = n/2$ if n is even. Note that $f(n)$ is positive when n is even and nonpositive when n is odd. Thus $f(n) = b$ for $b \in \mathbb{Z}$ has the unique solution $n = 2b$ when $b > 0$ and $n = -2b + 1$ when $b \leq 0$.   ∎

We formalize the notion of one-to-one correspondence using functions.

4.10. Definition. A function $f: A \to B$ is a **bijection** if for every $b \in B$ there is exactly one $x \in A$ such that $f(x) = b$.

4.11. Example. *Pairing up spouses.* Let M be the set of men at a party, and let W be the set of women. If the attendees consist entirely of married couples, then we can define a function $f: M \to W$ by letting $f(x)$ be the spouse of x. For each woman $w \in W$, there is exactly one $x \in M$ such that $f(x) = w$. Hence f is a bijection from M to W.   ∎

4.12. Example. *Linear equations in two variables.* Given constants $a, b, c, d \in \mathbb{R}$, let $f: \mathbb{R}^2 \to \mathbb{R}^2$ be defined by $f(x, y) = (ax + by, cx + dy)$.

Theorem 2.2 states that the pair of equations $ax + by = r$ and $cx + dy = s$ has a unique solution for each $(r, s) \in \mathbb{R}^2$ if and only if $ad - bc \neq 0$. In other words, the function f is a bijection if and only if $ad - bc \neq 0$. ∎

When discussing a bijection $f \colon A \to B$, we often speak more informally of a one-to-one correspondence *between* A and B. This emphasizes that we can view the paired elements of the domain and target from either direction. Each element of B is the image of exactly one element of A. Thus assigning to each element of B the element of A of which it is the image defines a function from B to A that "undoes" f.

4.13. Definition. If f is a bijection from A to B, then the **inverse** of f is the function $g \colon B \to A$ such that, for each $b \in B$, $g(b)$ is the unique element $x \in A$ such that $f(x) = b$. We write f^{-1} for the function g.

4.14. Example. The identity function on a set S is a bijection from S to S that is its own inverse.

The function $f \colon \mathbb{R} \to \mathbb{R}$ defined by $f(x) = 3x$ is a bijection. Its inverse is defined by $f^{-1}(b) = b/3$.

When $ad - bc \neq 0$, the inverse of f in Example 4.12 is the function that expresses the solution pair (x, y) in terms of the pair (r, s). ∎

4.15. Remark. *If f is a bijection and g is the inverse of f, then g is also a bijection and f is the inverse of g.* This follows from the interpretation of a bijection as a pairing up of sets; in one direction the map is f, and in the other it is g. Thus $(f^{-1})^{-1} = f$. ∎

4.16. Example. The formula for converting Celsius temperature to Fahrenheit temperature is $f(x) = (9/5)x + 32$; this defines a bijection from $\mathbb{R}$ to $\mathbb{R}$. The inverse function g is defined by $g(b) = (5/9)(b - 32)$. We have $g(f(x)) = x$ for $x \in \mathbb{R}$, and also $f(g(b)) = b$ for $b \in \mathbb{R}$. When g is applied to true physical temperatures, the domain is $\{b \in \mathbb{R} \colon b \geq -273.15\}$.

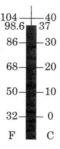

When interpreting physical measurements, care is needed in converting from one scale to another. It is commonly believed that "normal body temperature" is 98.6 degrees Fahrenheit, which equals 37 degrees

Celsius, exactly. Body temperature was first discussed using the Celsius scale. Perhaps "37 degrees" is the average body temperature accurate to the nearest degree Celsius. It is inappropriate to state the "normal" Fahrenheit body temperature to the accuracy suggested by 98.6.   ∎

Showing that $f: A \to B$ is a bijection means showing that for each $b \in B$, the equation $f(x) = b$ has a unique solution in A. Solving the equation to write a formula that determines x in terms of b obtains a formula for f^{-1}. We must check that the formula is valid on all of B.

4.17. Example. Define $f: \mathbb{R} \to \mathbb{R}$ by $f(x) = 5x - 2|x|$. We show that f is a bijection by solving $f(x) = b$ for each $b \in \mathbb{R}$, thereby obtaining f^{-1}. Observe that $f(x)$ has the same sign as x. When seeking $f^{-1}(b)$ for $b \in \mathbb{R}$, this enables us to assume that x has the same sign as b.

When $b \geq 0$, our equation becomes $b = 5x - 2|x| = 3x$, and $x = b/3$ is the unique solution. When $b < 0$, our equation becomes $b = 5x - 2|x| = 5x + 2x = 7x$, and $x = b/7$ is the unique solution.

For each $b \in \mathbb{R}$, we have shown that $f(x) = b$ has a unique solution. Thus f is a bijection. Its inverse is defined by $f^{-1}(b) = b/3$ if $b \geq 0$ and $f^{-1}(b) = b/7$ if $b < 0$.   ∎

A bijection transforms elements of one set into elements of another, allowing us to work in either context. For example, we can encode a subset S of $[n]$ by recording the presence or absence of element i as a 1 or 0 in position i of an n-tuple $m(S)$. An n-tuple with entries in $\{0, 1\}$ is a **binary n-tuple**; we call $m(S)$ the **binary encoding** of S. From a binary n-tuple b we will uniquely retrieve S such that $m(S) = b$. Thus binary encoding is a bijection from the power set of $[n]$ to the set of binary n-tuples. (Recall that the power set of T is the set of all subsets of T.)

4.18. Example. Given lights labeled $1, \ldots, n$, we can specify a subset of $[n]$ by turning the corresponding lights on. Binary encoding records in position i whether light i is on or off. Below we illustrate the correspondence when $n = 3$. The bijection for general n enables us to view subsets of $[n]$ as binary n-tuples and vice versa, transforming statements about one context into statements about the other.   ∎

lights on:	Ø	{1}	{2}	{3}	{1, 2}	{1, 3}	{2, 3}	{1, 2, 3}
image:	(0,0,0)	(1,0,0)	(0,1,0)	(0,0,1)	(1,1,0)	(1,0,1)	(0,1,1)	(1,1,1)

4.19. Proposition. Binary encoding establishes a bijection from the power set of $[n]$ to the set of binary n-tuples.

Proof: Let $m(S)$ be the binary encoding of S. We prove that for each binary n-tuple b, there is exactly one subset S of $[n]$ such that $m(S) = b$.

Let b be a binary n-tuple; we construct a set S such that $m(S) = b$.

For each $i \in [n]$, we put $i \in S$ if $b_i = 1$ and $i \notin S$ if $b_i = 0$. Applying m, we find that the ith position in $m(S)$ is b_i, and thus $m(S) = b$.

We also show that for each binary n-tuple b there is at most one solution to $m(S) = b$. We prove the contrapositive; distinct subsets S and T of $[n]$ have distinct images. Because $S \neq T$, some $i \in [n]$ belongs to one of $\{S, T\}$ but not the other. Hence position i has a 1 in one of $\{m(S), m(T)\}$ and a 0 in the other. Thus $m(S)$ and $m(T)$ cannot both equal b. ∎

Binary encoding provides a useful language for proving results about subsets. We can also interpret binary n-tuples as binary representations of integers from 0 to $2^n - 1$, using the bijection in Theorem 4.6.

INJECTIONS AND SURJECTIONS

The condition for a function to be a bijection is the combination of two conditions that we can consider independently. Often we prove that a function is a bijection by verifying these two conditions separately.

4.20. Definition. A function $f \colon A \to B$ is **injective** if for each $b \in B$, there is at most one $x \in A$ such that $f(x) = b$. A function $f \colon A \to B$ is **surjective** if for each $b \in B$, there is at least one $x \in A$ such that $f(x) = b$. The corresponding nouns are **injection** and **surjection**.

In Proposition 4.19 we proved first that binary encoding is surjective and then that it is injective.

4.21. Remark. *Geometric interpretation of injection and surjection.* A function $f \colon \mathbb{R} \to \mathbb{R}$ is injective if and only if every horizontal line intersects its graph at most once, and f is surjective if and only if every horizontal line intersects its graph at least once. ∎

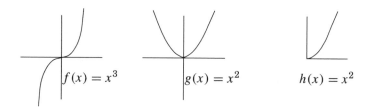

$f(x) = x^3$ $\qquad$ $g(x) = x^2$ $\qquad$ $h(x) = x^2$

4.22. Example. Graphing $f(x) = x^3$ and using Remark 4.21 suggests that this formula defines a bijection $f \colon \mathbb{R} \to \mathbb{R}$. We shall see that f is injective because it is an increasing function. The proof of surjectivity requires the existence of cube roots of real numbers.

The function $g \colon \mathbb{R} \to \mathbb{R}$ defined by $g(x) = x^2$ is neither injective nor surjective. Since $g(-1) = g(1)$, it is not injective. Since its image contains no negative numbers, it is not surjective. When $P = \{x \in \mathbb{R}: x > 0\}$, the function $h: P \to P$ defined by $h(x) = x^2$ *is* a bijection, as is the function defined by this rule with domain and target $[0, 1]$.   ∎

4.23. Example. Consider $f: \mathbb{R} \to \mathbb{R}$ defined by $f(x) = 1/(1 + x^2)$. Attempting to solve for x in $b = 1/(1 + x^2)$ yields $x = \pm\sqrt{(1/b) - 1}$. When $b \leq 0$ or $b > 1$, there is no solution. Here the inverse image of b is empty, and thus f is not surjective. When $0 < b \leq 1$, there is at least one solution. For $b \in (0, 1)$ there are two solutions, so f is not injective.

Let $A = \{x \in \mathbb{R}: x \geq 0\}$ and $B = \{x \in \mathbb{R}: 0 < x \leq 1\}$. If we use the formula $g(x) = 1/(1 + x^2)$ to define a function $g: A \to B$, then g *is a bijection*. We have chosen the target to ensure existence of a solution to $g(a) = b$ for each $b \in B$, and we have chosen the domain to ensure uniqueness of the solution to $g(a) = b$.   ∎

4.24. Remark. *Schematic interpretation of injection and surjection.* In the diagram of a function $f: A \to B$ suggested in Remark 1.22, each element of A is the tail of exactly one arrow; this follows from the definition of *function*. The function f is injective if each element of B is the head of at most one arrow, meaning that there is no "collapsing" of elements. The function is surjective if each element of B is the head of at least one arrow, meaning that no element of the target is "missed".

Reversing the arrows yields a function if and only if f is a bijection, in which case the resulting function from B to A is f^{-1}.   ∎

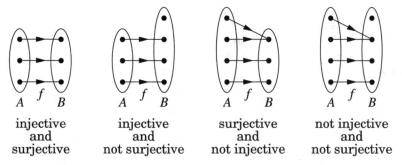

| injective and surjective | injective and not surjective | surjective and not injective | not injective and not surjective |

The geometric interpretation of injection (Remark 4.21) suggests that every increasing real-valued function is injective.

4.25. Proposition. Let f be a real-valued function defined on a subset of $\mathbb{R}$. If f is strictly monotone, then f is injective.

Proof: Given distinct x, y in the domain, we may assume by symmetry that $x < y$. If f is increasing, then $f(n) > f(y)$. If f is decreasing, then $f(x) > f(y)$. In either case, $x \neq y$ implies $f(x) \neq f(y)$, so f is injective. ∎

4.26. Example. *Exponentiation.* The function $f\colon \mathbb{R} \to \mathbb{R}$ defined by $f(x) = x^n$ is strictly increasing when n is an odd natural number. When n is even, f is not injective; in this case $x^n = y^n$ whenever $y = \pm x$. Exercise 18 requests the details.   ∎

4.27. Example. What are the solutions to the equation below?

$$x^4 + x^3 y + x^2 y^2 + xy^3 + y^4 = 0 \qquad\qquad (*)$$

Certainly $(x, y) = (0, 0)$ is a solution; we show that there are no others. First consider the solutions with $x = y$. Setting $x = y$ in $(*)$ yields $5x^4 = 0$, which implies that $x = 0$.

Next consider solutions with $x \neq y$. Using Lemma 3.13, we obtain

$$0 = (x^4 + x^3 y + x^2 y^2 + xy^3 + y^4)(x - y) = x^5 - y^5.$$

Exponentiation to an odd power is injective, so $x \neq y$ yields no solution. ∎

COMPOSITION OF FUNCTIONS

When we have a function whose target is contained in the domain of a second function, we can create a new function by applying the first and then the second. This yields a function from the domain of the first function into the target of the second.

4.28. Definition. If $f\colon A \to B$ and $g\colon B \to C$, then the **composition** of g with f is a function $h\colon A \to C$ defined by $h(x) = g(f(x))$ for $x \in A$. When h is the composition of g with f, we write $h = g \circ f$.

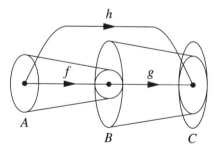

4.29. Example. If $f\colon \mathbb{R} \to \mathbb{R}$ and $g\colon \mathbb{R} \to \mathbb{R}$ are defined by $f(x) = x - 2$ and $g(x) = x^2 + x$, then $g \circ f$ is defined by

$$(g \circ f)(x) = g(f(x)) = (x - 2)^2 + (x - 2) = x^2 - 3x + 2.$$

On the other hand, $(f \circ g)(x) = f(g(x)) = x^2 + x - 2$.   ∎

The properties we have been studying are preserved by composition.

4.30. Proposition. The composition of two injections is an injection.
The composition of two surjections is a surjection.
The composition of two bijections is a bijection.
If f, g are bijections (so $g \circ f$ is a bijection), then $(g \circ f)^{-1} = f^{-1} \circ g^{-1}$
(the Inverse Composition Formula).

Proof: (Exercise 33).   ∎

4.31. Example. The function $h\colon \mathbb{R} \to \mathbb{R}$ defined by $h(x) = mx + b$ is a
bijection when $m \neq 0$. Its inverse l is defined by $l(y) = (y - b)/m$. Let f
be "multiplication by m", and let g be "addition of b". We have $h = g \circ f$
and $l = f^{-1} \circ g^{-1}$, thus illustrating the Inverse Composition Formula. ∎

If $f\colon A \to B$ is a bijection and $g = f^{-1}$, then $g \circ f$ is the identity
function on A, and $f \circ g$ is the identity function on B. Exercises 35–36 ask
whether f must be a bijection when $g \circ f$ or $f \circ g$ is an identity function.

Example 4.29 shows that $f \circ g$ need not equal $g \circ f$; composition of
functions from a set to itself is not generally commutative. On the other
hand, composition is always associative. We can form a composition $h \circ g \circ f$ by composing $h \circ g$ with f or by composing h with $g \circ f$. These always
yield the same function, which justifies dropping the parentheses.

4.32. Proposition. (Associativity of composition) If $f\colon A \to B$ and
$g\colon B \to C$ and $h\colon C \to D$, then $h \circ (g \circ f) = (h \circ g) \circ f$.

Proof: The two named functions have domain A and target D, so it suffices to show that they agree on each element of A. We evaluate each
function at an arbitrary element $x \in A$.

$$(h \circ (g \circ f))(x) = h((g \circ f)(x)) = h(g(f(x)))$$
$$((h \circ g) \circ f)(x) = (h \circ g)(f(x)) = h(g(f(x)))$$ ∎

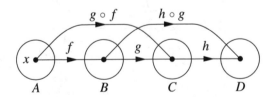

We close this section with several examples of ways to obtain new
functions from given functions. Such a procedure is a function whose domain and target are themselves sets of functions. To avoid confusion, we
use the word "operator" to describe a function defined on a set of functions. Operators that map a collection of functions to itself can be applied
successively, allowing us to discuss composition of operators.

The simplest example of an operator is the identity operator, mapping each function f to itself. We mention several others.

4.33. Example. *Translation and Scaling.* Given $f\colon \mathbb{R} \to \mathbb{R}$, let $T_a f\colon \mathbb{R} \to \mathbb{R}$ be the function defined by $(T_a f)(x) = f(x + a)$. The "machine" T_a takes a function as its input and returns a function as its output. When $a = 0$, the translation operator is the identity operator.

Similarly, the scaling operator S_b takes f to the function $S_b f$ defined by $(S_b f)(x) = f(bx)$. When $b = 1$, this is the identity operator.

Translation and scaling have natural interpretations in terms of the graph of the function f (Exercise 38). ∎

4.34. Example. *Sum and Product.* In Definition 1.25 we defined new functions from real-valued functions f, g with the same domain, calling these the *sum* $f + g$ and the *product* fg. We can understand this in another way. Define operators A and M by $A(f, g) = f + g$ and $M(f, g) = fg$. If W is the set of all real-valued functions with this domain, then the operators A and M are functions with domain $W \times W$ and target W. ∎

4.35. Example. Let S be the set of polynomials in one variable. Given the polynomial f defined by $f(x) = \sum_{i=0}^{k} a_i x^i$, let Df denote the polynomial whose value at x is $\sum_{i=1}^{k} a_i i x^{i-1}$. The operator D (the differentiation operator from calculus) is a function $D\colon S \to S$. It is surjective; the polynomial with coefficients $\{a_k\}$ is the image of the polynomial in which the coefficient of x^0 is 0 and the coefficient of x^k is a_{k-1}/k for $k \geq 1$. The operator D is not injective; the polynomials f, g defined by $f(x) = x + 1$ and $g(x) = x + 2$ have the same image.

We define another operator $J\colon S \to S$. For $f(x) = \sum_{i=0}^{k} a_i x^i$, let Jf denote the polynomial whose value at x is $\sum_{i=0}^{k} a_i x^{i+1}/(i + 1)$. If $Jf = Jg$, then term-by-term comparison of coefficients shows that $f = g$; hence J is injective. On the other hand, J is not surjective, because there is no polynomial f such that Jf is a nonzero polynomial of degree 0.

We can compose operators. We have $D(J(f)) = f$ for all $f \in S$, but $J(D(f))$ does not equal f when $f(0) \neq 0$. For example, if $f(x) = x^2 + 3$, then $J(D(f))$ is the function g defined by $g(x) = x^2$. ∎

CARDINALITY

Often we want to know how big a set is. The precise meaning of this involves bijections. The definition agrees with intuition, and we have been using it implicitly. We have the notation $[k] = \{1, 2, \ldots, k\}$ for $k \in \mathbb{N}$; we also define $[0] = \emptyset$. We need several preliminary notions.

4.36. Definition. A set A is **finite** if there is a bijection from A to $[k]$ for some $k \in \mathbb{N} \cup \{0\}$. A set is **infinite** if there is no such bijection.

Note that the empty set is considered a finite set.

4.37. Proposition. If there is a bijection $f \colon [m] \to [n]$, then $m = n$.
Proof: (Exercise 42).   ∎

4.38. Corollary. If A is finite, then for exactly one n there is a bijection from A to $[n]$.
Proof: By the definition of finiteness, such a number exists. Suppose that bijections $g \colon A \to [m]$ and $h \colon A \to [n]$ exist. Because the composition of two bijections is a bijection Proposition 4.30, the function $f = h \circ g^{-1}$ is a bijection from $[m]$ to $[n]$. By Proposition 4.37, $m = n$.   ∎

4.39. Definition. The **size** of a finite set A, written $|A|$, is the unique n such that there is a bijection from A to $[n]$. A set of size n is an *n-element set* or *n*-**set**.

4.40. Remark. *Size of finite sets.* The domain of the *size* function is the set of all finite sets; its target is the set of nonnegative integers. Corollary 4.38 states that this function is well-defined in the sense of Remark 1.24. When we write $A = \{a_1, \ldots, a_n\}$ with the a_i's distinct, we are specifying a bijection from $[n]$ to A and stating that the size of A is n.

The notation for size is the same as the notation for absolute value; size measures discrete distance to A from the empty set, and absolute value measures linear distance to a number from 0. Since size applies only to sets and absolute value applies only to numbers, the context indicates whether size or absolute value is being used.   ∎

The definition of size using bijections leads to many natural results.

4.41. Corollary. If A and B are disjoint finite sets, then
$$|A \cup B| = |A| + |B|.$$

Proof: Let $m = |A|$ and $n = |B|$. Given bijections $f \colon A \to [m]$ and $g \colon B \to [n]$, we define $h \colon A \cup B \to [m+n]$ by $h(x) = f(x)$ for $x \in A$ and $h(x) = g(x) + m$ for $x \in B$. Upon checking that h is a bijection (Exercise 44), the conclusion follows.   ∎

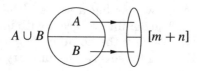

Deleting an element of an infinite set leaves another infinite set, but deleting an element of a nonempty finite set decreases its size by one. This enables us to prove statements about finite sets by induction on size.

4.42. Corollary. Every nonempty finite set of real numbers has both a maximum element and a minimum element.

Proof: We use induction on the size of the set. If $|A| = 1$, then the only element of A is both its maximum and its minimum. If $|A| = 2$, then the larger element is the maximum and the smaller is the minimum. If $|A| > 2$, choose $x \in A$. The induction hypothesis yields a maximum M and minimum L for $A - \{x\}$. Compare x with M to find the maximum and with L to find the minimum. ∎

In Chapter 5, we will study the counting of finite sets in much more depth. Meanwhile, we consider infinite sets. We do not extend | | to infinite sets. Nevertheless, we can use bijections to compare infinite sets.

4.43. Definition. An infinite set A is **countably infinite** (or **countable**) if there is a bijection from A to $\mathbb{N}$; otherwise A is **uncountably infinite** (or **uncountable**). Sets A and B **have the same cardinality** if there is a bijection from A to B.

Some authors allow "countable" to apply also to finite sets. We adopt the more common convention that a countable set has the same cardinality as $\mathbb{N}$ and hence is infinite.

We have seen that $\mathbb{Z}$ is countable (Example 4.9). Also $\mathbb{Q}$ is countable (Exercise 8.17), but $\mathbb{R}$ is not (Theorem 13.27).

When there is a bijection from A to a proper subset of B but no bijection from A to B, we think of B as being *larger* than A; thus infinite sets are larger than finite sets. Since $\mathbb{N}$ is a subset of $\mathbb{R}$ and $\mathbb{R}$ is uncountable, $\mathbb{R}$ is larger than $\mathbb{N}$. Since $\mathbb{Z}$ is countable, we do consider $\mathbb{Z}$ to be larger than $\mathbb{N}$; they have the same cardinality.

To show that a set S is countable, we place its elements in a sequence so that each element appears exactly once. This specifies a bijection from $\mathbb{N}$ to S. Using this approach, we show next that the Cartesian product of two countable sets is countable.

4.44. Theorem. The sets $\mathbb{N} \times \mathbb{N}$ and $\mathbb{N}$ have the same cardinality ($\mathbb{N} \times \mathbb{N}$ is countable).

Proof: View the ordered pairs $\{(i, j): i, j \in \mathbb{N}\}$ as points in the plane with positive integer coordinates. We list the ordered pairs in sequence by listing each successive diagonal in order. The pairs appear in order of increasing $i + j$, and the pairs with a fixed value of $i + j$ appear in increasing order by j, as illustrated below. ∎

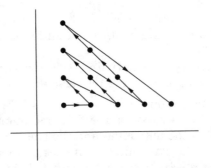

4.45. Example. *Another bijection from* $\mathbb{N} \times \mathbb{N}$ *to* $\mathbb{N}$. Define $f\colon \mathbb{N} \times \mathbb{N} \to \mathbb{N}$ by $f(m, n) = 2^{m-1}(2n - 1)$. By Proposition 3.32, every natural number is an odd number times a power of 2, so f is surjective. By the uniqueness part of Proposition 3.32, f is also injective. ∎

We have seen that an infinite set can have the same cardinality as another set that properly contains it. Here is another example.

4.46. Solution. *A bijection from* $(0, 1)$ *to* $\mathbb{R}$. Obtaining a bijection from one set to another proves that the two sets have the same cardinality. Consider the open interval $(0, 1)$ and the set $\mathbb{R}$. We can map $(0, 1)$ to an interval centered on 0 by subtracting $1/2$. We then want to stretch the first half of the interval onto the set of negative real numbers and the second half onto the set of positive real numbers. The graph of a function doing this crosses the horizontal axis at $x = 1/2$ and rises without bound as x approaches 1, but falls without bound as x approaches 0.

We construct an example whose injectivity and surjectivity we can verify without appealing to geometric intuition. Define $f\colon (0, 1) \to \mathbb{R}$ by

$$f(x) = \begin{cases} \frac{x - (1/2)}{x} & \text{if } x \leq 1/2 \\ \frac{x - (1/2)}{1 - x} & \text{if } x \geq 1/2. \end{cases}$$

Since $f(x) > 0$ when $x > 1/2$ and $f(x) < 0$ when $x < 1/2$, we may consider one side of $1/2$ at a time. If $x, x' < 1/2$ and $\frac{x - (1/2)}{x} = \frac{x' - (1/2)}{x'}$, then simplifying yields $x = x'$. The computation is similar when $x, x' > 1/2$.

When $y < 0$, we find $x < 1/2$ such that $y = f(x)$. From $y = \frac{x - (1/2)}{x}$, we solve for x to obtain $x = \frac{1}{2(1-y)}$. Since $1 - y > 1$, we have $x \in (0, 1/2)$. Similarly, when $y > 0$ we use $y = \frac{x - (1/2)}{1 - x}$ to obtain $x = \frac{y + (1/2)}{y + 1}$. Since $y > 0$, we obtain $x \in (1/2, 1)$. ∎

Consider sets A and B and functions $f\colon A \to B$ and $g\colon B \to A$. If A and B are finite and f and g are injections, then f and g must also be bijections (Exercise 46). When A and B are not finite, the conclusion that f and g are bijections need not hold. For example, let $A = (0, 1)$ and $B = [0, 1]$, and define $f\colon A \to B$ and $g\colon B \to A$ by $f(x) = x$ and $g(x) = (x + 1)/3$. Then f and g are both injections, but neither is a bijection. Nevertheless, the existence of injections f and g always implies that A and B have the same cardinality. This gives us a method for proving that two sets have the same cardinality without providing an explicit bijection.

4.47. Theorem. (Schroeder-Bernstein Theorem) If $f\colon A \to B$ and $g\colon B \to A$ are injections, then there exists a bijection $h\colon A \to B$, and hence A and B have the same cardinality.

Proof: (optional) We view A and B as disjoint sets, making two copies of common elements. For each element z of $A \cup B$, we define the *successor* of z to be $f(z)$ if $z \in A$, and $g(z)$ if $z \in B$. The *descendants* of z are the elements that can be reached by repeating the successor operation. We say that z is a *predecessor* of w if w is the successor of z. Because f and g are injective, every element of $A \cup B$ has at most one predecessor. The *ancestors* of z are the elements that can be reached by repeating the predecessor operation.

The *family* of z consists of z together with all its ancestors and descendants; call this $F(z)$. We use the structure of families to define a one-to-one correspondence between A and B. The successor operation defines a function f' on $A \cup B$; below we show several possibilities for families using a graphical description of f'.

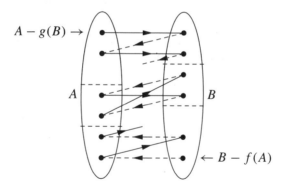

First suppose that z is a descendant of z. Because every element has at most one predecessor, in this case $F(z)$ is finite (repeatedly composing the successor function leads to a "cycle" of elements involving z). Applying f' alternates between A and B, and thus $F(z)$ has even size. For every $x \in A$ in $F(z)$, we pair x with $f(x)$; because $F(z)$ has even size, this is a one-to-one correspondence between $F(z) \cap A$ and $F(z) \cap B$.

Otherwise, $F(z)$ is infinite. In this case, the set $S(z)$ of ancestors of z may be finite or infinite. When $S(z)$ is finite, it contains an *origin* that has no predecessor (all elements of $F(z)$ have the same origin). If $S(z)$ has an origin in B, then for every $x \in A \cap F(z)$ we pair x with its predecessor

$g^{-1}(x)$; because B contains the origin, $g^{-1}(x)$ exists. When $S(z)$ is infinite or has an origin in A, we pair x with its successor $f(x)$.

Because every element has at most one predecessor, the pairing we have defined is a one-to-one correspondence between the elements of A and the elements of B within $F(z)$. Since the families are pairwise disjoint, it is also a one-to-one correspondence between A and B. In more technical language, we have defined the function $h: A \rightarrow B$ by $h(x) = g^{-1}(x)$ when the family of x has an origin in B, and $h(x) = f(x)$ otherwise. The function h is the desired bijection.   ∎

HOW TO APPROACH PROBLEMS

Functions and their properties are fundamental tools in all areas of mathematics. Although these concepts may seem abstract at first, they arise from familiar situations such as solving equations. The exercises emphasize understanding and application of definitions rather than subtle insights or inventiveness. We list a few helpful principles.

1) Definitions are your friends.
2) Standard examples can provide counterexamples or suggest proofs.
3) Bijections can transform problems into more convenient contexts.
4) Countability of a set amounts to placing its elements in a sequence.

The role of definitions.

The definitions in this chapter provide road maps for what must be done to answer a question. To show that a function is a bijection, one must show that the inverse image of each target element consists of exactly one domain element. The concepts of injection and surjection break this requirement into two pieces. Although we can interpret injections and surjections schematically or geometrically, we return to the definitions to write proofs. The table below summarizes the meanings of injective, surjective, and their negations. Given $f: A \rightarrow B$ and $b \in B$, recall that $I_f(b)$ denotes $\{x \in A: f(x) = b\}$ (Definition 1.35).

A proof that $f: A \rightarrow B$ is injective shows that f never maps two elements of A to a single element of B; every element of B is the image of at most one element under f. We consider pairs $x, y \in A$ and prove "$f(x) = f(y)$ implies $x = y$" or its contrapositive "$x \neq y$ implies $f(x) \neq f(y)$". To prove that f is not injective, it suffices to exhibit a pair $x, y \in A$ with $x \neq y$ and $f(x) = f(y)$.

A function $f: A \rightarrow B$ is surjective if its image is all of its target. For all $b \in B$, we must prove that there exists $x \in A$ such that $f(x) = b$. Usually, we construct an example of such an x (in terms of b); this means finding a solution to $f(x) = b$. To prove that f is not surjective, we show for some $b \in B$ that $f(x) = b$ has no solution, so b is not in the image of f.

Injective	Not Injective
$(\forall b \in B)[I_f(b)$ has at most one element]	$(\exists b \in B)[I_f(b)$ has at least two elements]
$(\forall x, x' \in A)[x \neq x' \Rightarrow f(x) \neq f(x')]$	$(\exists x, x' \in A)[x \neq x' \text{ and } f(x) = f(x')]$
$(\forall x, x' \in A)[f(x) = f(x') \Rightarrow x = x']$	

Surjective	Not Surjective
$(\forall b \in B)[I_f(b)$ is nonempty]	$(\exists b \in B)[I_f(b)$ is empty]
$(\forall b \in B)(\exists x \in A)[f(x) = b]$	$(\exists b \in B)(\forall x \in A)[f(x) \neq b]$

The best way to prove injectivity depends on the function. When f is defined by a formula, it may be easy to manipulate the equation given by the hypothesis $f(x) = f(y)$ to derive $x = y$; consider the example $f(x) = mx + c$ with $m \neq 0$. When f is defined using words, it may be more natural to show that $x \neq y$ implies $f(x) \neq f(y)$, as in Proposition 4.19.

Some authors use the terms "one-to-one" and "onto" for injective and surjective; we avoid these to eliminate confusion between "one-to-one function" and "one-to-one correspondence".

The prefix "sur" means "over" or "above". The word "surjection" suggests projecting the domain down onto the target; a simple example is the function $f: \mathbb{R}^2 \to \mathbb{R}$ defined by $f(x, y) = x$. On the other hand, the word "injection" suggests placing something inside something else; a simple example is the map $g: \mathbb{R} \to \mathbb{R}^2$ defined by $g(x) = (x, 0)$. These two examples should help the student remember which is which.

The usefulness of standard examples and graphs.

Many exercises in this chapter ask whether a statement about functions is true or false. Standard examples provide both insight into proofs and counterexamples for false statements. Such examples include polynomials, ratios of polynomials, absolute value, and other elementary functions. Also, the schematic diagrams in Remark 4.24 actually specify functions on finite sets.

We remark also on the use of graphs. Graphing a function suggests properties one might try to prove about it, but statements that depend on the visual interpretation of a picture require rigorous proof. For example, surjectivity of the function $f: \mathbb{R} \to \mathbb{R}$ defined by $f(x) = x^3$ depends on the ability to take cube roots of real numbers. We accept this informally now for convenience, but a rigorous proof requires the methods of Part IV.

Similarly, stating that the value of $x/(1 + x^2)$ is $1/2$ when $x = 1$ and "approaches" 0 as x "gets large" does not prove that every value between $1/2$ and 0 is achieved. Making this inference requires results about limits and continuity, which are not available to us until Part IV. Meanwhile we must rely on the definitions; showing that a particular value b is attained requires solving for x in terms of b (see Example 4.23 and Exercise 22).

Bijections as transformations.

Our first example of bijection gave us the numerical system of base q representation. Also important is the flexibility of interpreting natural numbers as lists. In Exercise 17, binary representation leads to a startling result about the game of Nim.

Many geometric operations can be interpreted as bijections; see Exercise 20, Exercise 30, and Exercise 38. For readers familiar with calculus, here we illustrate the use of bijections in making changes of variables.

4.48. Example. The function defined by $f(x) = e^x$ maps $\mathbb{R}$ to $\mathbb{R}$. It is increasing and hence injective, but it is not surjective; its image is the set of positive real numbers. By restricting the target of e^x to the set of positive real numbers, we obtain a bijection. The natural logarithm function, its inverse, is a bijection from the set of positive real numbers to $\mathbb{R}$. The function defined by $f(x) = \sin x$ from $\mathbb{R}$ to the interval $[-1, 1]$ is surjective but not injective. By restricting the domain of $\sin x$ to $[-\pi/2, \pi/2]$, we obtain a bijection to the interval $[-1, 1]$. ∎

4.49. Example. *Limits of integration under change of variables.* In calculus, we often compute a definite integral by changing variables. For example, consider $\int_0^2 (x^3 + 1)^5 3x^2 \, dx$. Let $f(x) = x^3 + 1$. The function f is a bijection from the interval $[0, 2]$ to the interval $[1, 9]$. Letting $y = f(x)$ leads to $\int_0^2 (x^3 + 1)^5 3x^2 \, dx = \int_1^9 y^5 \, dy = (1/6)(9^6 - 1)$. Similarly, because $y = \sin x$ defines a bijection from $[-\pi/2, \pi/2]$ to $[-1, 1]$, we compute $\int_{-\pi/2}^{\pi/2} \sin x \cos x \, dx = \int_{-1}^{1} y \, dy = 0$.

Changing variables by writing $y = g(x)$ requires that g be a bijection from the interval of integration in x to the interval of integration in y. As x varies from a to b, y varies from $g(a)$ to $g(b)$. ∎

Infinite sets and countability.

How do we show that a set is countable? Expressing A as $\{a_1, a_2, \cdots\}$ specifies a bijection from $\mathbb{N}$ to A. Since the inverse of a bijection is a bijection, proving that a set A is countable is equivalent to obtaining a sequence that contains each element of A *exactly once*. Example 4.9 presents such a sequence for $A = \mathbb{Z}$; this lists the values $f(n)$ in order: $0, 1, -1, 2, -2, 3, -3, \cdots$. Arguing that this sequence names every integer exactly once proves that $\mathbb{Z}$ is countable.

This is the fundamental technique for proving countability. To show that the union of a (countable) sequence of countable sets is countable (Exercise 44), construct a sequence that lists each element of the union exactly once. Note that one cannot list all of the first set first; one would never reach the second set. Between any two terms of a sequence there are only finitely many terms. An arbitrarily large finite set, no matter how large, is not infinite.

EXERCISES

4.1. (−) Let $120102_{(3)}$ and $110222_{(3)}$ be ternary representations of two natural numbers. Use base 3 arithmetic to add them. Check the answer by converting each to base 10, adding, and converting back to base 3.

4.2. (−) Which integer is bigger, $333_{(12)}$ or $3333_{(5)}$?

4.3. (−) Note that $(15)^2 = 225$, $(25)^2 = 625$, and $(35)^2 = 1225$. For $n \in \mathbb{N}$, prove that the square of the number given by appending 5 to the base 10 representation of n is the number given by appending 25 to the base 10 representation of $n(n+1)$.

4.4. (−) Consider a temperature scale T, where water freezes at 20 degrees and boils at 80 degrees. Suppose that there are constants a and b so that when the temperature on the Fahrenheit scale is x, the temperature on the T scale is $ax + b$. Determine what the Fahrenheit temperature is when the temperature is 50 on the T scale. (Hint: Solving the problem does not require finding a and b.)

4.5. (−) For which sets A does there exist a bijection from A to A that is different from the identity function on A?

4.6. (−) Let A be the set of days in the week. Let f assign to each day the number of letters in its English name. Does f define an injection from A to $\mathbb{N}$?

4.7. (−) For each of the three functions A, M, D defined in Example 1.37, determine whether it is injective and whether it is surjective.

4.8. (−) Let f and g be polynomials defined by $f(x) = x - 1$ and $g(x) = x^2 - 1$. Find formulas for $f \circ g$ and $g \circ f$.

4.9. (−) Decide whether the following statement is true or false; justify.
"If f and g are monotone functions from $\mathbb{R}$ to $\mathbb{R}$, then $g \circ f$ is also monotone."

4.10. Suppose that $f(x) = ax + b$ and $g(x) = cx + d$ for constants a, b, c, d with a and c not zero. Explain why f and g are injective and surjective. Show that the function h defined by $h = g \circ f - f \circ g$ is neither injective nor surjective.

4.11. (−) Explain why multiplication by 2 defines a bijection from $\mathbb{R}$ to $\mathbb{R}$ but not from $\mathbb{Z}$ to $\mathbb{Z}$.

4.12. Determine which of the following statements are true. Give proofs for the true statements and counterexamples for the false statements.
a) Every decreasing function from $\mathbb{R}$ to $\mathbb{R}$ is surjective.
b) Every nondecreasing function from $\mathbb{R}$ to $\mathbb{R}$ is injective.
c) Every injective function from $\mathbb{R}$ to $\mathbb{R}$ is monotone.
d) Every surjective function from $\mathbb{R}$ to $\mathbb{R}$ is unbounded.
e) Every unbounded function from $\mathbb{R}$ to $\mathbb{R}$ is surjective.

•   •   •   •   •

4.13. (!) Let n be an integer between 1 and 999. Written as three decimal digits, let n be abc (that is, $n = 100a + 10b + c$). Let the *reverse* of a three-digit decimal number with digits $\alpha\beta\gamma$ be the three-digit number with digits $\gamma\beta\alpha$.

Suppose that $a \neq c$ (either may be 0). Let x be the difference between n and its reverse. Prove that x and its reverse sum to 1089.

4.14. Prove that the method of Example 4.5 generates a base q representation for every natural number.

4.15. (!) Consider a balance scale plus k objects of known weights $1, 3, \ldots, 3^{k-1}$ (the first k powers of 3). Prove by induction on k that every unknown weight in the set $\{1, \ldots, (3^k - 1)/2\}$ can be balanced.

4.16. Consider a balance scale and objects with positive integer weights $w_1 \leq \cdots \leq w_k$. Prove that using these objects it is possible to balance each integer weight from 1 to $\sum_{j=1}^{k} w_j$ (as in Problem 4.1) if and only if $w_j \leq 1 + 2\sum_{i=1}^{j-1} w_i$ for $1 \leq j \leq k$. For example, everything from 1 to 10 is achievable using $\{1, 2, 7\}$, but with weights $\{1, 2, 8\}$ it is not possible to balance the integer 4.

4.17. (+) *The Game of Nim.* A position in Nim consists of some piles of coins. Two players alternate, with each move removing a portion of one pile. The winner is the player who takes the last coin.

Suppose that the starting piles have sizes $n_1, \ldots, n_k$. Prove that Player 2 has a winning strategy if and only if for every j, an even number of $n_1, \ldots, n_k$ have a 1 in position j in their binary representation. For example, when the sizes are $1, 2, 3$, the binary representations are $1, 10, 11$, and the condition holds.

4.18. Prove that exponentiation to a positive odd power defines a strictly increasing function. For $n \in \mathbb{N}$, find all solutions to $x^n = y^n$. (Hint: Consider the cases $x < 0 < y$, $0 < x < y$, and $x < y < 0$.)

4.19. For $k \in \mathbb{N}$, determine all ordered pairs (x, y) such that $\sum_{j=0}^{2k} x^{2k-j} y^j = 0$. (Hint: Generalize Example 4.27.)

4.20. Let $f \colon \mathbb{R}^2 \to \mathbb{R}^2$ be defined by $f(x, y) = (ax - by, bx + ay)$, where a, b are numbers with $a^2 + b^2 \neq 0$.
a) Prove that f is a bijection.
b) Find a formula for f^{-1}.
c) Give a geometric interpretation of f for the case $a^2 + b^2 = 1$. (Describe the effect f has on geometric figures in the plane.)

4.21. (!) Let A be the set of subsets of $[n]$ that have even size, and let B be the set of subsets of $[n]$ that have odd size. Establish a bijection from A to B, thereby proving that $|A| = |B|$. (Such a bijection is suggested below for $n = 3$.)

$$
\begin{array}{ccccc}
A & \varnothing & \{2, 3\} & \{1, 3\} & \{1, 2\} \\
B & \{3\} & \{2\} & \{1\} & \{1, 2, 3\}
\end{array}
$$

4.22. Verify that $f(x) = \frac{2x-1}{2x(1-x)}$ defines a bijection from the interval $(0, 1)$ to $\mathbb{R}$. (Hint: In the proof that f is surjective, use the quadratic formula.)

4.23. Determine which formulas below define injections from $\mathbb{R}$ to $\mathbb{R}$. Determine which define surjections. For each that does not define a bijection, find a nontrivial interval $S \subseteq \mathbb{R}$ (containing more than a single point) such that the formula defines a bijection from S to S.
a) $f(x) = x^3 - x + 1$.
b) $f(x) = \cos(\pi x/2)$.

4.24. Let f and g be surjections from $\mathbb{Z}$ to $\mathbb{Z}$, and let $h = fg$ be their product (Definition 1.25). Must h also be surjective? Give a proof or a counterexample.

4.25. Determine which formulas below define surjections from $\mathbb{N} \times \mathbb{N}$ to $\mathbb{N}$.
    a) $f(a, b) = a + b$.
    b) $f(a, b) = ab$.
    c) $f(a, b) = ab(b + 1)/2$.
    d) $f(a, b) = (a + 1)b(b + 1)/2$.
    e) $f(a, b) = ab(a + b)/2$.

4.26. (−) Given $f: \mathbb{R} \to \mathbb{R}$, suppose that there are positive constants c, α such that, for all $x, y \in \mathbb{R}$, $|f(x) - f(y)| \geq c\,|x - y|^{\alpha}$. Prove that f is injective.

4.27. Let $f: \mathbb{R} \to \mathbb{R}$ be a quadratic polynomial. Prove that f is not surjective. Find a cubic polynomial that is not injective; justify.

4.28. (+) Determine which cubic polynomials from $\mathbb{R}$ to $\mathbb{R}$ are injective. (Hint: This is easy if calculus is allowed. To avoid calculus, first use geometric arguments to reduce the problem to the case $x^3 + rx$. Comment: All cubic polynomials from $\mathbb{R}$ to $\mathbb{R}$ are surjective, but proving this requires the methods of Part IV.)

4.29. Consider three functions f, g, h mapping $\mathbb{R}$ to $\mathbb{R}$, defined by
$$f(x) = x/(1 + x^2), \qquad g(x) = x^2/(1 + x^2), \qquad h(x) = x^3/(1 + x^2).$$
    a) Determine which of these functions are injective.
    b) Prove that f and g are not surjective.
    c) Graph all three functions. (Comment: The graph of h should suggest that h is surjective, but proving this requires the methods of Part IV.)

4.30. (!) Given real numbers a, b, c, d, let $f: \mathbb{R}^2 \to \mathbb{R}^2$ be defined by $f(x, y) = (ax + by, cx + dy)$. Prove that f is injective if and only if f is surjective.

4.31. (!) Let $f: A \to B$ be a bijection, where A and B are subsets of $\mathbb{R}$. Prove that if f is increasing on A, then f^{-1} is increasing on B.

4.32. Let F be a field. Define f on F by $f(x) = -x$, and define g on $F - \{0\}$ by $g(x) = x^{-1}$. Prove that f is a bijection from F to F and that g is a bijection from $F - \{0\}$ to $F - \{0\}$.

4.33. (!) Prove the following statements about composition of functions.
    a) The composition of two injections is an injection
    b) The composition of two surjections is a surjection.
    c) The composition of two bijections is a bijection.
    d) If $f: A \to B$ and $g: B \to C$ are bijections, then $(g \circ f)^{-1} = f^{-1} \circ g^{-1}$. (Hint: Use associativity of composition to prove that the function $f^{-1} \circ g^{-1}$ must be the inverse of the function $g \circ f$.)

4.34. (!) Given $f: A \to B$ and $g: B \to C$, let $h = g \circ f$. Determine which of the following statements are true. Give proofs for the true statements and counterexamples for the false statements.
    a) If h is injective, then f is injective.
    b) If h is injective, then g is injective.
    c) If h is surjective, then f is surjective.
    d) If h is surjective, then g is surjective.

4.35. (!) Consider $f: A \to B$ and $g: B \to A$. Answer each question below by providing a proof or a counterexample.
    a) If $f(g(y)) = y$ for all $y \in B$, does it follow that f is a bijection?
    b) If $g(f(x)) = x$ for all $x \in A$, does it follow that $f(g(y)) = y$ for all $y \in B$?

4.36. Consider $f\colon A \to B$ and $g\colon B \to A$. Prove that if $f \circ g$ and $g \circ f$ both are identity functions, then f is a bijection. In particular, prove that
    a) If $f \circ g$ is the identity function on B, then f is surjective.
    b) If $g \circ f$ is the identity function on A, then f is injective.

4.37. Consider $f\colon A \to A$. Prove that if $f \circ f$ is injective, then f is injective.

4.38. Given $f\colon \mathbb{R} \to \mathbb{R}$, define the functions $T_a f$ and $S_b f$ by $(T_a f)(x) = f(x + a)$ and $(S_b f)(x) = f(bx)$. Determine how to modify the graph of f to obtain the graphs of $T_a f$ and $S_b f$. (Hint: For $S_b f$, consider the cases $b > 0$, $b = 0$, and $b < 0$.)

4.39. Let $f\colon \mathbb{R} \to \mathbb{R}$ be defined by $f(x) = a(x + b) - b$. Obtain an explicit formula for the function g that is obtained by n successive applications of f.

4.40. Suppose that $f\colon A \to B$ is a bijection and that $g\colon B \to B$. Let h be the composition $h = f^{-1} \circ g \circ f$, so $h\colon A \to A$. Derive a formula in terms of f and g for the function from A to A obtained by n successive applications of h.

4.41. For $f\colon A \to A$ and $n \in \mathbb{N}$, let f^n be defined by $f^1 = f$ and $f^n = f \circ f^{n-1}$ for $n > 1$. Let n and k be natural numbers with $k < n$. Prove that $f^n = f^k \circ f^{n-k}$.

4.42. Let f be a bijection from $[m]$ to $[n]$. Prove that $m = n$. (Hint: Use induction.)

4.43. Let B be a proper subset of a set A, and let f be a bijection from A to B. Prove that A is an infinite set. (Hint: Use Exercise 4.42.)

4.44. $(-)$ Prove that the function h in the proof of Corollary 4.41 is a bijection.

4.45. (!) Let f be a function from a finite set A to itself. Prove that f is injective if and only if f is surjective. Prove that this equivalence fails when A is infinite.

4.46. (!) Given finite sets A, B, consider a function $f\colon A \to B$.
    a) When f is injective, what is implied about the sizes of A and B?
    b) When f is surjective, what is implied about the sizes of A and B?
    c) Prove that if A and B are finite and $f\colon A \to B$ and $g\colon B \to A$ are injections, then $|A| = |B|$ and f and g are bijections.

4.47. Prove that the natural numbers, the even natural numbers, and the odd natural numbers form sets of the same cardinality (they are countable).

4.48. The proof of countability of $\mathbb{N} \times \mathbb{N}$ in Theorem 4.44 specifies a sequence containing every ordered pair (i, j). Determine the index of the ordered pair (i, j) in this sequence, as a function of i and j. (Comment: This defines the bijection $f\colon \mathbb{N} \times \mathbb{N} \to \mathbb{N}$ explicitly.)

4.49. (!) Let $A_1, A_2, \ldots$ be a sequence of sets, each of which is countable. Prove that the union of all the sets in the sequence is a countable set.

4.50. Let $A = (0, 1)$ and $B = \{y \in \mathbb{R}\colon 0 \le y < 1\}$. Define $f\colon A \to B$ and $g\colon B \to A$ by $f(x) = x$ and $g(y) = (y + 1)/2$. Obtain an explicit formula for the function h constructed by the proof of Theorem 4.47 for these functions f, g.

4.51. (!) Construct an explicit bijection from the open interval $(0, 1)$ to the closed interval $[0, 1]$.

PART II

PROPERTIES

OF NUMBERS

Chapter 5

Combinatorial Reasoning

Techniques for determining the sizes of finite sets have applications in probability, the analysis of computer procedures, and many other areas. In this chapter we study fundamental models for counting problems, both on their own and in relation to properties of functions.

5.1. Problem. *Summation of Integer Powers.* Given a correct formula for $\sum_{i=0}^{n} i^k$, we can prove it by induction. Lacking a formula, how can we discover it? ∎

5.2. Problem. *Comparison of Poker Hands.* A poker hand consists of five cards from an ordinary deck of cards. Why is "three-of-a-kind" a higher-ranked poker hand than "two-pair", and why is a "flush" ranked higher than a "straight"? ∎

5.3. Problem. *Nonnegative Integer Solutions.* Suppose that each resident of New York City has 100 coins in a jar. The coins come in five types (pennies, nickels, dimes, quarters, half-dollars). We consider two jars of coins to be "equivalent" if they have the same number of coins of each type. Is it possible that no two people have equivalent jars of coins? ∎

5.4. Problem. *The Drummer Problem.* At a party there are n married couples. Each woman is dancing with some man, but not necessarily with her spouse. The band has two drummers, who alternate songs. After each song, two of the women switch partners. During the last song, each woman dances with her husband. If we know only the initial dancing pairs and the initial drummer, can we determine which drummer is playing at the end? ∎

5.5. Problem. *Sorting by Transpositions.* Given a list of the numbers 1 through n in some order, how many exchanges of entries are needed to sort the numbers into the order $1, 2 \cdots, n$? ∎

ARRANGEMENTS AND SELECTIONS

In this section we count sets consisting of arrangements and selections of objects from finite sets. This introduces the factorial function and the binomial coefficients, which arise in many mathematical contexts.

Many problems can be solved by expressing them in terms of arrangements and selections. More complicated problems may involve combining several steps. We introduce two elementary rules for combining subproblems, the *rule of sum* and the *rule of product*.

5.6. Definition. A **partition** of a set A is a collection of pairwise disjoint subsets of A whose union is A. The **rule of sum** states that if A is finite and $B_1, \ldots, B_m$ is a partition of A, then $|A| = \sum_{i=1}^{m} |B_i|$.

The rule of sum follows from Corollary 4.41 by induction on m (Exercise 15). The rule of product is a bit more subtle. Often we can describe a set by building its elements in stages such that the *number* of choices available at the ith step does not depend on previous choices, although the actual choices available may depend on them.

5.7. Example. A music practice room is available for only one hour during each weekday. In how many ways can three students sign up to use the room during the week? The first student picks one of the five days. The choices for the second student depend on the choice the first student makes, but in each case four choices remain. Similarly, the third student always has three choices. There are therefore $5 \cdot 4 \cdot 3 = 60$ possibilities. ∎

5.8. Definition. Let T be a set whose elements can be described using a procedure involving steps $S_1, \ldots, S_k$ such that step S_i can be performed in r_i ways, regardless of *how* steps $S_1, \ldots, S_{i-1}$ are performed. The **rule of product** states that $|T| = \prod_{i=1}^{k} r_i$.

The rule of product follows from the rule of sum by induction on k (Exercise 16). Its most elementary application is $|A \times B| = |A| \cdot |B|$, determining the size of the Cartesian product of finite sets. Repeating this observation yields a useful example.

5.9. Example. *The number of q-ary n-tuples is q^n.* Consider lists of length n from a set of size q, such as $\{0, 1, \ldots, q - 1\}$. As q-ary representations, these yield the numbers 0 through $q^n - 1$; thus the set has size q^n.

The product rule counts this directly without using bijections. There are q choices for each position, regardless of the choices in other positions. By the product rule, there are q^n ways to form the n-tuple. ∎

A list of length k using elements of S specifies a function from $[k]$ to S. Allowing repeating elements yields all functions from $[k]$ to S. Forbidding repetitions restricts the functions to injections. Listing all the elements without repetition yields a bijection.

5.10. Definition. A **permutation** of a finite set S is a bijection from S to itself. The **word form** of a permutation of $[n]$ is the list obtained by writing the image of i in position i. We write $n!$, read as "n **factorial**", to mean $\prod_{i=1}^{n} i = n(n-1)\cdots 1$.

The word form of a permutation simply records the function; for example, $f\colon [3] \to [3]$ defined by $f(1) = 2$, $f(2) = 3$, and $f(3) = 1$ is the permutation with word form 231. We often use the term "permutation" for both the function and the word form; an n-tuple of elements of S is equivalent to a function from $[n]$ to S.

In counting problems, we use the word **arrangements** to refer to lists formed from a specified set. We generalize permutations by considering arrangements without repetition.

5.11. Theorem. An n-element set has $n!$ permutations (arrangements without repetition). In general, the number of arrangements of k distinct elements from a set of size n is $n(n-1)\cdots(n-k+1)$.

Proof: We count the injections from $[k]$ to S, where $n = |S|$. There is no injection when $k > n$, which agrees with the formula. We construct all injections by choosing images one by one; choosing the image of i is choosing the element in position i of the corresponding list.

There are n ways to choose the image of 1. For each way we do this, there are $n-1$ ways to choose the image of 2. In general, after we have chosen the first i images, avoiding them leaves $n-i$ ways to choose the next image, no matter how we made the first i choices. The rule of product yields $\prod_{i=0}^{k-1}(n-i) = n!/(n-k)!$ for the number of arrangements. ∎

By convention, we define $0! = 1$, so the general formula simplifies to $n!$ when we count permutations. This is consistent with saying that there is exactly one bijection from $\emptyset$ to $\emptyset$. This illustrates a general convention; the value of an empty sum is the additive identity, and the value of an empty product is the multiplicative identity. For example, we set $x^0 = 1$.

We have counted the arrangements of k distinct elements from S. We also consider selections of k elements from S, in which the order of the selected elements is unimportant.

5.12. Definition. A **selection** of k elements from $[n]$ is a k-element subset of $[n]$. The number of such selections is "n choose k", written as $\binom{n}{k}$.

If $k < 0$ or $k > n$, then $\binom{n}{k} = 0$; in these cases there are no selections of k elements from $[n]$. When $0 \le k \le n$, we obtain a simple formula.

5.13. Theorem. For integers n, k with $0 \le k \le n$,

$$\binom{n}{k} = \frac{n!}{k!(n-k)!}.$$

Proof: We relate selections to arrangements. We count the arrangements of k elements from $[n]$ in two ways. Picking elements for positions as in Theorem 5.11 yields $n(n-1) \cdot (n-k+1)$ as the number of arrangements.

Alternatively, we can select the k-element subset first and then write it in some order. Since by definition there are $\binom{n}{k}$ selections, the product rule yields $\binom{n}{k}k!$ for the number of arrangements.

In each case, we are counting the set of arrangements, so we conclude that $n(n-1)\cdots(n-k+1) = \binom{n}{k}k!$. Dividing by $k!$ completes the proof. ∎

We often interpret counting problems in the language of probability. We give a formal definition of probability in Chapter 9. Here we consider only experiments that have n equally likely possible outcomes. We describe the experiment as choosing one of these outcomes **at random**. When A is a subset of the set of outcomes, we define the **probability** of A (or of obtaining an outcome in A) to be $|A|/n$.

5.14. Example. Standard dice have faces showing the numbers 1 through 6. When we roll two different 6-sided dice, there are 36 possible outcomes, equally likely. In six of these, the total showing on the two dice is 7, so the probability of *rolling* 7 is $1/6$. ∎

5.15. Example. A record of n coin flips is a binary n-tuple, using 1 for heads and 0 for tails. We view the 2^n lists as equally likely. The probability that the number of heads is even is the fraction of the lists having an even number of 1s. Using binary encoding (Proposition 4.19), this is also the fraction of subsets of $[n]$ having even size. When $n > 0$, half the subsets of $[n]$ have even size (Exercise 4.21 or Exercise 27); we conclude that the probability of getting an even number of heads is $1/2$. ∎

5.16. Solution. *Comparison of Poker Hands.* A **standard deck of cards** consists of 52 cards. These come in 13 ranks of four cards each. They are also grouped into four *suits*, with one card of each rank in each suit.

When we choose five cards at random from a standard deck, there are $\binom{52}{5} = 2{,}598{,}960$ possible outcomes (*hands*); we view them as equally likely. The probability of a particular type is the number of hands of that type divided by $\binom{52}{5}$. In poker, rarer types are ranked higher. To rank types of hands, we compare the number of each type.

Three-of-a-kind means three cards of the same rank and one in each of two other ranks. This can occur in $\binom{13}{1}\binom{4}{3}\binom{12}{2}\binom{4}{1}\binom{4}{1} = 54{,}912$ ways, since we pick the special rank, pick three cards from it, pick two other ranks, and pick one card from each of those. The rule of product applies; the number of choices at each step does not depend on the earlier choices made.

Two-pair means two cards each in two ranks and the fifth card in some third rank. This can occur in $\binom{13}{2}\binom{4}{2}\binom{4}{2}\binom{44}{1} = 123,552$ ways; we pick the ranks for the pairs, pick the cards from those ranks, and pick the final card from the remaining ranks. The computation shows that three-of-a-kind is less than half as likely and hence is ranked higher.

A *flush* consists of five cards in one suit and occurs in $4\binom{13}{5} = 5,148$ ways. A *straight* consists of one card each in five consecutive ranks; the Ace can be considered either the lowest or the highest rank. A straight can begin at one of 10 possible ranks; thus it occurs in $10 \cdot 4^5 = 10,240$ ways. The flush is rarer. (Here we have counted in each type the hands that are both straights and flushes—see Exercise 23.) ∎

The numbers $\binom{n}{k}$ are called the **binomial coefficients** due to their appearance as coefficients in the nth power of a sum of two terms.

5.17. Theorem. (Binomial Theorem)

$$(x + y)^n = (x + y)(x + y) \cdots (x + y) = \sum_{k=0}^{n} \binom{n}{k} x^k y^{n-k}$$

Proof: The proof interprets the process of multiplying out the factors. To form a term in the product, we must choose x or y from each factor; some factors contribute x, some y. The number of factors contributing x is some integer k from 0 to n, and the remaining $n - k$ factors contribute y. The number of terms of the form $x^k y^{n-k}$ is the number of ways to choose k of the factors to contribute x. Summing over k accounts for all the terms. ∎

BINOMIAL COEFFICIENTS

We next discuss interpretations, properties, and applications of the binomial coefficients. These numbers satisfy many useful identities. We observe first that many statements can be proved in a variety of ways.

5.18. Lemma. $\binom{n}{k} = \binom{n}{n-k}$.

Proof: *Proof 1* (counting two ways). By definition, $[n]$ has $\binom{n}{k}$ subsets of size k. Another way to count selections of k elements is to count selections of $n - k$ elements to omit, and there are $\binom{n}{n-k}$ of these.

Proof 2 (bijections). The left side counts the k-element subsets of $[n]$, the right side counts the $n - k$-element subsets, and the operation of "complementation" establishes a bijection between the two collections.

Proof 3 (arithmetic). Having computed $\binom{n}{k} = \frac{n!}{k!(n-k)!}$ (Theorem 5.13), we observe that the formula is unchanged by switching k and $n - k$. ∎

We include counting arguments as another weapon in our arsenal of proof techniques. A proof that interprets a formula as the size of a finite set is a **combinatorial proof**. The technique of **counting two ways** allows us to establish equality between two formulas by proving that both count the same set. We have used this idea in Remark 3.10, Example 3.15, and Corollary 4.41. Counting two ways is closely related to proving equality of size by establishing a bijection. Combinatorial proofs may provide more information and deeper understanding than manipulation of formulas, but discovering them may require some cleverness.

We can phrase a combinatorial proof about binomial coefficients using selections or using one of several alternative models. In Proposition 4.19, we constructed a bijection ("binary encoding") from the set of subsets of $[n]$ to the set of binary n-tuples. Whenever we discuss k-element subsets of $[n]$, we could alternatively discuss binary n-tuples with k 1s. Yet another model interprets these as paths in the plane.

5.19. Definition. A **lattice path** in the plane is a path joining integer points via steps of unit length rightward or upward. Alternatively, it is a list of ordered pairs of integers, with each step increasing one coordinate by 1. The **length** of a path is the total number of steps.

5.20. Example. *Lattice paths and binary lists.* Typically we start lattice paths at the origin. Since each step increases a coordinate by 1, the length of the walk is the sum of the coordinates of the ending point.

We can encode a path by recording in position i a 1 when the ith step is rightward and a 0 when the ith step is upward. In a path of length n, the final location is determined by how many steps we take to the right; if there are k steps to the right, we reach the point $(k, n - k)$, and the encoding has k 1s.

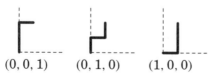

$(0, 0, 1)$ $(0, 1, 0)$ $(1, 0, 0)$

Furthermore, the actual path is determined by which steps are taken to the right. Thus the path is determined by the binary n-tuple. This establishes a one-to-one correspondence between the lattice paths reaching $(k, n - k)$ and the binary n-tuples with k ones. Hence the number of lattice paths to $(k, n - k)$ is $\binom{n}{k}$. The illustration below shows the number of paths to specified points. ∎

1					
1	5				
1	4	10			
1	3	6	10		
1	2	3	4	5	
1	1	1	1	1	1

5.21. Proposition. For nonnegative integers a, b, the number of lattice paths from the origin to the point (a, b) is $\binom{a+b}{a}$.

Proof: The discussions before and in Example 5.20 interpret lattice paths from the origin to (a, b) as selections of a elements from $[a + b]$. ∎

The lattice path or *block-walking* model suggests an inductive formula for the binomial coefficients. It permits inductive proofs of identities when a combinatorial proof doesn't come to mind. It is sometimes called *Pascal's Formula* in honor of Blaise Pascal (1623–1662). The triangular array of numbers in which row n consists of all the binomial coefficients with n "on top" (starting with row 0) is called **Pascal's Triangle**, though it was known to Chinese mathematicians much earlier.

$$
\begin{array}{ccccccccccc}
&&&&& 1 &&&&& \\
&&&& 1 && 1 &&&& \\
&&& 1 && 2 && 1 &&& \\
&& 1 && 3 && 3 && 1 && \\
& 1 && 4 && 6 && 4 && 1 & \\
1 && 5 && 10 && 10 && 5 && 1 \\
\end{array}
$$

5.22. Lemma. (Pascal's Formula) If $n \geq 1$, then

$$\binom{n}{k} = \binom{n-1}{k} + \binom{n-1}{k-1}.$$

Proof: These proofs use the same idea, phrased in different models.

   Proof 1. By Proposition 5.21, the number of lattice paths reaching $(k, n - k)$ is $\binom{n}{k}$. Each path arrives at $(k, n - k)$ from exactly one of the points $(k, n - k - 1)$ and $(k - 1, n - k)$. By Proposition 5.21 again, there are $\binom{n-1}{k}$ paths of the first type and $\binom{n-1}{k-1}$ paths of the second type.

   Proof 2. Using the subset model, we count the k-sets in $[n]$. There are $\binom{n-1}{k}$ such sets not containing n and $\binom{n-1}{k-1}$ such sets containing n.

   Proof 3. $(1 + x)^n = (1 + x)(1 + x)^{n-1}$. Using the Binomial Theorem, we expand both $(1 + x)^n$ and $(1 + x)^{n-1}$ to obtain

$$\sum_{k=0}^{n} \binom{n}{k} x^k = (1+x) \sum_{k=0}^{n-1} \binom{n-1}{k} x^k = \sum_{k=0}^{n-1} \binom{n-1}{k} x^k + \sum_{k=0}^{n-1} \binom{n-1}{k} x^{k+1}.$$

Shifting the index in the last summation yields $\sum_{k=1}^{n} \binom{n-1}{k-1} x^k$. Since $\binom{n-1}{n} = \binom{n-1}{-1} = 0$, we can add $\binom{n-1}{n}$ to the first sum and $\binom{n-1}{-1}$ to the second to obtain

$$\sum_{k=0}^{n} \binom{n}{k} x^k = \sum_{k=0}^{n} \left[\binom{n-1}{k} + \binom{n-1}{k-1} \right] x^k.$$

By Corollary 3.25, the corresponding coefficients must be equal. ∎

If we first derived Pascal's Formula by Proof 1 or Proof 2, we could then use it to prove the formula for $\binom{n}{k}$ by induction on n (Exercise 25).

We view n-tuples as arrangements with repetitions allowed. What happens when we consider selections with repetitions allowed? The next theorem permits us to solve Problem 5.3.

5.23. Theorem. With repetition allowed, there are $\binom{n+k-1}{k-1}$ ways to select n objects from k types. This also equals the number of nonnegative integer solutions to $x_1 + \cdots + x_k = n$.

Proof: Selections are determined by how many objects are chosen of each type. Let x_i be the number chosen of type i. This establishes a one-to-one correspondence between the selections and the nonnegative integer solutions to $x_1 + \cdots + x_k = n$.

We model these solutions as arrangements of n dots and $k - 1$ vertical separating bars. We represent selecting x_1 items of type 1 by recording x_1 dots and marking the end with a bar before continuing to the next type. Doing this for each type forms an arrangement of dots and bars. Below we illustrate the result when $x_1 = 5$, $x_2 = 2$, $x_3 = 0$, and $x_4 = 3$. Since we want $x_1 + \cdots + x_k = n$, we have n dots and $k - 1$ bars.

$$\bullet \ \bullet \ \bullet \ \bullet \ \bullet \ | \ \bullet \ \bullet \ | \ | \ \bullet \ \bullet \ \bullet$$

Given an arrangement of n dots and $k - 1$ bars, we can invert the process to obtain x_i; it equals the number of dots in the ith group. This establishes a one-to-one correspondence between solutions to $x_1 + \cdots + x_k = n$ and arrangements of n dots and $k - 1$ bars. These arrangements are determined by choosing the locations for the bars in a list of length $n + k - 1$, so there are $\binom{n+k-1}{k-1}$ of them. We have counted the solutions to the equation and hence also the selections of n objects from k types. ∎

This formula can also be written as $\binom{n+k-1}{n}$, so care must be taken to distinguish between the number of types and the number of elements being selected, whatever these happened to be named in an application. It may be safer to remember the proof than to remember the formula.

5.24. Solution. *Nonnegative Integer Solutions.* New York City has about 7 million residents. Suppose that each resident has 100 coins in a jar. Two jars are "equivalent" if they have the same number of pennies, same number of nickels, and similarly for dimes, quarters, and half-dollars. When x_i denotes the number of coins of type i, the number of pairwise inequivalent jars of coins is the number of solutions to $x_1 + x_2 + x_3 + x_4 + x_5 = 100$ using nonnegative integers. By Theorem 5.23, this equals $\binom{104}{4}$ = 4,598,126. Hence some two people must have equivalent jars. ∎

Selections with repetition also correspond to the terms in the expansion of a power of a sum of several terms.

5.25. Corollary. The expansion of $\left(\sum_{i=1}^{m} x_i\right)^d$ has $\binom{d+m-1}{m-1}$ terms.

Proof: The terms correspond to the solutions of $\sum_{i=1}^{m} d_i = d$. ∎

5.26. Example. *Monomials in a multinomial expansion.* Every monomial in the expansion of $(w + x + y + z)^3$ has total degree 3. Ignoring the coefficients, we list the monomials below. By Corollary 5.25, there are $\binom{3+4-1}{4-1} = 20$ of these. In Chapter 9 we compute the formula for the coefficients, called the *multinomial coefficients.* ∎

$$
\begin{array}{ccccc}
w^3 & w^2x & w^2y & w^2z & wxy \\
x^3 & x^2w & x^2y & x^2z & wxz \\
y^3 & y^2w & y^2x & y^2z & wyz \\
z^3 & z^2w & z^2x & z^2y & xyz
\end{array}
$$

Identities such as Lemma 5.18 and Pascal's Formula can be helpful in solving problems involving binomial coefficients. They also illustrate combinatorial techniques of proof. We prove two more.

5.27. Lemma. (The Chairperson Identity)

$$
k\binom{n}{k} = n\binom{n-1}{k-1}
$$

Proof: Each side counts the k-person committees with a designated chairperson that can be formed from a set of n people. On the left, we select the committee and then select the chair from it; on the right, we select the chair first and then fill out the rest of the committee. ∎

Combinatorial proofs of summation formulas often consist of defining a set whose size is the total and partitioning that set into subsets whose sizes are the terms in the sum; this again is "counting two ways".

5.28. Theorem. (The Summation Identity)

$$\sum_{i=0}^{n} \binom{i}{k} = \binom{n+1}{k+1}$$

Proof: The right side counts the binary $n + 1$-tuples with $k + 1$ ones. We can partition this set into disjoint subsets according to which position holds the rightmost 1. The number of ways to form the list so the rightmost 1 is in position $i + 1$ is $\binom{i}{k}$.   ∎

$$\underbrace{\hspace{2cm}}_{k \ \ \text{ones}} \quad 1 \ \ 0 \ \ 0 \ \ 0$$

$$\uparrow \hspace{2.5cm} \uparrow \ \uparrow \hspace{1.5cm} \uparrow$$
$$1 \ \text{positions} \ i \ \ i+1 \hspace{1cm} n+1$$

The block-walking version of this proof counts the paths to $(k, n - k)$ according to the height at which they take the last step to the right. Exercise 30 requests a proof by induction.

5.29. Solution. *Summation of integer powers.* Formulas for sums of powers (Problem 5.1) are easy to verify by induction but difficult to guess. The Summation Identity provides a method that automatically generates the answer and the proof. Notice that $i = \binom{i}{1}$. Therefore, the Summation Identity proves the summation formula for the first n natural numbers by $\sum_{i=0}^{n} i = \sum_{i=0}^{n} \binom{i}{1} = \binom{n+1}{2} = n(n + 1)/2$. End of proof! For the squares, we rewrite i^2 using binomial coefficients. Since $i^2 = 2\binom{i}{2} + i = 2\binom{i}{2} + \binom{i}{1}$,

$$\sum_{i=0}^{n} i^2 = 2 \sum_{i=0}^{n} \binom{i}{2} + \sum_{i=0}^{n} \binom{i}{1} = 2\binom{n+1}{3} + \binom{n+1}{2} = \frac{n(n+1)(2n+1)}{6}.$$

The last step extracts the common factor $n(n + 1)$ from the formulas for $\binom{n+1}{3}$ and $\binom{n+1}{2}$. This approach yields $\sum_{k=0}^{n} f(k)$ for any polynomial f. ∎

This method eliminates the guesswork but not the "grunt-work" to obtain the exact formula, as we must write i^k in terms of $\{\binom{i}{j}: 0 \le j \le k\}$ to apply the Summation Identity. Nevertheless, for all k the method shows that $\sum_{i=1}^{n} i^k$ is a polynomial of degree $k + 1$ in n. In Theorem 5.31, we obtain the two leading terms. Calculus provides another approach; see Example 5.46 and Exercise 17.32.

5.30. Remark. *Binomial coefficients and polynomials.* Viewed as a function of n, the binomial coefficient $\binom{n}{k} = \frac{1}{k!}n(n - 1) \cdots (n - k + 1)$ is a polynomial of degree k. The coefficient of n^k is $\frac{1}{k!}$, and the coefficient of n^{k-1} is $\frac{1}{k!}\sum_{j=1}^{k-1}(-j) = \frac{-1}{k!}\binom{k}{2} = \frac{-1}{2(k-2)!}$. When n is large, the contribution from lower terms is relatively unimportant; thus we may be content with knowing the leading term or first two terms of a polynomial.

In the next proof, we use $O(n^k)$ to indicate an unspecified polynomial of degree at most k. We therefore write $f(x) = 2x^k + O(x^{k-1})$ to mean that f is a polynomial of degree k with leading coefficient 2. (The "Big Oh" notation applies more generally to describe "order of growth" of functions; see Exercise 2.23.)

This use of the "Big Oh" notation allows us to write

$$k!\binom{n}{k} = n^k - \binom{k}{2}n^{k-1} + O(n^{k-2}).$$

Also, subtracting a polynomial of lower degree from a polynomial f does not change the leading term. Hence when f has degree k and g has degree $k-1$, we have $f(n) - g(n) = f(n) + O(n^{k-1})$. ∎

5.31. Theorem. For $k \in \mathbb{N}$, the value of $\sum_{i=1}^{n} i^k$ is a polynomial in n with leading term $\frac{1}{k+1}n^{k+1}$ and next term $\frac{1}{2}n^k$.

Proof: (optional) By Remark 5.30, there is a polynomial g such that

$$k!\binom{i}{k} = i(i-1)\cdots(i-k+1) = i^k - \binom{k}{2}i^{k-1} + g(i),$$

with g of degree at most $k-2$. Solving for i^k yields $i^k = k!\binom{i}{k} + \binom{k}{2}i^{k-1} - g(i)$.

We use induction on k. For $k = 1$, the formula $\sum_{i=1}^{n} i = \frac{1}{2}n^2 + \frac{1}{2}n$ agrees with the claim. For $k > 1$, we have

$$\sum_{i=1}^{n} i^k = k! \sum_{i=1}^{n} \binom{i}{k} + \binom{k}{2} \sum_{i=1}^{n} i^{k-1} - \sum_{i=1}^{n} g(i).$$

By the induction hypothesis, the term of degree j in $g(i)$ contributes a polynomial of degree $j+1$ to $\sum_{i=1}^{n} g(i)$. Thus $\sum_{i=1}^{n} g(i) = O(n^{k-1})$. Also, the induction hypothesis yields $\binom{k}{2} \sum_{i=1}^{n} i^{k-1} = \binom{k}{2}\frac{1}{k}n^k + O(n^{k-1})$, and the Summation Identity yields $k! \sum_{i=1}^{n} \binom{i}{k} = k!\binom{n+1}{k+1}$.

These three formulas yield $\sum_{i=1}^{n} i^k = k!\binom{n+1}{k+1} + \frac{k-1}{2}n^k + O(n^{k-1})$. We use Lemma 5.27 to replace $\binom{n+1}{k+1}$ with $\frac{n+1}{k+1}\binom{n}{k}$, and next we replace $\binom{n}{k}$ using the displayed expression in Remark 5.30. We obtain

$$\sum_{i=1}^{n} i^k = k!\frac{n+1}{k+1}\binom{n}{k} + \frac{k-1}{2}n^k + O(n^{k-1})$$

$$= k!\frac{1}{k+1}(n+1)\frac{1}{k!}\left[n^k - \binom{k}{2}n^{k-1}\right] + \frac{k-1}{2}n^k + O(n^{k-1})$$

$$= \frac{1}{k+1}n^{k+1} + \left\{\frac{1}{k+1}\left[1 - \binom{k}{2}\right] + \frac{k-1}{2}\right\}n^k + O(n^{k-1}).$$

To complete the induction step, we simplify the coefficient of n^k:

$$\frac{1}{k+1}\left[1 - \binom{k}{2}\right] + \frac{k-1}{2} = \frac{2 - k(k-1) + (k+1)(k-1)}{2(k+1)} = \frac{2 + k - 1}{2(k+1)} = \frac{1}{2}. \quad ∎$$

PERMUTATIONS

We view a permutation on $[n]$ both as an arrangement of $[n]$ (the *word form*) and as a bijection from $[n]$ to $[n]$. The solution of Problem 5.4 involves a closer study of the word form.

5.32. Definition. The **identity permutation** of $[n]$ is the identity function from $[n]$ to $[n]$; its word form is $1\,2\cdots n$. A **transposition** of two elements in a permutation switches their positions in the word form.

5.33. Solution. In the Drummer Problem, label the couples $1,\ldots,n$. To describe the arrangement at a given time, list for each i the index of the woman dancing with the ith man. This list is the word form of a permutation of $[n]$. At the last dance we reach the identity permutation.

Between two songs, two women switch, which transposes their indices in the permutation. The drummers also switch. The first drummer plays the last song if and only if we perform an even number of transpositions in reaching the identity permutation from the original permutation.

To study the parity of the number of transpositions, we define a number $P(f)$ for each permutation f of $[n]$. When f has word form $x_1,\ldots,x_n$, let $P(f) = \prod_{j>i}(x_j - x_i)$. We claim that $P(f)$ changes sign with each transposition. Exchanging x_k and x_l replaces $x_l - x_k$ with $x_k - x_l$ in the product. Also, $x_l - x_i$ and $x_i - x_k$ change sign for each i with $k < i < l$, but these two changes cancel. The other factors are unchanged. Thus the product changes sign.

When f^* is the identity permutation, $P(f^*)$ is positive. For the initial permutation f_0, the computation of $P(f_0)$ uses no information about transpositions made to reach f^*. No matter how f^* is reached, $P(f_0)$ is positive if and only if the number of transpositions made to reach f^* is even. Thus the first drummer is playing at the end if and only if $P(f_0)$ is positive, and we can compute this from the initial pairing. ∎

The analysis of Solution 5.33 distinguishes two types of permutations. A permutation f of $[n]$ is **even** when $P(f)$ is positive, and it is **odd** when $P(f)$ is negative. This categorization has important applications in algebra, matrix theory, and combinatorics. When $n = 1$, there is one even permutation of $[n]$ and no odd permutation. For $n \geq 2$, there are $n!/2$ even permutations and $n!/2$ odd permutations (Exercise 52).

Every permutation can be transformed to the identity permutation by applying transpositions. No matter what transpositions are used, an even permutation will take an even number of transpositions to reach the identity, while an odd permutation will take an odd number.

Problem 5.5 asks how many transpositions are needed to reach the identity from a known permutation; we solve this in the next section. In computer science, reaching the identity permutation via specified kinds of operations is called *sorting*. There the initial permutation is unknown and is gradually discovered by pairwise comparison of elements.

FUNCTIONAL DIGRAPHS

Viewing a permutation as a function from a set to itself allows us to compose permutations. Bijections have inverses; the inverse of a permutation is a permutation, and their composition is the identity permutation. Since composing bijections yields a bijection, the composition of two permutations of [n] is a permutation of n. We compose permutations as bijections but still name them by their word forms.

5.34. Example. Let f be the permutation 4123 of [4]. The function $f \circ f$ is the permutation 3412. The inverse of f is the permutation 2341.

The result of composing two permutations of [n] usually depends on which is applied first. Let g and h be the permutations 132 and 213, respectively. Now $h \circ g$ is 231, while $g \circ h$ is 312.   ∎

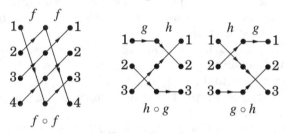

5.35. Definition. The *n*th **iterate** of $f\colon A \to A$ is the function f^n obtained by composing n successive applications of f.

Precisely, we set $f^1 = f$ and $f^n = f \circ f^{n-1}$ for $n > 1$. Since composition of functions is associative (Proposition 4.32), we also have $f^k \circ f^{n-k}$ whenever $0 \le k \le n$ (see Exercise 4.41).

5.36. Example. *Rotation by 90 degrees.* Let $f\colon \mathbb{R}^2 \to \mathbb{R}^2$ be the function that rotates the plane by 90 degrees counterclockwise. The formula for f is $f(x, y) = (-y, x)$. The fourth iterate of f is the identity function. When restricted to the four points $a = (1, 0)$, $b = (0, 1)$, $c = (-1, 0)$, and $d = (0, -1)$, the function f defines a permutation that maps a, b, c, d to b, c, d, a, respectively.   ∎

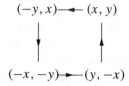

This suggests a visual way to study iteration.

5.37. Definition. The **functional digraph** of a function $f: A \to A$ consists of a point for each element of A and, for each $x \in A$, an arrow from the point representing x to the point representing $f(x)$. These points are **vertices**. A list of vertices $a_1, \ldots, a_k$ is a **cycle** of length k if there is an arrow from a_i to a_{i+1} for $1 \leq i \leq k-1$ and an arrow from a_k to a_1. A **loop** is a cycle of length 1.

Mathematicians use "graph" for various structures with pictorial representations; the word comes from the Greek for "picture".

The functional digraph of f differs from the picture after Remark 1.22 by using only one copy of A. By the definition of function, each vertex in a functional digraph is the tail of exactly one arrow. The function is injective if every vertex is the head of at most one arrow, and it is surjective if every vertex is the head of at least one arrow. A function from a set to itself has a fixed point if and only if its functional digraph has a loop.

5.38. Example. *The functional digraph for the Penny Problem.* In the Penny Problem (Application 1.14), we defined a function on the set of nondecreasing lists of positive integers. We proved that the fixed points of this function are the lists of the form $12 \cdots n$.

Since the function does not change the total number of pennies, we can study it on the subset S_n of lists with sum n. Below we illustrate the resulting functional digraph when $n = 5$. There is no fixed point among lists with sum 5, and thus the functional digraph has no loop. It does have a cycle of length 3. ∎

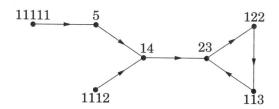

The functional digraph makes it easy to study what happens when we repeatedly compose a function with itself. In Example 5.38 we always reach the cycle of length 3.

The **2-line form** for a permutation f of $[n]$ lists $[n]$ in order on the top line and the corresponding elements on the bottom line; thus it lists the pairs $\{(x, f(x)): x \in [n]\}$. For example, $\left(\begin{smallmatrix} 1 & 2 & 3 & 4 \\ 4 & 3 & 2 & 1 \end{smallmatrix}\right)$ is the 2-line form of the permutation with word form 4 3 2 1. One advantage of the 2-line form is that it allows us to describe permutations of sets other than $[n]$.

5.39. Example. *The functional digraph of a permutation.* Consider the permutation $f : [9] \rightarrow [9]$ with 2-line form $\left(\begin{smallmatrix} 1 & 2 & 3 & 4 & 5 & 6 & 7 & 8 & 9 \\ 3 & 6 & 1 & 4 & 2 & 5 & 8 & 9 & 7 \end{smallmatrix}\right)$. Its visual presentation using the model in Remark 1.22 appears on the left below. On the right we draw the functional digraph. In the functional digraph of a permutation, each vertex is the tail of one arrow and is the head of one arrow, and thus the arrows group into cycles. We completely specify a permutation by its **cycle description**, which lists the cycles of elements formed by iteration. In this example the cycle description is $(789)(4)(265)(13)$. ∎

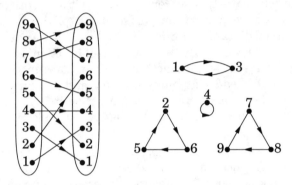

The cycle description of a permutation is an important tool for analyzing its structure. We use it to solve Problem 5.5.

5.40. Example. Consider the permutation f with word form 23416785 and 2-line form $\left(\begin{smallmatrix} 1 & 2 & 3 & 4 & 5 & 6 & 7 & 8 \\ 2 & 3 & 4 & 1 & 7 & 6 & 8 & 5 \end{smallmatrix}\right)$. The functional digraph appears below with solid arrows; the cycle description is $(1234)(5678)$. Transposing 3 and 5 in the word form yields f' with 2-line form $\left(\begin{smallmatrix} 1 & 2 & 3 & 4 & 5 & 6 & 7 & 8 \\ 2 & 5 & 4 & 1 & 7 & 6 & 8 & 3 \end{smallmatrix}\right)$. The functional digraph of f' appears below with dashed arrows; the cycle description is (12567834), a single cycle. Transposing 3 and 5 again returns f' to f. ∎

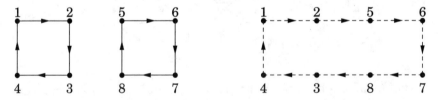

5.41. Solution. *The number of transpositions needed to sort the word form of a permutation of $[n]$ is $n - k$, where k is the number of cycles in its cycle description.*

We claim that transposing two elements lying on distinct cycles combines their elements into a single cycle, while transposing two elements in the same cycle splits it into two cycles. When $f(i) = x$ and $f(j) = y$, transposing x and y yields f' with $f'(i) = y$ and $f'(j) = x$; elsewhere f and f' agree. If x and y are in distinct cycles $x \cdots i$ and $y \cdots j$ of f, then the transposition yields the cycle $x \cdots i \; y \cdots j$ in f'. Other cycles are unchanged. If x and y are in the same cycle $\cdots i \; x \cdots j \; y \cdots$ of f, then f' has distinct cycles $x \cdots j$ and $y \cdots i$ on these elements.

By this claim, each transposition changes the number of cycles by one—up if the transposed elements were on the same cycle, otherwise down. The identity permutation consists of n cycles of length 1. Thus a permutation with k cycles needs at least $n - k$ transpositions to reach the identity. Furthermore, the identity can be reached in $n - k$ steps, because when there are fewer than n cycles we can find two elements on the same cycle and transpose them to increase the number of cycles. ∎

HOW TO APPROACH PROBLEMS

Combinatorial arguments are often simple to follow but hard to produce. Often it is easier to find a proof by induction or algebraic manipulation of formulas, but combinatorial proofs usually provide more information about the underlying structure of the problem. After establishing a bijection, one can study the images of subsets to obtain more detailed identities. Here we emphasize the basic techniques.

1) Understand the distinction between cases and stages to apply the rules of sum and product.

2) Interpret formulas for natural numbers as sizes of sets using geometric or combinatorial models.

3) Prove equalities by counting a set in two ways; a sum can be interpreted as counting the subsets in a partition of a set.

4) When lacking a combinatorial proof, consider established identities, induction, or other methods.

The rules of sum and product.

In applying these techniques, pay attention to whether order matters. When picking three cards from one suit and one from each of two others, picking the suits successively as $4\binom{13}{3}3\binom{13}{1}2\binom{13}{1}$ counts all hands twice, because the order of picking the suits for the last two cards does not matter.

Not every counting problem can be solved by building a set in successive stages. Although the classical models emphasize the rule of product, often the number of ways to do a particular step depends depends on how the previous steps were done. When this happens, we must consider cases. The consideration of cases is an application of the rule of sum, and within each case we hope to use the rule of product.

5.42. Example. How many ways are there to label the four corners of a square with roman letters such that adjacent corners get different letters? We try to build labelings in stages. There are 26 ways to label the top left corner. Whatever we choose, avoiding that label on the top right leaves 25 choices. Again we have 25 choices on the bottom right. On the lower left corner we must avoid both the label of the top left and bottom right, so we would like to multiply by 24 to get $26 \cdot 25 \cdot 25 \cdot 24$.

Unfortunately, we actually have 25 choices available at the last step if the top left and bottom right received the same color. Thus the simple rule of product does not work. After the first two steps, there are 24 ways to perform the third step so that the first three positions have distinct labels, and one way to do it so that it agrees with the first color. Thus a correct computation is $26 \cdot 25 \cdot 24 \cdot 24 + 26 \cdot 25 \cdot 1 \cdot 25$.   ■

Cases and subcases can become very complex. We try to organize arguments to avoid cases where possible, taking advantage of symmetry and the rule of product.

Interpreting formulas as sizes of sets.

An expression like k^n involving natural number exponentiation can be interpreted combinatorially using the set of k-ary n-tuples or geometrically using an n-dimensional grid of dots.

5.43. Example. Consider $m^2 = m(m-1) + m$. Algebraically, we apply the distributive law. Geometrically, we partition an m-by-m square; deleting one row leaves an m-by-$(m-1)$ rectangle, so both sides of the identity count the dots. Combinatorially, we can interpret m^2 as the number of ordered pairs (i, j) from $[m]$; there are $m(m-1)$ with $i \neq j$ and m with $i = j$.   ■

We can interpret $n!$ using permutations. We can interpret $\binom{n}{k}$ as k-element subsets of $[n]$, binary n-tuples with k 1s, lattice paths from the origin to $(k, n-k)$, etc. We can interpret a product of two terms as counting ordered pairs or counting a two-stage process. For example, in Lemma 5.27 we interpret $k\binom{n}{k}$ as the number of ways to form a committee size k from n people and then designate a chairperson. The quantity $n!\binom{n}{k}$ could represent ordered pairs consisting of a permutation of $[n]$ and a designation of k special positions in the permutation.

A combinatorial proof that a divides b can be given by grouping a set of size b into a sets of equal size or into sets of size a. This is the essence of the *multiple counting* argument used to derive the formula for $\binom{n}{k}$.

5.44. Example. $k - 1$ *divides* $k^n - 1$. We already have an algebraic proof of this using the geometric sum: $k^n - 1 = (k-1) \sum_{i=0}^{n-1} k^i$. Now we present a combinatorial proof.

Let $B = \{0, 1, \ldots, k-1\}$. The set B^n has size k^n, but we only want a set of size $k^n - 1$, so we discard the n-tuple that is all 0 to obtain a set S of size $k^n - 1$. Each n-tuple in S has a leftmost nonzero value; let A_i consist of those where that value is i.

If $|A_i| = |A_j|$ when $i \neq j$, then we have partitioned S into $k-1$ sets of equal size, which proves that $k-1$ divides $|S| = k^n - 1$. To prove that $|A_i| = |A_j|$, we define a bijection from A_i to A_j by changing the leftmost nonzero element of each n-tuple in A_i from i to j. We illustrate S when $(k, n) = (5, 2)$, grouped by these classes, with columns indicating corresponding elements under the bijection. ∎

$$
\begin{array}{ccccccc}
A_1 & 01 & 10 & 11 & 12 & 13 & 14 \\
A_2 & 02 & 20 & 21 & 22 & 23 & 24 \\
A_3 & 03 & 30 & 31 & 32 & 33 & 34 \\
A_4 & 04 & 40 & 41 & 42 & 43 & 44
\end{array}
$$

Counting two ways.

A familiar instance of counting two ways is the technique of interchanging the order of summation in a double summation: $\sum_i \sum_j f(i, j) = \sum_j \sum_i f(i, j)$. Here the summands are identified by ordered pairs, and we have the option of grouping the terms by the first index or by the second.

The proof of the Chairperson Identity (Lemma 5.27) is essentially of this type. We interpret $k\binom{n}{k}$ as a set formed by a two-step procedure, choosing a subset and distinguishing one element. To prove the identity, we counted the same set by performing the two steps of the construction in the other order, choosing the distinguished element first.

Summation formulas may require more delicate arguments. We are told a formula for the value of the sum, or perhaps we guess a formula from computations with small examples. Usually we then define a set whose size is this formula. The remainder of the proof consists of devising a way to cut this set into pieces so that the sizes of the pieces correspond to terms in the sum; see Theorem 5.28.

In Exercise 41, for example, the value of the sum is $\binom{n}{3}$, so we naturally consider the set of triples from $[n]$. How can we cut this set into pieces so that the number of ways to form a triple in the ith piece is $(i-1)(n-i)$? Exercise 40 is similar and easier. In Exercise 42, we consider the selections of k elements from a set of size $m+n$, and the form of the summands makes it rather easy to identify which ways of selecting k elements should be in the ith piece.

Other techniques.

Besides combinatorial arguments, techniques such as induction, algebraic manipulation, properties of polynomials, or even calculus may work. For example, we have shown combinatorially that $[n]$ has as many odd subsets as even subsets, but induction also easily proves this.

The results of earlier combinatorial arguments may also help; identities may permit substitutions that simplify computations.

5.45. Example. Suppose we want to evaluate $\sum_{k=0}^{n} k\binom{n}{k}$. From the Chairperson Identity, we know that this equals $n \sum_{k=0}^{n} \binom{n-1}{k-1}$. The nonzero terms
in the new sum count the subsets of $[n-1]$, grouped by their size. Hence
the sum is $n2^{n-1}$. Exercise 39 asks for a direct combinatorial proof. ∎

5.46. Example. Calculus quickly yields the leading term in Theorem
5.31. The interpretation of definite integrals using area (see Chapter
17) yields $\sum_{i=1}^{n-1} i^k \le \int_0^n x^k \, dx \le \sum_{i=1}^{n} i^k$. With $\int_0^n x^k \, dx = \frac{n^{k+1}}{k+1}$, we obtain
$\frac{n^{k+1}}{k+1} \le \sum_{i=1}^{n} i^k \le \frac{n^{k+1}}{k+1} + n^k$. This yields the leading term and suggests that
$n^k/2$ might be the next. Exercise 17.32 uses calculus to prove it. ∎

EXERCISES

In these problems, n denotes a positive integer. The phrase "count the" means
"determine the number of" and requires justification of the answer.

5.1. (−) When rolling n dice, what is the probability that the sum of the numbers
obtained is even?

5.2. (−) For each integer k between 2 and 12, find the probability of obtaining the
total k when rolling two fair dice (see Example 5.14).

5.3. (−) Many games involve rolling two dice with sides numbered 1 through 6.
Explain simply why x and $14 - x$ are equally likely to be the sum of the numbers
facing up on the two dice.

5.4. (−) A *word* is a string of letters from an alphabet. How many words of length
l can be formed from an alphabet of size m? How many can be formed in which
each letter is used at most once?

5.5. (−) Given n married couples, how many ways are there to form pairs consisting of one man and one woman who are not married to each other?

5.6. (−) Count the bijections from A to B, given that $|A| = |B| = n$.

5.7. (−) How many ways are there to pick two cards from a standard 52-card deck
such that the first card is a spade and the second card is not an Ace?

5.8. (−) Determine the coefficient of $x^4 y^5$ in the expansion of $(x + y)^9$.

•   •   •   •   •

5.9. Compute the probability that a random five-card hand has the following.
    a) At least three cards with the same rank.
    b) At least two cards with the same rank.

5.10. A fair coin is flipped exactly $2n$ times. Compute the probability of obtaining
exactly n heads. Evaluate the formula for $n = 10$.

5.11. The following problem appeared on a statewide exam for grade 10 in California. "A game involves two cubes with sides numbered 1 through 6. After throwing the two cubes, the smaller number is subtracted from the larger number to find the difference. If a player throws the cubes many times, what difference will probably occur most often? Provide a diagram and written explanation that you could use to explain to a friend."

5.12. We roll a fair six-sided die exactly three times. Determine the probability that the sum of the values rolled equals eleven.

5.13. We roll a fair six-sided die exactly four times. For $k \in \{0, 1, 2, 3, 4\}$, determine the probability that we roll a six exactly k times. Check your answer by verifying that these probabilities sum to one.

5.14. Consider a dial having a pointer that is equally likely to point to each of n regions numbered $1, 2, \ldots, n$. When we spin the dial three times, what is the probability that the sum of the selected numbers is n?

5.15. (−) Use Corollary 4.41 to prove that the size of the union of k pairwise disjoint finite sets is the sum of their sizes.

5.16. Use the rule of sum to prove the rule of product. (Hint: Use induction on the number k of steps used to form elements of the set T being counted, after expressing the elements of T as k-tuples.)

5.17. Suppose that $n! + m! = k!$, where $n, m, k \in \mathbb{N}$. Prove that $(n, m, k) = (1, 1, 2)$.

5.18. Count the sets of six cards from a standard deck of 52 cards that have at least one card in every suit.

5.19. (!) There are 999,999 natural numbers less than one million. For $1 \le k \le 6$, determine how many of these have k distinct digits in their decimal representations. Leading zeros count; treat 111 as 000111 and count it for $k = 2$.

5.20. (!) Prove that $(n^5 - 5n^3 + 4n)/120$ is an integer for all $n \in \mathbb{N}$.

5.21. Count the rectangles of all sizes formed using segments in a grid with m horizontal lines and n vertical lines. In the picture below, $m = 4$ and $n = 5$.

5.22. Let P be an n-sided polygon in the plane such that every segment joining pairs of vertices of P lies inside P; such segments are "diagonals" of P. Count the pairs of diagonals of P that cross.

5.23. In terms of binomial coefficients, count the (five-card) poker hands having
 a) One pair (two cards of equal rank and no others of equal rank).
 b) Full house (two cards of equal rank and three cards of another rank).
 c) Straight flush (five consecutive cards from the same suit).

5.24. (!) A *bridge hand* consists of 13 cards from a standard 52-card deck. Its *distribution* is the list in nonincreasing order of the number of cards in each suit.

Thus 5440 denotes a hand with five cards in one suit and four cards in each of two others. List the distributions, find their probabilities, and rank them. Explain intuitively why 4333 ranks so low.

5.25. Use Pascal's Formula to prove by induction on n that $\binom{n}{k} = \frac{n!}{k!(n-k)!}$ when $0 \le k \le n$ and $\binom{n}{k} = 0$ otherwise. Assume that $\binom{0}{0} = 1$ and that $\binom{0}{k} = 0$ when $k \ne 0$.

5.26. Use Pascal's Formula to prove the Binomial Theorem by induction on n.

5.27. Exercise 4.21 proves combinatorially that $[n]$ has exactly as many subsets with even size as subsets with odd size. Use the Binomial Theorem to give another proof. Does the conclusion remain true when $n = 0$?

5.28. Count the solutions in nonnegative integers $x_1, \ldots, x_k$ to $x_1 + \cdots + x_k \le n$.

5.29. Count the solutions in positive integers $x_1, \ldots, x_k$ to $x_1 + \cdots + x_k = n$.

5.30. Prove by induction that $\sum_{i=0}^{n} \binom{i}{k} = \binom{n+1}{k+1}$ for integers $k, n \ge 0$.

5.31. (!) Count the ways to group $2n$ distinct people into pairs. (The answer is 1 when $n = 1$ and is 3 when $n = 2$.)

5.32. By counting a set of dots in two different ways, give a combinatorial proof that $n^2 = 2\binom{n}{2} + n$.

5.33. *Summing the cubes.*
    a) Prove directly that $m^3 = 6\binom{m}{3} + 6\binom{m}{2} + m$.
    b) Use part (a) to prove that $\sum_{i=1}^{n} i^3 = (\frac{n(n+1)}{2})^2$ (without using induction).
    c) Prove part (a) by counting a set in two ways. (Hint: Count the ordered triples that can be formed from $[m]$.)

5.34. Use the Summation Identity to count the cubes of all integers sizes formed by an n by n by n assembly of unit cubes.

5.35. Consider a track meet with k^n contestants. In each round, the remaining contestants are placed in groups of size k. The winner in each group advances to the next round.
    a) Use this to give another combinatorial proof that $k - 1$ divides $k^n - 1$.
    b) How many races are run in the entire competition?

5.36. Let x be an element of a set A of size $2n$. Among the n-element subsets of A, count those containing x and those omitting x. Conclude that $\binom{2n}{n} = 2\binom{2n-1}{n-1}$.

5.37. (!) By counting a set in two ways, prove that $\binom{n}{k}\binom{k}{j} = \binom{n}{j}\binom{n-j}{k-j}$.

•   •   •   •   •

In Exercises 38–45, prove each summation formula by counting a set in two ways.

5.38. $\sum_{k=1}^{n} 2^{k-1} = 2^n - 1$.

5.39. $\sum_{k=0}^{n} k\binom{n}{k} = n2^{n-1}$.

5.40. $\sum_{i=1}^{n} (i - 1) = \binom{n}{2}$.

5.41. (!) $\sum_{i=1}^{n} (i - 1)(n - i) = \binom{n}{3}$.

5.42. $\sum_{i=0}^{k} \binom{m}{i}\binom{n}{k-i} = \binom{m+n}{k}$.

5.43. $\sum_{k=-m}^{n} \binom{m+k}{r}\binom{n-k}{s} = \binom{m+n+1}{r+s+1}$.

5.44. $\sum_{i=0}^{k} \binom{m+k-i-1}{k-i}\binom{n+i-1}{i} = \binom{m+n+k-1}{k}$. (Hint: Use selections with repetition.)

5.45. $\sum_{A\subseteq[n]} \sum_{B\subseteq[n]} |A \cap B| = n4^{n-1}$. (Hint: Consider the ordered triples (x, A, B) such that $A, B \subseteq [n]$ and $x \in A \cap B$; count this set in two ways.)

•   •   •   •   •

5.46. (+) Evaluate $\sum_{S\subseteq[n]} \prod_{i\in S} 1/i$.

5.47. (!) Consider $f_m\colon \mathbb{N} \to \mathbb{N}$ defined by $f_m(n) = \sum_{k=0}^{m} \binom{n}{k}$. Prove that $f_m(n) = 2^n$ when $n \le m$. Find an n such that $f_m(n) \ne 2^n$. (Hint: Count subsets.)

5.48. Count the ways to choose distinct subsets $A_0, A_1, \ldots, A_n$ of $[n]$ such that $A_0 \subset A_1 \subset \cdots \subset A_n$. What happens if repetitions are allowed in the list?

5.49. (−) Determine the parity and the inverse for each of the following permutations of $[9]$.
  a) 987654321   b) 135792468   c) 259148637

5.50. Consider three covered bins containing apples, oranges, and a mixture of apples and oranges, respectively. The three bins have labels Apples, Oranges, and Apples/Oranges, but the labels have been moved so that *all* the labels are wrong. We are allowed to reach into one bin and select one piece of fruit (without seeing the rest). Prove that by selecting the right bin to sample, we can determine the correct labeling of the bins. Explain how this relates to permutations.

5.51. Suppose that Problem 5.4 is changed by having three drummers who rotate. Prove that the final drummer cannot be determined from the initial permutation.

5.52. For $n > 1$, prove that the number of even permutations of $[n]$ equals the number of odd permutations of $[n]$. (Hint: Establish a one-to-one correspondence.)

5.53. Let $s(f)$ be the minimum number of transpositions needed to transform the permutation f to the identity permutation. Without considering cycle structure, give a direct procedure to sort a permutation using at most $n - 1$ permutations. Prove that the permutation $n\,n-1\cdots 1$ requires at least $n/2$ transpositions to sort.

5.54. Let $s^*(f)$ be the minimum number of transpositions of adjacent elements needed to transform the permutation f to the identity permutation. Prove that the maximum value of $s^*(f)$ over permutations of $[n]$ is $\binom{n}{2}$. Explain how to determine $s^*(f)$ by examining f.

5.55. (+) Let A_n be the set of permutations of $[n]$. Let B_n be the set of n-tuples $(b_1, \ldots, b_n)$ such that $1 \le b_i \le i$ for each $i \in [n]$. Construct a bijection from A_n to B_n. (Hint: Use induction on n, employing a bijection from A_{n-1} to B_{n-1} to construct a bijection from A_n to B_n. Below we illustrate this process for $n = 3$.)

A_3	321	231	213	312	132	123
B_3	111	112	113	121	122	123

5.56. Use induction to determine the positive integers n such that $n! \ge 2^n$. By forming permutations from subsets, prove that $n! \ge 2^n - n$ for all $n \in \mathbb{N}$.

5.57. For $n \in \mathbb{N}$, find and prove a formula for $\sum_{k=1}^{n} k \cdot k!$. (Comment: There are several proofs, including a combinatorial proof like that of Theorem 5.28.)

5.58. How many permutations of [4] have no fixed point? How many permutations of [5] have no fixed point?

5.59. For $f: A \to A$ and $n \in \mathbb{N}$, let f^n be defined by $f^1 = f$ and $f^n = f \circ f^{n-1}$ for $n > 1$. Let n and k be natural numbers with $k < n$. Prove that $f^n = f^k \circ f^{n-k}$.

5.60. Let S_n be the set of nondecreasing lists summing to n. Let $f: S_n \to S_n$ be the function defined on S_n by the operation in the Penny Problem (Application 1.14).
   a) Draw the functional digraph of f when $n = 6$.
   b) Determine all values of n such that f is injective. Determine all values of n such that f is surjective.

5.61. (+) Let a, b be nonzero real numbers, and define $f: \mathbb{R} \to \mathbb{R}$ by $f(x) = 1/(ax + b)$ for $x \neq -b/a$ and $f(-b/a) = (-1/b) - (b/a)$. Determine the set of ordered pairs (a, b) such that the functional digraph of f has a 3-cycle. Solve the analogous problem when $f(x) = \frac{cx+d}{ax+b}$ and $ad \neq bc$.

5.62. A **partition** of the integer n is a nonincreasing list of positive integers that sum to n. For example, the partitions of 4 are 4, 31, 22, 211, 1111. The elements of the list are the "parts" of the partition.
   a) List the partitions of 6.
   b) Prove that the number of partitions of n with k parts equals the number of partitions of n with largest part k. (Hint: View the parts as rows of dots.)

5.63. (+) By establishing a bijection, prove that the number of partitions of n into distinct parts equals the number of partitions of n into odd parts. For example, the partitions of 4 into distinct parts are 4 and 31, and the partitions of 4 into odd parts are 31 and 1111. (Hint: Consider Proposition 3.32.)

5.64. Let n and k be natural numbers. Prove that there is exactly one choice of integers $m_1, \ldots, m_k$ such that

$$0 \leq m_1 < m_2 < \cdots < m_k \text{ and } n = \binom{m_1}{1} + \binom{m_2}{2} + \cdots + \binom{m_k}{k}.$$

(Hint: Observe that $\binom{m}{k} = \sum_{i=1}^{k} \binom{m-i}{k+1-i}$. Comment: This is called the k-*nomial representation* of n, by analogy with the q-ary representation.)

5.65. (+) The goal of this problem is to determine which polynomials p with rational coefficients have the property that $p(n) \in \mathbb{Z}$ if $n \in \mathbb{Z}$. Let I be the set of polynomials with this property. Recall that the sum $p + q$ of two functions p, q on a set S is the function h such that $h(x) = p(x) + q(x)$. Similarly, the scalar multiple $n \cdot p$ is the function h such that $h(x) = n \cdot p(x)$.
   a) Show that if $p, q \in I$ and $n \in \mathbb{Z}$, then $p + q \in I$ and $n \cdot p \in I$.
   b) Show that $p_j \in I$, where $p_j(x) = \binom{x}{j}$, and that $\sum_{j=0}^{k} n_j \binom{x}{j} \in I$ for $\{n_j\} \subseteq \mathbb{Z}$.
   c) Let f be a polynomial of degree k with rational coefficients. Prove that f can be expressed as $f(x) = \sum_{j=0}^{k} b_j \binom{x}{j}$, where the b_j's are rational. (Hint: One way to prove this uses induction on the degree of the polynomial.)
   d) Prove that $f \in I$ if and only if $f(x) = \sum_{j=0}^{k} b_j \binom{x}{j}$, where the b_j's are integers. (Hint: Evaluate f at the integers in the set $\{0, \ldots, k\}$. Note that $\binom{0}{0} = 1$, by our convention that $0! = 1$.)

Chapter 6

Divisibility

Since ancient times, people have known that one cannot always separate n objects into k equal piles; this is possible only when n is divisible by k. In this chapter we study divisibility properties of the integers.

6.1. Definition. If $a, b \in \mathbb{Z}$ with $b \neq 0$, and $a = mb$ for some integer m, then a is **divisible by** b, and b **divides** a (written as $b|a$). We call b a **divisor** or **factor** of a. A natural number other than 1 is **prime** if its only positive factors are itself and 1.

The first few primes are $2, 3, 5, 7, 11$. We will prove that every natural number has a unique factorization into primes. To ensure uniqueness, we must declare that the number 1 is not prime. We will also study integer solutions to linear equations and solve the following problems.

6.2. Problem. How can we find the greatest common divisor of two large numbers without factoring them? ∎

6.3. Problem. *The Dart Board Problem.* Suppose a dart board has regions with values a and b, where a and b are natural numbers having no common divisor other than 1. What is the largest integer k that cannot be achieved by summing the values of thrown darts? We seek k such that $ma + nb = k$ has no solutions in nonnegative integers m, n, but $ma + nb = j$ does have such a solution whenever j is an integer larger than k. ∎

FACTORS AND FACTORIZATION

6.4. Definition. Integers a and b are **relatively prime** when they have no common factor greater than 1. When m and n are integers, the number $ma + nb$ is an **integer combination** of a and b.

The term "relatively prime" does not mean "somewhat prime"; it means "prime in relation to each other". Our first lemma allows us to express 1 as an integer combination of relatively prime integers.

6.5. Lemma. If a and b are relatively prime, then there exist integers m and n such that $ma + nb = 1$.

Proof: When $|a| = |b|$ or when $b = 0$, the numbers are not relatively prime, unless $|a| = 1$, in which case the conclusion holds with $(m, n) = (a, 0)$. We may thus assume that $|a| > |b|$. Since multiplying by -1 does not change the common factors, and since m, n may be positive or negative, we may assume also that a and b are nonnegative. We prove this case using strong induction on $a + b$. We have already proved the basis step, where $a + b = 1$.

For the induction step, suppose that $a + b \geq 2$. By symmetry, we may assume that $a > b$. We have considered the case $b = 0$. When $b > 0$, we will apply the induction hypothesis to the integers b and $a - b$. These integers are positive and have sum less than $a + b$. They also are relatively prime, since every common divisor of b and $a - b$ also divisor of b and a.

Hence we may apply the induction hypothesis to b and $a - b$, obtaining integers m', n' such that $m'b + n'(a - b) = 1$. The crucial computation rewrites this as $n'a + (m' - n')b = 1$. Setting $m = n'$ and $n = m' - n'$ now yields the desired integer combination of a and b. ∎

6.6. Proposition. If a and b are relatively prime and a divides qb, then a divides q.

Proof: Since a, b are relatively prime, Lemma 6.5 provides integers m, n such that $1 = ma + nb$. Thus $q = maq + nbq = mqa + nqb$. Since a divides each term on the right, a must also divide their sum, q. ∎

6.7. Proposition. If a prime p divides a product of k integers, then p divides at least one of the factors.

Proof: We use induction on k. The statement is trivial when $k = 1$. For $k \geq 2$, let $b_1, \ldots, b_k$ be k integers whose product is divisible by p, and let $n = \prod_{i=1}^{k-1} b_i$. Thus p divides nb_k. If p divides b_k, then the claim holds.

Otherwise, since p is prime, p and b_k are relatively prime. Now Proposition 6.6 implies that p divides n. By applying the induction hypothesis to n, we conclude that p divides one of $\{b_1, \ldots, b_{k-1}\}$. ∎

6.8. Definition. A **prime factorization** of n expresses n as a product of powers of distinct primes; the exponent on each prime is its **multiplicity**. We write a prime factorization of n as $n = \prod_{i=1}^{k} p_i^{e_i}$.

For example, we write $2^4 3^1 5^2$ as a prime factorization of 1200. A prime that does not divide n has multiplicity 0 in every factorization of n. The next theorem is the fundamental result about factorization.

6.9. Theorem. (Fundamental Theorem of Arithmetic) Every positive integer n has a prime factorization, which is unique except for reorderings of the factors.

Proof: We use strong induction on n. For $n = 1$, there are no prime factors. By convention, the product of the integers in an empty set is the multiplicative identity 1, so the basis step holds.

For the induction step, consider $n > 1$. Let S be the set of integers larger than 1 that divide n; this is nonempty, since $n \in S$. By the Well-Ordering Property, S has a smallest element p. Furthermore, p is prime; otherwise, p has a smaller prime factor that also divides n.

By Proposition 6.7, p appears in every list of primes (repetition allowed) whose product is n. Therefore, every prime factorization of n consists of p and a prime factorization of n/p. By the induction hypothesis, n/p has a unique prime factorization. Hence there is exactly one prime factorization of n, obtained by adding one to the multiplicity of p in the unique prime factorization of n/p. ∎

6.10. Corollary. If a, b are relatively prime and both divide n, then $ab|n$.
Proof: Exercise 28. ∎

When integers a and b are not relatively prime, they have a common divisor larger than 1. Often we need to know the largest such divisor.

6.11. Definition. Given integers a, b not both 0, the **greatest common divisor** $\gcd(a, b)$ is the largest natural number that divides both a and b. By convention, $\gcd(0, 0) = 0$.

If $d = \gcd(a, b) \neq 0$, then a/d and b/d are relatively prime. This enables us to describe all integer combinations of a and b.

6.12. Theorem. The set of integer combinations of a and b is the set of multiples of $\gcd(a, b)$.

Proof: Let $d = \gcd(a, b)$. The set of integer combinations of a and b is $S = \{ra + sb \colon r, s \in \mathbb{Z}\}$. Let T denote the set of multiples of d.

We first prove $S \subseteq T$. Since d divides both a and b, there are integers k and l such that $a = kd$ and $b = ld$. The distributive law now yields

$ma + nb = mkd + nld = (mk + nl)d$, and thus d also divides $ma + nb$. Since this holds for every integer combination, we have $S \subseteq T$.

To prove $T \subseteq S$, we express each multiple of d as an integer combination of a and b. Since the integers a/d and b/d are relatively prime, by Lemma 6.5 there exist integers m, n such that $m(a/d) + n(b/d) = 1$. Thus $ma + nb = d$. For $k \in \mathbb{Z}$, we now have $(mk)a + (nk)b = kd$. Thus $T \subseteq S$.   ∎

THE EUCLIDEAN ALGORITHM

Our applications of common divisors have been based on knowing the greatest common divisor. For two large numbers, the greatest common divisor is not immediately obvious; we need a procedure for computing it. We also want to know the integer combination m, n that yields $ma + nb = d$, where $d = \gcd(a, b)$. There is an efficient algorithm based on the idea underlying the proof of Lemma 6.5. (An **algorithm** is a procedure for performing a computation or construction.)

6.13. Proposition. If a, b, k are integers, then $\gcd(a, b) = \gcd(a - kb, b)$.

Proof: By the distributive law, every integer dividing a and b must also divide $a - kb$. Similarly, every integer dividing $(a - kb)$ and b must also divide a. Thus d is a common divisor of a and b if and only if d is a common divisor of $a - kb$ and b, and hence $\gcd(a, b) = \gcd(a - kb, b)$.   ∎

When $k = 1$, Proposition 6.13 allows us to subtract a smaller number from a larger number without changing the gcd. Doing this repeatedly will lead us to the gcd. If one number has 10 digits and the other has 100 digits, then we would have to do many subtractions before making much progress. We can speed up the procedure by performing many subtractions at once, as suggested by the "k" in Proposition 6.13. Finding the right number of subtractions is the role of division.

6.14. Proposition. If a and b are integers with $b \neq 0$, then there is a unique integer pair k, r such that $a = kb + r$ and $0 \leq r \leq |b| - 1$.

Proof: (Exercise 16).   ∎

The process of obtaining k and r in Proposition 6.14 is the **Division Algorithm**. The resulting r is the **remainder** of a under division by b; the remainder is 0 if and only if a is divisible by b. To express k, we define the **floor** of a real number x, written $\lfloor x \rfloor$, to be $\max\{z \in \mathbb{Z} : z \leq x\}$. When $a, b > 0$, we have $k = \lfloor a/b \rfloor$. The **ceiling** of x, written $\lceil x \rceil$, is $\min\{z \in \mathbb{Z} : z \geq x\}$.

We next describe an algorithm for computing greatest common divisors. This solves Problem 6.2. Exercise 12.26 considers the efficiency of the algorithm. Exercise 43 considers a similar algorithm.

6.15. Algorithm. (The Euclidean Algorithm)
   INPUT: A pair of nonnegative integers, not both 0.
   OUTPUT: The greatest common divisor of the input pair.
   INITIALIZATION: Set the *current pair* to be the input pair.
   ITERATION: If one element of the current pair is 0, then report the other element as the output, and stop. Otherwise, replace the maximum element of the current pair with its remainder upon division by the other element, and repeat using this new pair as the current pair.

   Note that if the current pair is (n, n), then the next pair is $(n, 0)$, after which the algorithm reports n as the gcd and stops.
   We must prove that the Euclidean Algorithm terminates and correctly reports the gcd. Keeping careful track of the quotients in the divisions also enables us to compute $\gcd(a, b)$ as an integer combination of a and b. Each step of the algorithm expresses the new value as an integer combination of the previous values, so expressing the final value in terms of the original values is a succession of substitutions.

6.16. Example. *The Euclidean Algorithm and integer combinations.* When (a, b) is the current pair, the new number is the remainder in $a = kb + r$, and we express the remainder in terms of the current pair as $r = a - kb$. At the end we work backward, undoing the subtractions by substitutions, to express the greatest common divisor as an integer combination of the original inputs. When applied to the pair $(154, 35)$, the algorithm takes three steps and finds 7 as the common divisor.

$$\begin{array}{ll}
(154, 35) & 14 = 154 - 4 \cdot 35 \\
(35, 14) & 7 = 35 - 2 \cdot 14 \\
(14, 7) & 0 = 14 - 2 \cdot 7 \\
(7, 0) &
\end{array}$$

$$7 = 35 - 2 \cdot 14 = 35 - 2(154 - 4 \cdot 35) = -2 \cdot 154 + 9 \cdot 35 \qquad \blacksquare$$

6.17. Theorem. Applied to integers a, b with $a \geq b \geq 0$ and $a \neq 0$, the Euclidean Algorithm reports $\gcd(a, b)$ as output. Furthermore, reversing the substitution steps of the algorithm yields an expression of $\gcd(a, b)$ as $ma + nb$ for some $m, n \in \mathbb{Z}$.

Proof: The proof is by strong induction on b, the smaller entry of the input pair. For the basis step, we have $b = 0$. In this case, the output is a. This equals $\gcd(a, 0)$, and $a = 1 \cdot a + 0 \cdot 0$ expresses the greatest common divisor as an integer combination of a and b.
   For the induction step, we have $a \geq b \geq 1$, and we assume that the Euclidean Algorithm computes the gcd whenever the smaller input is less than b. The result of the first step is a pair (b, c) with $b > c \geq 0$, satisfying

$a = kb + c$ for some $k \in \mathbb{N}$. Writing $c = a - kb$ shows that c is obtained from a by subtracting a multiple of b. By Proposition 6.13, $\gcd(a, b) = \gcd(b, c)$.

The remaining computations are the same as when applying the Euclidean Algorithm with the input (b, c). Since $b > c \geq 0$, the induction hypothesis tells us that continuing the Euclidean Algorithm with the pair (b, c) yields the output $(d, 0)$, where $d = \gcd(b, c) = \gcd(a, b)$.

The induction hypothesis also tells us that reversing the substitutions as in Example 6.16 expresses d as an integer combination of b and c. Let this be $d = m'b + n'c$, where $m', n' \in \mathbb{Z}$. Since we have c expressed in terms of a and b as $c = a - kb$, we can substitute to obtain

$$d = m'b + n'c = m'b + n'(a - kb) = n'a + (m' - n'k)b.$$

Thus setting $m = n'$ and $n = (m' - n'k)$ yields an expression for the greatest common divisor as an integer combination of a and b. ∎

The replacement operation used at each step in the Euclidean Algorithm is a function $E: S \to S$, where S is the set of nonnegative integer pairs (a, b) with $a \geq b \geq 0$ and $a \neq 0$. The Euclidean Algorithm iterates E until it produces a pair in which the second coordinate is 0. Thus Theorem 6.17 guarantees that $E^n(a, b) = (\gcd(a, b), 0)$ for some $n \in \mathbb{N}$.

The expression of $\gcd(a, b)$ as an integer combination of a and b can be used to solve linear equations in integers. An equation for which we seek integer solutions is called a **diophantine equation**, in honor of Diophantus (third century A.D.).

6.18. Example. *Impossibility of solutions.* The equation $6x + 15y = 79$ has no solution in integers. Such a solution would express 79 as an integer combination of 6 and 15. All such combinations are multiples of $\gcd(6, 15) = 3$, but 79 is not a multiple of 3. ∎

After finding one solution to the diophantine equation $ax + by = c$, we can easily find all solutions. We illustrate this by an example.

6.19. Example. *Description of all solutions.* What are the integer solutions of $6x + 15y = 99$? Let S denote this set. Since 99 is a multiple of $3 = \gcd(6, 15)$, Theorem 6.12 guarantees a solution. To find solutions, we first divide the equation by this gcd to obtain the **reduced** equation $2x + 5y = 33$; doing so does not change the set of solutions. Setting $x = -2$ and $y = 1$ produces 1 as an integer combination of the coefficients 2 and 5: $2(-2) + 5(1) = 1$. Had we not seen a solution to $2x + 5y = 1$, we could have used the Euclidean Algorithm to generate one.

Multiplying the solution to the reduced equation by 33 produces the solution $(x, y) = 33 \cdot (-2, 1) = (-66, 33)$ in S. We can generate other solutions by increasing x and decreasing y, or vice versa. We must increase $2x$

and decrease $5y$ by the same amount. Hence this amount must be a multiple of both 2 and 5. Since 2 and 5 are relatively prime, we find all other solutions by changing x by a multiple of 5 while we change y in the other direction by that multiple of 2. Thus $S = \{(-66 + 5k, 33 - 2k) : k \in \mathbb{Z}\}$. ∎

THE DART BOARD PROBLEM

Our analysis of diophantine equations allows negative integers. What happens when we forbid negative values in the solution? The Dart Board Problem is such a question. Its solution is also known as Sylvester's Theorem, for James Joseph Sylvester (1814–1897).

Let a, b, k be positive integers with a and b relatively prime. We begin with a geometric argument that suggests that $k = ma + nb$ must have a nonnegative integer solution (m, n) when k is large. Since a, b are relatively prime, the equation $k = ma + nb$ has integer solutions. We move from one to the next by adding b to m and subtracting a from n. Viewed as points in the plane, the solution pairs (m, n) lie along a line. There is a nonnegative integer solution for k if and only if the line for k contains an integer point in the first quadrant, which by definition is the set of points with both coordinates nonnegative.

The lines for distinct choices of k are parallel; they are the level sets of the function defined by $f(m, n) = ma + nb$. Below we sketch these lines for $(a, b) = (3, 5)$ and $k \in \{1, 2, 4, 7\}$. These k are the positive integers not expressible as nonnegative integer combinations of a and b. The dots indicate the integer points closest to the first quadrant on these lines. As k increases, the line crosses more of the first quadrant. Since the integer points have the same spacing on each line, making k large guarantees a solution. In terms of a and b, we ask how large k must be to guarantee the existence of a nonnegative integer solution.

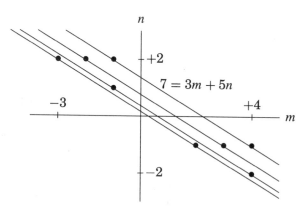

6.20. Solution. *The Dart Board Problem.* For relatively prime positive integers a and b, we prove that $ab - a - b$ is the largest integer not expressible as a nonnegative integer combination of a and b. Call k **achievable** if $k = ma + nb$ has a nonnegative integer solution. We prove that $ab - a - b$ is not achievable and that every larger number is achievable.

First we prove that $ab - a - b$ is not achievable, by the method of contradiction. If $ab - a - b$ is achievable, then $ab - a - b = ma + nb$ for some nonnegative integers m, n. Thus $ab = (m + 1)a + (n + 1)b$. Since a and b are relatively prime, this implies that a divides $(n + 1)$ and that b divides $(m + 1)$. Since $m, n \geq 0$, this in turn implies that $n + 1 \geq a$ and $m + 1 \geq b$. These inequalities yield the contradiction

$$ab = (m + 1)a + (n + 1)b \geq 2ab.$$

Next, we prove "$k > ab - a - b \Rightarrow k$ is achievable" by proving the contrapositive: "k not achievable $\Rightarrow k \leq ab - a - b$". Suppose that k is not achievable. Because $\gcd(a, b) = 1$, we can find integers r, s such that $1 = ra + sb$. Multiplying the equation by k yields $k = (kr)a + (ks)b$; this is an integer solution to $k = ma + nb$, but one coefficient is negative. Adding b to m and subtracting a from n produces another integer solution to the equation $k = ma + nb$. Since k is not achievable, there are no integer solutions in the first quadrant. Hence there are consecutive integer solutions with (m', n') in the second quadrant and $(m' + b, n' - a)$ in the fourth quadrant. Since these are integer solutions in these quadrants, they must satisfy $m' \leq -1$ and $n' - a \leq -1$. Now we can compute

$$k = m'a + n'b \leq (-1)a + (a - 1)b = ab - a - b. \qquad \blacksquare$$

Achievability of the numbers larger than $ab - a - b$ can also be proved directly. Consider the example $(a, b) = (3, 10)$ (Exercise 3.44). Checking successive numbers reveals that 17 is not achievable but that 18, 19, 20 are. All larger numbers are also achievable, since each larger number exceeds one of these by a positive multiple of 3.

The crucial property is having achievable numbers with all remainders under division by a. We next establish a condition for equally-spaced numbers to have distinct remainders. This yields another solution of the Dart Board Problem and will be applied in Chapter 7.

6.21. Theorem. When a, b are relatively prime and $x \in \mathbb{Z}$, the numbers $x, x + b, \ldots, x + (a - 1)b$ have distinct remainders upon division by a.

Proof: Suppose that $x + ib$ and $x + jb$ have the same remainder, which means that $x + ib = ka + r$ and $x + jb = la + r$ for some integers k, l, r with $0 \leq r \leq a - 1$. Subtracting the equations yields $(i - j)b = (k - l)a$. Since a divides $(k - l)a$, a must also divide $(i - j)b$. Since a and b are relatively prime, Proposition 6.6 implies that a must divide $(i - j)$. Since i and j are nonnegative integers less than a, this requires $i = j$. $\qquad \blacksquare$

6.22. Solution. *The Dart Board Problem, alternative proof.* We prove
that every integer larger than $ab - a - b$ is achievable. If such a number
x has the same remainder under division by a as a smaller achievable
number y, then also x is achievable, since we can increase the multiple of
a in $ma + nb = y$ to achieve x.

By using $m = 0$ and the nonnegative values of n, the numbers in $T =$
$\{0, b, 2b, \ldots, (a - 1)b\}$ are achievable. By Theorem 6.21, these numbers
have different remainders under division by a. Since $a < b$, we have
$(a - 2)b < ab - a - b$, and all of these numbers are less than $ab - a - b$
except $(a - 1)b$. However, $(a - 1)b - (ab - a - b) = a$, so $(a - 1)b$ is the
first number after $ab - a - b$ in its remainder class. Thus we have proved
that every integer larger than $ab - a - b$ has the same remainder under
division by a as a number in T that is no bigger and is achievable. ∎

$$ab - a - b$$

| 0 | b | $2b$ | $(a-2)b$ $(a-1)b$ | $\leftarrow T$ |

MORE ON POLYNOMIALS (optional)

In this section we consider the set of polynomials in one variable as
a mathematical system in its own right. This set has many properties
analogous to those of the integers. For example, the Division Algorithm,
the Euclidean Algorithm, and unique factorization into primes all apply.

Let $\mathbb{R}[x]$ denote the set of polynomials in one variable, and let $\mathbb{Z}[x]$
denote the subset consisting of those polynomials whose coefficients are
integers. We add and multiply polynomials in the natural way:

$$\sum_k a_k x^k + \sum_l b_l x^l = \sum_n (a_n + b_n)x^n$$

$$\sum_k a_k x^k \sum_l b_l x^l = \sum_n \left(\sum_{k=0}^n a_k b_{n-k}\right) x^n.$$

Since the sum and the product of two polynomials are polynomials,
and $\mathbb{Z}$ is closed under addition and multiplication, addition and multiplication are binary operations on $\mathbb{R}[x]$ and on $\mathbb{Z}[x]$ (a *binary operation* on a
set S is a function from $S \times S$ to S). The constant polynomials 0 and 1 are
additive and multiplicative identities, respectively.

We use letters such as a, b, q, r for polynomials. Since we treat them
as elements of $\mathbb{R}[x]$, we usually omit the indeterminate x in the notation
for a polynomial. Recall from Corollary 3.25 that a and b are the same
object in $\mathbb{R}[x]$ if and only if a and b are equal as functions on $\mathbb{R}$.

6.23. Theorem. (Division Algorithm for Polynomials) If $a, b \in \mathbb{R}[x]$, and $b \neq 0$, then there exist unique $q, r \in \mathbb{R}[x]$ such that $a = qb + r$ and either $r = 0$ or $\deg(r) < \deg(b)$.

Proof: For each b, we prove existence by strong induction on the degree of a. Let $m = \deg(b)$ and $n = \deg(a)$. If $n < m$, then the desired conclusion holds with $q = 0$ and $r = a$. Thus the result holds whenever $n < m$.

For the induction step, we consider a polynomial a with degree $n \geq m$, and we assume that the result holds for all polynomials of degree less than n. Let a_n and b_m be the leading coefficients of a and b. Let $h(x) = \frac{a_n}{b_m} x^{n-m}$. Note that $hb - a$ has no term of degree n. We can therefore write $a = hb + c$, where $c = 0$ or $\deg(c) < n$. If $c = 0$, then the conclusion holds with $q = h$ and $r = 0$. Otherwise the strong induction hypothesis allows us to write $c = Qb + R$, where $\deg(R) < m$. Thus,

$$a = Qb + R + hb = (Q + h)b + R.$$

This is the desired conclusion with $q = Q + h$ and $r = R$.

We leave the proof of uniqueness to Exercise 58.   ∎

6.24. Corollary. If p is a polynomial, then there is a polynomial q with

$$p(x) = (x - x_0)q(x) + p(x_0).$$

Proof: Apply the Division Algorithm with $b(x) = x - x_0$. The remainder must be a constant. Evaluating at $x = x_0$ determines the constant.   ∎

The statement and the proof of Theorem 6.23 parallel those of the Division Algorithm on $\mathbb{Z}$. The same will hold for results about greatest common divisors.

6.25. Definition. A nonempty subset I of $\mathbb{R}[x]$ is an **ideal** if it satisfies properties (a) and (b) below.

   a) $p, q \in I$ imply $p + q \in I$.

   b) $p \in I$ and $r \in \mathbb{R}[x]$ imply that $rp \in I$.

An ideal I is a **principal ideal** if there exists $g \in \mathbb{R}[x]$ such that $I = \{pg : p \in \mathbb{R}[x]\}$. The polynomial g is a **generator** of I.

It follows from property (b) that every ideal contains 0.

6.26. Theorem. Every ideal in $\mathbb{R}[x]$ is a principal ideal.

Proof: If I consists of only the zero polynomial, then the result is true with $g = 0$. Otherwise, let b be a nonzero polynomial in I of least degree. Let a be an arbitrary element of I. By the Division Algorithm we can write $a = qb + r$. Either $r = 0$ or the degree of r is less than the degree of b.

Since a and qb both belong to I, also $r = a - qb = a + (-q)b \in I$. Since b is an element of I with minimum degree, and $r \in I$, $\deg(r) < \deg(b)$ is impossible. We conclude that $r = 0$, and hence $a = qb$.

Thus every element of I is a multiple of b. By the definition of ideal, every multiple of b is in I. Thus I consists of all multiples of b. ∎

The notion of "ideal" extends to more general mathematical systems called *rings* and *integral domains*. These are sets endowed with operations of addition and multiplication whose properties resemble those of the integers $\mathbb{Z}$. Within $\mathbb{Z}$, for example, the set of integer combinations of integers a and b is by definition an ideal, and Theorem 6.12 shows that this ideal is a principal ideal with generator $\gcd(a, b)$. Exercise 55 is the analogue of Theorem 6.26 for $\mathbb{Z}$; every ideal in $\mathbb{Z}$ consists of the multiples of a single integer.

Given $d, a \in \mathbb{R}[x]$, we say that d **divides** a or is a **divisor** of a if there is a polynomial q such that $dq = a$. A **greatest common divisor** of a and b is a polynomial divisible by every common divisor of a and b.

6.27. Theorem. If a and b are elements of $\mathbb{R}[x]$, then a and b have a greatest common divisor d. Furthermore, there are polynomials s and t so that $d = as + bt$.

Proof: The set S of polynomials that can be written as $as + bt$ for some polynomials s and t is an ideal. Theorem 6.26 therefore implies that S consists of all multiples of some polynomial d. We have $a, b \in S$, since $a = 1a + 0b$ and $b = 0a + 1b$. Thus each is a multiple of d.

If q divides both a and b, then since $d = as + bt$, also q divides d. Thus d is a greatest common divisor of a and b. ∎

In abstract algebra, it is proved that whenever every ideal in an integral domain is principal, every element has a unique factorization into primes. This applies in particular to $\mathbb{R}[x]$, as we now show.

6.28. Definition. A polynomial $u \in \mathbb{R}[x]$ is a **unit** if it is a nonzero constant. A polynomial $a \in \mathbb{R}[x]$ is **reducible** if it can be expressed as $a = bc$ with neither b nor c being a unit. A nonconstant $a \in \mathbb{R}[x]$ is **irreducible** if $a = bc$ implies that b or c is a unit.

We call a nonzero constant a "unit" because it has a multiplicative inverse in $\mathbb{R}[x]$ (Exercise 63). We do not consider units to be irreducible, for the same reason that we do not consider 1 to be a prime number.

6.29. Lemma. If an irreducible polynomial p divides a product ab, then p divides a or b.

Proof: If p doesn't divide a, then $\gcd(p, a) = 1$. By Theorem 6.23, $1 = sp + ta$, and hence $b = spb + tab$. Now p divides both terms on the right and thus also divides b. ∎

6.30. Theorem. Every nonconstant $a \in \mathbb{R}[x]$ can be written as the product of irreducible polynomials. The factorization is unique up to order of the factors and multiplication of them by units.

Proof: (sketch) If a is irreducible, then the existence and uniqueness results are immediate. If a is reducible, then there exist polynomials b, c with degrees smaller than a such that $a = bc$. A formal proof of existence then follows either by the method of descent or by strong induction on the degree. The uniqueness follows by strong induction and Lemma 6.29 (see Exercise 59). ∎

EXERCISES

6.1. (−) Explain why the following makes no sense: "Let n be relatively prime."

6.2. (−) Let p be a prime number. Which integers are relatively prime to p?

6.3. (−) Determine which numbers are relatively prime to 0.

6.4. (−) Suppose that $\gcd(a, b) = 1$. Prove that $\gcd(na, nb) = n$.

6.5. (−) Let n be a natural number. What is the list of pairs produced when the Euclidean Algorithm is applied to the input $(5n, 2n)$?

6.6. (−) How many steps does it take the Euclidean Algorithm to reach $(1, 0)$ when the input is $(n + 1, n)$?

6.7. (−) Is 61 an integer combination of 9 and 15? Is 61 an integer combination of 9 and 16?

6.8. (−) For each pair below, use the Euclidean Algorithm to compute the greatest common divisor, and express the greatest common divisor as an integer combination of the two numbers.
  a) 126 and 224.   b) 221 and 299.

6.9. (−) For each diophantine equation below, find all solutions, if any exist.
  a) $17x + 13y = 200$.   c) $60x + 42y = 104$.
  b) $21x + 15y = 93$.   d) $588x + 231y = 63$.

6.10. (−) Show that the first 10 multiples of 7 end in different digits (in base 10), but the first 10 multiples of 8 do not. Explain the distinction.

•   •   •   •   •

American coins have values 1, 5, 10, 25, and 50 cents, called pennies, nickels, dimes, quarters, and half-dollars, respectively.

6.11. (−) A person has the same (nonzero) number of each type of American coin. The total amount she has is a whole number of dollars. Determine the smallest such nonzero amount. Answer the same question assuming she has no pennies. Answer the same question assuming she has no pennies and no nickels.

6.12. (−) A parking meter contains the same number of dimes and quarters, in total a nonzero whole number of dollars. What is the minimum number of coins?

6.13. (−) A parking meter can hold k quarters, $2k$ nickels, and $4k$ dimes. Find all k such that the total when the meter is full is a whole number of dollars.

6.14. (−) Suppose a parking meter accepts only dimes and quarters and has twice as many dimes as quarters. If the total amount of money is a nonzero whole number of dollars, what is the smallest possible number of quarters?

6.15. What is the smallest number of American coins (values may repeat) sufficient to make change equal to any value from 1 cent through 99 cents? Is there only one optimal solution? What is the answer when coins can be made with any desired value?

• • • • •

6.16. For $a, b \in \mathbb{Z}$, prove that there is exactly one pair $k, r \in \mathbb{Z}$ such that $0 \le r \le |b| - 1$ and $a = kb + r$.

6.17. Prove that $\gcd(a + b, a - b) = \gcd(2a, a - b) = \gcd(a + b, 2b)$.

6.18. Suppose that $\gcd(a, b) = 1$. Does this determine $\gcd(a^2, b^2)$? Does this determine $\gcd(a, 2b)$?

6.19. (!) Let n, k, j be natural numbers with $n > k > j$. Prove that $\binom{n}{k}$ and $\binom{n}{j}$ are not relatively prime. (Hint: Use Exercise 5.37 to relate these quantities.)

6.20. By counting an appropriate geometric arrangement of points, prove that $2 \sum_{i=1}^{q-1} \lfloor ip/q \rfloor = (p - 1)(q - 1)$ if p and q are relatively prime.

6.21. For $x \in \mathbb{R}$ and $k \in \mathbb{N}$, prove that $\lfloor -x \rfloor = -\lceil x \rceil$ and that $\lceil \frac{x-k+1}{k} \rceil = \lfloor \frac{x}{k} \rfloor$.

6.22. Find every integer k such that $k \ge 3$ and $k - 2$ divides $2k$.

6.23. Determine the values of n such that $\{n, n + 2, n + 4\}$ are all prime.

6.24. Prove that 3 divides $4^n - 1$, for every positive integer n. Prove that 6 divides $n^3 + 5n$, for every positive integer n.

6.25. Let $\langle a \rangle$ be a sequence such that $a_1 = 1$, $a_2 = 1$, and $a_{n+1} = a_n + 2a_{n-1}$ for $n \ge 2$. Prove that a_n is divisible by 3 if and only if n is divisible by 3.

6.26. For $n \in \mathbb{N}$, prove that $(n - 1)^3 + n^3 + (n + 1)^3$ is divisible by 9.

6.27. Let $f \colon \mathbb{N} \times \mathbb{N} \to \mathbb{N}$ be defined by $f(x, y) = 3^{x-1}(3y - 1)$. Show that f is not surjective. Explain why this differs from Example 4.45.

6.28. Suppose that $\gcd(a, b) = 1$ and that $a|n$ and $b|n$. Prove that $ab|n$.

6.29. The **least common multiple** (lcm) of natural numbers a and b is the least natural number divisible by both. Prove that $\mathrm{lcm}(a, b) \cdot \gcd(a, b) = a \cdot b$.

6.30. (!) Prove that $(2n)!/(2^n n!)$ is an odd number when n is a nonnegative integer.

6.31. (!) Let a, b, c be integers such that $a^2 + b^2 = c^2$.
 a) Is it always true that at least one of $\{a, b\}$ is even?
 b) For c divisible by 3, prove that a and b are both divisible by 3.

6.32. (!) A bear's cage has two jars of jelly beans, one with x beans and the other with y. Each jar has a lever. When a jar has at least two beans, pressing its lever will give the bear one bean from it and move one bean from it to the other jar; otherwise the lever has no effect. Obtain necessary and sufficient conditions on the pair x, y so that the bear can eat all the beans except one.

6.33. Let abc be a 3-digit natural number (written in base 10). Prove that the 6-digit number $abcabc$ has at least three distinct prime factors.

6.34. (!) Prove using contradiction that the set of prime numbers is not finite.

6.35. (!) Let n be a positive integer. Construct a set of n consecutive positive integers that are not prime. (Hint: Determine a positive integer x such that x is divisible by 2, $x + 1$ is divisible by 3, $x + 2$ is divisible by 4, etc.)

6.36. (!) *Primes and factorials.*
   a) Express the exponent of prime p in the factorization of $k!$ as a finite sum. In particular, compute the exponent of 5 in 250!.
   b) Use the answer to (a) to prove that N is divisible by $k!$ if N is the product of k consecutive natural numbers.
   c) Give another proof of (b) using a combinatorial argument.

6.37. (!) Let p be a prime number.
   a) Prove that p divides $\binom{p}{k}$ if $1 \le k \le p - 1$.
   b) Prove that $n^p - n$ is divisible by p for every $n \in \mathbb{N}$. (Hint: Use the Binomial Theorem and part (a) in a proof by induction.)

6.38. (!) Let x, y, k be nonnegative integers, with k not being a power of 2. Prove that $x^k + y^k$ is not prime. Conclude that if $2^n + 1$ is prime and n is not a power of 2, then n is prime.

6.39. Prove that if $2^n - 1$ is prime, then n is prime. (Hint: Prove the contrapositive; if n is not prime, then $2^n - 1$ is not prime. Comment: Primes of the form $2^n - 1$ are called **Mersenne primes**; 36 such primes are known—see `http://www.mersenne.org/2976221.htm`.)

6.40. (!) A natural number is **perfect** if it is the sum of the smaller natural numbers that divide it; 6 and 28 are the first two perfect numbers. Prove that if $2^n - 1$ is prime, then $2^{n-1}(2^n - 1)$ is perfect. (Hint: List the divisors and sum them. Comment: Euclid conjectured that all perfect numbers have this form. This is not known, but it is known that these are the only even perfect numbers.)

6.41. *Pólya's proof for infinitude of primes.* Let $a_n = 2^{2^n} + 1$. Prove by induction that a_n divides $a_m - 2$ if $n < m$. Conclude that a_n and a_m have no common factors if $n \ne m$. Use this to prove that there are infinitely many primes. (This method also proves that there are at least $\log_2 \log_2 N$ primes less than N.)

6.42. Let n be an integer. Let $f(n)$ denote the number of distinct digits that occur as the last digit in the base 10 representation of the numbers $n, 2n, 3n, \ldots, 10n$. Compute $f(n)$.

6.43. Let a and b be nonnegative integers. Prove that the following algorithm computes $\gcd(a, b)$. Each step of the algorithm replaces the current pair of numbers with a new pair or reports an output, according to the following rules.

1) When one number is 0 or they are equal, stop and report the maximum of the pair as output.

2) When both numbers are nonzero and at least one is even, divide the first even member of the pair by 2.

3) When both numbers are odd, replace the larger one with their difference. (Comment: This algorithm runs faster than the Euclidean Algorithm.)

6.44. Example 4.6 suggests a procedure for computing the base q representation of a natural number n. Prove that the following inductive procedure also works, and use it to compute the representation of 729 in base 5.

1) If $1 \leq n \leq q - 1$, then the base q representation of n is $a_0 = n$.

2) If $n \geq q$, then let $n = kq + r$, where $r \in \{0, \dots, q - 1\}$, and let $b_m, \dots, b_0$ be the base q representation of k. The base q representation of n is $a_{m+1}, \dots, a_0$, where $a_0 = r$ and $a_i = b_{i-1}$ for $i > 0$.

6.45. The royal treasury has 500 7-ounce weights, 500 11-ounce weights, and a balance scale. An envoy arrives with a bar of gold, claiming it weighs 500 ounces. Can the treasury determine whether the envoy is lying? If so, how? What if the weights are 6-ounce and 9-ounce weights?

6.46. $(-)$ Find all integer solutions to $70x + 28y = 518$. Determine how many solutions have both variables positive.

6.47. Find all integer solutions to $\frac{1}{60} = \frac{x}{5} + \frac{y}{12}$.

6.48. Given $a, b, c \in \mathbb{Z}$, let $d = \gcd(a, b)$, and suppose that d divides c. Prove that the set of integer solutions to $ax + by = c$ is nonempty. Express the set of solutions in terms of one given solution and the parameters a, b, d.

6.49. A jar contains some pennies, some nickels, and some dimes. Suppose that the total value of the coins in cents is s, and the total number of coins is t. Determine the smallest s that permits more than one solution for some t.

6.50. *A "reciprocal" dart board problem.*

a) Do there exist natural numbers m, n such that $7/17 = 1/m + 1/n$?

b) $(+)$ Let p be a prime number. For which $k \in \mathbb{N}$ do there exist $m, n \in \mathbb{N}$ such that $k/p = 1/m + 1/n$?

6.51. $(+)$ *The Coconuts Problem.* Five suspicious sailors spend the day gathering a pile of coconuts. Exhausted, they postpone dividing it until the next morning. Suspicious, each decides to take his share during the night. The first sailor divides the pile into five equal portions plus one extra coconut, which he gives to a monkey. He takes one pile and leaves the rest in a single pile. The second sailor later does the same; again the monkey receives one leftover coconut. The third, fourth and fifth sailors also do this; each time a remainder of one goes to the monkey. In the morning they split the remaining coconuts into five equal piles, and each sailor gets his "share". (Each knows some were taken, but none complains, since each is guilty!) What is the smallest possible number of coconuts in the original pile? (This problem appeared in the *Saturday Evening Post* on October 9, 1926.)

6.52. $(+)$ *The Postage Stamp Problem* (special case). The Post Office wants to issue stamps with two different values. Postage is one cent per ounce, and each envelope has space for s stamps. Correctly posting a one-ounce envelope requires that one of the stamp values be 1. The problem is to choose the other value m to maximize n such that all weights in $[n]$ can be correctly posted.

a) Prove that m should be at most $s + 1$.

b) Prove that for each m satisfying $2 \leq m \leq s+1$, the smallest integer weight that cannot be formed using at most s stamps of values 1 or m is $m(s + 3 - m) - 1$. (Hint: Prove that this value is one less than a multiple of m.)

c) Use part (b) to prove that the best choice of m is $\lceil s/2 \rceil + 1$. (Comment: The more general problem in which d different values are allowed is unsolved.)

6.53. (+) Consider cards labeled $1, \ldots, 2n$. The cards are shuffled and dealt to two players A and B so that each gets n of the cards. Let x be the sum of the labels on the cards that have been played; initially, $x = 0$. Starting with A, play alternates between the two players. At each play, a player adds one of his or her remaining cards to x. The first player who makes x divisible by $2n + 1$ wins. Prove that for every deal, player B has a strategy to win. (Hint: Prove that B can always make it impossible for A to win on the next move.)

6.54. (+) Let S be a set of three positive integers. If r, s are members, with $r \leq s$, then r, s can be replaced with $2r$ and $s - r$. Prove that every set S of three positive integers can be transformed by such operations into a set that contains 0. (Hint: If x is the smallest number in S and y is the next smallest, prove that y can be expressed as $y = (2^n + a)x + b$, where $a < 2^n$ and $b < x$. Use this to prove the claim by strong induction on the minimum value in S.)

6.55. (+) A set $S \subseteq \mathbb{Z}$ is an **ideal** in $\mathbb{Z}$ if S is nonempty and satisfies 1) if $a, b \in S$, then $a + b \in S$, and 2) if $a \in S$ and $n \in \mathbb{Z}$, then $na \in S$. Prove that every ideal in $\mathbb{Z}$ is the set of multiples of a single integer. (Comment: This strengthens Theorem 6.12, showing that every ideal in $\mathbb{Z}$ is a principal ideal—see Definition 6.25. The analogous result for $\mathbb{R}[x]$ is Theorem 6.26.)

6.56. For each pair of polynomials below, compute the greatest common divisor.

a) x^2 and $3x^3 + x + 1$.

b) $x^2 + x$ and $x^3 + 2x^2 + 2x + 1$

c) $x^3 - 3x - 2$ and $x^3 - x - 2x^2 + 2$.

6.57. (−) Show that $\deg(p+q) \leq \max(\deg(p), \deg(q))$. Under what circumstance does strict inequality hold?

6.58. Prove the uniqueness of the polynomials q, r in Theorem 6.23.

6.59. Complete the details of the proof of Theorem 6.30 using induction.

6.60. (!) Reprove Theorem 6.9 by first solving exercise 50 and then mimicking the logic of Theorem 6.30.

6.61. (!) Consider the set of polynomials $\mathbb{R}[x, y]$ in two variables. Show that there are ideals that are not principal.

6.62. Solve Exercise 6.18 when a and b are elements of $\mathbb{R}[x]$.

6.63. Show that if $ab = 1$ for $a, b \in \mathbb{R}[x]$, then a and b are constants.

6.64. Find a polynomial in $\mathbb{Z}[x]$ whose factors lie in $\mathbb{R}[x]$ but not in $\mathbb{Z}[x]$.

6.65. Consider $A, B, C \in \mathbb{R}$ with $A \neq 0$. Obtain a necessary and sufficient condition on A, B, C so that $Ax^2 + Bx + C$ is irreducible in $\mathbb{R}[x]$.

Chapter 7

Modular Arithmetic

In Chapter 6 we studied divisibility; now we study what is left after division. Parity describes the remainder when the divisor is 2; the odd integers are those having remainder 1. Considerations of parity are fundamental to atomic physics and computer science as well as to mathematics. We generalize parity by considering divisors other than 2; this leads to many applications and to another notion of arithmetic.

7.1. Problem. How can one easily determine, from the 0s and 1s in the binary representation of a natural number, whether the number is divisible by 3? ∎

7.2. Problem. *Chinese Remainder Problem.* A general in ancient China wanted to count his troops. Suppose that when his soldiers were split into three equal groups there was one soldier left over, when split into five equal groups there were two left over, and when split into seven equal groups there were four left over. What is the minimum number of soldiers that makes this possible? ∎

7.3. Problem. *The Newspaper Problem.* A math professor cashes a check for x dollars and y cents, but the teller inadvertently pays y dollars and x cents instead. After the professor buys a newspaper for 50 cents, the remaining money is twice the original value of the check. What was the value of the check? How does the solution change if the cost of the newspaper changes? ∎

7.4. Problem. *Primality Testing.* Is is possible to prove that a number is not prime without knowing any of its factors? ∎

RELATIONS

Comparison of objects is fundamental in mathematics. For example, we can compare two real numbers by asking whether $x \leq y$, we can compare two sets by asking whether $A \subseteq B$, and we can compare points in the plane by asking which is closer to the origin.

Given two objects s and t, not necessarily of the same type, we may ask whether they satisfy a given relationship. Let S denote the set of objects of the first type, and let T denote the set of objects of the second type. Some of the ordered pairs (s, t) may satisfy the relationship, and some may not. The next definition makes this notion precise.

7.5. Definition. When S and T are sets, a **relation** between S and T is a subset of the product $S \times T$. A **relation on** S is a subset of $S \times S$.

We usually define a relation R by stating a condition for pairs; the relation is the set of ordered pairs satisfying the condition.

7.6. Example. Let S be the set of students and T the set of teachers in a school. We define a relation R between S and T by letting R be the set of ordered pairs (x, y) in $S \times T$ such that x has taken a class from y. Each element of S or T may belong to many ordered pairs satisfying the relation. ∎

7.7. Example. If $f: \mathbb{R} \to \mathbb{R}$, then the graph of f is a relation on $\mathbb{R}$. It is the set of ordered pairs $\{(x, y) \in \mathbb{R}^2: y = f(x)\}$; each element of $\mathbb{R}$ is the first coordinate in exactly one such pair.

The conditions "$|x| = |y|$" and "$x^2 + y^2 \leq 1$" also define relations on $\mathbb{R}$. These relations are not the graphs of functions. ∎

7.8. Example. *Parity.* The condition "have the same parity" defines a relation on the set $\mathbb{Z}$. If x, y are both even or both odd, then (x, y) satisfies this relation; otherwise it does not. ∎

The parity relation satisfies several properties we now define. Together, they yield an important type of relation.

7.9. Definition. An **equivalence relation** on a set S is a relation R on S such that for all choices of distinct $x, y, z \in S$,
a) $(x, x) \in R$ (**reflexive property**).
b) $(x, y) \in R$ implies $(y, x) \in R$ (**symmetric property**).
c) $(x, y) \in R$ and $(y, z) \in R$ imply $(x, z) \in R$ (**transitive property**).

7.10. Example. For every set S, the **equality relation** $R = \{(x, x): x \in S\}$ is an equivalence relation on S. Echoing the notation for equality, we often write $x \sim y$ instead of $(x, y) \in R$ when R is an equivalence relation.

Let S be the set of students at a college. The condition "x and y have been in a class together" generally does not define an equivalence relation on S. It defines a relation that is reflexive and symmetric, but it need not be transitive. When x has been in a class with y, and y has been in a class with z, it does not follow that x has been in a class with z.

On the other hand, the condition "x and y were born in the same year" does define an equivalence relation on S. All three properties hold, because each person is born in only one year. ∎

7.11. Example. *Order relations.* The **divisibility relation** defined by $R = \{(m, n) \in \mathbb{N}^2 : m|n\}$ is not an equivalence relation. It is reflexive and transitive, but it is not symmetric. Indeed, it is **antisymmetric**: $(x, y) \in R$ and $(y, x) \in R$ together imply $x = y$. A relation that is reflexive, anti-symmetric, and transitive is an **order relation**.

Another example of an order relation is the **inclusion relation** on the set S of subsets of a set X. For all $A, B, C \in S$, we have $A \subseteq A$, $(A \subseteq B$ and $B \subseteq A)$ implies $A = B$, and $(A \subseteq B$ and $B \subseteq C)$ implies $A \subseteq C$. ∎

7.12. Example. Given a function $f : \mathbb{R}^2 \to \mathbb{R}$, we define a relation on $\mathbb{R}^2$ by putting $p \sim q$ when $f(p) = f(q)$. This relation is an equivalence relation; two points satisfy the relation if and only if they belong to the same level set of f. ∎

A topographic map of a region on the surface of the earth illustrates this. Let $f(p)$ be the height of the point p above sea level. Points in the same level set of f have the same height; points in different level sets have different heights. A hiker walking in a single level set does no work against gravity. The level sets partition the plane into subsets where an important measurement is constant. This leads us to a general definition.

7.13. Definition. Given an equivalence relation on S, the set of elements equivalent to $x \in S$ is the **equivalence class** containing x.

The equivalence classes of an equivalence relation on S form a parti-tion of S; elements x and y belong to the same class if and only if (x, y) satisfies the relation. The converse assertion also holds. If $A_1, \ldots, A_k$ is a partition of S, then the condition "x and y are in the same set in the partition" defines an equivalence relation on S (Exercise 12).

7.14. Example. *Cycles in a permutation.* Let f be a permutation of a fi-nite set A. Iterating f allows us to group the elements of A into "cycles" under f (Example 5.39). These cycles are the equivalence classes of a nat-ural equivalence relation on A; we put $(x, y) \in R$ when y can be obtained from x be repeatedly applying f. ∎

CONGRUENCE

In this chapter, we focus on an equivalence relation associated with divisibility, called "congruence". The notions of congruence and modular arithmetic, introduced by Karl Friedrich Gauss (1777–1855), are so fundamental that we have special terminology and notation for them.

7.15. Definition. *Congruence.* Given a natural number n, the integers x and y are **congruent modulo** n if $x - y$ is divisible by n. We write this as $x \equiv y \bmod n$. The number n is the **modulus**.

7.16. Theorem. For every $n \in \mathbb{N}$, congruence modulo n is an equivalence relation on $\mathbb{Z}$.

Proof: Reflexive property: $x - x$ equals 0, which is divisible by n.

Symmetric property: If $x \equiv y \bmod n$, then by definition $n | (x - y)$. Since $y - x = -(x - y)$, and since n divides $-m$ if and only if n divides m, we also have $n | (y - x)$, and hence $y \equiv x \bmod n$.

Transitive property: If $n | (x - y)$ and $n | (y - z)$, then integers a, b exist such that $x - y = an$ and $y - z = bn$. Adding these equations yields $x - z = an + bn = (a + b)n$, so $n | (x - z)$. Thus the relation is transitive. ∎

7.17. Definition. The equivalence classes of the relation "congruence modulo n" on $\mathbb{Z}$ are the **remainder classes** or **congruence classes** modulo n. The set of congruence classes is written as $\mathbb{Z}_n$ or $\mathbb{Z}/n\mathbb{Z}$.

7.18. Remark. *Remainder classes.* We show that there are n remainder classes modulo n; for $0 \leq r < n$, the rth class in $\mathbb{Z}_n$ is $\{kn + r : k \in \mathbb{Z}\}$. When $n = 10$, the last decimal digit determines the remainder class.

By definition, $a \equiv b \bmod n$ if and only if $a - b$ is divisible by n. The Division Algorithm yields unique integers k, r such that $a = kn + r$ and $0 \leq r < n$; here r is the remainder upon division by n. If $a = kn + r$ and $b = ln + s$ with remainders $r, s \in \{0, \ldots, n - 1\}$, then $n | (a - b)$ if and only if $r - s = 0$. Hence $a \equiv b \bmod n$ if and only if a and b have the same remainder "modulo n". This justifies our description of the classes. ∎

The next lemma is the property of the congruence relation that allows us to define arithmetic with congruence classes.

7.19. Lemma. If $a \equiv r \bmod n$ and $b \equiv s \bmod n$, then $a + b \equiv r + s \bmod n$ and $a \cdot b \equiv r \cdot s \bmod n$.

Proof: Since $a \equiv r \bmod n$ and $b \equiv s \bmod n$, there exist integers k, l such that $a = kn + r$ and $b = ln + s$. Adding these equations yields $a + b = (k + l)n + (r + s)$, and thus $a + b \equiv r + s \bmod n$. Multiplying the equations yields $a \cdot b = kln^2 + (ks + lr)n + r \cdot s$, and thus $a \cdot b \equiv r \cdot s \bmod n$. ∎

7.20. Example. Since $79 \equiv 4 \bmod 5$ and $23 \equiv 3 \bmod 5$, we can multiply the congruence classes to obtain $79 \cdot 23 \equiv 12 \bmod 5$. Since $12 \equiv 2 \bmod 5$, we can further reduce this to $79 \cdot 23 \equiv 2 \bmod 5$. ∎

Lemma 7.19 enables us to define arithmetic on congruence classes. The result of adding or multiplying two congruence classes will itself be a congruence class. When the modulus n is given, we use $\bar{a}$ to denote the congruence class containing a.

7.21. Definition. A **binary operation** on a set S is a function from $S \times S$ to S. On $\mathbb{Z}_n$, **addition** is the binary operation defined by letting the sum of the congruence classes $\bar{a}$ and $\bar{b}$ be the class containing the integer $a + b$. On $\mathbb{Z}_n$, **multiplication** is the binary operation defined by letting the product of $\bar{a}$ and $\bar{b}$ be the class containing the integer $a \cdot b$. In notation, $\bar{a} + \bar{b} = \overline{a + b}$ and $\bar{a} \cdot \bar{b} = \overline{a \cdot b}$.

In the formulas in Definition 7.21, the operations between classes are the operations being defined; the operations on the right are previously known operations on integers. Lemma 7.19 guarantees that these operations on $\mathbb{Z}_n$ are well-defined functions; when we choose integers a_1, a_2 in the class $\bar{a}$ and integers b_1, b_2 in the class $\bar{b}$, the numbers $a_1 + b_1$ and $a_2 + b_2$ lie in the same congruence class, as do $a_1 \cdot b_1$ and $a_2 \cdot b_2$. We can choose any elements from the classes to perform the computation; the result is always the same congruence class.

For this reason, we use the notation $\bar{a}$ only when we need to emphasize the congruence class as an object. The expression $6 + 6 \equiv 5 \bmod 7$ is both a statement about integers and a statement about congruence classes. The validity of addition and multiplication with congruence classes is one reason we use the "equality-like" notation ($\equiv$) for the congruence relation.

7.22. Example. *Binary arithmetic.* We have already used arithmetic modulo 2. The congruence classes are "even" (0 mod 2) and "odd" (1 mod 2). The addition table modulo 2 states that the sum of two integers with the same parity is even, and the sum of two integers with opposite parity is odd. The multiplication table says that the product of two integers is odd if and only if they are both odd. ∎

+	0	1		*	0	1
0	0	1		0	0	0
1	1	0		1	0	1

7.23. Example. *Clock arithmetic.* Minutes on a clock behave like arithmetic mod 60. If a 90-minute movie starts at "quarter-past", then it will end at "quarter-'til". This is independent of the hour, just as the sum of two odd numbers is even no matter which odd numbers we use. ∎

7.24. Remark. *Modular computation.* Lemma 7.19 holds for all integers r and s; they need not lie between 0 and $n - 1$. Thus when performing arithmetic operations in which we care only about the congruence class of the result modulo n, *we may at any time replace a number by a more convenient representative of its congruence class.* We may write the computation of Example 7.20 as

$$79 \cdot 23 \equiv 4 \cdot 3 \equiv 12 \equiv 2 \bmod 5.$$

Here the " mod 5" indicates that all four expressions belong to the same congruence class modulo 5.   ∎

7.25. Example. *"Casting out nines": An integer is divisible by 9 if and only if the sum of its decimal digits is divisible by 9.* Since 10 is congruent to 1 modulo 9, every nonnegative power of 10 is also congruent to 1 modulo 9. Therefore $\sum_{n \geq 0} a_n 10^n \equiv \sum_{n \geq 0} a_n 1^n \equiv \sum_{n \geq 0} a_n \bmod 9$. This is called "casting out nines" and was used as a check by clerks adding columns of figures before adding machines were invented.

To check the computation of $\sum c_i$, let s be the sum of the digits of the result, and for each i let b_i be the sum of the digits of c_i. If the addition is correct, then $\sum b_i$ must be congruent to s modulo 9. For example, suppose we add the numbers 123, 456, 789 and obtain 1268. The sums of the digits in the three numbers are 6, 15, 24, respectively, which sum to 45. The sum of the digits of 1268 is 17. Since 17 is not congruent to 45 modulo 9, we must have made a mistake. The correct sum is 1368.   ∎

7.26. Solution. *Divisibility by 3.* In binary representation, we write $n = \sum_0^m a_j 2^j$, where each a_j is 0 or 1. Thus n is divisible by 3 if and only if

$$0 \equiv \sum_0^m a_j 2^j \equiv \sum_0^m a_j (-1)^j \bmod 3.$$

Since each a_j is 0 or 1, n is divisible by three if and only if the number of 1s in even-indexed positions differs from the number of 1s in odd-indexed positions by a multiple of 3.

For example, suppose $n = 1010101111$ in binary. Since there are five 1s in even-indexed position and two 1s in odd-indexed positions, n is divisible by three. In base 10, $n = 687 = 3 \cdot 229$.   ∎

In arithmetic modulo n, the class $\overline{0}$ is an additive identity, and the class containing $-x$ is an additive inverse for the class containing x. The class $\overline{1}$ is a multiplicative identity, but multiplicative inverses do not always exist. The next lemma allows us to find multiplicative inverses when they exist. It also leads to solutions of Problems 7.2–7.4.

7.27. Lemma. If a and n are relatively prime integers, then multiplication by a defines a bijection from $\mathbb{Z}_n - \{0\}$ to itself; equivalently, multiplication by a permutes the nonzero congruence classes.

Proof: Since a and n are relatively prime, $0, a, 2a, \ldots, (n-1)a$ all have different remainders modulo n (Theorem 6.21). Since 0 has remainder 0, the others are nonzero. Since they are distinct, the list defines an injection from $\mathbb{Z}_n - \{0\}$ to itself. Since the set is finite, this injection is a bijection. ∎

7.28. Corollary. If a and n are relatively prime integers, then solutions to $ax \equiv 1 \bmod n$ exist and lie in a single congruence class. In the language of $\mathbb{Z}_n$, the class $\overline{x}$ is the **multiplicative inverse** of $\overline{a}$. ∎

APPLICATIONS

We use Corollary 7.28 first to present an *ad hoc* solution of Problem 7.2 and then to prove a theorem that provides another algorithm.

7.29. Solution. *Chinese Remainder Problem.* We seek a number x that is congruent to 1 mod 3, to 2 mod 5, and to 4 mod 7. Thus $x = 3n + 1$ for some integer n. Incorporating the second requirement, we have $3n + 1 \equiv 2 \bmod 5$, which becomes $3n \equiv 1 \bmod 5$. Since 3 and 5 are relatively prime, there is a unique congruence class modulo 5 as a solution; we have $n \equiv 2 \bmod 5$. Writing $n = 5m + 2$ yields $x = 3(5m + 2) + 1 = 15m + 7$.

Incorporating the third requirement, we have $15\dot{m} + 7 \equiv 4 \bmod 7$. Since $15 \equiv 1 \bmod 7$ and $7 \equiv 0 \bmod 7$, we obtain $m \equiv 4 \bmod 7$, so that $m = 7k + 4$ for some $k \in \mathbb{Z}$. Hence $x = 15(7k + 4) + 7 = 105k + 67$. The smallest positive number (of soldiers) is 67. ∎

This method can be combined with induction on the number of congruences to prove the next theorem. We present a short proof that yields another algorithm and avoids induction.

7.30. Theorem. (Chinese Remainder Theorem) If $\{n_i\}$ is a set of r natural numbers that are pairwise relatively prime, and $\{a_i\}$ are any r integers, then the system of congruences $x \equiv a_i \bmod n_i$ has a unique solution modulo $N = \prod n_i$.

Proof: If x and x' are solutions, then they must be congruent modulo N. To see this, note that $x \equiv x' \equiv a_i \bmod n_i$ for each i, so $n_i | (x - x')$. Since the n_i's are relatively prime, this yields $N | (x - x')$, by Corollary 6.10.

Now we construct a solution. For each i, let $N_i = N/n_i$. Since n_i is relatively prime to the other moduli, $\gcd(N_i, n_i) = 1$. By Corollary 7.28, there is exactly one congruence class modulo n_i, call it $\overline{y}_i$, such that $N_i y_i \equiv 1 \bmod n_i$. Set $x = \sum_{j=1}^{r} a_j N_j y_j$. When we consider this equation modulo n_i, the terms with $j \neq i$ are congruent to 0, since $n_i | N_j$ for $j \neq i$. Only the term with $j = i$ remains, and from $N_i y_i \equiv 1 \bmod n_i$ we obtain $x \equiv a_i N_i y_i \equiv a_i \bmod n_i$. Hence x satisfies all required congruences. ∎

7.31. Example. Suppose we seek x such that $x \equiv 2 \bmod 5$, $x \equiv 4 \bmod 7$, and $x \equiv 3 \bmod 9$. This yields $N = 315$ and $N_1, N_2, N_3 = 63, 45, 35$.

i	a_i	n_i	N_i	$N_i \bmod n_i$	y_i
1	2	5	63	3	2
2	4	7	45	3	5
3	3	9	35	-1	-1

By the Chinese Remainder Theorem, we obtain a solution by setting x to be $2 \cdot 63 \cdot 2 + 4 \cdot 45 \cdot 5 + 3 \cdot 35 \cdot (-1) = 1047$. (Hand computation using this procedure should check the solution at this stage!) All numbers congruent to 1047 modulo 315 are solutions. The one with smallest absolute value is $1047 - 3 \cdot 315 = 102$.   ∎

When the moduli are not pairwise relatively prime, there may be no solution, or it may be possible to modify the problem to use the Chinese Remainder Theorem anyway (Exercise 35).

The solution of Problem 7.3 uses a different equivalence relation.

7.32. Solution. *The Newspaper Problem.* In Problem 7.3, the check for x dollars and y cents is paid instead as y dollars and x cents. Note that x and y are between 0 and 99. After subtracting 50 cents for the newspaper, the remaining money is twice the original value of the check.

We can encode this information as $100y + x - 50 = 2(100x + y)$, which simplifies to $98y - 199x = 50$. This is a diophantine equation, which we can solve by the method of Example 6.19. After some calculation, we obtain $(x, y) = (-1650 + 98j, -3350 + 199j)$ for $j \in \mathbb{Z}$. To enforce $0 < x < 100$, we take $j = 17$ and obtain $(x, y) = (16, 33)$. This answer checks, since $\$33.16 - \$0.50 = 2 \times \$16.33$.

A natural equivalence relation leads to a uniform approach for all possible prices of the newspaper. Define an equivalence relation on integer pairs (r, s) representing r dollars plus s cents by saying that pairs are equivalent if they represent the same amount of money. Thus (a, b) and (a', b') are equivalent if $(a, b) = (a' + n, b' - 100n)$ for some $n \in \mathbb{Z}$.

The problem states that $(y, x - 50)$ and $(2x, 2y)$ are equivalent, each representing the amount of money remaining after buying the newspaper. After setting $y = 2x + n$ and $x - 50 = 2y - 100n$, we eliminate y to obtain $3x + 50 = 98n$. Since $x \geq 0$, we have $n \geq 0$. Since x is an integer, $98n - 50$ must be divisible by 3; this holds if we choose $n = 1$. With $n = 1$ we obtain $x = 16$ and $y = 2x + 1 = 33$, as before.

When the newspaper costs k cents, we obtain $3x + k = 98n$. For $k = 75$, this requires n to be positive and divisible by 3, but $n \geq 3$ implies $x \geq (98 \cdot 3 - 75)/3 = 73$. Since $y = 2x + n$, this yields $y \geq 149$, which violates the conditions of the problem. Hence there is no solution when $k = 75$. Each choice of n up to 99 yields solutions for various k. For example, when $n = 99$, we have the solution \$0.99 for the original check, if the newspaper costs \$97.02. See also Exercise 37. ∎

FERMAT'S LITTLE THEOREM

Let p be a prime number, and let a be a nonzero integer not divisible by p. The function $f_a \colon \mathbb{Z}_p \to \mathbb{Z}_p$ defined by $f_a(x) = ax$ is a bijection, by Lemma 7.27. In the functional digraph for $f_5 \colon \mathbb{Z}_{13} \to \mathbb{Z}_{13}$ shown below, all the cycles have the same length except the cycle consisting of 0 alone.

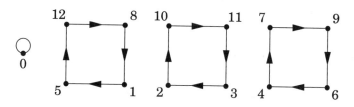

This observation holds in general and is the key to one of many proofs of Fermat's Little Theorem. Proved by Pierre de Fermat (1601–1665), the theorem states that $a^{p-1} \equiv 1 \bmod p$ when p is prime and a is not a multiple of p. Exercise 43 requests a proof using Lemma 7.27 and modular computation. Exercise 6.37 requests a proof using induction and binomial coefficients. Example 9.38 presents a proof using multinomial coefficients. We now present a proof due to Leonhard Euler (1707–1783).[†]

7.33. Definition. When some power of a is congruent to 1 modulo p, the **order** of a (in $\mathbb{Z}_p$) is the least k such that $a^k \equiv 1 \bmod p$.

7.34. Lemma. Let p be prime, and suppose that $a \not\equiv 0 \bmod p$. For $x \in \mathbb{Z}_p$, let $S_x = \{x, xa, xa^2, \ldots\}$. There is a positive integer k such that, for all $x \neq 0$, the set S_x consists of exactly k elements.

Proof: Since $\mathbb{Z}_p$ is finite, the positive powers of a cannot be distinct modulo p, and eventually some remainder repeats. If $a^m \equiv a^n \bmod p$ with $m > n$, then $a^{m-n} \equiv 1 \bmod p$. Thus the order of a is well-defined; call it k. Now $1, a, a^2, \ldots, a^{k-1}$ are distinct, and thereafter the list repeats. We have $|S_1| = k$.

By Lemma 7.27, multiplication by x permutes the elements of $\mathbb{Z}_p - \{0\}$. Thus $x, xa, xa^2, \ldots, xa^{k-1}$ are distinct. Since $xa^k \equiv x \bmod p$, thereafter the list repeats. We have proved that $|S_x| = k$ for all x. ∎

[†]For further reading about this theorem, see Andre Weil, *Number Theory: An Approach through History*, Birkhäuser (Boston, 1984).

The key idea is that the sets defined in Lemma 7.34 are the equivalence classes of an equivalence relation on $\mathbb{Z}_p$. Since multiplication by a defines a permutation of $\mathbb{Z}_p$ and these sets are the cycles of the permutation, Example 7.14 provides one way to establish this claim. Here we provide an algebraic proof.

7.35. Lemma. If R is the relation on $\mathbb{Z}_p$ defined by $(x, y) \in R$ if and only if $y \equiv xa^j \bmod p$ for some nonnegative integer j, then R is an equivalence relation.

Proof: Since $x \equiv xa^0 \bmod p$, R is reflexive. Let k be the order of a in $\mathbb{Z}_p$. When $y \equiv xa^j \bmod p$, we may assume that $0 \leq j \leq k - 1$. If $y \equiv xa^j \bmod p$, then $x \equiv ya^{k-j} \bmod p$, so R is symmetric. If $y \equiv xa^r \bmod p$ and $z \equiv xa^s \bmod p$, then $z \equiv xa^{r+s} \bmod p$, so R is transitive. ∎

7.36. Theorem. (Fermat's Little Theorem) If p is prime and a is not a multiple of p, then $a^{p-1} \equiv 1 \bmod p$.

Proof: Let k be the order of a in $\mathbb{Z}_p$. We prove that $p - 1$ is a multiple of k, and thus

$$a^{p-1} = a^{mk} = (a^k)^m \equiv 1^m \equiv 1 \bmod p.$$

In the equivalence relation R defined in Lemma 7.35, the equivalence class containing 0 is {0}. The remaining classes partition $\mathbb{Z}_p - \{0\}$. The equivalence class S_x containing x consists of all classes obtainable from x by multiplying by a power of a. By Lemma 7.34, S_x has size k. Thus R partitions $\mathbb{Z}_p - \{0\}$ into sets of size k, and $p - 1$ is a multiple of k. ∎

7.37. Example. We have seen that $5^4 \equiv 1 \bmod 13$. The smallest powers of 4, 3, 2 congruent to 1 modulo 13 are 4^6, 3^3, and 2^{12}. In each case, we obtain $a^{12} \equiv 1 \bmod 13$. ∎

7.38. Example. *Computation with Fermat's Little Theorem.* We can rapidly compute remainders for large numbers involving powers. For example, computation modulo the prime 31 yields

$$11^{902} = 11^{30 \cdot 30 + 2} = (11^{30})^{30} \cdot 11^2 \equiv 1^{30} \cdot 121 \equiv -3 \equiv 28 \bmod 31. \quad \blacksquare$$

7.39. Corollary. If p is prime and $a \in \mathbb{Z}$, then $a^p \equiv a \bmod p$. ∎

The contrapositive of this corollary of Fermat's Little Theorem enables us to solve Problem 7.4 most of the time.

7.40. Solution. *Primality testing.* Let a and p be integers. The contrapositive of Corollary 7.39 states that if a^p is not congruent to a modulo p, then p is not a prime number. Thus, finding such a number a proves that p is not prime without knowing any factors of p.

For example, suppose we want to test whether 341 is prime. If we choose $a = 7$, we have an easy computation. Because $7^3 = 343 \equiv 2$ mod 341 and $2^{10} = 1024 \equiv 1$ mod 341, we can compute

$$7^{341} = 7^{3 \cdot 113 + 2} \equiv 2^{113} 7^2 \equiv 2^{110+3} 7^2 \equiv 8 \cdot 49 \equiv 392 \equiv 51 \text{ mod } 341.$$

Since $51 \not\equiv 7$ mod 341, we conclude that 341 cannot be prime.

We can apply this test even without such a clever choice for a. Although it still takes some work, computing the congruence class of a^{341} never requires 341 multiplications. Using repeated squaring, we can compute the congruence classes of the numbers $\{a^{2^k}\}$. The binary representation of 341 tells us which of these to multiply together to compute a^{341}. For example, suppose $a = 3$. Repeated squaring yields

$$3^2 = 9 \qquad\qquad 3^8 = 81^2 = 6561 \equiv 82 \text{ mod } 341$$
$$3^4 = 81 \qquad\qquad 3^{16} \equiv 82^2 \equiv 245 \text{ mod } 341$$

and so on. The binary representation of 341 is 101010101. If we multiply together the congruence classes of 3^n for $n = 1, 4, 16, 64, 256$, we have the congruence class of 3^{341}. ∎

Modular multiplications are fast on computers. If n is not prime, then computing the congruence class of a^n for a few random choices of a is likely to prove that n is not prime, but this does not always work. There are some numbers such that $a^n \equiv a$ mod n for every a even though n is not prime. One such number is 561, which has prime factorization $3 \cdot 11 \cdot 17$. Such numbers are **Carmichael numbers**. We state a characterization of these numbers, without proof. A natural number n is a Carmichael number if and only if the following two properties hold: n has no repeated prime factors, and $(p - 1)|(n - 1)$ whenever p is a prime factor of n. For example, 2, 10, 16 all divide 560.

CONGRUENCE AND GROUPS (optional)

In the remainder of this chapter, we present a bit more formal discussion of the arithmetic properties of $\mathbb{Z}_n$. We have proved that addition and multiplication modulo n are well-defined. This enables us to specify addition and multiplication tables for the elements of $\mathbb{Z}_n$. Below we illustrate these for $\mathbb{Z}_6$ and $\mathbb{Z}_7$.

+	0 1 2 3 4 5
0	0 1 2 3 4 5
1	1 2 3 4 5 0
2	2 3 4 5 0 1
3	3 4 5 0 1 2
4	4 5 0 1 2 3
5	5 0 1 2 3 4

*	0 1 2 3 4 5
0	0 0 0 0 0 0
1	0 1 2 3 4 5
2	0 2 4 0 2 4
3	0 3 0 3 0 3
4	0 4 2 0 4 2
5	0 5 4 3 2 1

+	0 1 2 3 4 5 6
0	0 1 2 3 4 5 6
1	1 2 3 4 5 6 0
2	2 3 4 5 6 0 1
3	3 4 5 6 0 1 2
4	4 5 6 0 1 2 3
5	5 6 0 1 2 3 4
6	6 0 1 2 3 4 5

*	0 1 2 3 4 5 6
0	0 0 0 0 0 0 0
1	0 1 2 3 4 5 6
2	0 2 4 6 1 3 5
3	0 3 6 2 5 1 4
4	0 4 1 5 2 6 3
5	0 5 3 1 6 4 2
6	0 6 5 4 3 2 1

Since adding a multiple of n does not change a congruence class modulo n, the class 0 is an **identity element** for the addition modulo n. Furthermore, $(n - i) + i \equiv n \equiv 0 \bmod n$, so the class $n - i$ acts as an **additive inverse** of i. As verified earlier, the sum of two congruence classes is a congruence class. Also, $(a + b) + c \equiv a + (b + c) \bmod n$. These properties make $\mathbb{Z}_p$ a "group" under addition mod p.

7.41. Definition. A **group** is a set G together with a binary operation $\circ$ on $G^\dagger$ satisfying the following properties:
1) There is an element $e \in G$ such that for every $x \in G$, $x \circ e = x = e \circ x$ (e is the **identity element** of the group).
2) For every $x \in G$, there is an element $y \in G$ such that $x \circ y = e = y \circ x$ (y is the **inverse** of x).
3) For every $x, y, z \in G$, $(x \circ y) \circ z = x \circ (y \circ z)$ (**associative property**).

A fundamental example of a group is the set of permutations of $[n]$, where the binary operation is composition (Exercise 49).

The elements of a field (Definition 1.39) form a group under addition. The nonzero elements of a field form a group under multiplication. Whenever the binary operation in a group is written as $+$, we express the inverse of x as $-x$, write $y - x$ for $y + (-x)$, and name the identity element 0. We have done this for $\mathbb{Z}_n$ under addition mod n.

What about multiplication in $\mathbb{Z}_n - \{0\}$? We know that 1 is a multiplicative identity, but we soon run into trouble. When n is the product of integers a, b larger than 1, we have $a \cdot b \equiv 0 \bmod n$, so discarding 0 does not permit the remaining elements to form a group under multiplication (see the table for $\mathbb{Z}_6$).

When p is prime, $p | ab$ implies $p | a$ or $p | b$. Hence ab is not congruent to $0 \bmod p$ when a and b are not congruent to $0 \bmod p$. Thus multiplication is a binary operation on $\mathbb{Z}_p - \{0\}$. The associative property follows from the associative property of integer multiplication, since we can select any integers from these congruence classes to do the computation. Finally, we verify below that multiplicative inverses exist in $\mathbb{Z}_p - \{0\}$ whenever p is prime. The table before Definition 7.41 exhibits multiplicative inverses in $\mathbb{Z}_7 - \{0\}$; we have $6 \cdot 6 \equiv 1$, $5 \cdot 3 \equiv 1$, $4 \cdot 2 \equiv 1$, and $1 \cdot 1 \equiv 1$.

7.42. Corollary. When p is prime, $\mathbb{Z}_p - \{0\}$ is a group under multiplication.

Proof: We have verified all the needed properties except the existence of inverses. Consider $a \not\equiv 0$. Since a and p are relatively prime, Corollary 7.28 implies that there is some nonzero class $\bar{b}$ such that $\bar{a}\bar{b} \equiv \bar{1}$. The class $\bar{b}$ is the desired $(\bar{a})^{-1}$. Note also that $\bar{b}\bar{a} \equiv \bar{1}$, since multiplication modulo p is commutative. ∎

---

†The definition of a binary operation includes the property that $x \circ y \in G$ for all $x, y \in G$; this is the property of **closure** under $\circ$.

In our discussion, we have completed all the details of proving that $\mathbb{Z}_p$ is a field if (and only if) p is prime.

When can a number be its own multiplicative inverse?

7.43. Lemma. If p is prime and $a \in \mathbb{N}$, then $a^2 \equiv 1 \bmod p$ if and only if $a \equiv 1 \bmod p$ or $a \equiv -1 \bmod p$.

Proof: If $a^2 \equiv 1$, then p divides $a^2 - 1$, which equals $(a+1)(a-1)$. When a prime divides a product, it must divide one of the factors (Proposition 6.7). Hence p divides $a+1$ or $a-1$, yielding $a \equiv -1 \bmod p$ or $a \equiv 1 \bmod p$. Conversely, if a is in one of these classes, then p divides $(a+1)(a-1)$, and a^2 is in the same congruence class as 1. ∎

7.44. Theorem. (Wilson's Theorem) $(p-1)! \equiv -1 \bmod p$ for p prime.

Proof: This holds for $p = 2$ because $1 \equiv -1 \bmod 2$. Consider $p > 2$. For each $1 \le i \le p-1$, there is exactly one i' in $[p-1]$ such that $ii' = 1$, by Lemma 7.27. By Lemma 7.43, the numbers from 2 through $p-2$ form disjoint pairs of inverses. Hence $\prod_{i=2}^{p-2} i \equiv 1 \bmod p$, and $\prod_{i=1}^{p-1} i \equiv p-1 \equiv -1 \bmod p$. ∎

Wilson's Theorem was only conjectured by John Wilson (1741–1793); Joseph Louis Lagrange (1736–1813) gave the first proof in 1770.

EXERCISES

7.1. (−) Let a, b, x, n be positive integers. The following statement is not always true: "If $ax \equiv bx \bmod n$, then $a \equiv b \bmod n$." Provide a counterexample, and add a hypothesis on x and n to make the statement always true.

7.2. Does the last digit of an integer (written in base 10) determine whether the integer is divisible by 5? By 2? By 3?

7.3. Suppose that a person sleeps exactly 8 hours each day and goes to sleep at midnight on April 1. She always goes to sleep exactly 17 hours after she wakes up. Does she rise at each hour of the day during the month of April? What happens if instead she always goes to sleep exactly 18 hours after she wakes up? Explain.

7.4. (−) Prove that if two natural numbers have the same number of copies of each digit in their decimal representations, then they differ by a multiple of 9.

7.5. (−) What is the congruence class of 10^n modulo 11? Use this to determine the remainder when 654321 is divided by 11.

7.6. (!) Determine the last digit (the ones digit) in the base 8 expansion of 9^{1000}, 10^{1000}, and 11^{1000}.

7.7. (−) When the remainders modulo m of the numbers $1^2, 2^2, \ldots, (m-1)^2$ are listed in order, the list is symmetric around the center. Why?

7.8. (−) Let k be an odd number. Prove that $k^2 - 1$ is divisible by 8.

7.9. (−) Use Fermat's Little Theorem to find a number between 0 and 12 that is congruent to 2^{100} modulo 13.

•   •   •   •   •

7.10. Define a relation R on the set of humans on this planet by putting $(x, y) \in R$ if x and y are citizens of the same country. Is R an equivalence relation?

7.11. For each example below, determine whether the given relation R is an equivalence relation on the given set S.
   a) $S = \mathbb{N} - \{1\}$; $(x, y) \in R$ if and only if $\gcd(x, y) > 1$.
   b) $S = \mathbb{R}$; $(x, y) \in R$ if and only if there exists $n \in \mathbb{Z}$ such that $x = 2^n y$.

7.12. Let S be the union of disjoint sets $A_1, \ldots, A_k$. Let R be the relation consisting of pairs $(x, y) \in S \times S$ such that x, y belong to the same member of $\{A_1, \ldots, A_k\}$. Prove that R is an equivalence relation on S.

7.13. Let $\mathbf{P}$ be the power set of $[2n]$. Let R be a relation defined on $\mathbf{P}$ by $(A, B) \in R$ if and only if $A \cap [n] = B \cap [n]$. Determine whether R is an equivalence relation. What happens when $[n]$ and $[2n]$ are generalized to sets C and S such that $C \subseteq S$?

7.14. Given $f: \mathbb{R} \to \mathbb{R}$, let $O(f)$ be the set of functions g for which there exist positive constants $c, a \in \mathbb{R}$ such that $|g(x)| \le c \, |f(x)|$ for $x > a$ (see Exercise 2.23). Define a relation R on the set S of functions mapping $\mathbb{R}$ to $\mathbb{R}$ by putting $(g, h) \in R$ if and only if $g - h \in O(f)$. Prove that R is an equivalence relation on S.

7.15. Find the flaw in the following argument that the symmetric and transitive properties imply the reflexive property for a relation R on S: "Consider $x \in S$. If $(x, y) \in R$, then the symmetric property implies that $(y, x) \in R$. Now the transitive property applied to (x, y) and (y, x) implies that $(x, x) \in R$."

7.16. (!) Prove that every year (including leap years) has at least one Friday the 13th. What is the maximum number of Friday the 13ths in a year? (Hint: Use modular arithmetic to simplify the analysis.)

7.17. Give three proofs that $n^3 + 5n$ is divisible by 6 for every $n \in \mathbb{N}$.
   a) Use induction.
   b) Use modular arithmetic.
   c) Use an expression for $n^3 + 5n$ in terms of binomial coefficients.

7.18. Let p be an odd prime. Determine all solutions to $2n^2 + n \equiv 0 \bmod p$.

7.19. (!) For $m, n, p \in \mathbb{Z}$, suppose that 5 divides $m^2 + n^2 + p^2$. Prove that 5 divides at least one of $\{m, n, p\}$.

7.20. (−) Use modular arithmetic to prove that $k^n - 1$ is divisible by $k - 1$ for all $n, k \in \mathbb{N}$ with $k \ge 2$.

7.21. (!) Use modular arithmetic to prove that N is divisible by $k!$ if N is the product of k consecutive natural numbers.

7.22. (+) Prove that there are infinitely many primes of the form $4n + 3$ and infinitely many primes of the form $6n + 5$, where $n \in \mathbb{N}$. (Hint: Show first that the divisors of a number congruent to $-1 \bmod 4$ cannot all be congruent to 1 mod

4, and the divisors of a number congruent to -1 mod 6 cannot all be congruent to 1 or 3 mod 6. (Comment: Dirichlet proved more generally that if a and b are relatively prime, then there are infinitely many primes of the form $an + b$, but this is beyond the techniques we have available.)

7.23. (!) The base 10 representation of an integer is **palindromic** if the digits read the same when written forward or backward. Prove that every palindromic integer with an even number of digits is divisible by 11. More generally, prove that every integer whose base k representation is palindromic and has even length is divisible by $k + 1$.

7.24. Define $f: \mathbb{Z}_n \to \mathbb{Z}_n$ by $f(x) = x^2$. For which $n \in \mathbb{N}$ is f injective?

7.25. ($-$) Prove that the first six powers of 10 belong to distinct congruence classes modulo 7. (Comment: Gauss asked whether the powers of 10 yield $n - 1$ distinct congruence classes modulo n for infinitely many n; this remains unanswered. The moduli 5 and 13 fail, even though they are primes.)

7.26. Let n be a natural number whose base 10 representation is a permutation of the six digits $\{1, 2, 3, 4, 5, 6\}$. Suppose that for $1 \le i \le 6$, the i-digit number formed by the leftmost i digits is divisible by i. The integer 123456 fails, for example, because 1234 is not divisible by 4. Determine all possible values of n.

7.27. (+) Let n be a natural number whose base 10 representation is a permutation of the digits 0 through 9. Suppose that for $1 \le i \le 10$, the i-digit number formed by the leftmost i digits is divisible by i. Determine all possible values of n. (Hint: The divisibility requirements impose constraints; for example, the tenth digit must be 0, and then the fifth digit must be 5. Division is not needed.)

7.28. (!) *Test for divisibility by 7.*

a) Let $a_k \cdots a_0$ be the base 10 representation of n. We can determine whether n is divisible by 7 by treating n as $\sum a_i 10^i$ and reducing the powers of 10 modulo 7; we have discussed this approach to divisibility by 9. Apply this to check whether 7 divides 535801.

b) Given a positive integer n, let $f(n)$ be the integer formed by subtracting twice the last base 10 digit of n from the number formed by the remaining digits of n. For example, if $n = 154$, then $f(n) = 15 - 8 = 7$. Prove that $7|n$ if and only if $7|f(n)$. Apply this to check whether 7 divides 535801. (Hint: To prove that $7|n$ if and only if $7|f(n)$, prove first that $7|n$ if and only if $7|[10f(n)]$.)

7.29. *Test for divisibility by n* (generalizing Exercise 7.28). Let n be a positive integer. Let $f(n)$ be the result of subtracting j times the last base 10 digit of n from the number formed by the remaining digits of n ($j = 2$ in Exercise 7.28). Prove that if s is not divisible by 2 or 5 and $10j \equiv -1$ mod s, then n is divisible by s if and only if $f(n)$ is divisible by s. Describe the resulting tests for divisibility by 17 and by 19, and illustrate how they work on 323, which equals $17 \cdot 19$.

7.30. (!) *Primes and threes.*

a) Prove that the sum of the digits in the base 10 expansion of a natural number n is a multiple of 3 if and only if n is a multiple of 3.

b) Prove that $6|x$ when $x + 1$ and $x - 1$ are prime, with one exception.

c) Suppose that $x + 1$ and $x - 1$ are prime. Form a new number by concatenating the digits of one with the digits of the other. Thus $\{11, 13\}$ can become 1113 or 1311. Prove that the resulting number is not prime, with one exception.

7.31. (!) We say that k is a square modulo n if $k \equiv j^2$ mod n for some j. Suppose that $n = m^2 + 1$ for some $m \in \mathbb{N}$. Prove that if k is a square modulo n, then $-k$ is also a square mod n.

7.32. Suppose that $n \in \mathbb{N}$, $a, b \in \mathbb{Z}$, and $d = \gcd(a, n)$. Consider arithmetic modulo n. Prove that there is no congruence class $\overline{x}$ that solves the congruence equation $\overline{ax} = \overline{b}$ unless d divides b, in which case there are d solutions.

7.33. (!) 1500 soldiers arrive in training camp. A few soldiers desert the camp. The drill sergeants divide the remaining soldiers into groups of five and discover that there is one left over. When they divide them into groups of seven, there are three left over, and when they divide them into groups of 11, there are again three left over. Determine the number of deserters.

7.34. Find all integers that are congruent to 1 mod 7, 3 mod 8, and 5 mod 9. Which solution has the smallest absolute value?

7.35. Suppose that $x \equiv 3$ mod 6, $x \equiv 4$ mod 7, and $x \equiv 5$ mod 8. Explain why the Chinese Remainder Theorem does not apply to compute x. Transform the problem to an equivalent problem where the Chinese Remainder Theorem can be used. Compute the smallest positive solution for x. Give a precise (and concise) reason why there is no smaller positive number that works.

7.36. Derive a description of all integers congruent to x mod a, y mod b, and z mod c, given that n is such an integer.

7.37. (+) Analyze the Newspaper Problem in full (Solution 7.32). In particular, for which prices of the newspaper does the problem have a solution?

7.38. We form a necklace by placing distinguishable beads (numbered 1 through n) on a circular string. Two necklaces are indistinguishable if one can be rotated or flipped to make it look like the other. Prove that indistinguishability is an equivalence relation. Count the equivalence classes with n beads. (The beads are distinguished by their labels, so there is no problem of periodicity.)

7.39. Let n be prime. We are given k types of feathers, each in unlimited supply. We wish to place one feather at each corner of an n-cornered hat. Each such arrangement can rotate; unlike necklaces, hats cannot be worn upside-down. Two arrangements of feathers are indistinguishable if one can be rotated to look like the other. Prove that indistinguishability is an equivalence relation. Count the equivalence classes of hats with feathers.

7.40. We have a stick partitioned into n equal segments. We have k colors of paint and must paint each segment. A list of colors on the segments is indistinguishable from its reverse, because when tossed in the air the stick can land either way. How many distinguishable colorings of the stick are there?

7.41. Define f and g from $\mathbb{Z}_n$ to $\mathbb{Z}_n$ by $f(x) \equiv (x + a)$ mod n and $g(x) \equiv ax$ mod n.
    a) Give a complete description of the functional digraph of f.
    b) Draw the functional digraph of g for the case $(n, a) = (19, 4)$ and the case $(n, a) = (20, 4)$. Describe a property of the digraph that is true whenever n is prime and false whenever n is not prime.

7.42. (−) For all $a \in \mathbb{Z}_{13} - \{0\}$, find the least k such that $a^k \equiv 1$ mod 13. Also list the partition of [12] into cycles under multiplication by a.

7.43. (!) By Theorem 6.21, $\{a, 2a, \ldots, (p-1)a\}$ have distinct remainders modulo p when a and p are relatively prime. Use this to give a short proof of Fermat's Little Theorem.

7.44. Fermat's Little Theorem implies that p divides $2^p - 2$ if p is prime. Fermat conjectured that the converse is also true, meaning that p divides $2^p - 2$ only if p is prime, but he was wrong. Euler provided the counterexample of $p = 341$. Use Fermat's Little Theorem to prove that 341 is not prime and that 341 divides $2^{341} - 2$, verifying Euler's counterexample.

7.45. Let m be a positive integer. Give an example of a polynomial f with integer coefficients and leading coefficient 1 such that $f(x) \equiv 0 \bmod m$ for all $x \in \mathbb{Z}$. (Comment: Compare this with Corollary 3.25.)

7.46. (!) A **cyclic shift** of a p-tuple x is a p-tuple obtained by adding a constant (modulo p) to the indices of the elements of x; shifting x by $p+i$ positions produces the same p-tuple as shifting x by i positions. For $a \in \mathbb{N}$, let R be the relation on $[a]^p$ (the set of p-tuples with entries in $\{1, \ldots, a\}$) defined by putting $(x, y) \in R$ if the p-tuple y can be obtained from x by a cyclic shift.

  a) Prove that R is an equivalence relation on $[a]^p$.

  b) Use part (a) and Lemma 7.27 to prove that p divides $a^p - a$ when p is prime. (Hint: Partition a set of size $a^p - a$ into subsets of size p.)

  c) Use part (b) to prove Fermat's Little Theorem.

7.47. Let p be an odd prime. Prove that $2(p-3)! \equiv -1 \bmod p$. (Hint: Use Wilson's Theorem, Theorem 7.44.)

7.48. (−) Suppose that $p > 1$ and $(p-1)! \equiv -1 \bmod p$. Prove that p is prime. (Comment: This is the converse of Wilson's Theorem.)

7.49. Prove that the set of permutations of $[n]$, viewed as a set of functions from $[n]$ to $[n]$, forms a group under the operation of composition.

7.50. Prove that the polynomials of degree k with coefficients in $\mathbb{Z}_p$ form a group under addition modulo p.

7.51. Let G be a group under the binary operation $\circ$. Prove that for every $x \in G$ there is a unique y such that $y \circ x = 1$.

7.52. Let G be a group under the binary operation $\circ$. Define $f_y \colon G \to G$ by $f_y(x) = y \circ x$. Prove that f_y is a bijection.

7.53. (!) Let G be a finite group under multiplication, with identity element 1. Given $x \in G$, the least k such that $x^k = 1$ is the **order** of x. Prove that the order of x divides $|G|$.

Chapter 8

The Rational Numbers

Within the real number system, division by a nonzero number is always defined. In particular, when p and q are integers and $q \neq 0$, the quotient p/q is a real number. Such real numbers are called **rational**; the others are **irrational**. We denote the set of rational numbers by $\mathbb{Q}$. After relating rational numbers to geometry, we prove that some real numbers are irrational and characterize Pythagorean triples.

8.1. Problem. *The Billiard Problem.* Suppose that a square billiard table has corners at $\{(0,0), (1,0), (1,1), (0,1)\}$. A ball leaves the origin along a line with slope s. If the ball reaches a corner, it stops (or falls off the table). Whenever it hits the side of the table not at a corner, it continues to travel on the table, but the slope of the line is multiplied by -1. Does the ball reach a corner? When $s = 3/5$, the answer is "Yes". ∎

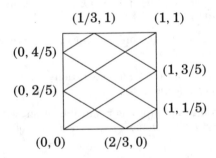

$(1/3, 1)$ $(1, 1)$

$(0, 4/5)$

$(1, 3/5)$

$(0, 2/5)$

$(1, 1/5)$

$(0, 0)$ $(2/3, 0)$

8.2. Problem. *Pythagorean triples.* What are the integer solutions to $a^2 + b^2 = c^2$? The positive solutions measure sides of right triangles. ∎

8.3. Problem. *Iterated averaging.* Starting with $\{0, 1\}$, what numbers can be found by iteratively averaging two numbers already found? ∎

RATIONAL NUMBERS AND GEOMETRY

The familiar word "fraction" is commonly used with many different (but related) meanings. To clarify understanding of the rational numbers, in this book we use one specific meaning.

8.4. Definition. A **fraction** is an expression consisting of an integer (the **numerator**), a division symbol, and a nonzero integer (the **denominator**). For integers a, b, we write the fraction as $\frac{a}{b}$ or a/b.

Since division by nonzero integers is well-defined in the real number system, a fraction represents a unique rational number, the result of the division. On the other hand, many fractions represent the same rational number; $\frac{1}{2} = \frac{2}{4} = \frac{3}{6}$, for example.

Since fractions represent unique rational numbers, we treat them in equalities and inequalities as the numbers they represent. We use the word "fraction" when we want to discuss the pair of numbers as an expression or as a particular representative of a rational number.

The elementary criterion for when fractions represent the same number uses only integer multiplication.

8.5. Remark. Fractions $\frac{a}{b}$ and $\frac{c}{d}$ represent the same rational number if and only if $ad = bc$. ∎

Remark 8.5 expresses a rational number as a set of fractions. From the perspective of the real numbers, a rational number is simply a real number that can be expressed as the quotient of two integers. In Appendix A we construct $\mathbb{Q}$ from $\mathbb{N}$ without mentioning $\mathbb{R}$, defining a rational number to be an equivalence class of fractions under the relation given in Remark 8.5 (see Chapter 7 for discussion of equivalence relations).

In Chapter 6, we proved that integers have unique prime factorizations. We thus can write the numerator and denominator of a fraction as products of prime powers. When they have a common factor, we can cancel it to obtain another representative of the same rational number.

8.6. Definition. A fraction a/b is **in lowest terms** if a and b have no common factors and $b > 0$.

8.7. Remark. A fraction is in lowest terms if and only if its denominator is the smallest positive number among the denominators of all representatives of the same rational number (Exercise 9). ∎

8.8. Remark. *Canonical "factorization" of rational numbers.* A rational number x has a unique representation as a fraction a/b in lowest terms. The prime factorizations of a and b use distinct primes. This yields a canonical representation of x using primes. We write $x = a/b = \prod p_i^{e_i}$, where e_i is positive for primes that divide a and negative for primes that divide b. The canonical representation simultaneously minimizes the absolute value of the exponent on each prime. ∎

Representing a rational number as a fraction in lowest terms is often convenient; see Theorem 2.3, for example. Nevertheless, this is not always the most appropriate representative. When adding two rational numbers, for example, one first writes them as fractions with a common denominator. This illustrates why it is useful to think of a rational number as the set of all its representations as fractions.

We next give a geometric interpretation of rational numbers.

8.9. Example. *Geometric interpretation of rational numbers.* Given a pair (a, b) of integers (not both zero), define the line $L(a, b)$ through the point $(0, 0) \in \mathbb{R}^2$ by $L(a, b) = \{(x, y) \in \mathbb{R}^2 : bx = ay\}$. The lines $L(a, b)$ and $L(c, d)$ are the same if and only if $ad = bc$. Observe that the integer point $(p, q) \in \mathbb{R}^2$ lies on the line $L(p, q)$.

This establishes a bijection between the rational numbers and the lines through the origin (other than the vertical line) that pass through integer points. The inverse of this bijection assigns to $L(a, b)$ the rational number b/a that is its slope. ∎

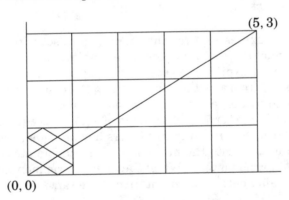

8.10. Solution. *The Billiard Problem.* Our ball starts at the origin and bounces off the walls of the unit square; let L be the line with slope s along which it starts. Vertical direction changes after each unit of vertical travel, but the magnitude of the vertical rate of travel remains the same. The same statement holds for horizontal motion.

Thus "reaching a corner" means simultaneously having traveled integer amounts m horizontally and n vertically. This occurs if and only if L contains the point (m, n). The path followed by the ball is then a folding of the segment from $(0, 0)$ to (m, n). The line L contains such a point if and only if s is rational. (Exercise 16 develops a stronger statement.) ∎

Often it is convenient to describe the set of points forming a line or curve in the plane using a single variable. The points in the set are ordered pairs, and each coordinate in the ordered pair is expressed as a function of the new variable. This variable is a **parameter**, and the two functions are called **parametric equations** for the set.

Parametric equations for a line are closely related to its slope. In the description below, the parameter describing position on the line is t; the pair (a, b) designates a particular line. For lines that don't contain the origin, see Exercise 6.

8.11. Proposition. Every line L in $\mathbb{R}^2$ that contains the origin and has rational slope is specified by an integer pair (a, b) (with $a \neq 0$) such that $(x, y) \in L$ if and only if $(x, y) = (at, bt)$ for some real number t.

Proof: If $(x, y) = (at, bt)$, then $bx - ay = 0$. Thus such pairs satisfy the equation of a line through the origin (see Definition 2.1). This line has rational slope b/a.

Conversely, suppose that the line L is defined by $Ax + By = 0$ for real numbers A and B that are not both 0. If $B = 0$, then the line is vertical and does not have rational slope. If $B \neq 0$, then the slope of the line is $-A/B$, which we have assumed is rational. Therefore we can write $-A/B = b/a$ for integers a, b. Now (x, y) lies on L if and only if $bx - ay = 0$, which is equivalent to $(x, y) = (at, bt)$ for some real number t. ∎

Parametric equations for a circle are more subtle. Later they will help us solve Problem 8.2. The **unit circle** is the set $\{(x, y) \in \mathbb{R}^2 : x^2 + y^2 = 1\}$.

8.12. Theorem. (Parametrization of the Unit Circle) If $x \neq -1$, then $x^2 + y^2 = 1$ if and only if there is a real number t such that

$$(x, y) = \left(\frac{1 - t^2}{1 + t^2}, \frac{2t}{1 + t^2} \right).$$

Furthermore, such a point (x, y) has rational coordinates if and only if t is a rational number.

Proof: For $t \in \mathbb{R}$, always $\frac{1-t^2}{1+t^2} \neq -1$. Thus if (x, y) has the specified form, then $x \neq -1$ and

$$x^2 + y^2 = \frac{(1 - t^2)^2}{(1 + t^2)^2} + \frac{4t^2}{(1 + t^2)^2} = \frac{1 + 2t^2 + t^4}{(1 + t^2)^2} = 1.$$

This proves that the condition is sufficient for $x^2 + y^2 = 1$ when $x \neq -1$.

To prove the converse, suppose that (x, y) satisfies $x \neq -1$ and $x^2 + y^2 = 1$. Consider the line containing the points $(-1, 0)$ and (x, y). Its equation is given by $y = t(x + 1)$, where t is its slope. Substituting this into the equation $x^2 + y^2 = 1$ yields a quadratic equation for x in terms of the parameter t, namely $x^2 + t^2(x + 1)^2 = 1$. We rewrite this as

$$(1+t^2)x^2 + 2t^2x + t^2 - 1 = 0.$$

One solution to this is $x = -1$, which is excluded for the point (x, y) we are studying. Since the product of the solutions is $\frac{t^2-1}{1+t^2}$ (see Exercise 1.20), the other solution is $x = \frac{1-t^2}{1+t^2}$. Using $y = t(x + 1)$ and simplifying yields $y = \frac{2t}{1+t^2}$. Thus (x, y) satisfies the claimed parametric equations.

For the final statement, note that if t is rational, then the parametric equations imply that x and y also are rational. Conversely, if $x \neq -1$ and x, y are rational, then $t = y/(x + 1)$ also is rational.   ∎

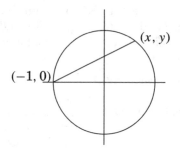

IRRATIONAL NUMBERS

The set $\mathbb{Q}$ of rational numbers forms an ordered field, but it does not satisfy the Completeness Axiom. This partly explains why some familiar equations such as $x^2 = 2$ have no rational solutions. In this section we study the existence of rational zeros of polynomials in one variable with integer coefficients. Earlier we found the rational solutions to the particular equations $bx + ay = 1$ and $x^2 + y^2 = 1$ in two variables.

We begin with the polynomial $x^2 - 2$. We proved in Theorem 3.31 that $\sqrt{2}$ is irrational. Here we give another proof and generalize both proofs.

8.13. Example. *Irrationality of $\sqrt{2}$.* We claim that the equation $x^2 = 2$ has no rational number x as a solution. Otherwise, we may consider a rational solution $x = a/b$ written in lowest terms. We have $a^2 = 2b^2$, so a^2 is even. Since the square of an odd integer is odd, a must be even. Now a^2 is divisible by 4, and hence $2b^2$ is divisible by 4, so b^2 is even. We conclude that b also is even. Now a, b are both divisible by 2, contradicting the choice of a/b in lowest terms.   ∎

This argument generalizes to prove that no prime number has a rational square root (Exercise 18), but stronger results are available. A somewhat different argument prohibits rational square roots for all natural numbers (except the squares of integers).

8.14. Theorem. The positive integer k has no rational square root if k is not the square of an integer.

Proof: We use contradiction. Suppose that $\sqrt{k}$ is rational and that m/n is a fraction representing it such that n is positive and minimal. If m/n is not an integer, then there is an integer q such that $m/n - 1 < q < m/n$. This is equivalent to $0 < m - nq < n$. Since $m - nq \neq 0$, we can write

$$\frac{m}{n} = \frac{m(m - nq)}{n(m - nq)} = \frac{m^2 - mnq}{n(m - nq)} = \frac{n^2 k - mnq}{n(m - nq)} = \frac{nk - mq}{m - nq}.$$

Since $0 < m - nq < n$, we have found a representation of m/n with smaller positive denominator, which contradicts the choice of n. Therefore, if the square root of k is rational, it must be an integer. ∎

8.15. Remark. The proof of Theorem 8.14 is actually an inductive proof phrased using the method of descent; we proved that there is no counterexample with smallest positive denominator. It is possible to prove statements about rational numbers by induction on the denominator of the representative in lowest terms.

Even though $\mathbb{Q}$ is countable (Exercise 17), we cannot use induction with respect to the usual order "$\leq$" to prove results about the nonnegative rational numbers. There is a place to start (0), but there is no "next" rational number; when x, y are distinct rational numbers, we can find another rational number between them, such as their average.

Selecting the representative of a rational number with smallest positive denominator is an example of **extremality**. The lack of more extreme examples helps shorten proofs; this is how the method of descent works. Another extremal description of fractions in lowest terms arises from the geometric interpretation of rational numbers: if $x = a/b$ in lowest terms, then (b, a) is the integer point with positive first coordinate that is closest to the origin on the line associated with x in Example 8.9. ∎

The proof of Theorem 8.14 does not use the prime factorization of integers and is generalized in Exercise 24. Other proofs appear in Exercise 19 and Exercise 22. We next generalize the argument in Example 8.13 to describe rational zeros of polynomials with integer coefficients.

8.16. Theorem. (Rational Zeros Theorem) Let $c_0, \dots, c_n$ be integers with $n \geq 1$ and $c_0, c_n \neq 0$, and let $f(x) = \sum_{i=0}^{n} c_i x^i$ for $x \in \mathbb{R}$. If r is a rational solution to the equation $f(x) = 0$, written as p/q in lowest terms, then p must divide c_0 and q must divide c_n.

Proof: When $f(r) = 0$, we can multiply both sides of $f(r) = 0$ by q^n to obtain $\sum_{i=0}^{n} c_i p^i q^{n-i} = 0$. Moving the term $c_n p^n$ to the other side yields

$$-c_n p^n = \sum_{i=0}^{n-1} c_i p^i q^{n-i} = q \sum_{i=0}^{n-1} c_i p^i q^{n-1-i}.$$

Since q divides one side of this equation, it must also divide the other side. Since q and p are relatively prime (because p/q was chosen in lowest terms), we conclude that q must divide c_n.

If we instead move the term $c_0 q^n$ to the other side, then we obtain

$$-c_0 q^n = \sum_{i=1}^{n} c_i p^i q^{n-i} = p \sum_{i=1}^{n} c_i p^{i-1} q^{n-i}.$$

Now p divides one side and hence the other. This implies that p divides c_0, since p and q are relatively prime.   ∎

8.17. Example. *No rational solutions.* If the equation $x^3 - 6 = 0$ has a rational solution r, written as p/q in lowest terms, then q must divide 1 and p must divide 6. The only possibilities are $r = \pm 1, \pm 2, \pm 3, \pm 6$, none of which work. Hence the cube root of 6 is irrational.   ∎

8.18. Example. *Solutions to quadratics.* The quadratic formula gives $(-b \pm \sqrt{b^2 - 4ac})/2a$ as the solutions to $ax^2 + bx + c = 0$. Even when a, b, c are integers, the solutions may be irrational. For example, $(1 + \sqrt{5})/2$ is a solution to the equation $x^2 - x - 1 = 0$. This number is not rational, because the Rational Zeros Theorem implies that the only possible rational solutions to this equation are ± 1, which do not satisfy the equation.   ∎

8.19. Remark. *Products of irrational numbers may be rational.* Since every nonzero real number has a reciprocal, every irrational number x has a reciprocal $1/x$ such that $x \cdot (1/x) = 1$. For example, the reciprocal of $(\sqrt{5} + 1)/2$ is $(\sqrt{5} - 1)/2$; their product is 1. In general, the product of the solutions of a quadratic equation with rational coefficients is always rational (Exercise 3).   ∎

PYTHAGOREAN TRIPLES

Why should we care that there is no rational solution to $x^2 = 2$? We believe there is a number $\sqrt{2}$ that is the ratio of two physical quantities. The length of the diagonal of a square with side-length 1 is a quantity that we believe is $\sqrt{2}$; it satisfies $x^2 = 2$. In elementary geometry, we construct right angles with straightedge and compass. We can thus construct a right triangle whose short sides have unit length. The length of the third side is $\sqrt{2}$, by the Pythagorean Theorem.

The ancients believed that all numbers were rational. It is said that therefore the person who discovered irrational numbers was murdered (by drowning). Are irrational numbers "crazy"? The Billiard Problem and decimal expansions (see Chapter 13) show that irrational numbers exhibit complicated behavior, but the term "irrational" does not arise from "crazy". The psychological meaning (lacking reason) and the mathematical meaning (not a ratio of integers) are related only in that "ratio" and "reason" come from the same Greek root. The Pythagoreans allowed only rational numbers in their reasoning.

8.20. Theorem. (Pythagorean Theorem) If a, b, c are the lengths of the sides of a right triangle, with c the length of the side opposite the right angle, then $a^2 + b^2 = c^2$.

Proof: (Sketch) We assume the notions of right angle, triangle, rectangle, and area. We assume that the area of a rectangle is the product of the lengths of two neighboring sides, that the area of a region is the sum of the areas of the regions formed by cutting it by line segments, and that the area of congruent regions is the same. Thus the area of a right triangle is half the product of the short sides, because the diagonal of a rectangle cuts it into two pieces of equal area.

The outer quadrilateral shown below is a square. By symmetry considerations, the four triangles are congruent and the inner quadrilateral also is a square (see Exercise 15). The area of the large square equals the area of the small square plus the areas of the triangles. This yields $(a + b)^2 = c^2 + 4(ab/2)$, which simplifies to $a^2 + b^2 = c^2$. ∎

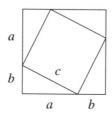

8.21. Example. *Pythagorean triples.* The integer solutions to $a^2 + b^2 = c^2$ are called **Pythagorean triples**. The most familiar example is the triple $(a, b, c) = (3, 4, 5)$, and integer multiples of this also work. Other examples where a, b, c have no common factor include $(5, 12, 13)$, $(8, 15, 17)$, $(7, 24, 25)$, $(20, 21, 29)$, and $(9, 40, 41)$. ∎

We prove that all Pythagorean triples can be generated using two independent integer parameters. Exercise 29 develops an alternative proof of this characterization.

8.22. Theorem. The Pythagorean triples are the integer multiples of triples of the form $(2rs, r^2 - s^2, r^2 + s^2)$ or $(r^2 - s^2, 2rs, r^2 + s^2)$, where r, s are integers.

Proof: Since $(2rs)^2 + (r^2 - s^2)^2 = (r^2 + s^2)^2$, the triples described are all Pythagorean triples. Multiplying such a triple by n multiplies the equality by n^2, so all integer multiples of these triples also satisfy $a^2 + b^2 = c^2$.

We must prove that every Pythagorean triple can be described in this way. For $c \in \mathbb{N}$, we seek integer solutions (a, b) to $a^2 + b^2 = c^2$. Letting $x = a/c$ and $y = b/c$ leads us to seek rational solutions to $x^2 + y^2 = 1$. Using Theorem 8.12 yields $\{(\frac{1-t^2}{1+t^2}, \frac{2t}{1+t^2}): t \in \mathbb{Q}\}$ as the set of rational solutions other than $(-1, 0)$. Since t is rational, we let $t = s/r$ in lowest terms and simplify to obtain

$$(a, b, c) = \frac{c}{r^2 + s^2}(r^2 - s^2, 2rs, r^2 + s^2). \tag{$*$}$$

Let $z = c/(r^2 + s^2)$. When z is an integer, $(*)$ expresses the triple in the desired form.

When z is not an integer, we show first that z is half an integer. Let $z = m/n$ in lowest terms. Since a and c are integers, $(*)$ implies that n divides both $r^2 - s^2$ and $r^2 + s^2$. Hence n also divides their sum $2r^2$ and difference $2s^2$. Since $\gcd(r, s) = 1$, also $\gcd(r^2, s^2) = 1$ (see Exercise 19). To divide both $2r^2$ and $2s^2$, n must therefore divide 2.

When z is half an integer, $r^2 + s^2$ must be even. Since $\gcd(r, s) = 1$, we conclude that both r and s are odd. Now the numbers $R = (r + s)/2$ and $S = (r - s)/2$ are integers. Note also that $r = R + S$ and $s = R - S$. This substitution yields $r^2 - s^2 = 4RS$, $2rs = 2(R^2 - S^2)$, and $r^2 + s^2 = 2(R^2 + S^2)$. Again we have (a, b, c) as an integer times a triple in the desired form:

$$(a, b, c) = \frac{2c}{r^2 + s^2}(2RS, R^2 - S^2, R^2 + S^2). \qquad \blacksquare$$

The famous **Fermat's Last Theorem** is the statement that $x^n + y^n = z^n$ has no solution in integers if $n \geq 3$. Fermat wrote this in the margin of a book in the seventeenth century, claiming to have a marvelous proof that would not fit in the margin, but he died without presenting a proof to anyone. Mathematicians labored for 350 years to find a proof. Andrew Wiles succeeded in 1994.

FURTHER PROPERTIES OF $\mathbb{Q}$ (optional)

We have seen examples of proofs about rational numbers using reduction to lowest terms and using the closure of the rational numbers under arithmetic operations. The next proof uses a different technique, analogous to the construction in Appendix A; first prove the statement for the natural numbers, then for the integers, then for the rational numbers.

8.23. Theorem. Suppose $f : \mathbb{Q} \to \mathbb{Q}$ satisfies $f(x + y) = f(x) + f(y)$ for all $x, y \in \mathbb{Q}$. Then $f(wx) = wf(x)$ for all $w, x \in \mathbb{Q}$.

Proof: First suppose $w = 1$; here the statement is a triviality. This provides the basis step for a proof by induction in the case $w \in \mathbb{N}$. For the induction step, suppose it is true when $w = n$. Then

$$f((n + 1)x) = f(nx + x) = f(nx) + f(x) = nf(x) + f(x) = (n + 1)f(x),$$

where we have used the distributive law, the defining property of f, the induction hypothesis, and the distributive law again. To prove the claim for $w = 0$, we need only show that $f(0) = 0$, which follows from $f(0) = f(0 + 0) = f(0) + f(0) = 2f(0)$. For $w = -1$, we use $0 = f(0) = f(x - x) = f(x) + f(-x)$, which implies $f(-x) = -f(x)$. Now we can prove the claim for $w \in \mathbb{Z}$: $f((-n)x) = f((-1)nx) = -f(nx) = -nf(x)$ for $n \in \mathbb{N}$.

Next suppose w is the reciprocal of the integer n. We have $f(x) = f(n(x/n)) = nf(x/n)$, and therefore $f(x/n) = (1/n)f(x)$. Note that at each stage we proved the statement for arbitrary $x \in \mathbb{Q}$, so these steps are justified. Now that we have the statement for all integers and for reciprocals of natural numbers, we can write $w \in \mathbb{Q}$ as a/b in lowest terms and conclude $f((a/b)x) = af((1/b)x) = (a/b)f(x)$. ∎

The statement of Theorem 8.23 is false when $\mathbb{Q}$ is replaced by $\mathbb{R}$; the conclusion then requires the additional hypothesis that f be continuous (continuity is discussed in Chapter 15).

8.24. Definition. A rational number is a **dyadic rational** if it can be expressed as a fraction whose denominator is a power of 2.

8.25. Solution. *Iterated averages and dyadic rationals.* We solve Problem 8.3: which numbers can be generated from $\{0, 1\}$ by iteratively taking the average (arithmetic mean) of two numbers already in the set?

Since the average of two rational numbers is rational, only rational numbers in the interval $[0, 1]$ can arise. In addition, the only numbers that can arise are dyadic rationals, since 0 and 1 are dyadic rationals, and the average of two dyadic rationals is also a dyadic rational.

To complete the solution, we prove that every dyadic rational in the interval $[0, 1]$ is generated. Except for 0 itself, each such rational is expressed in lowest terms as $(2j + 1)/2^k$ for some nonnegative integers j, k. We prove by induction on k that $(2j + 1)/2^k$ is achievable. For $k = 0$, the only such number is 1 itself. For the induction step, suppose $k > 0$, and consider the number $x = (2j + 1)/2^k$ in the interval $(0, 1)$. The number x is the average of $(2j)/2^k$ and $(2j + 2)/2^k$, which equal $j/2^{k-1}$ and $(j + 1)/2^{k-1}$ and lie in $[0, 1]$. Since one of $\{j, j + 1\}$ is even, one of these fractions is not in lowest terms (the numerator is an odd number times a power of 2 with positive exponent). After canceling these factors of 2, we have expressed x as the average of two dyadic rationals with smaller exponents in the denominator. By the induction hypothesis, each of these is achievable, so x also is achievable. ∎

In Chapter 13, we consider decimal and binary expansions of real numbers in the interval $[0, 1]$. The dyadic rationals are precisely the real numbers whose binary expansions terminate.

EXERCISES

8.1. (−) Suppose that x is rational and that a, b, c are irrational. Determine which statements below must be true. When true, provide a proof; when false, provide a counterexample.
    a) $x + a$ is irrational.
    b) xa is irrational.
    c) abc is irrational.
    d) $(x + a)(x + b)$ is irrational.

8.2. (−) Let f be a polynomial with rational coefficients. Prove that there is a polynomial with integer coefficients that has the same zeros as f.

8.3. (−) Consider $a, b, c \in \mathbb{Q}$ with $a \neq 0$. Suppose that $ax^2 + bx + c = 0$ has two solutions. Prove that their product is rational.

8.4. (−) Explain why we assume in Example 8.9 that a and b are not both 0.

8.5. (−) Find the image of the function $f: \mathbb{R} \to \mathbb{R}^2$ defined by $f(t) = \left(\frac{1-t^2}{1+t^2}, \frac{2t}{1+t^2} \right)$. (Hint: Consider Theorem 8.12.)

8.6. (−) Obtain parametric equations for a line of slope m through $(p, q) \in \mathbb{R}^2$.

8.7. (−) Show how the triples in Example 8.21 arise in the parametrization of Pythagorean triples.

•   •   •   •   •

8.8. (!) *How not to add fractions.* Find all $(x, y) \in \mathbb{R}^2$ such that $\frac{1}{x} + \frac{1}{y} = \frac{1}{x+y}$.

8.9. Prove that a fraction is in lowest terms if and only if its denominator is the smallest positive number among the denominators of all representatives of the same rational number.

8.10. (!) Let a/m and b/n be rational numbers expressed in lowest terms. Prove that $(an + bm)/(mn)$ is in lowest terms if and only if m and n are relatively prime.

8.11. Let x, y be real numbers such that $x/y = \sqrt{2}$. Simplify $(2y - x)/(x - y)$.

8.12. (!) Suppose that a, b, c, d are positive integers with $a/b < c/d$. Prove that $a/b < (a + c)/(b + d) < c/d$. Interpret this in terms of test scores or batting averages. Give a geometric interpretation using slopes of lines.

8.13. Suppose a, b, c, d are positive integers with $a \leq c \leq d$ and $c/d \leq a/b$. Prove that $b - a \leq d - c$. Prove that this conclusion does not always hold if $a \leq d < c$ and $c/d \leq a/b$.

8.14. Let $S = \{(x, y) \in \mathbb{R}^2 : x^2 - y^2 = 1\}$. Obtain parametric equations for the set of points in S with positive first coordinate. Graph this curve.

8.15. Discuss carefully how symmetry considerations are used in the proof of Theorem 8.20.

8.16. (!) In the Billiard Problem (Solution 8.10), for each corner of the square determine the condition on the slope s so that the process ends there.

8.17. (!) Prove that the set of rational numbers is countable.

8.18. (−) Generalize the proof in Example 8.13 to prove that the square root of every prime number is irrational (do not use the Rational Zeros Theorem).

8.19. (!) Prove that if r and s are relatively prime, then also r^2 and s^2 are relatively prime (see also Exercise 6.18). Use this to prove that the square root of an integer is irrational unless the root is an integer.

8.20. Let c be an integer, and let $f(x) = x^6 + cx^5 + 1$.
  a) Prove that the equation $f(x) = 0$ has rational solutions when $c = \pm 2$.
  b) Prove that the equation $f(x) = 0$ has no rational solutions when $c \neq \pm 2$.

8.21. Let $f(x) = 2x^3 + x^2 + x + 2$. Find all rational zeros of f, and then factor f to find the rest of the zeros. Graph f to check that your answer is reasonable.

8.22. Use the Rational Zeros Theorem to prove that the kth root of an integer is not a rational number unless it is an integer.

8.23. Suppose that $ax^2 + bx + c = 0$ has a rational solution, where a, b, c are integers and b is odd. Use the quadratic formula to prove that a and c cannot both be odd. (Comment: This provides another proof of Theorem 2.3.)

8.24. (+) Let f be a polynomial with integer coefficients and leading coefficient 1. Without using the Rational Zeros Theorem, prove that if $f(t) = 0$ for some $t \in \mathbb{Q}$, then $t \in \mathbb{Z}$. (Hint: If $t \notin \mathbb{Z}$, write $t = m/n$ with $n > 1$ and minimal. Let $q = t - \lfloor m/n \rfloor$, and use the numerators of $\{q^k\}$ to obtain a decreasing sequence of positive integers.)

8.25. (−) Give an example of a Pythagorean triple in increasing order that cannot be written in the form $(r^2 - s^2, 2rs, r^2 + s^2)$ for integers r, s. (Comment: By Theorem 8.22, the answer can be written in the form $(2rs, r^2 - s^2, r^2 + s^2)$. This shows that both forms are needed.)

8.26. Use the parametrization of Pythagorean triples to prove that every integer greater than two is a member of a Pythagorean triple not containing 0. (Hint: Give a construction when n is even and another construction when n is odd. The fact that $(k + 1)^2 - k^2 = 2k + 1$ may be useful.)

8.27. (!) Determine when the sum of two Pythagorean triples (under componentwise addition) is a Pythagorean triple. (The simple criterion needs no formula.)

8.28. Let x be an integer chosen at random from [20] (each with probability $1/20$). Let y be another integer, independently chosen in the same way.

a) Compute the probability that $x^2 + y^2$ is the square of an integer.

b) (+) Compute the probability that x and y belong to a Pythagorean triple.

8.29. (+) *Alternative proof of characterization of Pythagorean triples.* This exercise develops an alternative proof that every Pythagorean triple has the form described in Theorem 8.22. Let (a, b, c) be a Pythagorean triple such that a, b, c have no common factor (thus $\gcd(a, b) = \gcd(b, c) = \gcd(a, c) = 1$).

a) Prove that exactly one of a and b is even.

b) Let a be the even member of $\{a, b\}$. Prove that $(c + b)/2$ and $(c - b)/2$ are relatively prime and are squares of integers.

c) Given the result of part (b), let $(c + b)/2 = z^2$ and $(c - b)/2 = y^2$. Prove that $a = 2yz$, $b = z^2 - y^2$, and $c = z^2 + y^2$.

(Comment: The proof of Theorem 8.22 in the text emphasizes geometry and the properties of rational numbers. This proof emphasizes divisibility and primes.)

8.30. (+) *Solution of the general cubic equation.* Consider the equation $ax^3 + bx^2 + cx + d = 0$ with $a \neq 0$ and $a, b, c, d \in \mathbb{R}$.

a) Determine appropriate constants s, t so that the change of variables $x = s(y + t)$ reduces solving this equation to solving the equation $y^3 + Ay + B = 0$, where A, B are constants.

b) Determine a constant r such that the change of variables $y = z + r/z$ in the equation for y reduces it to a quadratic equation in z^3.

c) Solve the resulting quadratic equation for z^3, and use the solution to solve the general cubic equation for x. (Comment: This method is tedious even for easy cubic equations, and it uses complex numbers even when all the roots are real. Nevertheless, it does produce the solutions. There is no formula for solving a general polynomial equation of degree 5 or higher.)

8.31. Let $\mathbb{Q}^* = \mathbb{Q} - \{0\}$. Suppose that $f \colon \mathbb{Q}^* \to \mathbb{Q}^*$ and that f satisfies $f(x + y) = f(x)f(y)/[f(x) + f(y)]$ whenever $x, y \in \mathbb{Q}^*$. Suppose $c = f(1)$. Compute $f(x)$ in terms of c for every $x \in \mathbb{Q}^*$. (Hint: Consider the function $g = 1/f$.)

8.32. (+) A man has a watch with indistinguishable hands. An act of violence between midnight and the following noon simultaneously kills him and stops his watch. Can we always determine the time of death from this information if

a) the watch has hour, minute, and second hands?

b) the watch has only hour and minute hands?

PART III

DISCRETE
MATHEMATICS

Chapter 9

Probability

We give a precise definition of probability. Probability is an important tool for analysis of everyday events and for decision-making. We introduce the notions of conditional probability, independence, random variables and expectation, and multinomial coefficients. This enables us to solve the following problems.

9.1. Problem. *Bertrand's Ballot Problem.* Suppose that candidates A and B in an election receive a and b votes, respectively, with $a \geq b$, and that the votes are counted in random order. What is the probability that candidate A never trails? ∎

9.2. Problem. *Medical Testing.* Suppose 90% of patients with cancer test positive on a new test, and 4% of those without cancer test positive. Among the patients, 2% actually have cancer. Given that a randomly chosen patient tests positive, what is the probability that he or she has cancer?

9.3. Problem. *Bernoulli Trials.* Repeated performances of an experiment with a fixed probability of success are called **Bernoulli trials**, after Jakob Bernoulli (1654–1705). When we perform n trials of an experiment that has probability p of success, and the outcome of one trial cannot affect the outcome of any other trial, we expect to have about np successes. How can we make this intuition precise? ∎

9.4. Problem. *The Coupon Collector.* A restaurant gives one of five types of coupons with each meal, each with equal probability. A customer receives a free meal after collecting one coupon of each type. How many meals does a customer expect to need to buy before getting a free meal?

9.5. Problem. *Hitting for the cycle.* How often does a baseball player get a single, a double, a triple, and a home run in the same game? In Solution 9.40, we solve a special case for a particular batter. ∎

PROBABILITY SPACES

   In Chapter 5 we introduced an elementary model of probability to study experiments with n equally likely possible outcomes. For each set A of possible outcomes, the probability of obtaining an outcome in A is $|A|/n$.

9.6. Example. When we roll two different six-sided dice, there are 36 equally likely outcomes. In six of these outcomes, the two dice show the same number. Thus the probability of rolling doubles is 1/6.
   When we deal five cards from a standard deck, there are $\binom{52}{5}$ equally likely outcomes. The probability of a particular type of poker hand such as "full house" or "flush" is the proportion of the possible outcomes satisfying the specified criteria. In Solution 5.16, we computed these probabilities by counting the hands of the desired type.
   When we collect students' papers and return them *at random*, we are performing an experiment whose outcomes are permutations of the papers, with all permutations equally likely. The probability that no paper returned to its author is the proportion of the permutations that have no fixed point; we study this in Chapters 10 and 12.   ∎

   The notion of probability extends to settings where the outcomes are not equally likely. Let $S = \{a_1, \ldots, a_n\}$ be a finite set of outcomes. Associated with each outcome a_i we have a number p_i that we view as its probability. Our intuitive sense of probability requires that these numbers be nonnegative and sum to 1. Also, the probability that the outcome lies in some subset $T \subseteq S$ should equal $\sum_{a_i \in T} p_i$.
   Our formal model for probability has these properties. It also extends to situations where the set of outcomes is infinite.

9.7. Definition. A finite **probability space** is a finite set S together with a function P defined on the subsets of S (called **events**) such that
   a) For $A \subseteq S$, $0 \le P(A) \le 1$,
   b) $P(S) = 1$, and
   c) If A, B are disjoint subsets of S, then $P(A \cup B) = P(A) + P(B)$.

   Suppose that sets $B_1, \ldots, B_k$ form a *partition* of A, meaning that every element of A appears in exactly one of them (see Chapter 7). By induction on k, (c) implies that $P(A) = \sum_{i=1}^k P(B_i)$. In particular, when $k = |A|$ and each B_i contains exactly one element of A, we see that the probability of an event is the sum of the probabilities of the outcomes contained in it.

9.8. Proposition. (Elementary Properties) If A and B are events in a probability space S with probability function P, then
   a) $P(A^c) = 1 - P(A)$.

b) $P(\emptyset) = 0$.

c) $P(A \cup B) = P(A) + P(B) - P(A \cap B)$.

Proof: (a) We have $P(A) + P(A^c) = P(A \cup A^c) = P(S) = 1$.

(b) Apply (a) with $A = S$.

(c) As evident from a Venn diagram, the sets $A \cap B$, $A - B$, and $B - A$ partition $A \cup B$. Thus $P(A \cup B)$ is the sum of their probabilities. Since A is the disjoint union of $A - B$ and $A \cap B$ (similarly $B = (B - A) \cup (A \cap B)$), we obtain the desired formula from

$$P(A \cup B) = [P(A) - P(A \cap B)] + [P(B) - P(A \cap B)] + P(A \cap B). \qquad \blacksquare$$

9.9. Example. When we roll two dice, the probability of avoiding doubles is $1 - (1/6) = 5/6$; call this event A. The probability that the sum of the numbers is divisible by 4 is $1/4$, since this occurs on 9 of the 36 possible equally likely outcomes; call this event B. The probability that at least one of these two events occurs is not $1/6 + 1/4$; we have

$$P(A \cup B) = P(A) + P(B) - P(A \cap B) = \tfrac{6}{36} + \tfrac{9}{36} - \tfrac{3}{36} = \tfrac{1}{3}. \qquad \blacksquare$$

Our next example continues the theme of counting techniques for elementary discrete probability and features the use of complementation.

9.10. Solution. *Bertrand's Ballot Problem*, due to Joseph Louis Francois Bertrand (1822–1900). Candidates A and B receive a and b votes, respectively, and we assume that $a \geq b$. Counting the votes in random order could mean that the ballot box contains $a + b$ slips of paper, which can be removed from the box in $(a + b)!$ equally likely orders. Alternatively, it could mean that the lists of who received the ith vote are equally likely. A list such as $ABAABAB$ with final score (a, b) is determined by the positions of the As, which can be chosen in $\binom{a+b}{a}$ ways, so there are $\binom{a+b}{a}$ lists with final score (a, b).

In this problem, the two models give the same answer for the probability that A never trails. Consider the model with $(a + b)!$ outcomes. Changing the order of paper slips for one candidate does not change whether the candidate ever trails. Hence the $a!b!$ orderings of the paper slips that correspond to the list $ABAABAB$ give the same answer. Since each list corresponds to the same number of orderings of slips, the probability of each list is the same. Under either model, then, it suffices to count the lists in which A never trails and divide by $\binom{a+b}{a}$.

We use *election* to mean a list of A's and B's. An election is *good* if A never trails; otherwise it is *bad*. To count the good elections with final score (a, b), we count the bad elections with final score (a, b) and subtract them from the total. An election is bad if there is a k such that the score reaches $(k, k + 1)$. The minimal such k is the first time A trails. Modify the election after this time by changing every A to a B and every B to an A. Now A gets $b - k - 1$ additional votes and B gets $a - k$ additional votes, so the final score of the new election is $(b - 1, a + 1)$.

Every election with final score $(b - 1, a + 1)$ is won by B, since $a \geq b$. Therefore, in such an election there is a least k when the score is $(k, k+1)$. Switching the votes after this point as done previously generates an election with final score (a, b). The second map is the inverse of the first, and this establishes a bijection between the set of bad elections with final score (a, b) and the set of all elections with final score $(b - 1, a + 1)$. Hence there are $\binom{a+b}{a+1}$ bad elections. To obtain the probability of the good elections, we compute

$$\frac{\binom{a+b}{a} - \binom{a+b}{a+1}}{\binom{a+b}{a}} = 1 - \frac{b}{a+1} = \frac{a-b+1}{a+1}. \qquad \blacksquare$$

9.11. Remark. *Lattice paths and Catalan numbers.* The switching argument in Solution 9.10 is due to Antoine Désiré André (1840–1917). Graphing the successive vote totals as points in the plane yields a lattice path to (a, b). The path never steps above the diagonal if and only if candidate A never trails in the election. We can translate the switching argument into the language of lattice paths; the bijection maps the bad elections into lattice paths reaching $(b - 1, a + 1)$, via a reflection of the portion after $(k, k + 1)$ through the line with equation $y = x + 1$.

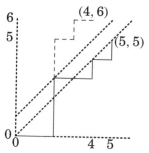

In the special case where $a = b = n$, the number of good paths is $\frac{1}{n+1}\binom{2n}{n}$. These numbers are known as the **Catalan numbers**; we shall see that they provide the solution to many counting problems (Exercises 36–39, Problem 12.4, Exercises 12.37–12.40). $\qquad \blacksquare$

In Solution 9.10, we were lucky; the two possible definitions of the probability space gave the same results. Our next example, also from Bertrand, begins to suggest the care that is needed. This example has uncountably many outcomes. In such a situation, it is not possible to assign probabilities to individual outcomes. Nevertheless, the notion of probability space does extend, as long as we define the probability function P only on an appropriately chosen family of subsets of S.

9.12. Example. *Bertrand's Paradox.* Choose a chord of the unit circle at random; what is the probability that its length exceeds $\sqrt{3}$? The answer depends on the meaning of "at random". As illustrated, the length of the chord exceeds $\sqrt{3}$ if and only if its midpoint is inside an inscribed equilateral triangle.

We could place a "spinner" at the center and spin it twice to select two points on the circumference to be the endpoints of the chord. In this model, the probability that the chord length exceeds $\sqrt{3}$ is 1/3.

Alternatively, we could choose the midpoint of the chord by throwing a dart at the circle that is equally likely to land in regions of equal area. The midpoint of a chord uniquely determines the chord, and in this model the probability that the length exceeds $\sqrt{3}$ is 1/4. Other reasonable models yield other values for the probability (Exercise 12). ∎

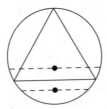

CONDITIONAL PROBABILITY

It has been remarked that probability theory is the area of mathematics in which an expert is most likely to blunder. The explanation may be that formulating and solving these questions requires precise language, but the problems often are stated informally and then misinterpreted. We have already seen that the expression "at random" may have more than one interpretation. Ambiguous language can cause difficulties.

9.13. Example. Consider the question "The Smiths have two children, and at least one is a boy; what is the probability that both are boys?" The correct answer depends on the procedure by which the information "At least one is a boy" is obtained. We assume that when we list the older child first, the four possibilities Boy-Boy, Boy-Girl, Girl-Boy, Girl-Girl are equally likely. Hence 1/3 of the families having at least one boy have two boys. On the other hand, the speaker may have encountered only the older child, noted that it was a boy, and said "at least one is a boy". If the information arose in this way, then the answer is 1/2. ∎

Care must be used in stating what is "given" when a probability is to be computed. The given information is a *condition* that defines a restricted experiment, and this leads to the notion of *conditional probability*.

9.14. Definition. Let A and B be events in a probability space S. When $P(B) \neq 0$, we define the **conditional probability** of A given B to be $\frac{P(A \cap B)}{P(B)}$; we write this as $P(A|B)$.

The conditional probability of A given B is the result of normalizing probabilities to restrict the probability space to the subset B. Given that B occurs, the probability of B should be 1, and the probability for each subset of this event is scaled up accordingly.

Conditional probability explains the confusion in Example 9.13. By stating that "[given that] at least one [child] is a boy", we are using conditional probability. The answer depends on whether the given event is "the first child is a boy" or "the children are not both girls". Understanding of conditional probability is needed in Exercises 8–11.

9.15. Definition. Events A and B are **mutually exclusive** if $A \cap B = \emptyset$. They are **independent** if $P(A \cap B) = P(A) \cdot P(B)$.

The definition of probability space yields $P(A \cup B) = P(A) + P(B)$ when A and B are mutually exclusive. Independence is related to conditional probability. When A and B are independent and $P(B) \neq 0$, the computation below shows that the probability of A is unaffected by knowing whether B occurs; this justifies the term "independent".

$$P(A|B) = \frac{P(A \cap B)}{P(B)} = \frac{P(A)P(B)}{P(B)} = P(A)$$

9.16. Example. *Bernoulli trials.* We flip a coin n times, where the probability of heads is p on each toss. The probability space consists of the n-tuples from $\{H, T\}$. The events H_i and H_j that correspond to heads on the ith flip and jth flip are independent when $i \neq j$. Thus the probability of a particular list with k heads and $n - k$ tails is $p^k(1 - p)^{n-k}$.

The lists of length n are the elements of the space; they yield mutually exclusive events. The event "obtaining k heads" is the union of the events for particular lists with k heads. Its probability is the sum of their probabilities. Since there are $\binom{n}{k}$ lists having k heads, the probability of the event "obtaining k heads" is $\binom{n}{k}p^k(1 - p)^{n-k}$. ∎

The next example underscores the distinction between "the fraction of students in each year that are math majors" and "the fraction of math majors that are in each year". We use it to motivate a general formula about conditional probability.

9.17. Example. Suppose that all college students are freshmen, sopho-mores, juniors, or seniors. Given the fractions of each year's students that are math majors, can we determine the fraction of math majors that are seniors? We can if we also know the number of students in each year.

In the four classes, suppose that the fractions of students that are math majors are 1/3, 1/4, 1/5, 1/6 and that the populations in the four classes are 1500, 1400, 1250, 1200, respectively. We can now compute the populations in the table below. The total number of math majors is 1300, and thus the fraction of math majors that are seniors is 2/13.   ∎

	Fr	So	Jr	Sr	Total
Math	500	350	250	200	1300
Other	1000	1050	1000	1000	4050
Total	1500	1400	1250	1200	5350

The method in Example 9.17 generalizes to a common situation in-volving conditional probability. We want to condition on an event A (in this case being a math major), but we are not given the probability of A. On the other hand, we *are* given the conditional probability of A with re-spect to each set in some partition $B_1, \ldots, B_k$ of the probability space (in this case the four school years).

9.18. Proposition. (Bayes' Formula) Let $B_1, \ldots, B_k$ be mutually exclu-sive events whose probabilities $b_i = P(B_i)$ are known and sum to 1. If A is an event such that the conditional probabilities $a_i = P(A|B_i)$ are known, then

$$P(B_i|A) = \frac{a_i b_i}{\sum_j a_j b_j}.$$

Proof: The definition of conditional probability yields $P(A \cap B_j) = P(A|B_j)P(B_j) = a_j b_j$. We then compute

$$P(B_i|A) = \frac{P(B_i \cap A)}{P(A)} = \frac{P(A \cap B_i)}{\sum_j P(A \cap B_j)} = \frac{a_i b_i}{\sum_j a_j b_j}. \qquad ∎$$

9.19. Solution. Problem 9.2 describes a typical scenario in the interpre-tation of medical testing. To compute the conditional probability, we need the proportion of patients testing positive that actually have cancer. Let "+" and "C" denote the events of testing positive and having cancer; we seek $P(C|+)$. Bayes' Formula provides the computation from the data of Problem 9.2. The probability of having cancer given a positive test is surprisingly small, because the healthy population is so large.

$$P(C|+) = \frac{P(+ \text{ and } C)}{P(+)} = \frac{P(+|C)P(C)}{P(+|C)P(C) + P(+|\neg C)P(\neg C)}$$

Using the data given, we thus have $P(C|+) = \frac{.90 \times .02}{.90 \times .02 + .04 \times .98} = \frac{.018}{.0572} \approx .315$. This answer is reasonable. Of 100 people, about 4 without cancer test positive, and about 2 with cancer test positive, so among the roughly 6 testing positive, about 1/3 have cancer. ∎

We provide one more example to show the care that must be exercised when making statements about ratios and probabilities.

9.20. Example. *Simpson's Paradox.* Many people believe that if A performs better than B in every category, then the overall performance of A must be better than that of B. We present a counterexample using airline performance.[†] The phenomenon is called **Simpson's Paradox**.

Destination	Alaska		America West	
	% on time	# arrivals	% on time	# arrivals
Los Angeles	88.9	559	85.6	811
Phoenix	94.8	233	92.1	5255
San Diego	91.4	232	85.5	448
San Francisco	83.1	605	71.3	449
Seattle	85.8	2146	76.7	262
Total	86.7	3775	89.1	7225

In 1987, airlines in the United States had to report the percentage of their flights that arrived on time at each of the nation's 30 busiest airports. Alaska Airlines served only five of these airports and performed better than America West at every one of them, but America West had a higher overall on-time average at these airports.

The explanation is that on-time performance depends on weather. Alaska Airlines serves primarily Seattle, with habitual bad weather; America West serves sunny Phoenix. Although Alaska always did better under comparable conditions, the overall statistic for America West is dominated by service under easy conditions, while Alaska is judged mostly by the airport where weather prevents good performance. ∎

Other instances of Simpson's Paradox appear in Exercises 19–20.

RANDOM VARIABLES AND EXPECTATION

We often associate a number with each outcome of an experiment; this defines a function on the probability space. We think of its "expected" value as the average of its values over many trials.

[†] A. Barnett, How numbers can trick you, *Technology Review* (1994), 38–45.

9.21. Definition. Let S be a finite probability space. A **random variable** is a function $X: S \to \mathbb{R}$. Each level set $I_X(k)$, the subset of S on which X takes the value k, is an event. We write $P(X = k)$ for the probability of this event. The **expectation** or **expected value** of X, written $E(X)$, is $\sum_k k \cdot P(X = k)$. In terms of individual points in the space, we can also write $E(X) = \sum_{a \in S} X(a) P(a)$.

9.22. Example. *Average grade.* We select a student at random from a class of n students, with each student having probability $1/n$ of being chosen. Suppose that the numbers of students receiving grades A, B, C, D, F are a, b, c, d, f, respectively, where $a+b+c+d+f = n$. Let X be the random variable whose value is the numerical value of the chosen student's letter grade. The average grade of the students is $(4a+3b+2c+1d+0f)/n$, which is the expectation $E(X)$.   ∎

9.23. Solution. *The binomial distribution.* We flip a coin n times, independently; the probability of heads is p on each toss. Let X be the number of heads obtained; this is a random variable with the possible values $0, \ldots, n$. As computed in Example 9.16, the probability that X equals k is $\binom{n}{k} p^k (1-p)^{n-k}$, since there are $\binom{n}{k}$ arrangements with k heads, each having probability $p^k (1-p)^{n-k}$. To compute the expectation, we have

$$E(X) = \sum_{k=0}^{n} k \binom{n}{k} p^k (1-p)^{n-k} = \sum_{k=1}^{n} n \binom{n-1}{k-1} p^k (1-p)^{n-k}$$

$$= np \sum_{k=1}^{n} \binom{n-1}{k-1} p^{k-1} (1-p)^{(n-1)-(k-1)} = np[p + (1-p)]^{n-1} = np.$$

Here we dropped the term for $k = 0$ (it equals 0), used Lemma 5.27 to extract the factor of n, and applied the Binomial Theorem.   ∎

There is a simpler way to compute $E(X)$ in Solution 9.23, using a fundamental and intuitive property of expected value.

9.24. Example. The expected total number of newspapers sold daily at newsstands in New York is the sum of the expected number sold at each individual newsstand. We can compute the total sales for a year in two ways. We can sum the sales for each day, or we can sum the sales for each newsstand. We then divide by the number of days to obtain the expectation.   ∎

9.25. Proposition. (Linearity of Expectation) Let X and $X_1, \ldots, X_n$ be random variables on a finite probability space.
  a) For $c \in \mathbb{R}$, $E(cX) = cE(X)$.
  b) $E(X_1 + \cdots + X_n) = E(X_1) + \cdots + E(X_n)$.

Proof: Both computations use the definition of expectation and the distributive law. For (a),

$$E(cX) = \sum_{a \in S} cX(a) = c \sum_{a \in S} X(a) = cE(X).$$

For (b), we also interchange the order of summation to compute

$$E(X) = \sum_{a \in S} X(a)P(a) = \sum_{a \in S} \left[\sum_{i=1}^{n} X_i(a) \right] P(a)$$

$$= \sum_{i=1}^{n} \left[\sum_{a \in S} X_i(a)P(a) \right] = \sum_{i=1}^{n} E(X_i) \quad \blacksquare$$

9.26. Solution. *The binomial distribution, revisited.* When we perform n trials with success probability p, we can define a variable X_i for the ith trial, with $X_i = 1$ if this trial is a success, and $X_i = 0$ if it is not. Letting X be the total number of successes, we have $X = \sum X_i$. Each X_i has expectation p, since $E(X_i) = 1 \cdot p + 0 \cdot (1 - p) = p$. By the linearity of the expectation, $E(X) = \sum E(X_i) = np$. $\quad \blacksquare$

The conclusion in Solution 9.26 does not require independence for the various events H_i of getting heads on the ith trial. This simpler computation thus gives a stronger result than the computation in Solution 9.23, because there the formula for the probabilities of the sample points depends on independence of the trials.

The random variables X_i in Solution 9.26 are called **indicator variables** because their value (0 or 1) indicates whether a particular event happens. Their use often simplifies the computation of expectation.

9.27. Application. Suppose that A, B, and n other people line up in random order. What is the expected number of people between A and B?

For each i, let $X_i = 1$ if the ith person stands between A and B, and let $X_i = 0$ otherwise. The expected number of people between A and B is then $E(X)$, where $X = \sum X_i$. Since $E(X_i) = 1/3$ for each i, we have $E(X) = \sum E(X_i) = n/3$. $\quad \blacksquare$

Suppose we are performing Bernoulli trials with independent success probability p and stop the experiment when we first obtain a success. This leads to an infinite probability space. Let X be the number of trials performed in the experiment. The probability that X takes the value k when we perform the experiment is $p(1 - p)^{k-1}$.

We assume that the experiment always has an outcome; in other words, the probability of *never* obtaining a success is 0. This is consistent with the statement that $\sum_k p(1 - p)^{k-1} = 1$.

9.28. Definition. Given a number p with $0 < p < 1$, let X be a random variable defined on $\mathbb{N}$ by $P(X = k) = p(1 - p)^{k-1}$. We say that X is a **geometric random variable** with parameter p.

The sum $\sum_k p(1 - p)^{k-1}$ has infinitely many terms. In Chapter 14 we discuss the precise meaning of infinite summation. Here we treat the issue informally to discuss the expectation of a geometric random variable X. In light of Definition 9.21, we define $E(X)$ by the infinite series

$$E(X) = \sum_{k=0}^{\infty} k P(X = k) = \sum_{k=0}^{\infty} k p (1 - p)^{k-1}.$$

Our informal argument to evaluate this sum uses properties of infinite series from Chapter 14 (such as the distributive law). The discussion assumes that the sum has a value; the methods of Chapter 14 provide several proofs of this.

9.29. Proposition. In Bernoulli trials with success probability p, the expected number of trials to obtain the first success is $1/p$.

Proof: (sketch) The desired value is $E(X)$ for a geometric random variable X with parameter p. We want to show that $\sum_{k=0}^{\infty} k p (1 - p)^{k-1} = 1/p$.

Consider $(1 - 2x + x^2) \sum_{k=1}^{\infty} k x^{k-1}$. We write the result of the multiplication by collecting the terms for each power of x. For x^0, the coefficient is 1. For x^1, the coefficient is $-2 \cdot 1 + 1 \cdot 2 = 0$. For $k \geq 2$, the coefficient is $1(k + 1) - 2(k) + 1(k - 1) = 0$. Everything cancels except the constant term, and therefore the product is 1.

From $(1 - x)^2 \sum_{k=1}^{\infty} k x^{k-1} = 1$, we obtain $\sum_{k=1}^{\infty} k x^{k-1} = \frac{1}{(1-x)^2}$, when $x \neq 1$. Setting $x = 1 - p$ and multiplying by p completes the proof. ∎

9.30. Solution. *The Coupon Collector.* We must obtain all n coupons, first one, then another, then a third, etc.

Let X_k be a random variable giving the number of meals eaten to obtain the next coupon when k coupons remain. The probability of getting a new coupon at the next meal is k/n. Thus the wait for a new coupon is the wait for the first success among Bernoulli trials with success probability k/n. The number of trials, X_k, has the geometric distribution. By Proposition 9.29, $E(X_k) = n/k$.

The total number of meals to obtain all coupons is $X = \sum_{k=1}^{n} X_k$. By the linearity of the expectation, $E(X) = n \sum_{k=1}^{n} 1/k$. When $n = 5$, for example, the expectation exceeds 11. ∎

Our final application of expectation uses probability in a different way. Instead of computing a probability in a given experiment, we are free to choose a probability to get the best result.

In many games, including professional sports, we have a set of options. We define a probability space on these options, assigning each some probability unknown to our opponent. Suppose that we have two options, with payoffs a and b. If we assign the probabilities x and $1 - x$ to the options, our expected payoff is $ax + b(1 - x)$. We want to choose x to maximize the expected payoff. The difficulty arises when the payoffs a and b depend on the opponent's choices.

9.31. Application. *The Odd/Even Finger Game.* This game has two players, A and B. On each play of the game, each player shows 1 or 2 fingers. The payoffs from A to B appear below for the four possible outcomes. The payment is the sum of the fingers showing; A wins it when the total is even, and B wins it when the total is odd.

	A shows 1	A shows 2
B shows 1	-2	$+3$
B shows 2	$+3$	-4

Although this seems like a fair game, it favors B. If B always shows 1 or always shows 2, then A can use this information to win. Hence B should show 1 with some probability x and show 2 with probability $1 - x$. Before each play, B privately performs an experiment with success probability x to determine how many fingers to show. Player A may know the strategy x, but A does not know what B will show on a particular play.

Solution 1. Knowing the strategy x, A can compute the expected payment to B under each option and play only the column with the smaller expectation. Hence x guarantees for B the minimum of the expectations in the two columns. These are $-2x + 3(1 - x)$ and $3x - 4(1 - x)$, which simplify to $3 - 5x$ and $7x - 4$. Player B chooses x to maximize their minimum. Since their graphs cross, the minimum is maximized when they are equal; $3 - 5x = 7x - 4$ yields $x = 7/12$. By choosing $x = 7/12$, B guarantees average payoff per game of at least $1/12$.

On the other hand, Player A can limit the average payoff to $1/12$. When A plays column 1 with probability y, the expected payoff to B is at most $\max\{-2y + 3(1 - y), 3y - 4(1 - y)\}$. Player A chooses y to minimize this. The minimum occurs at $y = 7/12$, where the two values equal $1/12$.

Solution 2. We treat the players symmetrically, with B using strategy x and A using strategy y. The expected payoff (to B) is

$$-2xy + 3y(1 - x) + 3x(1 - y) - 4(1 - y)(1 - x),$$

which equals $7x - 4 + y(7 - 12x)$. Again the value is $1/12$ when $x = 7/12$. If $x < 7/12$, then A does best with $y = 0$, and the expectation $7x - 4$ is then less than $1/12$. If $x > 7/12$, then A does best with $y = 1$ to make the payoff $3 - 5x$, again less than $1/12$. We have proved that $x = 7/12$ is the optimal choice for B. Choosing $x = 7/12$ makes the value independent of y. This corresponds to making the column expectations equal in Solution 1.

We can rewrite the expectation as $7y - 4 + x(7 - 12y)$. By a similar analysis, choosing $y = 7/12$ is optimal for Player A. ∎

MULTINOMIAL COEFFICIENTS

Many counting problems that involve two options generalize naturally to questions about m options, where $m \in \mathbb{N}$. In the Ballot Problem, there are $\binom{a+b}{b}$ elections that reach the final score (a, b). How does this generalize when we have m candidates and the final score is $(a_1, \ldots, a_m)$? We have seen that the binomial coefficient $\binom{n}{k}$ counts n-tuples having k entries with one value and $n - k$ entries with another. We have also seen that it is the coefficient of $x^k y^{n-k}$ in the expansion of $(x + y)^k$. We generalize these questions to m candidates or m types of letters or polynomials with m variables.

9.32. Definition. Suppose $k_1, \ldots, k_m$ are nonnegative integers summing to n. The **multinomial coefficient**, written $\binom{n}{k_1, \ldots, k_m}$, is the number of ways to arrange n objects of m types in a row, where there are k_i objects of type i.

The binomial coefficient counts arrangements of two types of objects. Suppose we have three types of objects. If our objects are a, b, b, c, c, then after placing a in one of five possible locations, there are six ways to complete an arrangement, since we choose the positions for the two b's from the remaining four positions. Hence $\binom{5}{1,2,2} = 30$. Generalizing this, we can choose positions for the first type of object, then positions for the second type, and so on, to obtain the formula $\binom{n}{k_1, \ldots, k_m} = \binom{n}{k_1}\binom{n-k_1}{k_2}\binom{n-k_1-k_2}{k_3} \cdots$. In Theorem 9.33, we provide a direct argument for a simpler formula.

9.33. Theorem. If $k_1, \ldots, k_m$ are nonnegative integers summing to n, then
$$\binom{n}{k_1, \ldots, k_m} = \frac{n!}{k_1! \cdots k_m!}.$$

Proof: Let M be the number of arrangements consisting of k_i letters of the ith type, for each i. We can turn such an arrangement into an arrangement of distinct objects by putting labels (e.g., subscripts) on the k_i letters of type i, for each i. For each i in a particular arrangement, we can assign the labels in $k_i!$ ways. Hence in total we have formed $M \prod_{i=1}^{m} k_i!$ arrangements of n distinct letters. Since we have made the letters distinct, this must equal $n!$, the total number of arrangements of n distinct letters. Hence $M = n!/\prod_{i=1}^{m} k_i!$. ∎

9.34. Example. Roll a balanced six-sided die 21 times. What is the probability of rolling exactly one 1, exactly two 2s, and so on, up to exactly six 6s? Answer: $\binom{21}{1,2,3,4,5,6}(1/6)^{21} = 0.0000935969$. ∎

The name "multinomial coefficient" arises from the expansion of polynomials with several variables.

9.35. Corollary. The number $\binom{n}{k_1,\dots,k_m}$ is the coefficient of $x_1^{k_1} \cdots x_m^{km}$ in the expansion of $(x_1 + \cdots + x_m)^n$.

Proof: The monomial $x_1^{k_1} \cdot \cdots \cdot x_m^{km}$ arises once in the expansion of $(x_1 + \cdots + x_m)^n$ for each way to arrange a set consisting of k_i copies of x_i for each i. Each such arrangement corresponds to a term in the expansion of the product. The jth position in the arrangement corresponds to the term chosen from the jth factor in $(x_1 + \cdots + x_m) \cdots (x_1 + \cdots + x_m)$. ∎

9.36. Example. *Trinomial expansion.*

$$(x + y + z)^3 = x^3 + y^3 + z^3 + 3x^2y + 3x^2z + 3y^2z + 3y^2x + 3z^2x + 3z^2y + 6xyz$$ ∎

9.37. Corollary. If p is prime and $\sum_{i=1}^m k_i = p$ with $0 \le k_i < p$, then p divides $\binom{p}{k_1,\dots,k_m}$.

Proof: Because the multinomial coefficient is the size of a finite set, Theorem 9.33 implies that $M = p!/\prod_{i=1}^m k_i!$ is an integer. Writing this as $p! = M \prod_{i=1}^m k_i!$, we observe that the left side is divisible by p. The factorials on the right side do not have p as a factor. Since p is prime, this implies that p divides M. ∎

Corollary 9.37 yields a remarkably short proof of Fermat's Little Theorem, due to Gottfried Wilhelm Leibniz (1646–1716). Exercise 6.37 requests a related proof using the Binomial Theorem.

9.38. Example. *Fermat's Little Theorem.* To prove that $a^{p-1} \equiv 1 \bmod p$ when p is prime and a is an integer not divisible by p, we prove that $a^p \equiv a \bmod p$. Modular computation allows us to assume that a is positive. Expressing a as $\sum_{i=1}^a 1$, we consider the expansion of $(1 + \cdots + 1)^p$, which we treat as $(x_1 + \cdots + x_a)^p$ with each x_i equal to 1. By Corollary 9.35, the coefficient of $x_1^{k_1} \cdots x_a^{k_a}$ is $\binom{p}{k_1,\dots,k_a}$. For each term x_i^p, in which the exponents on the variables other than x_i are 0, the coefficient equals 1; there are a of these. By Corollary 9.37, all the other coefficients are divisible by p. Hence $a^p = (1 + \cdots + 1)^p \equiv a \bmod p$. ∎

9.39. Proposition. (The Multinomial Distribution) Suppose an experiment has m possible outcomes, with p_j being the probability of the jth outcome and $\sum_{j=1}^m p_j = 1$. If we perform n independent trials, then the probability that for each j the jth outcome occurs exactly k_j times is $\binom{n}{k_1,\dots,k_m} p_1^{k_1} \cdots p_m^{km}$.

Proof: There are m^n possible lists of outcomes for n successive trials. Since the trials are independent, the probability of each particular list in which the jth outcome occurs precisely k_j times is $\prod_{j=1}^m p_j^{k_j}$. The number of lists of this type is the number of ways to arrange these outcomes (k_j of type j for all j) in a row, which equals the multinomial coefficient $\binom{n}{k_1,\ldots,k_m}$. The different lists are mutually exclusive events, so the probability is the number of lists of this type times the probability of each one. ∎

9.40. Solution. *Hitting for the cycle.* Our baseball player bats randomly, meaning that the at-bats are independent trials. An at-bat produces a single with probability .15, a double with probability .06, a triple with probability .02, a home run with probability .07, and otherwise an out. This describes a good hitter, whose batting average is .300 and "slugging average" (expected number of bases per at-bat) is .610. "Hitting for the cycle" means getting at least one hit of each type in a single game. What is the probability that this player hits for the cycle if he bats exactly five times in a game?

There are five ways this can occur. There can be one of each hit and one out, or two of one type of hit and one each of the three others. We use the multinomial distribution to compute each probability:

one of each hit, one out: $5!(.15)(.06)(.02)(.07)(.70) = 0.0010584$
two singles: $(5!/2)(.15)^2(.06)(.02)(.07) = 0.0001134$
two doubles: $(5!/2)(.15)(.06)^2(.02)(.07) = 0.00004536$
two triples: $(5!/2)(.15)(.06)(.02)^2(.07) = 0.00001512$
two home runs: $(5!/2)(.15)(.06)(.02)(.07)^2 = 0.00005292$.

These events are mutually exclusive, so we add the probabilities, obtaining the answer 0.0012852, which is about 1 in 800. Actually, this overestimates the player's probability of hitting for the cycle in a given game (see Exercise 31). ∎

EXERCISES

Exercises 1–6 consider events A and B in a probability space S. In each exercise, determine whether the statement is true or false. If true, provide a proof; if false, provide a counterexample.

9.1. If $A \subset B$, then $P(A) \le P(B)$.

9.2. If $P(A)$ and $P(B)$ are not zero, and $P(A|B) = P(B|A)$, then $P(A) = P(B)$.

9.3. If $P(A)$ and $P(B)$ are not zero, and $P(A|B) = P(B|A)$, then A and B are independent.

9.4. If $P(A) > 1/2$ and $P(B) > 1/2$, then $P(A \cup B) > 0$.

9.5. If A, B are independent, then A and B^c are independent.

9.6. If A, B are independent, then A^c and B^c are independent.

•   •   •   •   •

9.7. Determine when an event and its complement are independent.

9.8. (−) A man goes to his favorite restaurant often. With probability 1/2, he orders the pasta special. With probability 1/2, he orders the fish special. When he orders the pasta special, the probability is 1/2 that it is out of stock. When he orders the fish special, the probability is 1/2 that it is out of stock. What is the probability that the dish he orders is out of stock? Generalize the problem using variables for the probabilities.

9.9. Each of three containers has two marbles; one contains two red marbles, one contains two black marbles, and one contains one red and one black. A container is selected at random (each equally likely), and one of the two marbles inside is selected at random (each equally likely). Given that the selected ball is black, what is the probability that the other ball in its container is black?

9.10. We roll two dice, one red and one green. Under each assumption below, what is the probability that the roll is double-sixes?
   a) The red die shows a six.
   b) At least one of the dice shows a six. Does the method of obtaining this information affect the answer?

9.11. In a famous game show on television, a prize is placed behind one of three doors, with probability 1/3 for each door. The contestant chooses a door. The host then opens one of the other doors and says "As you can see, the prize is not behind this door. Do you want to stay with your original guess or switch to the remaining door?" When the contestant has chosen a wrong door, the host opens the other wrong door. When the contestant has chosen the right door, the host opens one of the two wrong doors, each with probability 1/2.
   Show that the contestant should switch.

9.12. (+) *Bertrand's Paradox.* In Example 9.12 (generating a random chord of the unit circle), let p be the probability that the length of the chord exceeds $\sqrt{3}$.
   a) Suppose the endpoints of the chord are generated by two random spins on the circumference of the circle. Prove that $p = 1/3$. (Assume that spinner points to an arc with probability proportional to the length of the arc.)
   b) Suppose the midpoint of the chord is generated by throwing a dart at the circle. Prove that $p = 1/4$. (Assume that the probability the dart lands in a region is proportional to the area of the region.)
   c) Devise a model for generating the chord that yields $p = 1/2$.

9.13. From n equally spaced points on a circle, a triple of three distinct points is chosen at random. What is the probability that they form an equilateral triangle? An isosceles triangle? A triangle with sides of distinct lengths?

9.14. In Bertrand's Ballot Problem (Problem 9.1), suppose the outcome is (a, b), with $a > b$, and the votes are counted in random order. What is the probability that A is always ahead of B (after the beginning)? What is the probability that the score is tied at some point during the election after the beginning?

9.15. (+) Let m 0s and n 1s be placed in some order around a circle. A position with a 0 is *good* if every arc of the circle extending clockwise from there contains more 0s than 1s. Prove that every arrangement of the elements on the circle has exactly $m - n$ good positions. Apply this to solve Bertrand's Ballot Problem.

9.16. Let X_1, X_2, X_3 be random variables such that $P(X_i = j) = 1/n$ for $(i, j) \in$ $[3] \times [n]$. Compute the probability that $X_1 + X_2 + X_3 \leq 6$ given that $X_1 + X_2 \geq 4$. Assume that $P(X_1 = a_1, X_2 = a_2, X_3 = a_3) = P(X_1 = a_1)P(X_2 = a_2)P(X_3 = a_3)$.

9.17. The fraction of the games that a tennis player wins against each of her four opponents is .6, .5, .45, .4, respectively. Suppose that she plays 30 matches against each of the first two and 20 matches against each of the last two. Given that she wins a particular match, what is the conditional probability that it is against the ith opponent, for $i \in \{1, 2, 3, 4\}$?

9.18. Half the females and one-third of the males in a class are smokers. Also, two-thirds of the students are male. What fraction of the smokers are female?

9.19. In baseball, "batting average" is defined as the fraction "Hits/(At-bats)". Consider two players A and B. Suppose their performance in day games and night games is as follows:

	Day		Night	
	A	B	A	B
Hits	a	c	w	y
At-bats	b	d	x	z

Find values for a, b, c, d, w, x, y, z so that A has a higher batting average than B in both day games and night games but B has a higher batting average overall.

9.20. (!) Consider universities H and Y, each having 100 professors. Construct an example where, in each of the categories "assistant professors", "associate professors", and "full professors", the proportion who are women is higher at H than at Y, and yet Y has more female professors than H.

9.21. In bowling, a *strike* occurs when the bowler knocks down all the pins in one roll. A *perfect game* consists of 12 consecutive strikes. Suppose that on each roll a bowler has probability p of rolling a strike. What must be the value of p so that the probability of a perfect game is .01? Use a calculator to estimate the answer.

9.22. Suppose that A, B, and n other people stand in a line in random order. For each k with $0 \leq k \leq n$, find the probability that exactly k people stand between A and B. Check that the sum of these probabilities equals one.

9.23. Beginning with A, players A and B alternate flipping a coin that has probability p of showing heads. The first player to get heads wins. Let x be the probability that A wins. Determine x as a function of p. Evaluate the formula in the special case of a fair coin, $p = .5$. (Hint: Use conditional probability to obtain an equation for x.)

9.24. Consider a dial having a pointer that is equally likely to point to each of n regions numbered $1, 2, \ldots, n$ in cyclic order. When the selection is k, the gambler receives 2^k dollars.

   a) What is the expected payoff per spin of the dial?

   b) Suppose that the gambler has the following option. After the spin, the gambler can accept that payoff or flip a coin to change it. If the coin shows heads, the pointer moves one spot counterclockwise; if tails, it moves one spot clockwise. When should the gambler flip the coin? What is the expected payoff under the optimal strategy?

9.25. (+) Consider n envelopes with amounts $a_1, \ldots, a_n$ in dollars, where $a_1 \leq \cdots \leq a_n$. A gambler is presented two successive envelopes, with the probability being p_i that the envelopes contain a_i and a_{i+1} dollars, for $1 \leq i \leq n-1$. He opens one of these two envelopes at random and sees what it contains. He can then either keep that amount or switch to the other envelope. Suppose that he sees a_k dollars. In terms of the data of the problem, determine whether he should switch.

9.26. Suppose X is a random variable that takes values only in $[n]$. Prove that $E(X) = \sum_{k=1}^{n} P(X \geq k)$.

9.27. A drunk has n keys, and only one will open the door. He tries keys at random. Under each model below, what is the expected number of selections until he opens the door?

a) He selects keys in a random order (without replacement) until one works.

b) After each mistake, he replaces the key and selects randomly again.

9.28. (!) Suppose that n pairs of socks are put into the laundry, with each sock having one mate. The laundry machine randomly eats socks; a random set of k socks returns. Determine the expected number of complete pairs of returned socks. (Hint: Use the linearity of expectation.)

9.29. Suppose that $2n$ people are partitioned into pairs at random, with each partition being equally likely. If the set consists of n men and n women, what is the expected number of male-female couples?

9.30. How many arrangements are there of the letters in $MISSISSIPPI$?

9.31. (For baseball enthusiasts.) Explain why the computation in Solution 9.40 overestimates the probability that the batter hits for the cycle in a given game.

9.32. (!) Find one polynomial p such that $p(n) = 3^n$ for $n = 0, 1, 2, 3, 4$. (Hint: Express 3^n as $(1+1+1)^n$ and use Theorem 9.33, letting $k_3 = n - k_1 - k_2$.)

9.33. Consider an experiment in which all the monomials in k variables with total degree n are equally likely to occur (0 is allowed as an exponent).

a) Determine the probability that all k variables have positive exponent in the chosen monomial.

b) For $(n, k) = (10, 4)$, determine the probability that the exponents are different. (Here 0 is allowed as an exponent.)

9.34. We have six dice; faces are equally likely to appear when a die is rolled. Each die has three red faces, two green faces, and one blue face. We roll the six dice. Derive a formula for the probability that we get a red face on three dice, a green face on two dice, and a blue face on one. (Hint: The answer reduces to a fraction with denominator 36.)

9.35. Determine the coefficient of $x^4 y^{16}$ in the expansion of $(x + xy + y)^{16}$. Determine the coefficient of $x^4 y^8$ in the expansion of $(x^2 + xy + y^2)^6$.

9.36. Let A be the set of lattice paths from $(0, 0)$ to (n, n) that do not move above the line given by $y = x$. Let B be the set of nondecreasing functions $f : [n] \to [n]$ such that $f(i) \leq i$ for all i. Establish a bijection from A to B.

9.37. Let a_n denote the number of lattice paths of length $2n$ that never step above the diagonal (these end at some point $(k, 2n - k)$ with $k \geq n$). Prove that $a_n = \binom{2n}{n}$.

9.38. A **ballot list** of length $2n$ is a binary $2n$-tuple $(b_1, \ldots, b_{2n})$ such that for each i, the number of 1s in $\{b_1, \ldots, b_i\}$ is at least as large as the number of 0s. In the language of Solution 9.10, ballot lists are equivalent to "good elections" with total score (n, n). Establish bijections from the set of ballot lists of length $2n$ to each of the sets below.

a) $2n + 1$-tuples of nonnegative integers in which consecutive entries differ by 1 and $a_1 = a_{2n+1} = 0$.

b) Arrangements of $2n$ people in 2 rows of length n so that heights are increasing in each row and column. (Example: $\binom{1247}{3568}$ is such an arrangement, where the people are 1 through $2n$ in increasing order of height.)

9.39. (+) Place $2n$ points on the boundary of a circle. Establish a bijection to prove that the number of ways to pair up the points by drawing noncrossing chords equals the number of ballot lists of length $2n$.

9.40. *The Finger Game (Application 9.31).*

a) Which values of x in the interval $[0, 1]$ guarantee a positive expectation for B no matter what A does?

b) We have seen that when each player shows one finger with probability $7/12$, B expects to win an average of $1/12$ dollars per game. With these strategies, what *proportion* of the games does B expect to win?

9.41. Let $\binom{a\ b}{c\ d}$ record the payoffs from A (column player) to B (row player) in a game in which each player has two options. Determine the conditions on a, b, c, d so that playing each option with probability $1/2$ will be optimal for each player.

9.42. Let $\binom{a\ b}{c\ d}$ record the payoffs from A (column player) to B (row player) in a game in which each player has two options. In terms of a, b, c, d, determine the maximum amount that B can guarantee receiving by choosing $x \in [0, 1]$ and playing the first row with probability x and the second with probability $1 - x$. (Hint: There are several cases, depending on the relative values of a, b, c, d.)

Chapter 10

Two Principles of Counting

In this chapter, we study two proof techniques in discrete mathematics, the Pigeonhole Principle and the Inclusion-Exclusion Principle. We consider these together because both are fairly easy to prove but have elegant applications that may require some cleverness to discover. They also can make it possible to avoid lengthy analysis by cases.

The Inclusion-Exclusion Principle is used to solve counting problems. The Pigeonhole Principle is a principle of counting in the sense that it considers cardinalities of sets, but its applications are to existence problems and extremal problems rather than to enumerative problems.

THE PIGEONHOLE PRINCIPLE

"Of three ordinary people, two must have the same sex."[†] The Pigeonhole Principle is also called the "Dirichlet drawer principle" in honor of Peter Gustav Lejeune-Dirichlet (1805–1859). It implies that extracting $n + 1$ shoes from a closet containing n pairs of shoes must produce a matched pair of shoes; they cannot all come from different pairs.

We proved a version of the Pigeonhole Principle in Chapter 2: in any set of real numbers, some number must be at least as large as the average. We have made arguments already using the essence of the principle (see Exercise 4.44, Solution 5.24, Theorem 6.21, and Lemma 7.27). The principle itself is elementary; the subtlety arises in the applications.

10.1. Theorem. (Pigeonhole Principle) Placing more than kn objects into n classes puts more than k objects into some class.

[†]This observation is attributed to Professor D. J. Kleitman of the Massachusetts Institute of Technology.

Proof: We prove the contrapositive. If no class has more than k objects, then the total number of objects is at most kn. This uses the property, proved by induction in Proposition 3.12, that the n inequalities $m_i \le k$ can be summed to obtain the inequality $\sum_{i=1}^{n} m_i \le kn$.   ■

To apply the Pigeonhole Principle, we must determine what should play the role of the objects and what should play the role of the classes. Sometimes the Pigeonhole Principle pops up in proof by contradiction.

10.2. Example. *Existence of multiplicative inverses modulo p.* If a and p are relatively prime, then there exists some $b \in \{1, \ldots, p - 1\}$ such that $ab \equiv 1 \bmod p$. Otherwise, $a, 2a, \ldots, (p - 1)a$ fall into the $p - 2$ nonzero congruence classes other than $\bar{1}$. By the Pigeonhole Principle, two fall in the same class. If ia and ja fall in the same class, then $ia \equiv ja \bmod p$ yields $p|(i - j)a$. Since a and p are relatively prime, this implies $p|(i - j)$ (by Proposition 6.6), which implies $i = j$ (since they are less than p), which is a contradiction.   ■

10.3. Example. *A society of friends.* Suppose that "being friends" is a symmetric relation. We prove that in any set S of people with $|S| \ge 2$, there must be two people that have the same number of friends in S. If $|S| = n$, then each person in S has between 0 and $n - 1$ friends in S. We cannot have a person with 0 friends and a person with $n - 1$ friends, however, because a person with $n - 1$ friends is a friend of everyone else. Hence at most $n - 1$ distinct numbers of friends arise among the n people, and some pair must have the same total.   ■

10.4. Example. *Midpoints between integer points.* Given five integer points in the plane, the midpoint of the segment joining some pair of them is also an integer point (an **integer point** is one with integer coordinates).
    The midpoint of the segment between integer points (a, b) and (c, d) is $(\frac{a+c}{2}, \frac{b+d}{2})$. This is an integer point if and only if a and c have the same parity (both odd or both even) and b and d have the same parity. This suggests putting the integer points into four classes by the parity of their x and y coordinates: (odd, odd), (odd, even), (even, odd), and (even, even). With five points, we must have two in the same class, and then the segment joining them has an integer midpoint. With four points, we can have one in each class and avoid having an integer midpoint.   ■

10.5. Example. *Forcing divisible pairs.* If S is a set of $n + 1$ numbers in $[2n]$, then S contains two numbers such that one divides the other. This result is best possible in that the set of n numbers $\{n + 1, n + 2, \ldots, 2n\}$ has no such pair. To apply the Pigeonhole Principle, we partition $[2n]$ into n classes such that for every pair of numbers in the same class, one divides the other. Recall that every natural number has a unique representation as an odd number times a power of two. For each k, the set $\{(2k-1)2^{j-1}: j \geq 1\}$ has the desired property; the smaller of any pair in this set divides the larger. Since there are only n odd numbers less than $2n$, we get the right number of classes. Explicitly, the kth class consists of those numbers in $\{2^{j-1}(2k-1): j \in \mathbb{N}\}$ that are at most $2n$. ∎

The preceding examples could have been phrased as extremal problems: What is the largest number of integer points in the plane such that no segment joining two of them has an integer midpoint? What is the largest size of a subset of $[2n]$ such that no element divides another? The Pigeonhole Principle establishes a bound, and a construction shows that the bound is best possible.

In such a problem, it does not suffice to present the construction and show that no element can be added; this does not forbid larger configurations constructed in other ways. For example, to build a large set that avoids divisible pairs in Example 10.5, it would be reasonable to choose primes. When $n = 5$, we pick the set $\{2, 3, 5, 7\}$ in this way, at which point we cannot add any more elements from $[10]$ without creating a divisible pair. That does not prove that the largest such set has size 4, and indeed $\{6, 7, 8, 9, 10\}$ is a larger example. To solve an extremal problem, our proof must show that all possible examples satisfy the bound.

10.6. Example. *Longest monotone sublist.* Consider a list of $n^2 + 1$ distinct numbers. A subset of the positions forms a **monotone sublist** if the numbers in those positions form an increasing list or a decreasing list when taken in order. For example, in the list $(3, 2, 1, 6, 5, 4, 9, 8, 7, 10)$, the numbers $3, 6, 9, 10$ form an increasing sublist of length 4. Erdős and Szekeres proved in 1935 that every list of $n^2 + 1$ distinct numbers contains a monotone sublist of length at least $n + 1$. Let $a_1, \ldots, a_{n^2+1}$ be the list. For each k, let x_k be the maximum length of an increasing sublist ending with a_k, and let y_k be the maximum length of a decreasing sublist ending with a_k. For the example above, the values of these parameters are

k	1	2	3	4	5	6	7	8	9	10
a_k	3	2	1	6	5	4	9	8	7	10
x_k	1	1	1	2	2	2	3	3	3	4
y_k	1	2	3	1	2	3	1	2	3	1

If there is no monotone sublist of length $n + 1$, then x_k and y_k never exceed n, and there are only n^2 possible pairs (x_k, y_k). Since there are n^2+1 values of k, the Pigeonhole Principle implies that two pairs are the same: $(x_i, y_i) = (x_j, y_j)$ for some $i < j$. If $a_i < a_j$, then $x_j > x_i$; if $a_i > a_j$, then $y_j > y_i$. This contradiction implies that a number exceeding n must appear in one of the pairs. Since there is a list of n^2 distinct numbers having no monotone sublist of length $n+1$ (Exercise 17), the result is best possible. ∎

10.7. Example. *A domino tiling problem.* A six by six checkerboard with 36 squares can be covered exactly by 18 dominoes consisting of two squares each; this is a **tiling** of the checkerboard by dominoes. We prove that every such tiling can be cut between some pair of adjacent rows or adjacent columns without cutting any dominoes. In the picture below, the tiling can be cut along the middle horizontal line.

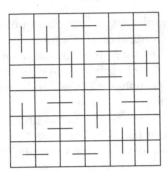

Consider a tiling. Every domino cuts one line between two adjacent rows or between two adjacent columns. There are 18 dominoes and 10 lines that may be cut, so the average number of cuts per line is 1.8. Since every set of numbers contains a number that is at most the average, some line is cut at most once. This is not strong enough to prove the claim, as it leaves the possibility that every line is cut at least once.

To complete the proof, we observe that every line is cut by an even number of dominoes; this implies that a line cut by at most one domino is not cut at all. The observation is easy: having an odd number of dominoes crossing a line would leave an odd number of squares on each side to be paired up by dominoes that don't cross the line, but each set of dominoes covers an even number of squares. ∎

10.8. Example. *The Chess Player Problem.* A chess player wants to practice for a championship match over a period of 11 weeks. She wants to play at least one game per day but at most 132 games in total. No matter how she schedules the games, there must be a period of consecutive days on which she plays a total of *exactly* 22 games.

We can study the total played on consecutive days by considering partial sums. Let a_i be the total number of games played on days 1 through i, and set $a_0 = 0$. Then $a_j - a_i$ is the total number of games played on days $i+1$ through j. We seek an i and a j such that $a_i + 22 = a_j$. This suggests considering both $\{a_j: 1 \le j \le 77\}$ and $\{a_i + 22: 0 \le i \le 76\}$. Since there is at least one game each day, the numbers in $\{a_j\}$ are distinct, as are the numbers in $\{a_i + 22\}$. Hence a duplication among these 154 numbers implies the desired result. Since $a_{77} \le 132$, and $a_{76} + 22 \le 153$, we have 154 numbers in [153], and some number must repeat. Because $a_{76} + 23$

could be as large as 154, this argument does not work to force a period of consecutive days with exactly k games if $k \geq 23$.   ∎

We often use the Pigeonhole Principle to prove existence results. Earlier we proved such statements by building an example of the desired object. The Pigeonhole Principle provides nonconstructive proofs of existence statements; it can also be an effective way to avoid case analysis.

Our examples suggest several remarks about using the Pigeonhole Principle. The classes may have different sizes. Partial sums may help with problems involving order or sums. An example showing that a claim is best possible can suggest classes and objects for applying the Pigeonhole Principle to prove the claim. Finally, the Pigeonhole Principle can be combined with proof by contradiction or other techniques.

THE INCLUSION-EXCLUSION PRINCIPLE

The rules of sum and product used to solve elementary counting problems are not helpful for counting problems involving forbidden conditions, because they lead to lengthy analysis by cases. In contrast, the Inclusion-Exclusion Principle leads quickly to formulas that solve such problems. The principle is based on the inclusion relation on the collection of subsets of a finite set.

10.9. Problem. *Derangements.* A professor collects homework papers from n students and returns them at random for peer grading. In this case, "at random" means that each of the $n!$ permutations is equally likely. A permutation in which no student receives his or her own paper is a **derangement**. What is the probability that a random permutation is a derangement?   ∎

10.10. Problem. *Dice-rolling.* We roll a six-sided die until each of the numbers one through five have appeared at least once. What is the probability that we succeed in the first n rolls?   ∎

10.11. Problem. *Euler totient.* Given a positive integer m, let $\phi(m)$ be the number of elements of $[m]$ that are relatively prime to m. The function $\phi \colon \mathbb{N} \to \mathbb{N}$ is the **Euler totient function**. How can we compute $\phi(m)$?   ∎

We begin by discussing the Euler totient for numbers with few prime factors. Recall that m and r are relatively prime if and only if $\gcd(m, r) = 1$. If m is a power of a prime p, then all numbers in $[m]$ are relatively prime to m except the multiples of p. There are m/p such multiples, so $\phi(m) = m - m/p$.

Next suppose that m has precisely two prime factors, p and q. We eliminate all m/p multiples of p and all m/q multiples of q in $[m]$, but this means that we have eliminated all the multiples of pq twice. We add these back in to correct the count and obtain $\phi(m) = m - m/p - m/q + m/pq$.

In general, m will have prime factors $p_1, \ldots, p_n$. Initially we include all of $[m]$. Excluding the multiples of each prime factor discards more than once every element divisible by more than one of the prime factors. When we then include the sets divisible by two prime factors, we will have included too often the elements divisible by more than two of them. Eventually the process of including and excluding will count each element the proper number of times.

Before presenting the general formula for the totient function, we develop the general setting for the Inclusion-Exclusion Principle. Consider a universe U of objects, in which we want to count the objects that appear in none of the n subsets $A_1, \ldots, A_n$. Each such subset corresponds to a forbidden condition.

In the totient problem, the universe is $[m]$, and the set A_i is the set of multiples of the ith prime factor. In the derangements problem, the universe U is the collection of all permutations of $[n]$. In order to count those with no fixed point, we will let $A_i \subseteq U$ be the set of permutations fixing i, and then the derangements are precisely the permutations that appear in none of these sets.

Let $N_\emptyset$ denote the number of elements of U that appear in none of the n specified subsets $A_1, \ldots, A_n$ of U. If $n = 1$, then $N_\emptyset$ counts the elements outside A_1, so $N_\emptyset = |U| - |A_1|$.

For $n = 2$, consider the Venn diagram below. We don't want to count elements in A_1 or A_2, so we subtract those from the total. This subtracts the elements of $A_1 \cap A_2$ twice, so we add those in again, obtaining $N_\emptyset = |U| - |A_1| - |A_2| + |A_1 \cap A_2|$. The Venn diagram makes it apparent that every element outside $A_1 \cup A_2$ makes a net contribution of 1, and every element inside $A_1 \cup A_2$ makes a net contribution of 0. (To count the elements belonging to *at least one* of the sets, the formula is $|A_1| + |A_2| - |A_1 \cap A_2|$. Readers who have studied Chapter 9 should note that dividing this by $|U|$ yields the probability of $A_1 \cup A_2$ when choosing a random element of U.)

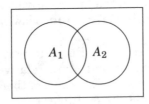

Before deriving the general formula, we also discuss the case $n = 3$ in detail. The reader may use the Venn diagram below to keep track of the "including" and "excluding" as we describe it.

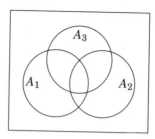

Again we start with all of U. To omit the elements belonging to each $\{A_i\}$, we subtract $|A_1| + |A_2| + |A_3|$ from $|U|$. Any element belonging to more than one of the sets has been deleted more than once, so we add $|A_1 \cap A_2| + |A_2 \cap A_3| + |A_1 \cap A_3|$ to correct this. Now an element in none of the sets contributes 1 to the count, an element in exactly one set contributes $1 - 1 = 0$, and an element in exactly two sets contributes $1 - 1 - 1 + 1 = 0$, but an element in all three sets contributes $1 - 1 - 1 - 1 + 1 + 1 + 1 = 1$. We subtract $|A_1 \cap A_2 \cap A_3|$ as a final correction. Thus the inclusion-exclusion formula for $N_\emptyset$ when there are three forbidden sets is

$$|U| - (|A_1| + |A_2| + |A_3|) + (|A_1 \cap A_2| + |A_2 \cap A_3| + |A_1 \cap A_3|) - |A_1 \cap A_2 \cap A_3|.$$

In general, for each subset S of the indices $1, \ldots, n$, we weight $\left|\bigcap_{i \in S} A_i\right|$ negatively if $|S|$ is odd and positively if $|S|$ is even. The count arising when $S = \emptyset$ is $|U|$, because each element is in every one of no sets. Just as a sum over no terms is the additive identity 0 and a product over no factors is the multiplicative identity 1, so an intersection over no sets is the "intersective identity" U.

10.12. Theorem. (Inclusion-Exclusion Principle) Given a universe U of items and subsets $A_1, \ldots, A_n$ of the items, the number $N_\emptyset$ of items belonging to none of the subsets is given by

$$N_\emptyset = \sum_{S \subseteq [n]} (-1)^{|S|} \left|\bigcap_{i \in S} A_i\right|.$$

Proof: We need only show that each item belonging to none of the sets contributes 1 to the total and that all other items contribute 0. An item x in none of the sets appears only in the term for $S = \emptyset$, so its contribution is 1. Otherwise, let $T \subseteq [n]$ be the nonempty set of indices i such that $x \in A_i$. In the formula, the item x is counted in the term for each subset of T. It contributes $+1$ for each $S \subseteq T$ of even size and -1 for each $S \subseteq T$ of odd size. Hence the total contribution for x is $\sum_{S \subseteq T} (-1)^{|S|} = \sum_{k=0}^{|T|} (-1)^k \binom{|T|}{k}$.

There are many ways to show that this sum is 0. We can treat it as a special case of $\sum_{k=0}^{t} \binom{t}{k} y^k$, with y set to -1. By the Binomial Theorem, the sum is $(1 + y)^t$, and when $y = -1$ it equals 0 since $t > 0$.

We also give a proof using a bijection. Choose $x \in T$. Let $A = \{R \subseteq T: |R| \text{ is odd}\}$ and $B = \{R \subseteq T: |R| \text{ is even}\}$. Given $R \in A$, let $f(R) = R - \{x\}$ if $x \in R$, and let $f(R) = R \cup \{x\}$ if $x \notin R$. Always $f(R)$ has even size. Furthermore, the same definition yields the inverse of f. Hence f is a bijection, and $|A| = |B|$. ∎

The Inclusion-Exclusion Principle is useful when (1) we can model our problem as counting the elements outside some sets $A_1, \ldots, A_n$, and (2) the quantities $\left| \bigcap_{i \in S} A_i \right|$ are easy to compute.

10.13. Solution. *Euler totient.* Suppose that m has n distinct prime factors $p_1, \ldots, p_n$. Within the universe $U = [m]$, we define the set A_i to be the multiples of p_i. The numbers relatively prime to m are the elements in none of $A_1, \ldots, A_n$. To apply the Inclusion-Exclusion Formula, we need the sizes of intersections of these sets. For the intersection of the sets indexed by the elements of $S \subseteq [n]$, we have $\left| \bigcap_{i \in S} A_i \right| = m / \prod_{i \in S} p_i$. By the Inclusion-Exclusion Principle, we thus have

$$\phi(m) = N_\emptyset = \sum_{S \subseteq [n]} (-1)^{|S|} \left| \bigcap_{i \in S} A_i \right| = \sum_{S \subseteq [n]} (-1)^{|S|} \frac{m}{\prod_{i \in S} p_i}.$$

For example, $60 = 2^2 \cdot 3 \cdot 5$, so we compute

$$\phi(60) = 60 - \frac{60}{2} - \frac{60}{3} - \frac{60}{5} + \frac{60}{6} + \frac{60}{10} + \frac{60}{15} - \frac{60}{30} = 16.$$

Another formula for $\phi(m)$ appears in Exercise 30. ∎

10.14. Solution. *Derangements.* We can model Problem 10.9 by writing the numbers $1, \ldots, n$ (for the papers) in the positions $1, \ldots, n$ (for the students). We want to count the permutations of $[n]$ such that no i is in position i. An instance of i in position i is a fixed point; the derangements are the permutations with no fixed points.

Within the universe U of permutations of $[n]$, let A_i be the set of permutations that leave i fixed. Because derangements have no fixed points, $D_n = N_\emptyset$. Consider a set $S \subseteq [n]$ with $|S| = k$. A permutation lies in all sets indexed by S if and only if it fixes $\{i : i \in S\}$. It can permute the other elements in any way (including fixing them), so $\left| \bigcap_{i \in S} A_i \right| = (n - k)!$. There are $\binom{n}{k}$ choices of S with size k, and we weight these contributions by $(-1)^{|S|}$, so the formula is

$$D_n = \sum_{k=0}^{n} (-1)^k \binom{n}{k} (n - k)! = n! \sum_{k=0}^{n} (-1)^k / k!.$$

Dividing by $n!$ yields $\sum_{k=0}^{n}(-1)^k/k!$ for the probability of a derangement. Surprisingly, the probability is almost independent of n and tends to a nonzero limit as n grows. The alternating sum converges rapidly to $1/e$, where $e = 2.71828\cdots$ (see Chapter 14). We discuss further properties of derangements in Chapter 12.   ∎

In the derangements computation, the size of $\bigcap_{i \in S} A_i$ depends only on $|S|$. This allows us to combine the terms for all sets S with $|S| = k$. The factor $\binom{n}{k}$ appears, multiplied by the size of each $\cap - i \in SA_i$ with $|S| = k$. We obtain a summation with $n + 1$ terms instead of a summation with 2^n terms. This simplification occurs often.

There are n^k functions from a k-set A to an n-set B. Using combinatorial arguments, we counted the injective ones in Proposition 5.11. These correspond to listing k distinct elements of B in order and assigning them to $a_1, \ldots, a_k$, and there are $n!/(n - k)!$ ways to do this. We use the Inclusion-Exclusion Principle to count the surjective functions.

10.15. Example. *Surjective functions.* How many functions from A to B are surjective? Let A_i be the set of functions that omit the ith element of $B = \{b_1, \ldots, b_n\}$. Given a set $S \subseteq [n]$ of indices, $\bigcap_{i \in S} A_i$ is the set of functions that omit the corresponding $|S|$ elements of B. There are $(n - |S|)^k$ of these functions, since we can map A onto the remaining elements without restriction (possibly missing more elements). When we combine the $\binom{n}{j}$ terms with $|S| = j$, for each j, we obtain $\sum_{j=0}^{n}(-1)^j \binom{n}{j}(n - j)^k$ for the number of surjective functions.   ∎

10.16. Solution. *Dice-rolling.* The Inclusion-Exclusion Principle applies to events in a finite probability space as well as to sets in a finite universe. When we normalize so that the total count of the universe is 1, we can interpret counting probability outside certain events as counting elements outside certain sets.

We roll a fair six-sided die n times and want to know the probability that each of the values $1, 2, 3, 4, 5$ appears during the experiment. If A_i is the event that i does not appear, we want the probability $P(\emptyset)$ outside all these events. The probability that we do not see one particular value is $(5/6)^n$. The probability that k such events from $\{A_i\}$ occur, meaning that k values fail to occur, is $[(6 - k)/6]^n$. This holds for each of the $\binom{n}{k}$ choices of k values, so the inclusion-exclusion formula yields

$$P(\emptyset) = 1 - 5(\tfrac{5}{6})^n + 10(\tfrac{4}{6})^n - 10(\tfrac{3}{6})^n + 5(\tfrac{2}{6})^n - (\tfrac{1}{6})^n$$

For $n = 5, 10, 15, 20$, the probabilities are .015, .356, .698, .873, respectively. The probability first exceeds .5 when $n = 12$.   ∎

EXERCISES

The first 23 problems are related to the Pigeonhole Principle, the others to Inclusion-Exclusion. The answers to most problems using the Inclusion-Exclusion Principle must be left as summations.

10.1. (−) Suppose that during a major league baseball season there are 140,000 at-bats and 35,000 hits. Which of the following must be true?
   a) There is some player who hits exactly .250.
   b) There is some player who hits at least .250.
   c) There is some player who hits at most .250.

10.2. Each year, the Grievance Committee consists of three professors. How many professors must there be in the department to avoid having the same committee in a period of eleven years?

10.3. Let S be a subset of $\{1, 2, \ldots, 3n\}$ having size $2n + 1$. Prove that S must contain three consecutive numbers. Show that this is best possible by exhibiting a set of size $2n$ for which the conclusion is false.

10.4. Let S be a set of $n + 1$ numbers in $[2n]$. Prove that S contains a pair of relatively prime numbers. Show that this is best possible by exhibiting a set of size n for which the conclusion is false.

10.5. (!) Prove that every set of seven distinct integers contains a pair whose sum or difference is a multiple of 10.

10.6. Suppose that the numbers 1 through 10 appear in some order around a circle. Prove that some set of three consecutive numbers sums to at least 17.

10.7. (!) The numbers 1 through 12 have fallen off the face of a clock and have been replaced in some random order. Prove that some set of three consecutive numbers has sum at least 20. Prove that some set of five consecutive numbers has sum at least 33. For three consecutive numbers, use more detailed analysis to determine whether it is possible for all the sums to be 19 or 20.

10.8. (!) Prove that every set of five points in the square of area 1 has two points separated by distance at most $\sqrt{2}/2$. Prove that this is best possible by exhibiting five points with no pair less than $\sqrt{2}/2$ apart. (Warning: Studying perturbations of the set found for the second part does not solve the first part.)

10.9. *Pigeonhole generalization.* Let $p_1, \ldots, p_k$ be natural numbers. Determine the minimum n such that for every way of distributing n objects into classes $1, \ldots, k$, there is some i such that class i receives at least p_i objects.

10.10. On a field 400 yards long, ten people each mark off football fields of length 100 yards. Prove that some point belongs to at least four of the fields.

10.11. (!) The **fractional part** of x is the amount by which it exceeds $\lfloor x \rfloor$. For $x \in \mathbb{R}$ and $n \in \mathbb{N}$, let $S = \{x, 2x, \ldots, (n-1)x\}$.
   a) Prove that if some pair of numbers in S have fractional parts that differ by at most $1/n$, then some number in S is within $1/n$ of an integer.
   b) Use part (a) to prove that some number in S is within $1/n$ of an integer.

10.12. Let S be a set of n integers. Prove that S has a nonempty subset whose sum is divisible by n. Show that this is best possible by exhibiting a set of $n - 1$ integers that has no nonempty subset whose sum is divisible by n.

10.13. (+) Consider a collection S of n positive integers summing to k. Call S "full" if for every $i \in [k]$, S has a subset with sum i. Prove that if $k \leq 2n - 1$, then S must be full. Show that this is best possible by exhibiting a set S of n numbers summing to $2n$ that is not full.

10.14. Six students come to class. Prove that among the six there must be three who all know each other or three who all don't know each other.

10.15. Use congruence classes to determine the maximum size of a subset of [99] that has no two numbers differing by 3.

10.16. (!) Given $n, k \in \mathbb{N}$, use congruence classes to determine the maximum size of a subset of $[n]$ that has no two numbers differing by k.

10.17. (!) Prove that the Erdős-Szekeres result is best possible by constructing for each n (with proof) a list of n^2 distinct numbers having no monotone sublist of length $n + 1$.

10.18. Consider an exam with three true/false questions, in which every student answers each question.

a) How many students are needed to guarantee that no matter how they answer the questions, some two students agree on every question?

b) How many students are needed to guarantee that no matter how they answer the questions, some two students agree on at least two questions? (Comment: Parts (a) and (b) each require a proof for the upper bound and an example for the lower bound.)

10.19. (!) *The Key Problem.* A private club has 90 rooms and 100 members. Keys must be given to members such that each set of 90 members can be assigned to 90 distinct rooms whose doors they can open. Each key opens one door. The management wants to minimize the total number of keys. Prove that the minimum number of keys is 990. (Hint: Consider the scheme where 90 of the members have one key, and the remaining 10 members have keys to all 90 rooms. Prove that this works and that no scheme with fewer keys works.)

10.20. (+) Generalize Example 10.8 to a player who plays on d consecutive days for a total of at most b games. We want to know whether there must be a total of exactly k games over some period of consecutive days, regardless of the schedule. Determine a formula $f(d, b)$ such that the argument of Example 10.8 works to prove the answer is "Yes" if $k \leq f(d, b)$.

10.21. (+) In Example 10.8, the chess player plays at most 132 days over 77 days. The argument there guarantees existence of a period of consecutive days with a total of exactly k games if $k \leq 22$. Using the Pigeonhole Principle and congruence classes modulo k, prove that there are also periods with exactly k games for $k \in \{23, 24, 25\}$. Construct a 77-day schedule of games such that no period of consecutive days has a total of exactly 26 games.

10.22. Given $m \geq 2n$, let S be a set of m points on a circle with no two diametrically opposite. Say that $x \in S$ is "free" if fewer than n points of $S - x$ lie in the semicircle clockwise from x. Prove that S has at most n free points. (Hint: Reduce the problem to the case $m = 2n$.)

10.23. Consider an n by n grid of dots at positions $\{(i, j) : 1 \leq i \leq n, 1 \leq j \leq n\}$ in the plane. Each dot is colored black or white. How large must n be such that for every way to color the dots, there is a rectangle whose four corners all have the same color? (Comment: The answer requires a proof for the upper bound and an example for the lower bound.)

10.24. How many ways are there to place 10 distinct people within three distinct rooms? How many ways are there to place 10 distinct people within three distinct rooms so that every room receives at least one person?

10.25. How many decimal n-tuples contain at least one each of $\{1, 2, 3\}$?

10.26. (!) Say that an integer is "full" if its base 10 representation contains at least one of each digit $0, 1, \ldots, 9$. For this problem, a representation with fewer digits is considered a representation with m digits by adding leading 0s. Derive a summation formula for the number of full m-digit integers.

10.27. A *bridge hand* consists of 13 cards from a standard deck of 52 cards. What is the probability that a bridge hand has at least one card in each suit? What is the probability that it has no cards (a void) in at least one suit?

10.28. How many natural numbers less than 252 are relatively prime to 252?

10.29. How many natural numbers less than 200 have no divisor in $\{6, 10, 15\}$?

10.30. (!) Let $\phi(m)$ denote the Euler totient function (the number of elements of $[m]$ that are relatively prime to m). If p, q are distinct prime numbers, prove that $\phi(pq) = \phi(p)\phi(q)$. In general, when $P(m)$ denotes the set of distinct prime factors of m, prove that $\phi(m) = m \prod_{p \in P(m)}(1 - 1/p)$.

10.31. Let $A_1, \ldots, A_n$ be subsets of a universe U. Let $T \subseteq [n]$ be a collection of indices, and let $N(T)$ be the number of elements of U that belong to the sets indexed by T but to no others among $A_1, \ldots, A_n$. By defining a new universe, prove the following generalization of the inclusion-exclusion formula:

$$N(T) = \sum_{T \subseteq S \subseteq [n]} (-1)^{|S|-|T|} \left| \bigcap_{i \in S} A_i \right|.$$

10.32. How many permutations of $[n]$ have no odd number as a fixed point?

10.33. (!) A math department has n professors and $2n$ courses, each professor teaching two courses each semester. How many ways are there to assign the courses in the fall semester? How many ways are there to assign the courses in the spring semester such that no professor teaches the same pair of courses in the spring as in the fall? If all the assignments are equally likely, what is the probability of this event?

10.34. (!) Given the five types of coins (pennies, nickels, dimes, quarters, half-dollars), how many ways can one select n coins so that no coin is selected more than 4 times? (Hint: Use inclusion-exclusion and selections with repetition.)

10.35. (!) Consider a set of n boys and n girls. In each situation below, use inclusion-exclusion to derive formulas for the number of ways to partition them

into pairs so that for each i the ith tallest boy is not paired with the ith tallest girl. (Leave answers as summations.)
    a) Same-sex pairs are allowed.
    b) Each pair has one person of each sex.

10.36. Given two each of n types of letters, how many distinguishable permutations are there such that no two consecutive letters are the same?

10.37. (!) How many ways are there to seat the people in n married couples around a merry-go-round so that no person sits next to his or her spouse? (Rotations of the seating arrangement are not distinguishable from each other.)

10.38. Let D_n count the permutations of $[n]$ with no fixed points. Let E_n^k count the permutations of $[n]$ with exactly k fixed points, for $0 \le k \le n$.
    a) Derive a formula for E_n^k in terms of $\{D_j : 0 \le j \le n\}$.
    b) Derive a formula for $n!$ in terms of $\{D_j : 0 \le j \le n\}$.

10.39. Use inclusion-exclusion to prove that $\sum_{k=0}^{n}(-1)^k \binom{n}{k} = 0$ if $n > 0$. What happens if $n = 0$?

10.40. Use inclusion-exclusion to prove that $\sum_{k=0}^{n}(-1)^k \binom{n}{k} 2^{n-k} = 1$. (Do not use the Binomial Theorem.)

10.41. Use inclusion-exclusion and selections with repetition to prove that

$$\sum_{k=0}^{n}(-1)^k \binom{n}{k}\binom{n-k+r-1}{r} = \binom{r-1}{n-1}.$$

Chapter 11

Graph Theory

The "graphs" of graph theory differ from the graphs of functions. Informally, a graph consists of "vertices" and connecting "edges". For example, we can think of people as vertices and join two people by an edge if they have met. Graph theory helps answer questions about acquaintance, chemical bonding, electrical networks, transportation networks, binary vectors, etc., along with those stated below. The techniques include induction, parity, extremality, counting two ways, the Pigeonhole Principle, Inclusion-Exclusion, and even the Dart Board Problem.

11.1. Problem. *The Königsberg Bridge Problem.* Some say that graph theory was born in the city of Königsberg in 1736. Located on the Pregel river, the parts of the city were linked by seven bridges as shown on the left below. The citizens wondered whether they could leave home, cross every bridge exactly once, and return home. This reduces to traversing the figure on the right, with heavy dots representing land masses and curves representing bridges. ∎

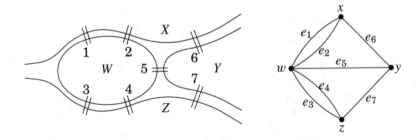

11.2. Problem. *The Marriage Problem.* Suppose there are n girls and n boys at a party, and each girl likes some subset of the boys. Under what conditions is it possible to match up the girls with the boys so that each girl is matched to a boy whom she likes? ∎

202

11.3. Problem. *The Platonic Solid Problem.* A Platonic solid has congruent regular polygons as faces and has the same number of edges meeting at each corner. The tetrahedron, cube, and octahedron appear below. The dodecahedron and icosahedron are the only other Platonic solids. Why are these five the only ones? ∎

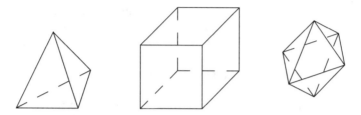

11.4. Problem. *The Art Gallery Problem.* A modern art gallery has the shape of a simple polygon in the plane, meaning a closed curve consisting of segments that meet only at successive vertices. What is the maximum number of stationary guards that may be needed to watch an art gallery with *n* corners? ∎

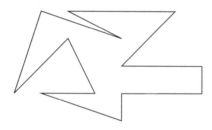

THE KÖNIGSBERG BRIDGE PROBLEM

To model the Königsberg Bridge Problem, we represent the land masses W, X, Y, Z by a set $V = \{w, x, y, z\}$. We represent the seven bridges by a set $E = \{e_1, e_2, e_3, e_4, e_5, e_6, e_7\}$. We encode the information about which land masses lie at the ends of each bridge by associating with each $e_i \in E$ a pair of elements of V. The relation between bridges and land masses permits us to answer the specific question of Problem 11.1 even before we formally define "graph".

11.5. Solution. *The Königsberg Bridge Problem.* The Swiss mathematician Leonhard Euler (1707–1783) observed in 1736 that Königsberg had no desired traversal. Every traversal passes through a land mass on the way from each bridge to the next. Each time we visit a land mass, we enter along one bridge and exit along another bridge. If we start and end in the same place, then we can also pair the first exit from that land mass with the last entrance to it. Thus the desired traversal requires that the number of bridges at each land mass is even. This condition fails in the Königsberg example, so the traversal does not exist. ∎

If in Problem 11.1 we add bridge 8 from W to Y and bridge 9 from X to Z, then there will be an even number of bridges at each land mass. Now $1, 2, 3, 4, 5, 6, 9, 7, 8$ is a traversal of the desired form. We will prove that having an even number of bridges at each land mass, together with being able to reach each bridge from every other, is sufficient for traversability. First we introduce the fundamental terminology for graphs.

11.6. Definition. A **graph** G is a triple consisting of a finite **vertex set** $V(G)$, a finite **edge set** $E(G)$, and a function h_G that assigns to each edge $e \in E(G)$ an unordered pair of vertices. When $h_G(e) = \{u, v\}$, we say that u and v are the **endpoints** of e and that e is **incident** to u and v. A graph G is **simple** if the function $h_G(e)$ is injective. In this case, we write $e = uv$ instead of $h_G(e) = \{u, v\}$.

11.7. Example. *The Königsberg graph.* The graph G in Problem 11.1 has vertex set $\{w, x, y, z\}$ and edge set $\{e_i : 1 \le i \le 7\}$. The endpoints of e_i for $1 \le i \le 7$ are $\{x, w\}$, $\{x, w\}$, $\{z, w\}$, $\{w, z\}$, $\{y, w\}$, $\{x, y\}$, $\{y, z\}$, respectively. This graph is not simple; $h_G(e_1) = h_G(e_2)$ and $h_G(e_3) = h_G(e_4)$. ∎

There are more general models of graphs. Our model is finite and does not permit directed edges or loops (edges with equal endpoints). Definition 5.37 (functional digraph) does allow these possibilities. In this chapter we consider only the model of *graph* in Definition 11.6.

The terms "vertex" and "edge" come from using graphs to model 3-dimensional solids. We visualize graphs by drawing them in the plane. To each vertex we assign a point; to each edge we assign a curve that joins the points assigned to its vertices. We take this as an informal aid for visualization; in Definition 11.60 we define drawings more precisely.

11.8. Definition. The **degree** of a vertex $x \in V(G)$, written $d(x)$, is the number of edges in G incident to x. A **subgraph** of a graph G is a graph H such that $V(H) \subseteq V(G)$ and $E(H) \subseteq E(G)$; we also require $h_H(e) = h_G(e)$ for $e \in E(H)$. When H is a subgraph of G, we write $H \subseteq G$ and say "G contains H".

11.9. Example. In the graph G of Problem 11.1, the degrees of w, x, y, z are $5, 3, 3, 3$, respectively. We have noted that we can make all the vertex degrees even by adding two edges sharing no endpoints. We can also do this by adding three edges with one common endpoint to obtain the graph H shown below; G is a subgraph of H.

We also show a subgraph F of G that has vertex degrees 2, 1, 1, 0 and is a simple graph; F illustrates the Handshake Problem (Solution 3.26) for two couples, with handshakes as edges. ∎

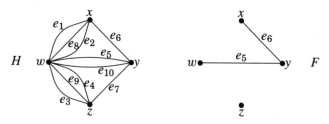

The full solution of the Königsberg Bridge Problem uses vertex degrees and a precise notion of "traversal". We must travel each bridge only once, but we may visit land masses more than once.

11.10. Definition. A **trail** in a graph G is a list $v_0, e_1, v_1, e_2, \ldots, e_k, v_k$ that alternates between vertices and edges, such that (1) $h_G(e_i) = v_{i-1}v_i$ for all i, and (2) edges $e_1, \ldots, e_k$ are distinct. The **length** of a trail is the number of edges. A u, v-**trail** is a trail with first vertex u and last vertex v; these are its **endpoints**.

A trail is **closed** if its endpoints are equal or if it has length 0. A trail in a graph is **maximal** if it is not a sublist of a longer trail. A graph is **Eulerian** if it has a closed trail containing all its edges.[†]

11.11. Example. In graph H of Example 11.9, $w, e_1, x, e_6, y, e_7, z, e_4, w$ is a closed trail of length 4. It is not a maximal trail, since it we can insert e_5, w, e_{10}, y between y and e_7 to enlarge it. Note that trails can repeat vertices, but not edges. This graph is Eulerian. ∎

We have shown that all vertex degrees must be even when a graph is Eulerian. Also necessary is that each edge be reachable from every other edge, meaning that there is a trail containing both. Euler remarked that these conditions are also sufficient, although no proof was published until 1871. To prove this, we use a lemma about maximal trails.

11.12. Lemma. If every vertex of a graph G has even degree, then every maximal trail in G is closed.

Proof: Since a trail contributes degree two when it passes through a vertex, a non-closed trail uses an odd number of edges at each endpoint. If the endpoint has even degree, then a non-closed trail can be extended. We have proved the contrapositive of the claim. ∎

[†]The name "Euler" is pronounced as "oiler", because it is a Germanic name like "Freud", not a Greek name like "Euclid".

11.13. Theorem. A graph G is Eulerian if and only if each vertex has even degree and each edge is reachable from every other.

Proof: We have argued that the conditions are necessary; we prove that they are also sufficient.

Suppose that G satisfies the conditions. Let T be a maximal trail in G; by Lemma 11.12, T is closed. If T does not include all of $E(G)$, let G' be the subgraph obtained from G by deleting $E(T)$. Since every edge of G is reachable from every other, there is a trail in G that starts with an edge of T and contains an edge of G'; let e be the first edge of G' on this trail, and let v be the vertex it follows.

Since T has even degree at every vertex, every vertex also has even degree in G'. Let T' be a maximal trail in G' beginning from v along e. By Lemma 11.12, T' is closed and ends at v. Hence we may incorporate T' to obtain a trail properly containing T. This contradicts the maximality of T, so we conclude that T already contains all edges of G.   ■

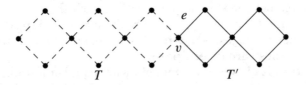

The proof of Theorem 11.13 uses extremality; we choose a maximal trail. Choosing an extremal example is a basic technique of proof. Induction proofs often amount to showing that a statement has no smallest counterexample. Our proofs of Fermat's Little Theorem (Theorem 7.36) and the Ballot Problem (Solution 9.10) illustrate other uses of extremality.

We close this section with further important observations about vertex degrees. First, the vertex degrees and the number of edges satisfy a simple equation proved by a counting argument.

11.14. Theorem. (Degree-sum Formula) If G is a graph with m edges, then $m = \frac{1}{2} \sum_{v \in V(G)} d(v)$.

Proof: Summing the degrees counts each edge twice, since each edge has two endpoints and contributes to the degree of each endpoint.   ■

11.15. Example. *The d-dimensional cube Q_d.* The cube Q_d is a simple graph with 2^d vertices that are the d-tuples of 0s and 1s. Two vertices of Q_d form an edge if and only if they differ in exactly one coordinate. Since each coordinate of a binary d-tuple can be changed in exactly one way, each vertex has degree d. By the Degree-sum Formula, Q_d thus has $d2^{d-1}$ edges. We show Q_2 and Q_3 below.   ■

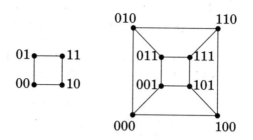

By the next corollary, the number of people in the world who have met an odd number of people is even. Applications of this and the Degree-sum Formula appear in Exercises 6–8 and Solutions 11.68 and 11.69.

11.16. Corollary. Every graph has an even number of vertices of odd degree.

Proof: By the Degree-sum Formula, the sum of the degrees is even. Hence the sum must have an even number of odd contributions. ∎

ISOMORPHISM OF GRAPHS

Consider the four cities {New York, Chicago, San Francisco, Champaign}. There are direct flights between Chicago and each of the other three cities, and direct flights between San Francisco and New York, but no direct flights between Champaign and either New York or San Francisco. We summarize this information by the graph below whose vertices are the four cities, and whose edges represent direct service.

Consider also the four integers {7, 10, 15, 42}. We define a graph with these integers as vertices, with two vertices forming an edge when they have a common factor larger than 1.

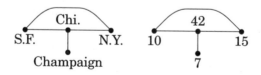

The picture shows that these two graphs have the same structure. They are not the same graph, since their vertices have different names. In order to treat them as the same object, we define a relation on the set of graphs, prove that it is an equivalence relation, and observe that these two graphs are in the same equivalence class.

To avoid complications, we define this relation only for simple graphs. In a simple graph, we name each edge by its endpoints and treat the edge set as a set of vertex pairs.

11.17. Definition. An **isomorphism** from a simple graph G to a simple graph H is a bijection $f\colon V(G) \to V(H)$ such that $uv \in E(G)$ if and only if $f(u)f(v) \in E(H)$. We say "G **is isomorphic to** H", written $G \cong H$, if there is an isomorphism from G to H. The set of pairs (G, H) such that G is isomorphic to H is the **isomorphism relation**.

When $G \cong H$, also $H \cong G$, so we may say "G and H are isomorphic". The adjective "isomorphic" applies only to pairs of graphs; the phrases "G is isomorphic" and "an isomorphic graph" have no meaning.

11.18. Example. The two 4-vertex graphs below are isomorphic. Consider mapping $1, 2, 3, 4$ to a, d, b, c, respectively. This changes the edges $12, 23, 34$ into ad, db, bc, respectively. Since these are the edges of the second graph, the vertex bijection is an isomorphism. Another isomorphism maps $1, 2, 3, 4$ to c, b, d, a, respectively.   ∎

We can describe isomorphism for simple graphs concisely using a relation on the vertex set.

11.19. Definition. Vertices u and v in a graph G are **adjacent** and are **neighbors** if they are the endpoints of an edge. The **adjacency relation** of G (defined on $V(G)$) is the set of ordered pairs (u, v) such that u and v are adjacent.

The adjacency relation is symmetric, and every symmetric relation is the adjacency relation of a graph. In the language of adjacency, simple graphs G and H are isomorphic if and only if there is a bijection $f\colon V(G) \to V(H)$ that preserves the adjacency relation.

11.20. Proposition. The isomorphism relation is an equivalence relation on the set of simple graphs.

Proof: The identity map on $V(G)$ is an isomorphism from G to itself. If $f\colon V(G) \to V(H)$ is an isomorphism from G to H, then f^{-1} is an isomorphism from H to G. If $f\colon V(F) \to V(G)$ and $g\colon V(G) \to V(H)$ are isomorphisms, then $g \circ f$ is a bijection from $V(F)$ to $V(H)$ that preserves the adjacency relation and hence is an isomorphism from F to H. Thus the isomorphism relation is reflexive, symmetric, and transitive.   ∎

11.21. Definition. An **isomorphism class** of graphs is an equivalence class of graphs under the isomorphism relation.

11.22. Remark. *Isomorphism classes.* Comments about the structure of a graph G also apply to every graph isomorphic to G. Some authors use the informal expression "unlabeled graph" instead of "isomorphism class of graphs". The vertices of a graph drawn on paper are named by their physical location; hence drawing a graph to illuminate its structure is choosing a convenient member of its isomorphism class.

A graph represents its isomorphism class just as a fraction represents a rational number. Asking whether a given graph "is" G is asking whether it is isomorphic to G. Similarly, we use the phrase "H is a subgraph of G" to mean that H is isomorphic to a subgraph of G. In this sense, the 2-dimensional cube Q_2 is a subgraph of the 3-dimensional cube Q_3 (see Exercise 10), even though the 2-tuples used as vertices in Q_2 are shorter than the 3-tuples used as vertices in G_3.   ∎

We usually prove that two graphs are isomorphic by presenting a bijection f and checking that it preserves the adjacency relation. Since structural properties are determined by the adjacency relation, we can prove that G and H are not isomorphic by finding some structural property of one that fails for the other. They may have different vertex degrees, different subgraphs, etc. Exhibiting a difference in structure proves that no vertex bijection preserves the adjacency relation.

11.23. Example. *Testing isomorphism.* An isomorphism from G to H must map every vertex $v \in V(G)$ to a vertex of H whose degree in H is $d_G(v)$. Hence the lists of vertex degrees of isomorphic graphs must be the same. For example, a graph whose vertices have degrees $1, 1, 1, 3$ cannot be isomorphic to a graph whose vertices have degrees $1, 1, 2, 2$, even though each has four vertices and three edges.

Nevertheless, two graphs may have the same list of vertex degrees and not be isomorphic. In each graph below, each vertex has degree 3. Only graph C has three vertices that are pairwise adjacent, so it cannot be isomorphic to any of the others. The others are pairwise isomorphic.

To show that $A \cong B$, we can verify that the bijection mapping u, v, w, x, y, z to $1, 3, 5, 2, 4, 6$, respectively, is an isomorphism. Sending u, v, w, x, y, z to $6, 4, 2, 1, 3, 5$ yields another isomorphism.

Graphs A and D have the same vertex set but different adjacency relations; $xw \in E(A)$, but $xw \notin E(D)$. Thus they are different graphs. They are isomorphic, though, by an isomorphism that maps u, v, w, x, y, z in $V(A)$ to u, v, z, x, y, w in $V(D)$, respectively.   ∎

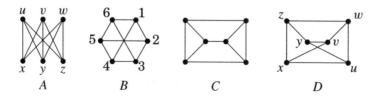

$$A \qquad\qquad B \qquad\qquad C \qquad\qquad D$$

When two simple graphs have many edges and have corresponding vertex degrees, looking at the nonadjacent pairs of vertices may make it easier to tell whether the graphs are isomorphic.

11.24. Definition. The **complement** $\overline{G}$ of a simple graph G is the graph with vertex set $V(G)$ and edge set $\{\{u, v\}: uv \notin E(G)\}$.

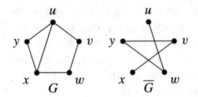

11.25. Example. Two graphs are isomorphic if and only if their complements are isomorphic (Exercise 13). The vertices of the graphs below have degree 5; in the complements the vertices have degree 2. The complement of the graph on the left has two closed trails of length 4. The complement of the other graph has no closed trail of length 4. Therefore, the graphs are not isomorphic.   ∎

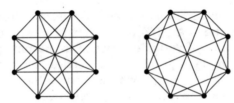

11.26. Example. *Counting graphs.* The number of pairs of distinct vertices in a set of size n is $\binom{n}{2}$. Each vertex pair may or may not form an edge; $\binom{n}{2}$ choices must be made to specify the adjacency relation. Thus there are $2^{\binom{n}{2}}$ simple graphs having a fixed set of n vertices.

For example, there are 64 graphs having a fixed set of four vertices. These fall into only 11 isomorphism classes. Representatives of these classes appear below; each graph is the complement of the other graph in its column. Only one of these is isomorphic to its complement.   ∎

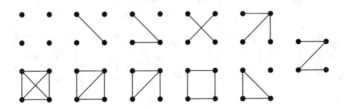

11.27. Remark. (optional) *Isomorphism* in mathematics generally describes a map between "equivalent" mathematical structures. An isomorphism between structures defined on sets S and T is a bijection between S and T that preserves the essential properties of the structure. For graphs, the sets S and T are the vertex sets, and the essential property is the adjacency relation.

We can define isomorphism for general graphs. In a graph G, the **multiplicity** of an unordered pair $\{u, v\} \in V(G)$ is the number of edges in G with endpoints $\{u, v\}$. Two graphs G, H are **isomorphic** if there is a bijection $f : V(G) \to V(H)$ that preserves multiplicity.

This agrees with our earlier definition for simple graphs, since uv is an edge in a simple graph if and only if $\{u, v\}$ has multiplicity 1. Describing a graph by its vertex set and multiplicities ignores the names of edges but includes all information about the structure of the graph.

In the language of Definition 11.6, isomorphism requires two bijections $f : V(G) \to V(H)$ and $\tilde{f} : E(G) \to E(H)$ such that for all $v \in V(G)$ and $e \in E(G)$, e is incident to v if and only if $\tilde{f}(e)$ is incident to $f(v)$. ∎

CONNECTION AND TREES

For most concepts in this chapter, the distinction between simple graphs and general graphs is unimportant. We now confine our attention to simple graphs, viewing an edge set as a set of unordered pairs of vertices.

11.28. Definition. A **path** is a simple graph whose vertices can be listed in an order so that two vertices are adjacent if and only if they are consecutive in the list. The **endpoints** of a path are the first and last vertices in such a list. A u, v-**path** is a path with endpoints u and v.

A **cycle** is a simple graph whose vertices can be place at distinct points of a circle so that two vertices are adjacent if and only if they appear consecutively on the circle.

The **length** of a path or cycle is its number of edges. We use P_n and C_n, respectively, to denote any representative of the isomorphism class that is a path or cycle with n vertices.

The definitions of P_n and C_n make sense because paths with n vertices are pairwise isomorphic, as are cycles with n vertices. The graph in Example 11.18 is P_4. We specify a path or a cycle (or a trail) as a subgraph of a simple graph by listing its vertices in order, since a simple graph has (at most) one edge with specified endpoints v_{i-1} and v_i. When we say that the subgraph is a cycle, we do not need to repeat the last vertex. This is consistent with the specification of cycles in permutations and functional digraphs (see Definition 5.10 and Definition 5.37).

11.29. Example. *Paths and cycles.* The 3-dimensional cube Q_3 (Example 11.15) contains subgraphs that are paths of lengths 0 through 7 and subgraphs that are cycles of lengths 4, 6, and 8. The graph G below contains three cycles. For each pair $s, t \in V(G)$, there is an s, t-path.   ■

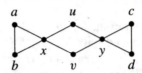

11.30. Definition. A graph G is **connected** if for all $u, v \in V(G)$, there is a u, v-path in G (otherwise, G is **disconnected**). A **component** of G is a connected subgraph of G that is not contained in any other connected subgraph. An **isolated vertex** is a vertex with degree 0.

11.31. Example. A connected graph, like that of Example 11.29, has one component. The graph below has three components, and one of these is an isolated vertex. The vertex sets of the components are $\{r\}$, $\{s, t, u, v, w\}$, and $\{x, y, z\}$. The subgraph consisting of the two components that are not isolated vertices is a disconnected graph with no isolated vertex.   ■

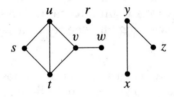

When studying paths in graphs, we say "u is connected to v" or "u and v are connected" when G has a u, v-path. The **connection relation** on $V(G)$ is the set of ordered pairs (u, v) such that G has a u, v-path. For the stronger statement that u and v are adjacent, we say "u and v are joined by an edge", *not* "u and v are connected".

11.32. Example. A u, v-path and a v, w-path together need not form a u, w-path. The concatenation of the u, v-path u, x, y, v and the v, w-path v, z, y, w is the trail u, x, y, v, z, y, w, which is not a path. Nevertheless, this trail **contains** the u, w-path u, x, y, w.   ■

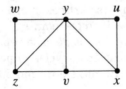

11.33. Proposition. If P is a u, v-path and P' is a v, w-path, then P and P' together contain a u, w-path.

Proof: We use extremality. At least one vertex of P appears in P', since both contain v. Let x be the first vertex of P that appears in P'. Following P from u to x and then P' from x to w yields a u, w path, since no vertex of P before x belongs to P'. ∎

11.34. Proposition. Let G be a graph. The connection relation on $V(G)$ is an equivalence relation, and its equivalence classes are the vertex sets of the components of G. If G has paths from one vertex to all others, then G is connected.

Proof: Reflexive property: v is connected to v by a path of length 0. Symmetric property: If P is a u, v-path, then reversing P yields a v, u-path. Transitive property: This is proved in Proposition 11.33.

Two vertices are in the same equivalence class if and only if they belong to a path; a path is a connected subgraph and hence appears in one component. If all vertices have paths to v, then Proposition 11.33 yields paths connecting all pairs of vertices. ∎

We often discuss subgraphs obtained by deleting an edge or a vertex.

11.35. Definition. The subgraph of G obtained by deleting an edge e is $G - e$. The subgraph obtained by deleting a vertex v and all edges containing v is $G - v$. The subgraph obtained by keeping all vertices but deleting the edges of a subgraph H is $G - E(H)$.

For example, if G is a cycle of length n with $e \in E(G)$ and $v \in V(G)$, then $G - e$ is a path of length $n - 1$, and $G - v$ is a path of length $n - 2$.

11.36. Lemma. If e is an edge of a connected graph G, then $G - e$ is connected if and only if e belongs to a cycle in G.

Proof: Suppose $e = xy \in E(G)$, and let $G' = G - e$. If $G - e$ is connected, then x and y belong to the same component in G', so G' contains an x, y-path, which completes a cycle with e in G.

Conversely, suppose e belongs to a cycle C. Choose $u, v \in V(G)$. Being connected, G has a u, v-path P. If P does not contain e, then P also exists in G'. If P contains e, suppose by symmetry that P reaches x before y when traveled from u to v. Since G' contains a u, x-path along P, an x, y-path along C, and a y, v-path along P, the transitivity of the connection relation implies that $G - e$ has a u, v-path. Since u, v were chosen arbitrarily from $V(G)$, we have proved that $G - e$ is connected. ∎

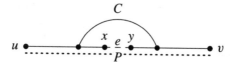

A **maximal** object of a particular type is one that is not contained in any other object of that type. Earlier we studied maximal trails in graphs, and components are maximal connected subgraphs. A maximal path in a graph is one that cannot be extended by adding a vertex at either end. Every path of maximum length is a maximal path, but maximal paths need not have maximum length: in Example 11.29, a, x, b is a maximal path that does not have maximum length. Considering maximal paths can lead to short proofs.

11.37. Lemma. If every vertex of G has degree at least two, then G contains a cycle.

Proof: Since $V(G)$ is finite, we can choose a maximal path P. Let v be an endpoint of P. Since $d(v) \geq 2$, v has a neighbor u that is not a neighbor of v on P. Since we cannot extend P to reach a new vertex from v, the vertex u already belongs to P, and the edge vu completes a cycle with the u, v-portion of P. ∎

If we allowed infinite vertex sets, this proposition would not hold. Consider $V(G) = \mathbb{Z}$ and $E(G) = \{xy : y - x = 1\}$. This infinite graph contains no cycle (it is a single "path" that extends infinitely in both directions), but every vertex has degree 2.

How many edges must a graph with n vertices have in order to be connected? Because deleting an edge of a cycle cannot disconnect a graph (Lemma 11.36), the minimal connected graphs have no cycles.

11.38. Definition. A **tree** is a connected graph with no cycles. A **leaf** is a vertex of degree 1. A **spanning tree** of a graph G is a subgraph of G that is a tree containing all vertices of G.

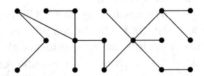

Gustav Kirchhoff (1824–1887) introduced spanning trees in connection with his work on electrical networks. Every connected graph has a spanning tree. This follows from Lemma 11.36: if G is connected, then deleting edges of cycles until no cycles remain produces a subgraph of G that is connected, has no cycles, and contains all vertices of G.

11.39. Lemma. Every tree with at least two vertices has a leaf, and deleting a leaf from a tree yields a tree with one less vertex.

Proof: Let G be a tree with n vertices, where $n \geq 2$. By the contrapositive of Lemma 11.37, a graph with no cycles has a vertex of degree less than two. Since G is connected and has more than one vertex, it has no vertex of degree 0, so it has a leaf x. Let $G' = G - x$.

We claim that G' is a tree with $n - 1$ vertices. We cannot create a cycle by deleting a vertex, so we need only show that G' is connected. Consider distinct vertices $u, v \in V(G')$. Because G is connected, there is a u, v-path P in G. Since internal vertices along a path have degree at least 2, P cannot contain x. Hence P is contained in G'. ∎

11.40. Theorem. Every tree with n vertices has $n - 1$ edges.

Proof: We use induction on n. A tree with 1 vertex has no edges. For the induction step, we consider $n > 1$ and assume that trees with $n - 1$ edges have $n - 2$ vertices. If G is a tree with n vertices, then Lemma 11.39 yields a leaf x and a tree $G' = G - x$ with $n - 1$ vertices. By the induction hypothesis, G' has $n - 2$ edges. Since x appears in only one edge, we conclude that G has $n - 1$ edges. ∎

Since deleting a leaf yields a smaller tree, each tree with $n + 1$ vertices arises from some tree with n vertices by adding an edge to a new vertex. This allows us to write an inductive proof about trees by "growing a leaf" (from an arbitrary vertex); all trees of the larger size will be considered.

BIPARTITE GRAPHS

Our next class of graphs includes all trees and d-dimensional cubes.

11.41. Definition. A set $S \subseteq V(G)$ is an **independent set** in a graph G if $uv \notin E(G)$ for all $u, v \in S$ (S may be empty). A **bipartite graph** with **bipartition** X, Y is a graph G such that $V(G) = X \cup Y$ and X, Y are disjoint (possibly empty) independent sets. We call X and Y the **partite sets** or **parts** of the bipartition.

11.42. Example. The d-dimensional cube Q_d is bipartite. Let X be the set of vertices whose encoding as a binary d-tuple has an odd number of ones. Let Y consist of those with an even number of ones. In each edge of Q_d, the parity of the number of ones in the encoding is different at the two endpoints. Hence X and Y are independent sets. ∎

11.43. Proposition. Every tree is bipartite.

Proof: We use induction on the number of vertices. A tree with one vertex has a bipartition with one set empty. For the induction step, suppose that every tree with n vertices is bipartite, and let T be a tree with $n + 1$ vertices. By Lemma 11.39, T has a leaf x such that $T - x$ is a tree T' with n vertices. Let y be the neighbor of x in T. By the induction hypothesis, we can partition $V(T')$ into two independent sets X and Y, with $y \in Y$. Placing x in X yields the desired bipartition of $V(T)$, since the only neighbor of x is in Y.   ∎

A disconnected bipartite graph has more than one bipartition, but a connected bipartite graph has only one. The parts or partite sets of a bipartition are not themselves called "partitions", just as the teams in a sports league are not themselves called "leagues".

Bipartite graphs have a simple structural characterization using an obvious necessary condition that we prove is also sufficient.

11.44. Theorem. A graph is bipartite if and only if it contains no cycle of odd length.

Proof: We use "odd cycle" for "cycle of odd length". To prove that the condition is necessary, consider a bipartite graph G. Every trail in a bipartite graph alternates between the two partite sets of a bipartition. Hence returning to the original partite set (or original vertex) happens only after an even number of steps. In particular, G has no odd cycle.

For sufficiency, consider a graph G with no odd cycles. We prove that each component H is bipartite. Since H is connected, it has a spanning tree T. By Proposition 11.43, T is bipartite with some bipartition X, Y. Consider $uv \in E(H)$. If u and v belong to the same partite set in T, then T has an u, v-path, since T is connected. This path has even length, since it alternates between X and Y. Edge uv would complete an odd cycle in G, contradicting the hypothesis. Hence every edge of H joins vertices from X and Y, and the bipartition of T is a bipartition of H.   ∎

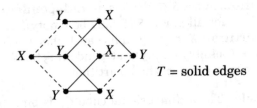

T = solid edges

11.45. Definition. A **complete graph** is a simple graph in which every pair of vertices forms an edge. We use K_n to denote the isomorphism class of complete graphs with n vertices (or its representatives).

11.46. Application. *The Airline Problem.* Suppose that an air traffic system has k airlines and n cities, and direct service between two cities includes flights in both directions. Suppose that each pair of cities has direct service from some airline, and suppose that no airline can offer a cycle through an odd number of cities. As a function of k, what is the maximum possible value of n?

The answer is 2^k. By Theorem 11.44, the direct flights for each airline form the edge set of a bipartite graph, and the question asks for the largest n such that all edges of K_n can be obtained using k bipartite subgraphs. Given such a solution, let X_i, Y_i be a partition of the ith subgraph G_i. We may assume that $X_i \cup Y_i$ contains all n vertices, since adding isolated vertices does not introduce odd cycles.

For each vertex v, define a binary k-tuple a by setting $a_i = 0$ if $v \in X_i$ and $a_i = 1$ if $v \in Y_i$. There are only 2^k binary k-tuples. If there are more than 2^k vertices, then the Pigeonhole Principle implies that two vertices receive the same k-tuple. Hence these two vertices belong to the same partite set in each bipartite subgraph, and the edge between them belongs to none of the subgraphs. This contradiction implies that $n \le 2^k$.

Conversely, when $n \le 2^k$ we can assign distinct binary k-tuples to the n vertices. Let $E(G_i)$ consist of all edges between vertices whose ith coordinate is 0 and vertices whose ith coordinate is 1. This constructs k bipartite subgraphs. Since distinct k-tuples differ in some coordinate, every edge belongs to some G_i, and we have constructed $G_1, \ldots, G_k$ covering the edges of K_n (illustrated below for $k = 2$ and $n = 4$). Hence the value $n = 2^k$ is achievable. ∎

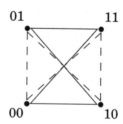

We cannot prove the upper bound in Application 11.46 by showing that one successful construction with 2^k vertices cannot accept another city. That argument does not consider all possible schedules for a system with $2^k + 1$ cities; it considers only those containing the special schedule with 2^k cities. We must consider all possible schedules (see Exercise 33).

Next we consider the Marriage Problem (Problem 11.2). We first develop an obvious necessary condition. Let $X = \{x_1, \ldots, x_n\}$ be the set of girls, and let A_i be the set of boys liked by x_i. Giving x_i a partner requires $|A_i| \ge 1$, for each i. We also require $|A_i \cup A_j| \ge 2$ when $i \ne j$, because two girls cannot have the same boy as a partner. In general, for each set of

k girls, a solution selects k distinct partners from the union of their sets; this requires $\left|\bigcup_{i \in J} A_i\right| \geq |J|$ for every set of indices $J \subseteq [n]$.

This condition is called **Hall's Condition**; it is also sufficient. Let the set of boys be $Y = \{y_1, \ldots, y_n\}$. We can form a bipartite graph G with bipartition X, Y by putting $x_i y_j \in E(G)$ if and only if $y_j \in A_i$. A selection of n distinct boys as partners for the girls corresponds to a selection of n pairwise disjoint edges G. In the example below, the pairing that results is $\{x_1 y_1, x_2 y_3, x_3 y_4, x_4 y_2\}$.

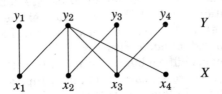

$$A_1 = \{y_1, y_2\}, \quad A_2 = \{y_2, y_3\}, \quad A_3 = \{y_2, y_3, y_4\}, \text{quad } A_4 = \{y_2\}$$

11.47. Definition. A **matching** in a graph is a set of pairwise disjoint edges; a **perfect matching** includes edges incident to each vertex.

11.48. Theorem. (Hall's Theorem) Given sets $A_1, \ldots, A_n$, there exist distinct elements $z_1, \ldots, z_n$ such that $z_i \in A_i$ for each i if and only if $\left|\bigcup_{i \in J} A_i\right| \geq |J|$ for every $J \subseteq [n]$.

Proof: We have observed that Hall's Condition is necessary. For sufficiency, suppose that the condition holds. We prove that the corresponding bipartite graph has a perfect matching. Let X, Y be the two partite sets, and put $x_i y_j \in E(G)$ if and only if $y_j \in A_i$. We must rephrase Hall's Condition using this bipartite graphs. Given a set $S \subseteq X$, let $J(S) = \{i : x_i \in S\}$, and let $N(S) = \bigcup_{i \in J(S)} A_i$. Hence $N(S)$ is the set of vertices in Y that have neighbors in S. Hall's Condition states that $|N(S)| \geq |S|$ for all $S \subseteq X$.

We prove by induction on n that this condition is sufficient; the statement is obvious for $n = 1$. For the induction step, consider $n > 1$, and suppose that Hall's Condition is sufficient in earlier cases. If $|N(S)| > |S|$ for every nonempty proper subset $S \subset X$, then we choose an arbitrary partner y for x_1 from A_1 and form $G' = G - x_1 - y$. Since this deletes at most one vertex of $N(S)$ for each $S \subseteq (X - \{x_1\})$, the graph G' satisfies Hall's Condition. By the induction hypothesis, G' has a perfect matching, which combines with $x_1 y$ to form a perfect matching in G.

Hence we may assume that $|N(S)| = |S|$ for some nonempty $S \subset X$. For all $S' \subseteq S$, we have $N(S') \subseteq N(S)$. Therefore, the subgraph consisting of S, $N(S)$, and the edges between them satisfies Hall's Condition. By the induction hypothesis, it has a perfect matching. It suffices to show that the graph G' obtained by deleting S and $N(S)$ also has a perfect matching (see illustration below).

Let $X' = X - S$ and $Y' = Y - N(S)$, so X', Y' is the bipartition of G'. For $T \subseteq X'$, let $N'(T)$ be the set of vertices in Y' with neighbors in T; we have $N'(T) = N(T) - N(S)$. By the induction hypothesis, it suffices to show that $\left|N'(T)\right| \geq |T|$ for all $T \subseteq X'$. Note that T and S are disjoint, as are $N'(T)$ and $N(S)$. From this and the three formulas

$$N(T \cup S) = N'(T) \cup N(S), \qquad |N(S)| = |S|, \qquad |N(T \cup S)| \geq |T \cup S|,$$

we compute

$$\left|N'(T)\right| = |N(T \cup S)| - |N(S)| \geq |T \cup S| - |S| = |T| \qquad \blacksquare$$

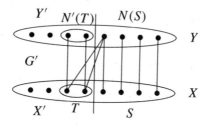

COLORING PROBLEMS

We use graph coloring to model questions about avoiding conflicts.

11.49. Problem. *Scheduling.* Suppose we want to schedule committee meetings in the Senate. Each committee needs one time period during the week, but we cannot assign two committees to the same time period if some Senator belongs to both of them. How can we determine the minimum number of time periods needed? ■

11.50. Definition. A k-**coloring** of a graph G is a function $f : V(G) \to S$, where S is a set of k elements called **colors** (the colors need not all be used). A k-coloring is **proper** if no pair of adjacent vertices receive the same color. The **chromatic number** of G, written $\chi(G)$, is the minimum k such that G has a proper k-coloring.

We use the Greek letter "chi" for chromatic number because it begins the Greek word for color.

11.51. Example. *Coloring of bipartite graphs, cycles, and complete graphs.* No two vertices receiving a given color can be adjacent, so $\chi(G)$ equals the minimum number of independent sets whose union is $V(G)$. Hence G is 2-colorable if and only if G is bipartite.

Thus odd cycles have chromatic number at least 3. We can also see this as follows. If C_{2k+1} were 2-colorable, the two colors would have to alternate as we follow the cycle. Since the number of vertices is odd, we wind up with two adjacent vertices of the same color. Changing one of these to a third color produces a 3-coloring.

A proper coloring of a complete graph must give the vertices distinct colors, and distinct colors suffice. Hence $\chi(K_n) = n$. Furthermore, if $K_n \subseteq G$, then $\chi(G) \geq n$. Always $\chi(G)$ is at least the size of its largest complete subgraph, but odd cycles show that equality need not hold. ∎

Finding time slots to schedule committee meetings is a graph coloring problem. Introducing a vertex for each committee, we let two vertices form an edge if their committees have a common member, since this means they cannot have the same time slot. The chromatic number of the resulting graph is the number of time slots needed.

This provides a mathematical model for the scheduling problem but not a solution; we have no general procedure to compute the chromatic number. We can compute it for graphs in some special classes. To prove that $\chi(G) = k$, we provide a proper k-coloring of G (this proves $\chi(G) \leq k$), and we prove that G is not $k - 1$-colorable (this proves $\chi(G) \geq k$). We consider a class of graphs that includes all cycles.

11.52. Example. *Coloring of generalized cycles.* Place n points around a circle. For $k \leq \lceil n/2 \rceil$, let $G_{n,k}$ be the graph obtained by making each point adjacent to the $k - 1$ nearest points in each direction. The graph $G_{n,1}$ is an independent set, which is 1-colorable. The graph $G_{n,2}$ is the ordinary cycle C_n, which has chromatic number 2 when n is even and 3 when n is odd. The graph $G_{8,3}$ below contains K_3, so $\chi(G_{8,3}) \geq 3$. In fact, $G_{8,3}$ is not 3-colorable. ∎

11.53. Theorem. If $n \geq k(k - 1)$, then the chromatic number of the generalized cycle $G_{n,k}$ is given by

$$\chi(G_{n,k}) = \begin{cases} k & \text{if } k \text{ divides } n \\ k+1 & \text{if } k \text{ does not divide } n. \end{cases}$$

Proof: Because every k consecutive points on the circle form a complete subgraph, we know that $\chi(G_{n,k}) \geq k$. If $G_{n,k}$ has a proper k-coloring, then each set of k consecutive points must receive k distinct colors. By symmetry, we may assume that the first k labels are $1, \ldots, k$ in order. Since the next point is adjacent to the $k - 1$ most recent points as we proceed, it must have a color different from those. If we use only k colors, the $k + 1$st point must have color 1, and as we continue the colors must cycle through $[k]$ in order repeatedly. This will be a proper coloring if and only if the last k vertices have colors $1, \ldots, k$ before restarting with color 1. Hence $\chi(G_{n,k}) = k$ if and only if $k | n$.

To complete the computation of $G_{n,k}$ for $n \geq k(k - 1)$, we need only prove that $G_{n,k}$ is $k + 1$-colorable. If we can partition the points around the circle into consecutive sets of sizes k and $k + 1$, then we can use stretches of colors $1, \ldots, k$ and $1, \ldots, k + 1$ to complete a proper coloring. Thus it suffices to express n as $mk + l(k + 1)$ for nonnegative integers m, l. The numbers $a = k$ and $b = k + 1$ are relatively prime. Solution 6.20 (the Dart Board Problem) guarantees that a solution exists when $n \geq ab - a - b + 1$, which here means $n \geq (k - 1)k$.

Alternatively, we can give an explicit formula for l and m. The Division Algorithm yields $n = qk + r$, with $0 \leq r < k$. With $n \geq k(k - 1)$, we have $q \geq k - 1$. We set $l = r \geq 0$ and $m = q - r \geq 0$.   ∎

When $n < k(k - 1)$, more than $k + 1$ colors are needed (Exercise 42).

In general, computing $\chi(G)$ is difficult. It is the smallest positive integer k such that the number of proper k-colorings is nonzero. We can consider the more general problem of counting the proper k-colorings. We will show for each graph G that this is a polynomial in k. Our discussion of this uses the Inclusion-Exclusion Principle and is optional.

11.54. Definition. Let $\chi(G; k)$ be the number of proper k-colorings of G. As a function of k, this is the **chromatic polynomial** of G.

11.55. Example. *Chromatic polynomials of complete graphs, their complements, and trees.* For some graphs, we can compute chromatic polynomials by using the rules of sum and product (Definitions 5.6 and 5.8) to count the ways of constructing proper colorings. When G is an independent set of size n, we can choose colors independently at the vertices; with k colors available, we have $\chi(G; k) = k^n$. For the complete graph K_n, the colors must be distinct. Choosing colors for vertices 1 through n in turn yields $\chi(K_n; k) = k(k - 1) \cdots (k - n + 1)$. We count k-colorings that differ by permuting the labels of the colors as distinct colorings.

Every tree arises from K_1 by iteratively adding a new vertex with one edge to an old vertex. If we color the vertices in the order added, then the color on the first vertex can be chosen in k ways. Subsequently there are $k - 1$ ways to choose the color for the each new vertex, no matter how the choices have been made so far. By the rule of product, the chromatic polynomial of a tree with n vertices is $k(k - 1)^{n-1}$.   ∎

We next obtain a formula for the chromatic polynomial of an arbitrary graph. It does not provide a good algorithm to compute $\chi(G)$, because there are too many subsets of the edge set. It expresses $\chi(G; k)$ as an integer combination of terms of the form k^c with $c \leq n$, where $n = |V(G)|$. Hence $\chi(G; k)$ is a polynomial in k of degree $|V(G)|$.

11.56. Theorem. Let $c(G)$ denote the number of components of a graph G. Given a set $S \subseteq E(G)$ of edges in G, let G_S denote the subgraph of G with vertex set $V(G)$ and edge set S. The number $\chi(G; k)$ of proper k-colorings of G is given by

$$\chi(G; k) = \sum_{S \subseteq E(G)} (-1)^{|S|} k^{c(G_S)}.$$

Proof: We want to count k-colorings that violate no edges, where an edge is "violated" if both endpoints have the same color. This suggests inclusion-exclusion. We define $|E(G)|$ subsets of the k-colorings; the set corresponding to edge e contains the colorings that violate edge e. Using the Inclusion-Exclusion Formula (Theorem 10.12), it remains only to show that $k^{c(G_S)}$ is the number of k-colorings that violate the edges in S. To violate all the edges in S, every vertex we can reach from x by a path of edges in S must have the same color as x. Hence all the vertices within a component of G_S must have the same color, which we can pick in k ways. The choices for the various components are independent; by the rule of product, there are $k^{c(G_S)}$ ways to make all the choices.   ∎

11.57. Example. *A chromatic polynomial.* When we apply Theorem 11.56 to a graph with n vertices and m edges, every subset with 0, 1, or 2 edges yields a subgraph with n, $n - 1$, or $n - 2$ components, respectively, so the contributions from the inclusion-exclusion sum always begin $k^n - mk^{n-1} + \binom{m}{2}k^{n-2}$.

When $|S| = 3$, the number of components is again $n - 2$ if the three edges form a triangle; otherwise it is $n-3$. The graph drawn below has two triangles, and the remaining $\binom{5}{3} - 2 = 8$ sets of three edges yield one component. All subgraphs with four or five edges have only one component. Hence the inclusion-exclusion computation is

$$\chi(G; k) = k^4 - 5k^3 + 10k^2 - (2k^2 + 8k^1) + 5k - k = k^4 - 5k^3 + 8k^2 - 4k.$$

By ad hoc counting, one can also see that $\chi(G; k) = k(k - 1)(k - 2)(k - 2)$. This is 0 when k is 1 or 2, but $\chi(G; 3) = 6$.   ∎

PLANAR GRAPHS

Three hermits A, B, C are living in the woods. We must cut paths from each house to three utilities (traditionally gas, water, and electricity). Can we do this without crossing paths? We will see that we cannot.

11.58. Example. *The Four Color Problem.* Can the regions of every map drawn in the plane (or on the surface of a globe) be colored with four colors so that neighboring regions receive different colors?

This question about a map M becomes a question about a graph G when we create a vertex for each region and join two vertices by an edge if the corresponding regions share a boundary of nonzero length. The number of colors needed for M is $\chi(G)$. The famous Four Color Conjecture, posed in 1852, was that four colors would suffice for every map. Kenneth Appel and Wolfgang Haken proved this in 1976 at the University of Illinois, assisted by a computer.

When we view the vertex for each region as the "capital" and draw roads from the capital to the midpoints of borders with neighboring regions, the roads combine to draw G in the plane with no crossing edges. Below we draw a map with five regions separated by dashed boundaries. The resulting graph G with solid edges is 3-colorable.   ■

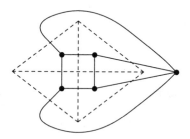

Until now, we have treated edges as abstract pairs of vertices. When we consider geometric properties of drawings of graphs, we treat edges as curves in the plane. We assume some intuitive geometric properties about regions and curves in the plane, without precise definitions. In particular, we merely think of a "continuous function from [0, 1] to $\mathbb{R}^2$" as a function whose image can be traced by a pencil without lifting it from the paper.

11.59. Definition. A **curve** from u to v in $\mathbb{R}^2$ is the image of a continuous function $f : [0, 1] \to \mathbb{R}^2$ such that $f(0) = u$ and $f(1) = v$; it is **simple** if f is injective, with the exception that $u = v$ is allowed in a simple curve. When $u = v$, the curve is **closed**.

11.60. Definition. A **drawing** of a graph G is a graph H isomorphic to G such that each vertex of H is a point in $\mathbb{R}^2$ and each edge e of H

with endpoints u, v is a simple curve from u to v. In a drawing of a graph, distinct edges e_1 and e_2 **cross** if they intersect other than at a common endpoint. A **planar graph** is a graph that has a drawing without crossings. Such a drawing is called a **plane graph**.

A plane graph $\hat{G}$ that is a drawing of a planar graph G is a convenient representative of the isomorphism class of G.

11.61. Definition. A set $R \subseteq \mathbb{R}^2$ is **path-connected** if for $u, v \in R$, there is a curve contained in R with endpoints u, v. A **face** F of a plane graph G is a maximal path-connected subset of $\mathbb{R}^2$ that intersects no edge or vertex of G. An edge e is in the **boundary** of F if some segment with an endpoint in F crosses e and no other edge of G.

11.62. Example. A plane graph with one vertex and no edges has one face. More generally, every plane graph G that is a tree has one face. If p, q are points in $\mathbb{R}^2$ that are not vertices and are not contained in edges of G, then there is a curve from p to q that intersects no edge or vertex of G (Exercise 44).   ∎

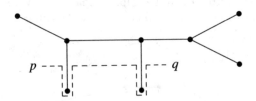

Our intuitive understanding of geometry in the plane suggests the next theorem, whose proof is surprisingly difficult and will be omitted. We use this theorem only in the proof of Theorem 11.64.

11.63. Theorem. (Jordan Curve Theorem) Every simple closed curve in $\mathbb{R}^2$ partitions its complement into two regions known as the **interior** and the **exterior**. Equivalently, every plane graph isomorphic to a cycle has two faces, one bounded and one unbounded.

11.64. Theorem. (Euler's Formula) If G is a connected plane graph with **v** vertices, **e** edges, and **f** faces, then $\mathbf{v} - \mathbf{e} + \mathbf{f} = 2$.

Proof: We use (strong) induction on the number of cycles in G. If G is connected and has no cycles, then G is a tree and $\mathbf{f} = 1$ (Example 11.62). Since $\mathbf{e} = \mathbf{v} - 1$ for a tree, we have $\mathbf{v} - \mathbf{e} + \mathbf{f} = 2$.

Otherwise, G has a cycle C containing an edge e. Because C is a cycle, the Jordan Curve Theorem implies that e is on the boundary of two faces in G; one inside C and one outside C. In the plane graph $G' = G - e$, these faces merge (including the points vacated by e) to form a single face.

Since e belongs to a cycle, Lemma 11.36 implies that G' is connected. Furthermore, G' has fewer cycles than G, since every cycle in G' appears in G, but C appears in G and not G'. Hence we can apply the induction hypothesis to G'; if its numbers of vertices, edges, and faces are $\mathbf{v}', \mathbf{e}', \mathbf{f}'$, this yields $\mathbf{v}' - \mathbf{e}' + \mathbf{f}' = 2$. Since $\mathbf{v} = \mathbf{v}'$, $\mathbf{e} = \mathbf{e}' + 1$, and $\mathbf{f} = \mathbf{f}' + 1$, we conclude that $\mathbf{v} - \mathbf{e} + \mathbf{f} = 2$. ∎

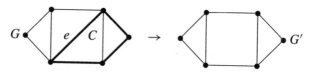

The statement and proof of Euler's Formula allows multiple edges. Our first application yields a necessary condition for planar graphs.

A **complete bipartite graph** is a simple bipartite graph in which two vertices are adjacent if and only if they belong to different partite sets. We use $K_{r,s}$ to denote such a graph with partite sets of sizes r and s. Example 11.23 shows three drawings of $K_{3,3}$. We will prove that K_5 and $K_{3,3}$ (the "gas-water-electricity" graph) are not planar.

11.65. Theorem. Every simple planar graph with $n \geq 3$ vertices has at most $3n - 6$ edges. Every simple planar graph with $n \geq 3$ vertices and no 3-vertex cycle has at most $2n - 4$ edges.

Proof: For $n = 3$, both statements are true by inspection. Consider a maximal simple plane graph G with $n \geq 4$. We may assume that G is connected, since otherwise we can add at least one edge. We can now use Euler's Formula to relate $\mathbf{v}$ and $\mathbf{e}$ if we can dispose of $\mathbf{f}$. Since G is simple, every face has at least three edges in its boundary. Every edge lies in the boundary of at most two faces. Summing the numbers of edges in the boundaries of the faces thus yields the inequality $2\mathbf{e} \geq 3\mathbf{f}$. Substituting this into $n - \mathbf{e} + \mathbf{f} = 2$ yields $\mathbf{e} \leq 3n - 6$.

For the second statement, we check the graphs with $n = 4$ individually (see Example 11.26). If also G has no C_3 and $n \geq 5$, then each face of G has at least four edges in its boundary. In this case the inequality becomes $2\mathbf{e} \geq 4\mathbf{f}$, and we obtain $\mathbf{e} \leq 2n - 4$. ∎

11.66. Example. K_5 *and* $K_{3,3}$. Theorem 11.65 implies that K_5 and $K_{3,3}$ are not planar graphs. For K_5, we have $\mathbf{e} = 10 > 9 = 3n - 6$. For the bipartite graph $K_{3,3}$, we have $\mathbf{e} = 9 > 8 = 2n - 4$. Each graph can be drawn using only one crossing. ∎

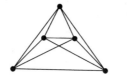 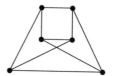

11.67. Remark. *Kuratowski's Theorem.* Example 11.66 yields the easy half of a characterization of planar graphs. Replacing an edge with a path having the same endpoints (and passing through new vertices) does not affect whether a graph is planar. Thus a graph containing a subgraph obtained from K_5 or $K_{3,3}$ by replacing edges with paths cannot be planar. Kuratowski's Theorem states that a graph is planar if and only if it contains no such subgraph.   ■

We next apply Euler's Formula to the Platonic Solid Problem. We describe solids and their relationship to planar graphs informally. A solid S is bounded by planes in space. The portion of the boundary of S belonging to one of these planes is a *face* of S. Intersections of neighboring faces are edges of S, and common endpoints of edges are vertices of S. The vertices and edges of S form a graph called the *skeleton* of S.

To draw the skeleton in the plane, we move S to rest on one of its faces in the plane. We puncture another face and gradually spread the surface out into the plane. We obtain a plane graph with the same intersection relationships as S among vertices, edges, and faces; it is a drawing of the skeleton of S. Each face bounded by l edges in S becomes a face of G whose boundary is a cycle of length l.

11.68. Solution. *The Platonic Solid Problem.* By definition, a Platonic solid has a skeleton in which every vertex has the same degree k and every face has the same length l. Also the physical properties require $k, l \geq 3$. We show that there are only five such solids by showing that there are only five planar graphs with these properties.

Consider a planar drawing of such a graph. By the Degree-sum Formula, $2e = vk$. Since every edge belongs to exactly two faces, we also have $2e = fl$. By Euler's Formula, we have $v - e + f = 2$. Substituting for v and f into Euler's Formula, we have $e(\frac{2}{k} - 1 + \frac{2}{l}) = 2$. Since e and 2 are positive, the other factor must also be positive, which yields $(2/k) + (2/l) > 1$, and hence $2l + 2k > kl$.

This inequality is equivalent to $(k - 2)(l - 2) < 4$. Since $k, l \geq 3$, we find that there are only five integer solution pairs. Once we specify the vertex degrees and the cycle lengths, there is essentially only one way to form the planar graph (we omit the details of this). Hence there are no more than the five known Platonic solids.   ■

k	l	$(k-2)(l-2)$	e	v	f	name
3	3	1	6	4	4	tetrahedron
3	4	2	12	8	6	cube
4	3	2	12	6	8	octahedron
3	5	3	30	20	12	dodecahedron
5	3	3	30	12	20	icosahedron

Euler's Formula applies to some geometric counting problems.

11.69. Solution. *Regions in a Circle.* We prove that the chords among n points on a circle cut the interior into $1 + \binom{n}{2} + \binom{n}{4}$ regions when no three chords have a common intersection. We obtain a plane graph G by treating the n points on the circle and the crossings of chords as vertices. Since each crossing of chords is determined by four points on the circle and each set of four points determines one pair of crossing chords, G has $\binom{n}{4} + n$ vertices. Since each internal vertex has degree 4 and each vertex on the circle has degree $n + 1$, the Degree-sum Formula yields $|E(G)| = \frac{1}{2}4\binom{n}{4} + \frac{1}{2}n(n+1)$. By Euler's Formula, the number of faces is $2 + \mathbf{e} - \mathbf{v} = 2 + \binom{n}{2} + \binom{n}{4}$. Subtracting 1 for the unbounded face leaves $1 + \binom{n}{2} + \binom{n}{4}$ regions inside the circle. ∎

We return to the issue of coloring planar graphs. The Four Color Theorem is difficult, but 6-colorability is not hard (Exercise 46). Also we can solve the Art Gallery Problem using an easier theorem about coloring a special class of planar graphs.

11.70. Definition. A graph is **outerplanar** if it has a drawing in the plane with every vertex on the boundary of the unbounded face.

11.71. Theorem. Every outerplanar graph is 3-colorable.

Proof: We use induction on the number of vertices; every graph with at most three is 3-colorable. For $n > 3$, let G be a plane graph with n vertices, all on the unbounded face. Every subgraph of G is also outerplanar.

Suppose first that G has a vertex x such that $G - x$ is disconnected. Let $G_1, \ldots, G_k$ be the components of $G - x$, and let G_i' be the graph consisting of G_i together with x and the edges from x to $V(G_i)$. Each G_i' is outerplanar and has fewer vertices than G. By the induction hypothesis, each G_i' is 3-colorable. We can permute the names of colors to make the colorings agree at x and obtain a proper 3-coloring of G.

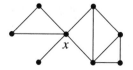
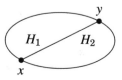

If G has no such vertex, then the unbounded face of G is bounded by a cycle C (drawn as a simple closed curve). If C is all of G, then G is a cycle and is 3-colorable. Otherwise, C has a chord xy. Let H_1 be the subgraph consisting of the cycle D formed by xy and an x, y-path on C, plus all chords of D. Let H_2 be the subgraph formed in this way using the other x, y-path on C. By the induction hypothesis, H_1 and H_2 are 3-colorable. Again we can permute the names of the colors in these 3-colorings to agree on $\{x, y\}$; this yields a proper 3-coloring of G.   ∎

11.72. Solution. *The Art Gallery Problem.* We prove that $\lfloor n/3 \rfloor$ guards suffice for any art gallery with n walls (this is best possible—Exercise 49). View the art gallery as a simple polygon in the plane. By adding chords between vertices, we obtain an outerplanar graph in which every bounded face is a triangle. By Theorem 11.71, this graph is 3-colorable. Below we triangulate and color the art gallery in Problem 11.4.

Given a proper 3-coloring, the Pigeonhole Principle implies that one of the colors is used on at most $\lfloor n/3 \rfloor$ vertices (color 1 in the example below). We claim that guards placed at the vertices with that color can watch the entire gallery. Because every bounded face is a triangle, each bounded face receives all three colors on its vertices and hence receives a guard at one vertex. This guard can watch this entire triangle.   ∎

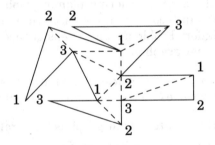

EXERCISES

11.1. Let G be the graph with vertex set [12] in which vertices u, v are adjacent if and only if u and v are relatively prime. Count the edges of G.

11.2. Let G be the graph with vertex set $\mathbb{Z}_n$ in which vertices u, v satisfy the adjacency relation if and only if u and v differ by 6. For each $n \geq 1$, determine the number of components of G.

11.3. Prove or disprove: There is no Eulerian graph with an even number of vertices and an odd number of edges.

11.4. (!) Let G be a connected non-Eulerian graph. Prove that the minimum number of trails that together traverse each edge of G exactly once is half the number of vertices having odd degree. (Hint: Transform G into a new graph G' by adding edges and/or vertices.)

11.5. Can the vertices of a simple graph have distinct degrees?

11.6. In a league with two divisions of 11 teams each, is it possible to schedule a season with each team playing seven games within its division and four games against teams in the other division?

11.7. (!) Let G be a connected graph in which every vertex has even degree. Prove that G has no edge whose deletion leaves a disconnected subgaph.

11.8. Let l, m, n be nonnegative integers with $l + m = n$. Find necessary and sufficient conditions on l, m, n such that there exists a connected n-vertex simple graph with l vertices of even degree and m vertices of odd degree.

11.9. Let G be a simple graph. Prove or disprove:
a) Deleting a vertex of maximum degree cannot raise the average degree.
b) Deleting a vertex of minimum degree cannot reduce the average degree.

11.10. Describe an inductive construction of the d-cube Q_d. Use it to prove that Q_d has $d2^{d-1}$ edges and has a cycle containing all its vertices (if $d \geq 2$).

11.11. Count the cycles of lengths 4 and 6 in the d-dimensional cube. (Hint: There is more than one "type" of 6-cycle.)

11.12. (−) Among the graphs below, which pairs are isomorphic?

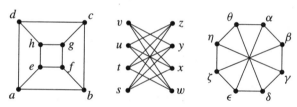

11.13. (−) For simple graphs G and H, prove that $G \cong H$ if and only if $\overline{G} \cong \overline{H}$.

11.14. What is the smallest value of n such that there are two n-vertex simple graphs with the same list of vertex degrees that are not isomorphic? (Hint: Use the list in Example 11.26.)

11.15. Prove that there are exactly two isomorphism classes of 7-vertex simple graphs in which every vertex has degree 4. (Hint: Consider the complements.)

11.16. Let G be a simple graph isomorphic to its complement $\overline{G}$. Prove that the number of vertices in G is congruent to 0 or 1 modulo 4.

11.17. (!) The **Petersen graph** is the graph on the left below. Prove that the graphs below are pairwise isomorphic and thus all represent the Petersen graph.

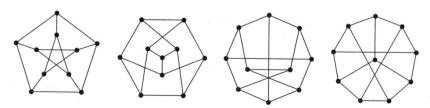

11.18. (!) Prove that there are exactly $2^{\binom{n-1}{2}}$ simple graphs with vertex set $v_1, \ldots, v_n$ in which every vertex has even degree. (Hint: Establish a bijection from this set to the set of all simple graphs with vertex set $\{v_1, \ldots, v_{n-1}\}$.)

11.19. (!) Give combinatorial proofs of the statements below by using simple graphs. (Hint: Interpret each quantity as counting something involving graphs.)
    a) If $n, k \in \mathbb{N}$ with $0 \le k \le n$, then $\binom{n}{2} = \binom{k}{2} + k(n-k) + \binom{n-k}{2}$.
    b) If $n_1, \ldots, n_k \in \mathbb{N}$ with $\sum_{i=1}^{k} n_i = n$, then $\sum_{i=1}^{k} \binom{n_i}{2} \le \binom{n}{2}$.

11.20. (!) *Common neighbors.* Let G be a simple graph with n vertices.
    a) Let x and y be nonadjacent vertices of degree at least $(n+k-2)/2$. Prove that x and y have at least k common neighbors.
    b) Prove that if every vertex has degree at least $\lfloor n/2 \rfloor$, then G is connected. Show that this bound is best possible whenever $n \ge 2$ by exhibiting a disconnected n-vertex graph where every vertex has at least $\lfloor n/2 \rfloor - 1$ neighbors.

11.21. (!) Prove that a graph G is connected if and only if for every partition of $V(G)$ into nonempty sets S, T, there is an edge xy with $x \in S$ and $y \in T$.

11.22. Consider three buckets with integer capacities $l > m > n$ in gallons. Initially, the largest bucket is filled. We need to measure out k gallons, but there are no markers on the buckets, so the only operation we can perform is to pour as much water from one bucket into another as will fit; in this way, we can keep track of how much water is in each bucket. Use graph theory to describe a method for determining whether it is possible to measure k gallons.

11.23. Let G be a graph in which every vertex has degree at least k, where k is an integer at least 2. Prove that G has a path of length at least k and a cycle of length at least $k + 1$. (Hint: Consider a maximal path.)

11.24. Let k be the maximum length of a path in a connected graph G. If P, Q are paths of length k in G, prove that P and Q have a common vertex.

11.25. (!) Let G be a simple graph having n vertices and no 3-vertex cycle. Prove that G has at most $n^2/4$ edges. (Hint: Consider the subgraph consisting of the neighbors of a vertex of maximum degree and the edges among them.)

11.26. (!) Prove that every graph with n vertices and $n - k$ edges has at least k components.

11.27. (!) Let G be a graph with n vertices and $n - 1$ edges. Prove that G is connected if and only if G has no cycles. (Comment: Compare with Theorem 11.40.)

11.28. (!) Prove that a graph G is a tree if and only if for all $x, y \in V(G)$, there is exactly one x, y-path in G.

11.29. ($-$) Prove that every tree with maximum degree k has at least k leaves.

11.30. Prove that a connected graph with n vertices has exactly one cycle if and only if it has exactly n edges.

11.31. Let $d_1, \ldots, d_n$ be n natural numbers. Prove that there exists a tree with n vertices that has these as its vertex degrees if and only if $\sum d_i = 2n - 2$. (Hint: Two implications are needed; use induction for one of them. Comment: It is not true that every n-vertex graph whose degrees sum to $2n - 2$ is a tree.)

11.32. (!) Let T be a tree with m edges, and let G be a simple graph in which every vertex has degree at least m. Prove that G contains T as a subgraph. (Hint: Use induction on m.)

11.33. Use induction on k (for both implications) to prove that $E(K_n)$ can be covered by k bipartite graphs if and only if $n \leq 2^k$. (Comment: This repeats the result of Application 11.46 (the Airline Problem).)

11.34. Let G be a graph having no cycle of even length. Prove that every edge of G appears in at most one cycle.

11.35. Prove that every tree has at most one perfect matching.

11.36. (!) Let G be a bipartite graph with bipartition X, Y in which every vertex of G has degree k.
    a) Prove that $|X| = |Y|$.
    b) Prove that G has a perfect matching.

11.37. $(-)$ How many perfect matchings does $K_{n,n}$ contain? How many cycles of length $2n$ does $K_{n,n}$ contain?

11.38. $(-)$ The **wheel** with n vertices consists of a cycle with $n - 1$ vertices and one additional vertex adjacent to all the vertices on that cycle. Determine the chromatic number of the wheel with n vertices.

11.39. Prove that if G does not have two disjoint odd cycles, then $\chi(G) \leq 5$.

11.40. (!) Suppose that every vertex of a graph G has degree at most k. Prove that $\chi(G) \leq k + 1$. For each k, construct a graph where the maximum vertex degee is k and the chromatic number equals $k + 1$.

11.41. Given a collection of lines in the plane with no three intersecting at a point, form a graph G whose vertices are the intersection points of the lines and whose edges are the segments along the lines that join intersection points. Prove that $\chi(G) \leq 3$.

11.42. Let $G_{n,k}$ be the generalized cycle defined in Example 11.52. Use the Pigeonhole Principle to prove that $\chi(G_{n,k}) > k + 1$ when $n = k(k - 1) - 1$.

11.43. Prove that the coefficients of $\chi(G; k)$ sum to 0 unless G has no edges.

11.44. Without using Euler's Formula, prove that a plane graph that is a tree has one face. (Hint: Use induction on the number of vertices.)

11.45. $(-)$ Let G be a simple planar graph with at least 11 vertices. Prove that $\overline{G}$ is not planar.

11.46. (!) Prove that every simple planar graph has a vertex of degree at most 5, and use this to prove that every planar graph has chromatic number at most 6.

11.47. Let G be an n-vertex simple planar graph with no cycle of length less than k. Prove that G has at most $(n - 2)k/(k - 2)$ edges, and use this to prove that the Petersen graph (Exercise 11.17) is not planar.

11.48. (!) Use Euler's Formula to prove that an outerplanar graph with n vertices has at most $2n - 3$ edges.

11.49. (+) For each n, construct an art gallery with n walls to prove that the bound of $\lfloor n/3 \rfloor$ in Solution 11.72 is best possible. (Hint: Use groups of three vertices to build "rooms" such that no guard can see into more than one room.)

Chapter 12

Recurrence Relations

Consider a row of n lights, and let a_n be the number of on/off settings of the lights. We can express a_n in terms of a_{n-1}. When we add a light, the new light may be on or off. The resulting $2a_{n-1}$ lists are distinct, and every list of length n arises in this way. Therefore, $a_n = 2a_{n-1}$. Since $a_1 = 2$, we see inductively that $a_n = 2^n$.

This discussion treats the counting of binary lists in two ways. The formula gives a_n explicitly as a function of n; the recursive definition also determines a_n. A recursive definition may be easier to find and may lead to a formula.

In this chapter, we consider combinatorial problems whose analysis leads to recursive definitions of sequences. We also develop techniques for obtaining explicit formulas from these recursive definitions. Recurrence relations enable us to analyze problems inductively, particularly when it is difficult to see a general pattern.

12.1. Problem. *The Tower of Hanoi.* The French mathematician Edouard Lucas (1842–1891) constructed a puzzle with three pegs and seven rings of different sizes that could slide onto the pegs. Legend has it that an order of monks had a similar puzzle with 64 large golden disks. Starting with all rings on one peg in order by size, the problem is to transfer the pile to another peg subject to two conditions: rings are moved one by one, and no ring is ever placed on top of a smaller ring. The monks supposedly believed that the world would crumble when the job was finished. How many moves are required? ∎

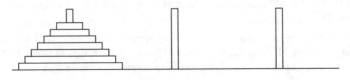

12.2. Problem. *The Fibonacci numbers.* Suppose that n spaces are available for parking along a curb. We can fill the spaces using Rabbits, which take one space, and/or Cadillacs, which take two spaces. In how many ways can we fill the spaces? In other words, how many lists of 1s and 2s sum to n? The answer arises in many natural phenomena. ∎

12.3. Problem. *Triangles in a triangular grid.* How many triangles are there in the equilateral triangular grid T_n with side length n? Below we illustrate the grid for $n = 3$. This has nine triangles of size 1, three of size 2, and one of size 3, for a total of 13. ∎

12.4. Problem. *The Polygon Problem.* A **triangulation** of a convex n-sided polygon cuts it into triangles by adding $n - 3$ noncrossing diagonals between corners. A triangle has 1 triangulation, a quadrilateral has 2, a pentagon has 5, and a hexagon has 14. How many triangulations does a convex n-gon have? ∎

GENERAL PROPERTIES

Recall that a sequence $\langle a \rangle$ of real numbers is a function from $\mathbb{N}$ or $\mathbb{N} \cup \{0\}$ to $\mathbb{R}$. We write a_n for the value at the integer n. In this chapter we start sequences with the 0th term. We can specify a sequence $\langle a \rangle$ by a formula for a_n or by a recursive expression for a_n.

12.5. Example. The sequence defined by the formula $a_n = 3(-1)^n$ can be defined recursively by $a_0 = 3$ and $a_n = -a_{n-1}$ for $n \geq 1$. It is easy to prove by induction that both definitions produce the same sequence. ∎

12.6. Definition. A **recurrence relation** or simply **recurrence** for a sequence $\langle a \rangle$ is an expression for a_n in terms of n and the values $a_0, \ldots, a_n$. If the expression for a_n depends only on n and $a_{n-k}, \ldots, a_{n-1}$ (and is used only when $n \geq k$), then the recurrence is of **order** k.

The recurrence in Example 12.5 is a first-order recurrence. Many claims about solutions to kth-order recurrences can be proved by induction on the index of the term. A recurrence of order k cannot be used to compute terms with index less than k. Thus the argument for the induction step is not valid there, and the basis step must explicitly consider the first k terms. We saw this phenomenon for second-order recurrences in Solution 3.27 and in Exercises 3.55–3.57. The proof of the induction step uses the induction hypothesis for the preceding k values.

12.7. Example. Suppose that $a_n = 4a_{n-1} - 4a_{n-2}$ for $n \geq 2$, with $a_0 = 1$ and $a_1 = 4$. The solution is $a_n = (n+1)2^n$. Checking the formula for $n = 0$ and $n = 1$ means verifying that $1 \cdot 2^0 = 1 = a_0$ and $2 \cdot 2^1 = 4 = a_1$. In the induction step we verify the formula when n is at least 2, given that it holds for $n - 1$ and $n - 2$. Using the recurrence for $\langle a \rangle$, we compute

$$a_n = 4a_{n-1} - 4a_{n-2} = 4n2^{n-1} - 4(n-1)2^{n-2} = 2n2^n - (n-1)2^n = (n+1)2^n.$$

The validity of this computation when $n = 2$ depends on having checked the formula for both $n = 0$ and $n = 1$.   ∎

Our first general example of this phenomenon is the statement of when a recurrence uniquely determines a sequence.

12.8. Proposition. A kth-order recurrence relation for $\langle a \rangle$, together with known initial values $a_0, \ldots, a_{k-1}$, uniquely determines a_n for all $n \geq 0$.

Proof: We use induction on n.
    Basis step: The hypothesis specifies the values of $a_0, \ldots, a_{k-1}$.
    Induction step (for $n \geq k$): The induction hypothesis states that $a_{n-k}, \ldots, a_{n-1}$ have been uniquely determined. The recurrence then determines a_n uniquely in terms of those values.   ∎

The notion of linearity is used throughout mathematics. If $x_1, \ldots, x_k$ are objects, and $c_1, \ldots, c_k$ are constants, then $\sum_{i=1}^{k} c_i x_i$ is a **linear combination** of $x_1, \ldots, x_k$ with **coefficients** $c_1, \ldots, c_k$. A polynomial is a linear combination of monomials. In Chapter 6, we studied linear combinations of integers with integer coefficients. In Chapter 9, we studied the expectation of linear combinations of random variables.
    This notion motivates the definition of linear recurrence relations. We want all linear combinations of sequences that satisfy a linear recurrence relation also to satisfy it. We make the definition special enough to emphasize the recurrence relations we study in this chapter.

12.9. Definition. A kth-order recurrence relation for $\langle a \rangle$ is **linear** if there exist functions f and $h_1, \ldots, h_k$ such that $a_n = f(n) + \sum_{i=1}^{k} h_i(n) a_{n-i}$ for $n \geq k$. The expression $f(n)$ is the **inhomogeneous term**. If $f(n) = 0$ for all n, then the relation is **homogeneous**.

12.10. Lemma. (Linearity) If both $\langle x \rangle$ and $\langle y \rangle$ satisfy a kth order homogeneous linear recurrence relation, and A and B are constants, then $\langle z \rangle$ defined by $z_n = Ax_n + By_n$ also satisfies the recurrence.

Proof: We have $x_n = \sum_{i=1}^{k} h_i(n)x_{n-i}$ and $y_n = \sum_{i=1}^{k} h_i(n)y_{n-i}$ for $n \geq k$. We multiply the first by A and the second by B and then add them. ∎

12.11. Example. The recurrence relations $a_n = 2a_{n-1}$ and $b_n = 2b_{n-1} + 1$ are linear. With initial values $a_0 = s$ and $b_0 = t$, their solutions are $a_n = s2^n$ and $b_n = (1+t)2^n - 1$, again proved by induction. ∎

FIRST-ORDER RECURRENCES

We begin with the simplest case: first-order recurrences.

12.12. Example. *Recurrences that become summations.* Consider the first-order linear recurrence $a_n = a_{n-1} + f(n)$ for $n \geq 1$. We "iterate" the recurrence to express a_n as a summation. This yields

$$a_n = a_{n-1} + f(n) = a_{n-2} + f(n) + f(n-1) = \quad \cdots \quad = a_0 + \sum_{i=1}^{n} f(i).$$

When we can evaluate the sum, we obtain a formula for a_n. Consider the example $f(n) = n$. Knowing a_0, the solution is $a_n = a_0 + \sum_{i=1}^{n} f(n) = a_0 + (n+1)n/2$ for $n \geq 0$. ∎

We consider two combinatorial problems that lead to first-order linear recurrence relations. There are two steps in solving a problem using recurrence relations. First we derive the recurrence relation, then we solve the recurrence to obtain a formula.

12.13. Example. *Regions in the plane.* A **configuration** of lines is a finite collection of lines in the plane such each pair of lines has one common point and no three lines have a common point. Let a_n be the number of regions created by a configuration of n lines. It is not obvious that every configuration of n lines creates the same number of regions; this follows inductively when we establish a recurrence for a_n.

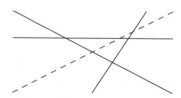

We begin with no lines and one region, so $a_0 = 1$. We prove that $a_n = a_{n-1} + n$ if $n \geq 1$. Consider a configuration of n lines, with $n \geq 1$, and let L be one of these lines. The other lines form a configuration of $n - 1$ lines. We argue that adding L increases the number of regions by n. The intersections of L with the other $n - 1$ lines partition L into n portions. Each of these portions cuts a region into two. Thus adding L increases the number of regions by n. Since this holds for all configurations we have $a_n = a_{n-1} + n$ for $n \geq 1$. By Proposition 12.8, this determines a unique sequence starting with $a_0 = 1$, and hence every configuration of n lines creates the same number of regions.

We discussed this recurrence in Example 12.12. Now the initial value is $a_0 = 1$, and the solution is $a_n = 1 + (n + 1)n/2$. ∎

12.14. Solution. *The Tower of Hanoi.* Consider the Tower of Hanoi problem for n rings; let a_n be the number of moves required to move the pile to another peg. To move the pile, we must move the bottom ring. In order to move the bottom ring, we must first move the top $n - 1$ rings to another peg. By the definition of a_{n-1}, this requires a_{n-1} moves. Then we move the bottom ring to the empty peg and must solve the smaller problem again to put the other rings back on top of it.

This yields the recurrence $a_n = 2a_{n-1} + 1$ for $n \geq 1$. We have argued that this many moves are needed, and also this many moves suffice. For the initial value we have $a_0 = 0$, since it takes no moves to move no rings. In Example 12.11, we found the solution $a_n = 2^n - 1$. Do not worry about the world crumbling; $2^{64} - 1$ seconds is more than 10^{11} years. ∎

Up to this point, we have presented solutions to recurrence relations and proved them by induction. Computing several terms of a sequence sometimes suggests a general formula, but we might guess wrong.

12.15. Example. *A cautionary remark.* Consider n points on a circle so that no three of the segments connecting them intersect at a single point. Let a_n be the number of regions into which such segments cut the region enclosed by the circle. For $n \geq 1$, the sequence begins 1, 2, 4, 8, 16, but $a_6 = 31$. Thus the answer is not the powers of 2 (see also Solution 11.69, Exercise 5.47, and Exercise 13). ∎

We cannot rely on guessing the answer, so we need systematic methods for solving recurrences. We have seen that $a_n = a_{n-1} + f(n)$ can be solved by evaluating $\sum_{i=1}^{n} f(i)$. We can also solve $a_n = ca_{n-1} + f(n)$ when c is a constant and f is a polynomial. The recurrence we derived for the Tower of Hanoi problem has this form.

12.16. Theorem. If $\langle a \rangle$ satisfies $a_n = ca_{n-1} + f(n)$ for $n \geq 1$, where c is a constant and f is a polynomial of degree d, then there is a constant A and a polynomial p such that $a_n = Ac^n + p(n)$. If $c \neq 1$, then p has degree d. If $c = 1$, then p has degree $d + 1$. The polynomial p is independent of the initial value; the initial value determines A.

Proof: When $c = 1$, we have $a_n = a_0 + \sum_{i=1}^{n} f(i)$, as in Example 12.12. When f is a polynomial of degree d, the value of this sum is a polynomial in n of degree $d + 1$, by Theorem 5.31.

When $c \neq 1$, by Proposition 12.8 only one sequence satisfies the recurrence and the initial value. Thus it suffices to find such a sequence that also satisfies $a_n = Ac^n + p(n)$. Satisfying the recurrence requires

$$Ac^n + p(n) = c[Ac^{n-1} + p(n-1)] + f(n).$$

Since Ac^n cancels, we may use any A, and it suffices to find p such that $p(n) = cp(n-1) + f(n)$. By Corollary 3.25, this equation holds for all $n \in \mathbb{N}$ if and only if corresponding coefficients of powers of n are equal. Given $p(n) = \sum_{k=0}^{d} b_k n^k$ and $f(n) = \sum_{k=0}^{d} c_k n^k$, the recurrence requires

$$\sum_{k=0}^{d} b_k n^k = c \sum_{k=0}^{d} b_k (n-1)^k + \sum_{k=0}^{d} c_k n^k.$$

Equating coefficients of n^k yields, for $0 \leq k \leq d$,

$$b_k = c \sum_{i=k}^{d} b_i \binom{i}{k} (-1)^{i-k} + c_k.$$

When $c \neq 1$, we can use the kth equation to express b_k in terms of $b_{k+1}, \ldots, b_d$. Thus we can compute b_d, then b_{d-1}, and so on.

After solving these equations to determine p, we make the sequence also match the initial value by requiring $a_0 = Ac^0 + p(0)$. This yields $A = a_0 - p(0)$. Now $a_n = Ac^n + p(n)$ satisfies the recurrence and the initial value, so this is the formula for the desired sequence. $\blacksquare$

In the proof, the point is that when $c \neq 1$ the equations are solvable to determine $b_d, \ldots, b_0$. The general formulas for the solution are unimportant; in practice, one simply solves the system of linear equations in terms of the data of the problem. The process of finding the coefficients of an unknown polynomial by requiring the polynomial to satisfy an equation is the **method of undetermined coefficients**.

12.17. Example. Suppose that $a_n = 2a_{n-1} + n^2 - 1$ for $n \geq 1$, with $a_0 = 1$. We have proved that the solution has the form $a_n = A2^n + p(n)$, where p has degree 2. We write $p(n) = b_0 + b_1 n + b_2 n^2$ and substitute the resulting formula for a_n into the recurrence to determine the coefficients.

$$A2^n + b_0 + b_1 n + b_2 n^2 = 2A2^{n-1} + 2b_0 + 2b_1(n-1) + 2b_2(n-1)^2 + n^2 - 1$$

We equate coefficients of corresponding powers of n in this equation (see Corollary 3.25).

exponent on n	left side	right side
0	b_0	$2b_0 - 2b_1 + 2b_2 - 1$
1	b_1	$2b_1 - 4b_2$
2	b_2	$2b_2 + 1$

The solution is $b_2 = -1$, $b_1 = -4$, $b_0 = -5$. We determine A to satisfy the initial value, $1 = A \cdot 2^0 + (-5)$. Thus $a_n = 6 \cdot 2^n - 5 - 4n - n^2$. ∎

The reader may wonder why the polynomial in Theorem 12.16 has higher degree when $c = 1$. The recurrence specifies $\langle a_n \rangle$ using $d + 3$ constants: the $d + 1$ coefficients from f, the constant c, and the initial value a_0. Thus the solution should involve $d + 3$ constants. When $c \neq 1$, these are c, A, and the $d + 1$ coefficients in p. When $c = 1$, the constant term in p is absorbed by the coefficient A. To have $d + 1$ independent coefficients in p, we need the coefficients of $n, \ldots, n^{d+1}$.

The techniques we have developed in this section apply more generally. Often a substitution reduces a recurrence to a simpler form. The introduction of α_n in the proof of Theorem 12.16 illustrates this technique; we close this section with another example.

12.18. Corollary. Let f be a polynomial of degree d. The solution of the recurrence $a_n = ca_{n-1} + f(n)\beta^n$ for $n \geq 1$ has the form $Ac^n + p(n)\beta^n$, where p is a polynomial. If $c \neq \beta$, then p has degree d. If $c = \beta$, then p has degree $d + 1$. The polynomial p is independent of the initial value; the initial value determines A.

Proof: We define b_n by setting $a_n = \beta^n b_n$. By substituting into the recurrence for $\langle a \rangle$ and canceling β^n, we obtain $b_n = (c/\beta)b_{n-1} + f(n)$. This recurrence has the form specified in Theorem 12.16. We take the solution given by Theorem 12.16 and multiply by β^n to obtain the formula for a_n. We leave the details to Exercise 14. ∎

SECOND-ORDER RECURRENCES

In this section, we present the most famous example of a second-order recurrence and then develop a solution method.

12.19. Example. *Fibonacci numbers.* The recurrence $a_n = a_{n-1} + a_{n-2}$ was studied by Leonardo of Pisa (1170?–1250), known as Fibonacci. The **Fibonacci numbers** F_n are the numbers determined by the **Fibonacci recurrence** $F_n = F_{n-1} + F_{n-2}$ with the initial values $F_0 = F_1 = 1$. (The initial values $F_0 = 0$ and $F_1 = 1$ are used by many authors; this merely shifts the index of the sequence by one.)

The Fibonacci sequence occurs in many applications; the mathematical journal *The Fibonacci Quarterly* is devoted to its study. Fibonacci studied a model of a rabbit farm where rabbits mature and reproduce rapidly. Suppose that every pair of rabbits produces a new pair of rabbits every month as soon as it is two months old. If the farm starts with one pair of rabbits born at time 0, then there is also one pair at time 1, and beginning at time 2 the rabbits alive two months earlier all give birth. Hence the number of pairs alive at time n are all those alive at time $n - 1$ plus those newly born, which is the number alive two months earlier. Hence the total number satisfies the recurrence $F_n = F_{n-1} + F_{n-2}$, with the initial values $F_0 = F_1 = 1$. ∎

12.20. Solution. *Rabbits and Cadillacs.* Suppose that n spaces in a row are available for parking. We want to count the ways to fill the spaces with Rabbits, which take one space, or Cadillacs, which take two spaces; let a_n be the number of ways to do this. There is one way when $n = 0$ or $n = 1$. When $n \geq 2$, we split the configurations into two types; those ending with a Rabbit and those ending with a Cadillac. By the definition of the sequence, there are a_{n-1} of the former and a_{n-2} of the latter. Since this accounts for all the configurations, there are $a_{n-1}+a_{n-2}$ ways to fill the n spaces. Thus a_n satisfies the same recurrence as the Fibonacci sequence, with the same initial values; hence $a_n = F_n$. In Solution 12.23, we obtain an explicit formula for F_n. ∎

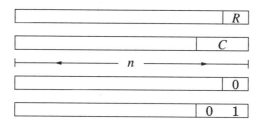

12.21. Example. *Fibonacci models.* In Solution 12.20 we have a canonical combinatorial model for the Fibonacci numbers; the Fibonacci number F_n counts the (ordered) lists of 1s and 2s that sum to n.

Another model for the Fibonacci numbers arises from binary n-tuples. Let a_n be the number of binary n-tuples with no consecutive 1s. There is one such list of length 0 and two of length 1. For $n > 1$, consider the last item in the list. If it is 0, then the first $n - 1$ digits can be any binary $n - 1$-tuple with no consecutive 1s. If the last item is 1, then position $n - 1$ must be 0, but the first $n - 2$ digits can be any list of length $n - 2$ with no consecutive 1s. Hence the number a_n of ways to build such a list of length n satisfies the recurrence $a_n = a_{n-1} + a_{n-2}$ for $n \geq 2$. With the initial conditions $a_0 = 1$ and $a_1 = 2$, we conclude by induction that $a_n = F_{n+1}$. ∎

The model of 1, 2-lists summing to n yields combinatorial arguments for identities involving the Fibonacci numbers, just as the subset-selection or block-walking models yield combinatorial arguments concerning binomial coefficients. It can also solve counting problems directly. For example, we can prove that there are F_{n+1} binary n-tuples with no consecutive 1s by establishing a bijection between the set of these n-tuples and the set of 1, 2-lists summing to n (Exercise 20).

Having obtained a recurrence for the Fibonacci numbers, we want to solve it to obtain an explicit formula for F_n. More generally, we consider second-order linear recurrence relations of the form $a_n = c_1 a_{n-1} + c_2 a_{n-2}$.

12.22. Theorem. Suppose that $a_n = c_1 a_{n-1} + c_2 a_{n-2}$, with $a_0 = s$ and $a_1 = t$. If the equation $x^2 - c_1 x - c_2 = 0$ has distinct solutions α, β, then there exist constants A, B such that $a_n = A\alpha^n + B\beta^n$. If $x^2 - c_1 x - c_2 = (x - \alpha)^2$, then there exist constants A, B such that $a_n = A\alpha^n + Bn\alpha^n$. In either case, the constants are determined by the initial values.

Proof: Suppose first that $x^2 - c_1 x - c_2 = 0$ has distinct solutions α, β. For $\gamma \in \{\alpha, \beta\}$, we have $\gamma^2 = c_1\gamma + c_2$. For $n \geq 2$, we can multiply this by γ^{n-2} to obtain $\gamma^n = c_1\gamma^{n-1} + c_2\gamma^{n-2}$. Hence the sequences defined by $a_n = \alpha^n$ and $a_n = \beta^n$ satisfy the recurrence. By linearity (Lemma 12.10), $a_n = A\alpha^n + B\beta^n$ also satisfies the recurrence.

By Proposition 12.8, it suffices to show that we can choose A, B to make this formula satisfy the initial values. Since $a_0 = s$ and $a_1 = t$, we require $s = A + B$ and $t = \alpha A + \beta B$. Since $\alpha \neq \beta$, this system of two linear equations in two unknowns has the solution $A = \frac{t - \beta s}{\alpha - \beta}$ and $B = \frac{t - \alpha s}{\beta - \alpha}$.

Now suppose that $x^2 - c_1 x - c_2 = (x - \alpha)^2$. The argument that α^n is a solution is as above. For $n\alpha^n$, we observe that the factorization requires $c_1 = 2\alpha$ and $c_2 = -\alpha^2$, and hence $c_1\alpha + 2c_2 = 0$. Thus,

$$c_1(n-1)\alpha^{n-1} + c_2(n-2)\alpha^{n-2} = n(c_1\alpha + c_2)\alpha^{n-2} - (c_1\alpha + 2c_2)\alpha^{n-2}.$$

Since $c_1\alpha + c_2 = \alpha^2$, the first term becomes $n\alpha^n$; since $c_1\alpha + 2c_2 = 0$, the second term is 0. Hence $c_1(n-1)\alpha^{n-1} + c_2(n-2)\alpha^{n-2} = n\alpha^n$, and $n\alpha^n$ satisfies the recurrence. By linearity, $a_n = A\alpha^n + Bn\alpha^n$ also satisfies it.

Again we can choose A, B to make this formula satisfy the initial values. From $n = 0$ and $n = 1$, we obtain the two requirements $s = A$ and $t = \alpha A + \alpha B$. We satisfy these by setting $A = s$ and $B = (t/\alpha) - s$. ∎

12.23. Solution. *Formula for the Fibonacci numbers.* Factoring the characteristic polynomial $x^2 - c_1 x - c_2$ of the recurrence introduces irrational numbers; the roots are $\alpha = \frac{1}{2}(1 + \sqrt{5})$ and $\beta = \frac{1}{2}(1 - \sqrt{5})$. We use $F_0 = 1$ and $F_1 = 1$ to determine A, B in the general solution $F_n = A\alpha^n + B\beta^n$, solving the linear equations as described in the proof of the theorem. The resulting formula (see Exercise 25) is

$$F_n = \frac{1}{\sqrt{5}}(\frac{1+\sqrt{5}}{2})^{n+1} - \frac{1}{\sqrt{5}}(\frac{1-\sqrt{5}}{2})^{n+1}$$

Since the recurrence produces a sequence of integers, the value of this strange formula involving irrational numbers is an integer for every nonnegative integer n.   ∎

GENERAL LINEAR RECURRENCES

Combining the ideas in Theorem 12.16 and Theorem 12.22 leads to solution techniques for kth-order linear recurrence relations. To determine the sequence $\langle a \rangle$, a recurrence of order k must be supplied with k **initial values** $a_0, \ldots, a_{k-1}$. As in Proposition 12.8, the recurrence for $n \geq k$ and the k initial values together specify a unique sequence.

We can write a kth-order linear recurrence relation in the form $a_n - h_1(n)a_{n-1} - h_2(n)a_{n-2} - \cdots - h_k(n)a_{n-k} = f(n)$, valid for $n \geq k$. Here we consider only the case of constant coefficients, where each h_i is constant. The **characteristic equation method** is a solution technique for linear recurrence relations with constant coefficients. It extends what we have done for $k = 1$ and $k = 2$.

12.24. Definition. Let $a_n - c_1a_{n-1} - c_2a_{n-2} - \cdots - c_ka_{n-k} = f(n)$ be a kth-order linear recurrence relation with constant coefficients. Its **characteristic polynomial** is the polynomial p defined by $p(x) = x^n - \sum_{i=1}^{k} c_i a_{n-i}$, and its **characteristic equation** is $p(x) = 0$.

For the homogeneous case, we extend the solution from $k = 2$. For the recurrence $a_n = c_1a_{n-1} + c_2a_{n-2}$, the characteristic polynomial was the polynomial defined by $x^2 - c_1x - c_2$.

12.25. Theorem. Suppose that $\langle a \rangle$ satisfies a kth-order homogeneous linear recurrence relation with constant coefficients. If the characteristic polynomial factors as $p(x) = \prod_{i=1}^{r}(x - \alpha_i)^{d_i}$ for distinct $\alpha_1, \ldots, \alpha_r$, then the solution of the recurrence is $a_n = \sum_{i=1}^{r} q_i(n)\alpha_i^n$, where each q_i is a polynomial of degree $d_i - 1$. The k coefficients of these polynomials are determined by the initial values.

Proof: (Exercise 28).   ∎

Before satisfying the initial conditions, we write the solution in Theorem 12.25 using unknown variables for the coefficients in each q_i. This expression is the **general solution** to the homogeneous recurrence relation. In the case $k = 2$, the general solution was the formula $A\alpha^n + B\beta^n$ or $A\alpha^n + Bn\beta^n$. The constants A, B are the k coefficients to be determined by the initial values.

These coefficients are determined by solving k linear equations obtained from the initial conditions; the value of the expression must agree with the initial value a_i for $0 \leq i \leq k - 1$.

Solving the characteristic equation is also the first step in solving an inhomogeneous relation. We combine this with a **particular solution** $\langle y \rangle$ of the inhomogenous relation (ignoring initial values) to obtain the general solution of the inhomogeneous relation.

12.26. Theorem. If $\langle y \rangle$ satisfies a kth-order linear recurrence relation with coefficients $c_1, \ldots, c_k$ and inhomogeneous term $f(n)$, then every sequence $\langle a \rangle$ that satisfies this recurrence relation has the form $a_n = x_n + y_n$, where $\langle x \rangle$ satisfies the homogeneous linear recurrence relation with coefficients $c_1, \ldots, c_k$.

Proof: (sketch) Summing the recursive expressions for x_n and y_n shows that every sequence of this form satisfies the recurrence. Conversely, a sequence $\langle a \rangle$ satisfying the recurrence is determined by k initial values. It suffices to show that the coefficients in the general solution to the homogeneous relation can be chosen so that the resulting $\langle x \rangle + \langle y \rangle$ agrees with the initial values $a_0, \ldots, a_{k-1}$. ∎

The remaining task is finding a particular solution $\langle y \rangle$. When the inhomogeneous term is $q(n)\beta^n$ for some polynomial q and constant β, the method of Theorem 12.16 and Corollary 12.18 yields a particular solution of the form $p(n)\beta^n$, where p is a polynomial. Furthermore, the degree of p exceeds the degree of q by the multiplicity of β as a root of the characteristic polynomial. When $f(n) = f_1(n) + f_2(n)$, we can find particular solutions for the inhomogeneous terms $f_1(n)$ and $f_2(n)$ separately and then sum them (Exercise 29); this is the **superposition principle**.

We apply these techniques to solve Problem 12.3.

12.27. Solution. *Triangles in a triangular grid.* Let a_n be the number of triangles (of all positive integer sizes) in the equilateral triangular grid T_n with side length n, shown below for $n = 3$. Note that $a_1 = 1$, $a_2 = 5$, and $a_3 = 13$. We first derive a recurrence for $\langle a \rangle$.

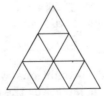

The grid T_n contains three copies of T_{n-1}. Each triangle T in T_n occurs in at least one of these copies of T_{n-1}, unless it touches all three sides. Let $f(n)$ be the number of triangles touching all three sides. Always there is one upright triangle touching all three sides (the full region), and when n

is even there is one inverted triangle with sides of length $n/2$ that touches all three sides. Thus $f(n) = 2$ when n is even and $f(n) = 1$ when n is odd, which we can write as $f(n) = \frac{3}{2} + \frac{1}{2}(-1)^n$.

The intersections of two or three of the copies of T_{n-1} are copies of T_{n-2} or T_{n-3}. Using the Inclusion-Exclusion Principle (Theorem 10.12) and the definition of $\langle a \rangle$, the number of triangles not contained in any of the three copies of T_{n-1} is $a_n - 3a_{n-1} + 3a_{n-2} - a_{n-3}$. This is another computation of $f(n)$, so we have

$$a_n - 3a_{n-1} + 3a_{n-2} - a_{n-3} = \frac{3}{2} + \frac{1}{2}(-1)^n \qquad \text{for } n \geq 3,$$

with initial values $a_0 = 0$, $a_1 = 1$, $a_2 = 5$.

The characteristic polynomial $x^3 - 3x^2 + 3x - 1 = (x - 1)^3$ has 1 as a triple root. Thus the general solution to the homogeneous recurrence is a quadratic polynomial in n. To find the particular solution, we sum solutions for the term $\frac{3}{2}$ and the term $\frac{1}{2} \cdot (-1)^n$.

Since -1 is not a characteristic root, the particular solution corresponding to the term $\frac{1}{2}(-1)^n$ has the form $A(-1)^n$. Satisfying the recurrence requires $A(-1)^n - 3A(-1)^{n-1} + 3A(-1)^{n-2} - 1A(-1)^{n-3} = \frac{1}{2}(-1)^n$. This simplifies to $A + 3A + 3A + A = \frac{1}{2}$, and hence $A = \frac{1}{16}$.

Since 1 is a characteristic root of multiplicity three and the inhomogenous term $\frac{3}{2}$ is 1^n times a polynomial of degree 0, the solution corresponding to this term is a polynomial of degree three. We need only determine the leading coefficient, because the lower-order terms belong to the homogeneous solution. We thus let $y_n = Bn^3$ and have

$$Bn^3 - 3B(n-1)^3 + 3B(n-2)^3(-1)^B(n-3)^3 = \frac{3}{2}.$$

The coefficients of the nonzero powers of n on the left (must!) cancel. The constant term yields $3B - 24B + 27B = 3/2$, with solution $B = 1/4$.

Now we determine coefficients in the general solution to satisfy the initial conditions. We evaluate

$$a_n = C_0 + C_1 n + C_2 n^2 + \frac{1}{16}(-1)^n + \frac{1}{4}n^3,$$

at $n = 0, 1, 2$ to obtain

$$0 = C_0 + \frac{1}{16}$$
$$1 = C_0 + C_1 + C_2 - \frac{1}{16} + \frac{1}{4}$$
$$5 = C_0 + 2C_1 + 4C_2 + \frac{1}{16} + 2$$

The solution of this system is $C_0 = -\frac{1}{16}$, $C_1 = \frac{1}{4}$, $C_2 = \frac{5}{8}$. The resulting formula for a_n is

$$a_n = \frac{1}{16}[4n^3 + 10n^2 + 4n - 1 + (-1)^n]. \qquad \blacksquare$$

Straightforward inductive arguments verify the solution of fixed-order linear constant-coefficient recurrences when the characteristic roots are known and the inhomogeneous term has the specified form. Essentially, we have guessed a formula and verified that it works. The method of generating functions provides a deeper explanation and applies in more general situations (see Application 12.40).

OTHER CLASSICAL RECURRENCES

We begin with a famous nonlinear recurrence relation. The Catalan numbers $C_n = \frac{1}{n+1}\binom{2n}{n}$ are named for Eugène Charles Catalan (1814–1894). In 1838 he discovered that C_n counts the ways to multiply together $n+1$ factors by a non-associative binary product (Exercise 37). Euler had already encountered them in 1758 in the solution of Problem 12.4. We have encountered them in Solution 9.10 as the number of good elections in the Ballot Problem. We use the ballot model to obtain a nonlinear recurrence for $\{C_n\}$ that is not a recurrence of fixed order; the value of C_n depends on all values $C_0, \dots, C_{n-1}$.

12.28. Example. *The Ballot Problem.* The **ballot paths** to (n, n) are the lattice paths of length $2n$ that do not step above the diagonal. Let there be a_n of these; we derive a recurrence for a_n. Every ballot path to (n, n) has some first return to the diagonal; suppose it occurs at position (k, k).

The first portion of the path steps right, does not rise above $y = x - 1$ until it reaches $(k, k - 1)$, and then steps up. Hence the possible initial portions of the path correspond to ballot paths of length $2(k - 1)$.

The portion of the path from (k, k) to (n, n) is a translation of a ballot path of length $2(n - k)$. Hence the number of ballot paths of length $2n$ that first return to the diagonal at (k, k) is $a_{k-1}a_{n-k}$. Summing over the choices for k yields $a_n = \sum_{k=1}^{n} a_{k-1}a_{n-k}$ for $n \geq 1$, with the initial value $a_0 = 1$. ∎

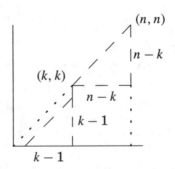

From Solution 9.10, we know that a_n in Example 12.28 equals the Catalan number $C_n = \frac{1}{n+1}\binom{2n}{n}$. We next prove that the solution to Problem 12.4 satisfies the same recurrence relation and has the same initial value.

Hence it must also be the Catalan sequence. In Exercises 37–40, we consider additional counting problems that lead to the same recurrence and hence to the Catalan numbers.

12.29. Solution. *The Polygon Problem.* Let a_n count triangulations of a convex polygon with $n+2$ sides, and define $a_0 = 1$. Starting with $n = 0$, the sequence begins $1, 1, 2, 5, 14, \cdots$. This agrees with the Catalan numbers.

Consider a convex polygon with $n + 2$ sides and vertices $v_0, \ldots, v_{n+1}$ in order. In every triangulation, the edge $v_{n+1}v_0$ lies on some triangle; let v_k be its third corner. To complete the triangulation, we must triangulate the polygon formed by $v_0, \ldots, v_k$ and the polygon formed by $v_k, \ldots, v_{n+1}$, which have $k + 1$ sides and $n - k + 2$ sides, respectively.

These smaller polygons can be triangulated in a_{k-1} and a_{n-k} ways, respectively. Summing over the choices of k yields $a_n = \sum_{k=1}^{n} a_{k-1}a_{n-k}$ for $n \geq 1$, with $a_0 = 1$. Hence $\langle a \rangle$ satisfies the same recurrence and initial value as the Catalan sequence. We conclude that there are $\frac{1}{n+1}\binom{2n}{n}$ ways to triangulate a convex polygon with $n + 2$ sides. ∎

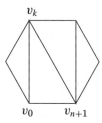

$$v_k$$
$$v_0 \quad\quad v_{n+1}$$

How would we solve the Catalan recurrence if we did not already have a formula? We introduce a technique for this in the next section. Meanwhile, we discuss another classical recurrence that can be solved by the **method of substitution**. We begin with an easy example. (The remainder of this section is optional.)

12.30. Example. Suppose that $a_n = (1 - \frac{1}{n+1})a_{n-1}$ for $n \geq 1$, with $a_0 = 1$. Multiplying the recurrence by $n+1$ yields $(n + 1)a_n = na_{n-1}$. This suggests substituting $b_n = (n + 1)a_n$ to simplify the recurrence. The new sequence $\langle b \rangle$ then satisfies $b_n = b_{n-1}$ for $n \geq 1$, with $b_0 = (0 + 1)a_0 = 1$. The solution to this recurrence is $b_n = 1$ for all $n \geq 0$. Undoing the substitution yields $a_n = b_n/(n + 1) = 1/(n + 1)$ for $n \geq 0$. ∎

We illustrate the techniques of substitution and reduction of order by giving another solution of the Derangements Problem (Problem 10.9). We obtain a second-order linear recurrence relation with nonconstant coefficients. (Exercise 34 obtains another recurrence for this.)

12.31. Solution. *Derangements.* Recall that D_n counts the permutations of $[n]$ without fixed points. With $D_1 = 0$ and $D_2 = 1$, we derive a second-order recurrence valid for $n \geq 3$. We classify the derangements of $[n]$ by the position $k \in [n-1]$ where n is placed. For each such k, we consider whether k appears in position n. If k appears in position n, then we can complete the derangement in D_{n-2} ways by deranging the other $n-2$ objects. Hence there are D_{n-2} derangements of this type for each choice of k.

n	j
k	n

If k does not appear in position n, then some element j other than k appears there, and we can interchange the elements in positions n and k to obtain a derangement of $[n-1]$, with n appended at the end. Conversely, given any derangement of $[n-1]$ with n appended at the end, making the same interchange restores the original configuration. Hence the number of derangements of this type (element n in position k, and element k not in position n) is D_{n-1}, for each choice of k.

Since there are $n-1$ choices for the position k where n is placed, we obtain the recurrence $D_n = (n-1)(D_{n-1} + D_{n-2})$ for $n \geq 3$. We can make the recurrence valid also for $n = 2$ by defining $D_0 = 1$. This is natural; there is one permutation with no elements, and it has no fixed points.

To solve this recurrence, we make successive substitutions to simplify its form until we obtain a recurrence that we can solve. First we want to eliminate the nonlinear factor $n-1$. We try setting $f_n = D_n/n!$. Note that $f_0 = 1$ and that f_n is the probability that a random permutation is a derangement. Substitution yields

$$n! f_n = (n-1)(n-1)! f_{n-1} + (n-1)! f_{n-2},$$

and then we divide by $n!$ to obtain $f_n = (1 - \frac{1}{n}) f_{n-1} + (\frac{1}{n}) f_{n-2}$.

The second substitution reduces the order of the recurrence. We rewrite the recurrence for f_n as

$$f_n - f_{n-1} = (-\tfrac{1}{n})(f_{n-1} - f_{n-2}).$$

After doing this, it becomes natural to define $g_n = f_n - f_{n-1}$ for $n \geq 1$ and obtain the recurrence $g_n = (-\frac{1}{n}) g_{n-1}$. Since we defined g_n for $n \geq 1$, and $g_1 = f_1 - f_0 = -1$, we can iterate the recurrence for g_n to write $g_n = (-1)^{n-1} g_1/n! = (-1)^n/n!$.

The final step is to reverse the substitutions. Retrieving f_n uses $\sum_{k=1}^n g_k = \sum_{k=1}^n (f_k - f_{k-1}) = f_n - f_0$. This is called a **telescoping sum**, because all the intermediate terms cancel out or "collapse" as a telescope collapses. Since $f_0 = 1$, we obtain

$$f_n = f_0 + \sum_{k=1}^n g_k = 1 + \sum_{k=1}^n \tfrac{(-1)^k}{k!} = \sum_{k=0}^n \tfrac{(-1)^k}{k!}.$$

Thus $D_n = n! \sum_{k=0}^n (-1)^k/k!$.   ∎

GENERATING FUNCTIONS (Optional)

The last method we discuss for solving recurrence relations uses a technique that combines combinatorics and algebra. Before applying it to recurrences, we develop the basic ideas.

We associate with a sequence $\langle a \rangle$ the expression $a_0 x^0 + a_1 x^1 + \cdots$. Here we do not treat x^n as a number; it is merely a place-holder for the term a_n in the sequence. We do not call this an "infinite series" or seek a sum; Chapters 14 and beyond address that interpretation.

12.32. Definition. A **formal power series** is an expression of the form $\sum_{n=0}^{\infty} a_n x^n$ in which x is treated as a formal variable, not as a number. The formal power series $\sum_{n=0}^{\infty} a_n x^n$ is the **generating function** for the sequence $\langle a \rangle$. The **sum** of two formal power series $\sum_{n=0}^{\infty} a_n x^n$ and $\sum_{n=0}^{\infty} b_n x^n$ is the formal power series with $a_n + b_n$ as the coefficient of x^n. Their **product** is the formal power series with $\sum_{j=0}^{n} a_j b_{n-j}$ as the coefficient of x^n.

By the definition, two formal power series are equal if and only if they have the same sequence of coefficients. The definitions of sum and product of formal power series agree with our experience in multiplying polynomials. These definitions allow us to build generating functions to solve counting problems; we want the coefficient of x^n to be the number a_n of solutions when the parameter is n. The name *generating function* emphasizes "generating" the values a_n.

Consider subsets of an r-element set. When $r = 1$, there is one way to choose the element and one way not to choose it, so the generating function is $1 + x$. To build the generating function for a fixed r, we take r such factors, because we make such a choice for each element. The contributions to the coefficient of x^n correspond to the subsets of size n. We argued in Chapter 5 that therefore the coefficient of x^n must be $\binom{r}{n}$. This principle applies more generally.

12.33. Lemma. Consider sets A, B, C with $C = A \times B$. For $n \in \mathbb{N} \cup \{0\}$, let a_n, b_n, c_n be the number of elements of "size" n in A, B, C, respectively. If the size of each $\gamma = (\alpha, \beta) \in C$ with $\alpha \in A$ and $\beta \in B$ is the sum of the sizes of α and β, then the generating function for $\langle c \rangle$ is the product of the generating functions for $\langle a \rangle$ and $\langle b \rangle$.

Proof: Choose $\gamma = (\alpha, \beta)$ with $\alpha \in A$ and $\beta \in B$. If z has size n, then for some k the sizes of α and β are k and $n - k$. Any element of A having size k can be paired with any element of B having size $n - k$, since the coordinates of elements in a Cartesian product are chosen independently. The subsets using a particular size from A are disjoint, so we sum over the possible values of k to obtain $c_n = \sum_{k=0}^{n} a_k b_{n-k}$. Hence the generating functions satisfy the definition of product. ∎

12.34. Application. *Selections with repetition.* We want to select objects from r types, with no restriction on the number of objects of each type. How many we select of one type is independent of how many we select of any other type. Hence the overall set of selections is an r-fold Cartesian product of the sets of selections involving each individual type.

The generating function is the formal power series in which the coefficient of x^n is the number of elements in which the total number of objects selected is n. Using Lemma 12.33, this generating function is the product of r generating functions, one for each type of object. When we have only one type of object, there is one way to choose n copies of it, so the generating function is $\sum_{n=0}^{\infty} x^n$. We conclude that the generating function for selections from r types of objects is $(\sum_{n=0}^{\infty} x^n)^r$.

We solved this problem in Theorem 5.23 by bijective arguments. We know that the number of ways to select n objects is $\binom{n+r-1}{r-1}$, and by definition this equals the coefficient of x^n in the generating function. Hence we have given a combinatorial proof of an algebraic identity about formal power series: $(\sum_{n=0}^{\infty} x^n)^r = \sum_{n=0}^{\infty} \binom{n+r-1}{r-1} x^n$. ∎

The formal power series 1 (the expression $1x^0 + 0x^1 + 0x^2 + \cdots$) is a multiplicative identity. Hence we may study multiplicative inverses.

12.35. Theorem. For $r \in \mathbb{N}$, the formal power series expansion of the generating function $(1-x)^{-r}$ is $\sum_{n=0}^{\infty} \binom{n+r-1}{r-1} x^n$.

Proof: When we multiply the two formal power series $1-x$ and $\sum_{n=0}^{\infty} x^n$, we obtain 1. Hence we write $\sum_{n=0}^{\infty} x^n = (1-x)^{-1}$ as formal power series. When we raise the series to the rth power, we obtain $\sum_{n=0}^{\infty} \binom{n+r-1}{r-1} x^n$ by Application 12.34. ∎

The special case $r = 1$ is the formal sum of the geometric series: $(1-x)^{-1} = \sum_{n=0}^{\infty} x^n$. Exercises 48–57 involve generating functions for combinatorial problems.

We next discuss the use of generating functions to solve recurrence relations. The recurrence for $\langle a \rangle$ leads to an equation for its generating function. We try to solve the equation to find the generating function explicitly. Extract the coefficients may yield a formula for a_n. The technique applies to both linear and nonlinear recurrences. We illustrate it by sketching an alternative derivation of the Fibonacci numbers.

12.36. Example. *Generating function for Fibonacci numbers.* Let $F(x) = \sum_{n=0}^{\infty} F_n x^n$. We multiply the recurrence by x^n and sum over the values of n where the recurrence is valid. This yields

$$\sum_{n=2}^{\infty} F_n x^n = \sum_{n=2}^{\infty} F_{n-1} x^n + \sum_{n=2}^{\infty} F_{n-2} x^n.$$

With $F_0 = F_1 = 1$, this becomes $F(x) - 1 - x = x(F(x) - 1) + x^2 F(x)$. Hence $F(x) = 1/(1 - x - x^2)$. The denominator factors as $(1 - \alpha x)(1 - \beta x)$, where $\{\alpha, \beta\} = (1 \pm \sqrt{5})/2$. The method of partial fractions then yields constants A, B (see Exercise 25) such that

$$F(x) = \frac{A}{1 - \alpha x} + \frac{B}{1 - \beta x}.$$

The geometric series ($r = 1$ in Theorem 12.35) now yields $F_n = A\alpha^n + B\beta^n$, the same formula obtained previously for the Fibonacci numbers. ∎

12.37. Application. We sketch the **generating function method** for solving kth order linear recurrence relations with constant coefficients. After multiplying the recurrence by x^n and summing over the region of validity ($n \geq k$), we obtain the generating function $A(x)$ for the sequence $\langle a \rangle$ as the ratio of two polynomials. The coefficients of the denominator are the coefficients of the characteristic polynomial in reverse. We have $(x - \alpha)$ as a factor of the characteristic polynomial if and only if $(1 - \alpha x)$ is a factor in the denominator polynomial of $A(x)$, with the same multiplicity.

When the characteristic polynomial factors as $p(x) = \prod_{i=1}^{r}(x - \alpha_i)^{d_i}$ for distinct $\alpha_1, \ldots, \alpha_r$, the denominator of $A(x)$ factors as $\prod_{i=1}^{r}(1 - \alpha_i x)^{d_i}$. Using partial fractions, we write $A(x) = \sum_{i=1}^{r} q_i(x)/(1 - \alpha_i x)^{d_i}$, where q_i is a polynomial of degree less than d_i. The polynomials q_i are determined by the initial values for $\langle a \rangle$.

Because $(1 - \alpha x)^{-d} = \sum_{n=0}^{\infty} \binom{n+d-1}{d-1} \alpha^n x^n$ (Theorem 12.35), we now obtain a formula for a_n of the form claimed in Theorem 12.26. This formula emerges from the generating function expansion without guesses.

Furthermore, suppose that the inhomogeneous term of the recurrence is β^n times a polynomial of degee d in n. Because $\sum_{n=0}^{\infty} \beta^n x^n = 1/(1 - \beta x)$, this adds d factors of $(1 - \beta x)$ to the denominator of $A(x)$. Thus the particular solution also emerges automatically. ∎

Finally, we use the generating function method to solve the Catalan recurrence, obtaining the formula for C_n yet again. We only sketch the steps, leaving the details to Exercise 47.

12.38. Solution. *Solution of the Catalan recurrence.* Our sequence $\langle C \rangle$ satisfies $C_n = \sum_{k=1}^{n} C_{k-1} C_{n-k}$ for $n \geq 1$, with $C_0 = 1$. Let $A(x) = \sum_{n=0}^{\infty} C_n x^n$ be the generating function. Multiplying the recurrence by x^n and summing over the region $n \geq 1$ where the recurrence is valid yields

$$A(x) - C_0 = x \sum_{n=1}^{\infty} \sum_{l=0}^{n-1} C_l C_{n-1-l} x^{n-1} = x \sum_{m=0}^{\infty} \sum_{l=0}^{m} C_l C_{m-l} x^m = x[A(x)]^2.$$

The resulting equation for A is $xA^2 - A + 1 = 0$. Using the quadratic formula, we have $A(x) = (1 \pm (1 - 4x)^{1/2})/2x$. The formula for the coefficients of $A(x)$ can be extracted by using an extended form of the binomial

theorem: $(1 - 4x)^{1/2} = \sum_{n=0}^{\infty} \binom{1/2}{n}(-4x)^n$. The extended binomial coefficient $\binom{u}{k}$ is defined to be $u(u - 1) \cdots (u - n + 1)/n!$ (see Remark 5.30); we
set $u = 1/2$.

In the formula for $A(x)$, we choose the negative square root because
by definition the coefficient of x^{-1} in $A(x)$ is 0. Thus the Catalan number
for $n \geq 1$ is the coefficient of x^n in $-(1 - 4x)^{1/2}/2x$. We obtain

$$C_n = -\binom{1/2}{n+1}(-4)^{n+1}/2 = \frac{1}{n+1}\binom{2n}{n}. \qquad \blacksquare$$

EXERCISES

"Obtain" requests proof, and obtaining a recurrence relation includes specifying the initial values. Solve these recurrences only when asked.

In Exercises 1–5, find a formula for a_n given the stated recurrence relation
and initial values.

12.1. $a_n = 3a_{n-1} - 2$ for $n \geq 1$, with $a_0 = 1$.

12.2. $a_n = a_{n-1} + 2a_{n-2}$ for $n \geq 2$, with $a_0 = 1$ and $a_1 = 8$.

12.3. $a_n = 2a_{n-1} + 3a_{n-2}$ for $n \geq 2$, with $a_0 = a_1 = 1$.

12.4. $a_n = 5a_{n-1} - 6a_{n-2}$ for $n \geq 2$, with $a_0 = 1$ and $a_1 = 3$.

12.5. $a_n = 3a_{n-1} - 1$ for $n \geq 1$, with $a_0 = 1$.

$$\bullet \qquad \bullet \qquad \bullet \qquad \bullet \qquad \bullet$$

12.6. Suppose that $\langle a \rangle$ satisfies the recurrence $a_n = -a_{n-1} + \lambda^n$. Determine the
values of λ such that $\langle a \rangle$ can be unbounded.

12.7. Let $a_n = n^3$. Find a constant-coefficient first-order linear recurrence relation
satisfied by $\langle a \rangle$. Does there exist a homogeneous constant-coefficient first-order
linear recurrence relation satisfied by $\langle a \rangle$? Why or why not?

12.8. Obtain a recurrence to count the pairings of $2n$ people.

12.9. Consider n circles in the plane such that each intersects every other and
no three circles meet at a point. Obtain a recurrence for the number of regions
formed. Solve the recurrence.

12.10. Use Euler's formula (Theorem 11.64) to count the regions determined by a
configuration of n lines in the plane, where no three lines have a common point and
no two are parallel. (Hint: Add a circle that encloses all the points of intersection,
and count vertices and edges.)

12.11. At the start of each year, \$100 is added to a savings account. At the end of
each year, interest equal to 5% of the amount in the account is added by the bank.
Let a_n be the amount of money in the account after the interest payment in the
nth year. Obtain a recurrence for a_n and solve it.

12.12. On a particular \$50,000 mortgage, interest is calculated each year as 5% of the unpaid amount, and afterward a payment of \$5,000 ends the year. Obtain a recurrence for the amount outstanding at the end of the nth year. Using a calculator, determine the number of years needed to pay off the mortgage. What happens if the interest rate is 10% instead of 5%?

12.13. Consider the sequence $\langle a \rangle$ defined in Example 12.15, where a_n is the number of regions inside the circle when all $\binom{n}{2}$ chords are drawn among n points on a circle and no three chords have a common intersection.

    a) Obtain the recurrence relation $a_n = a_{n-1} + f(n)$ for $n \geq 1$, where $f(n) = n - 2 + \sum_{i=1}^{n-1}(i - 1)(n - 1 - i)$, with initial value $a_0 = 1$.

    b) Using the methods of Chapter 5, solve the recurrence of part (a) to obtain an explicit formula for a_n.

12.14. Complete the proof of Corollary 12.18, solving the recurrence $a_n = ca_{n-1} + f(n)\beta^n$, where f is a polynomial and β is a constant.

12.15. Obtain a recurrence relation to count the ways to move a marker exactly n spaces, given that the number of spaces moved at one time can be 1 or 2 or 3.

12.16. Obtain a recurrence relation for the number of ways to fill up the parking along a curb with n spaces, given that there are three types of cars, of which one type takes one space and two types take two spaces.

12.17. Obtain a recurrence relation for the number of ways to tile a two-by-n checkerboard using n identical dominoes (see Example 10.7 for definitions).

12.18. A shopkeeper makes change for n cents by placing one coin at a time on the counter, keeping a running total; pennies, nickels, and dimes are available. Let a_n be the number of ways to make change for n cents. For example, $a_6 = 3$, by the lists $111111, 51$, and 15. Obtain a recurrence relation for a_n.

12.19. Prove by induction that the Fibonacci numbers satisfy the following relations. (For all problems about Fibonacci numbers, we use the sequence $\{F_n\}$ defined by $F_0 = F_1 = 1$ and $F_n = F_{n-1} + F_{n-2}$ for $n \geq 2$.)

    a) $\sum_{i=0}^{n} F_i^2 = F_n F_{n+1}$.

    b) $\sum_{i=0}^{n} F_{2i} = F_{2n+1}$.

    c) $\sum_{i=0}^{2n-1}(-1)^i F_{2n-i} = F_{2n-1}$.

12.20. Establish a bijection between the set of 1,2-lists that sum to n and the set of 0,1-lists of length $n - 1$ that have no consecutive 1s.

12.21. Prove by induction and by combinatorial argument that $1 + \sum_{i=0}^{n} F_i = F_{n+2}$.

12.22. Prove by induction and by combinatorial argument that $F_n = \sum_{i=0}^{n}\binom{n-i}{i}$.

12.23. Prove bijectively (using lists of 1s and 2s) that $F_{m+n} = F_m F_n + F_{m-1}F_{n-1}$. Conclude for each $k \in \mathbb{N}$ that F_{n-1} divides F_{kn-1}.

12.24. Prove that every natural number can be written as a sum of distinct numbers that are Fibonacci numbers.

12.25. (−) Complete the details of computing the formula for the Fibonacci numbers in Solution 12.23. Also complete the details of computing the formula for the Fibonacci numbers in Example 12.36.

12.26. *Fibonacci numbers and the Euclidean algorithm* (Algorithm 6.15). We say that (a_0, a_1) with $a_0 > a_1$ *takes* k *steps* when the Euclidean algorithm applied to it produces $(a_0, a_1), (a_1, a_2), \ldots, (a_k, 0)$ with $a_0 > a_1 > \cdots > a_k > 0$. For example, $(3, 2)$ takes two steps, and $(5, 3)$ takes three steps. For $k \geq 2$, prove that $a_0 + a_1 \geq F_{k+2}$ when (a_0, a_1) takes k steps. Prove also that this is best possible: for each $k \geq 2$ there is a pair with sum F_{k+2} that takes k steps.

12.27. (+) Consider cards $1, \ldots, n$ in a pile. When the top card is m, we reverse the order of the first m cards. The process stops only when card 1 is at the top. Prove that the process always stops, regardless of the initial order of cards. Prove that the process always takes at most $F_n - 1$ steps when there are n cards. (Hint: Prove inductively that if k distinct cards appear at the top during the process, then there are at most $F_k - 1$ steps.)

12.28. Prove Theorem 12.25, describing the general solution to a kth-order homogeneous linear recurrence relation with constant coefficients.

12.29. Let $\langle b \rangle$ and $\langle d \rangle$ be solutions to the inhomogeneous recurrences $x_n = f(n) + \sum_{i=1}^{k} h_i(n)x_{n-i}$ and $x_n = g(n) + \sum_{i=1}^{k} h_i(n)x_{n-i}$, respectively. Prove that $\langle b \rangle + \langle d \rangle$ is a solution to the recurrence $x_n = f(n) + g(n) + \sum_{i=1}^{k} h_i(n)x_{n-i}$.

12.30. Suppose that $\langle a \rangle$ is a solution of $x_n = c_1 x_{n-1} + c_2 x_{n-2} + c\alpha^n$, where $c_1, c_2, c, \alpha \in \mathbb{R}$. Prove that $\langle a \rangle$ and $C\alpha^n$ are solutions of the homogeneous third-order recurrence $x_n = (c_1 + \alpha)x_{n-1} + (c_2 - \alpha c_1)x_{n-2} - \alpha c_2 x_{n-3}$.

12.31. Suppose that $a_n = a_{n-1} + a_{n-2} + a_{n-3}$ for $n > 3$. Prove that $a_n \leq 2^{n-2}$ if $a_i = 1$ for $i \in \{1, 2, 3\}$, and $a_n < 2^n$ if $a_i = i$ for $i \in \{1, 2, 3\}$.

12.32. Solve the recurrence $a_n = \frac{2}{3}(1 + \frac{2}{3^n+1})a_{n-1}$ for $n \geq 1$, with $a_0 = 1$. (Hint: Substitute $b_n = (3^n + 1)a_n$.)

12.33. The following algorithm finds the extreme numbers (largest and smallest) among n numbers. If $n = 2$, compare the two numbers. If $n > 2$, (1) split the numbers into sets of sizes $\lfloor n/2 \rfloor$ and $\lceil n/2 \rceil$, (2) apply the algorithm inductively to determine the extreme numbers in each subset, (3) use the resulting numbers to compute the extreme numbers in the original set. Let a_n be the number of comparisons used on a set of size n. Obtain a recurrence for a_n. Use the substitution method to obtain a formula for a_n when n is a power of 2.

12.34. Consider the recurrence $D_n = (n-1)(D_{n-1} + D_{n-2})$ for derangements (with $D_0 = 1$). Substitute $f_n = D_n - nD_{n-1}$ to obtain the first-order recurrence $D_n = nD_{n-1} + (-1)^n$. Use this and induction to prove that $D_n = n! \sum_{k=0}^{n}(-1)^k/k!$.

12.35. Let B_n be the number of equivalence relations on n elements; this equals the number of partitions of the set $[n]$. Prove that $B_n = \sum_{k=1}^{n} \binom{n-1}{k-1} B_{n-k}$ for $n \geq 1$, with the initial value $B_0 = 1$. (Comment: These are the **Bell numbers**.)

12.36. Let a_n be the number of sets $\{x, y, z\} \subseteq \mathbb{N}$ such that x, y, z are the lengths of the sides of a triangle with perimeter n. Obtain a recurrence relation for a_n (the formula depends on the parity of n).

12.37. When numbers are combined using a non-associative binary operation, the order of operations matters. There is one way to combine two numbers, but a list of three can be combined using $a(bc)$ or $(ab)c$. With four numbers, there are five ways: $a(b(cd))$, $a((bc)d)$, $(a(bc))d$, $((ab)c)d$, $(ab)(cd)$. Each such grouping is a **parenthesization**. Let a_n be the number of parenthesizations of an ordered list of $n+1$ elements. Prove recursively that a_n equals the Catalan number C_n.

12.38. Obtain a bijection between the parenthesizations of $n+1$ distinct elements and the triangulations of a convex $n+2$-gon. (Hint: Consider the figure below.)

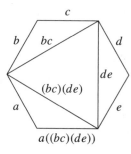

$$a((bc)(de))$$

12.39. Let a_n be the number of ways to pair up $2n$ points on a circle by noncrossing chords; note that $a_3 = 5$. Prove recursively that a_n equals the Catalan number C_n.

12.40. An arrangement of pennies is built on a row of n pennies. Each penny not in the base row rests on two pennies immediately below it, as illustrated below. Prove that the number of arrangements that can be built on a row of n pennies is the Catalan number C_n.

12.41. Let f be a polynomial of degree n. The **first difference** of f is the function $g = \Delta f$ defined by $g(x) = f(x+1) - f(x)$. The k**th difference** of f is the function $g^{(k)}$ defined inductively by $g^{(0)} = f$ and $g^{(k)} = \Delta g^{(k-1)}$ for $k \geq 1$. Obtain a formula for the nth difference of f.

12.42. Let $s(n, k)$ be the number of surjective functions from an n-element set to a k-element set. Derive a recurrence relation for $s(n, k)$ in terms of $s(n-1, k)$ and $s(n-1, k-1)$. Do not forget to specify initial values.

12.43. Let G_n be the graph consisting of a path with n vertices plus one vertex adjacent to each vertex of the path. Let a_n be the number of spanning trees in G_n.
  a) Prove that $a_n = a_{n-1} + \sum_{i=0}^{n-1} a_i$ for $n \geq 2$, where $a_0 = a_1 = 1$.
  b) Prove that $a_n = 3a_{n-1} - a_{n-2}$ for $n \geq 3$.

12.44. Let G_n be the graph on $2n$ vertices and $3n - 2$ edges pictured below, for $n \geq 1$. Prove that the chromatic polynomial of G_n is $(k^2 - 3k + 3)^{n-1}k(k - 1)$.

12.45. *Gambler's Ruin.* Two people gamble by flipping a fair coin until one goes broke. If the flip is heads, then A pays B \$1; otherwise B pays A \$1. Suppose that A starts with r dollars and B starts with s dollars. Let $a_n(r, s)$ be the probability that A goes broke on the nth flip. Obtain a recurrence relation for $a_n(r, s)$. (There are three parameters; be careful about the initial values.)

12.46. The **partitions** of an integer n are the nonincreasing lists of positive integers that sum to n. Let $p_{n,k}$ be the number of partitions of n having k parts (the partitions of 5 with three parts are 311 and 221, so $p_{5,3} = 2$). Prove that $p_{n,k} = p_{n-1,k-1} + p_{n-k,k}$ in general, and specify initial values that enable the recurrence to determine all values of $p_{n,k}$.

12.47. Add the details in Solution 12.38 to complete the derivation of the formula for the Catalan numbers using the method of generating functions.

12.48. Build the generating function for the Fibonacci sequence directly, using the model that F_n is the number of 1,2-lists that sum to n.

12.49. Suppose that $\langle a \rangle$ satisfies the recurrence $a_n = c_1 a_{n-1} + c_2 a_{n-2}$, with initial values a_0, a_1. Express the generating function for a as a ratio of two polynomials (see Example 12.36 for a special case).

12.50. Use generating functions to obtain a formula for the number of jars of 12 American coins (five types) having between 2 and 6 coins of each type. (Hint: Let a_n be the number of ways to do this selecting n coins instead of 13, and obtain the generating function for the sequence $\langle a \rangle$.)

12.51. Use generating functions to prove that $\sum_{k=0}^{n} \binom{n}{k}^2 = \binom{2n}{n}$.

12.52. Use generating functions to solve Exercises 5.42–5.44.

12.53. Suppose that $b_n = \sum_{k=0}^{n} a_k$ and that $A(x)$ is the generating function for the sequence $\langle a \rangle$. Obtain the generating function for $\langle b \rangle$ in terms of $A(x)$.

12.54. Let a_n be the number of ways to select $r \in \mathbb{N}$, roll a six-sided die r times, and obtain a sum of n. Obtain an expression for the generating function of $\langle a \rangle$.

12.55. Let a_n be the number of partitions of n using parts of size at most k (see Exercise 12.46). Obtain an expression for the generating function of $\langle a \rangle$.

12.56. Let a_n be the number of partitions of n (Exercise 12.46) using distinct parts. Let b_n be the number of partitions of n using odd parts. Derive expressions for the generating functions of $\langle a \rangle$ and $\langle b \rangle$.

12.57. (+) Establish a bijection to prove that the numbers a_n and b_n defined in Exercise 12.56 are equal. (Hint: Proposition 3.32 states that every natural number has a unique expression as an odd number times a power of 2.)

PART IV

CONTINUOUS

MATHEMATICS

Chapter 13

The Real Numbers

We now begin to study the Completeness Axiom and its consequences. Our main objective in this chapter is to understand the relationship between decimal expansions of real numbers and convergence of sequences.

13.1. Problem. In Chapter 8, we proved that there is no *rational* number whose square is 2. Nevertheless, we argued geometrically that there should be a *real* number whose square is two. How can we prove this? ∎

13.2. Problem. Is the set of real numbers countable? ∎

THE COMPLETENESS AXIOM

The Completeness Axiom is implicit in our understanding of real numbers. Recall that a sequence of real numbers is a function from $\mathbb{N}$ to $\mathbb{R}$. We name such a function using angled brackets and write its successive values using subscripts: $\langle x \rangle = \{x_1, x_2, \cdots\}$.

It is easy to generate numbers whose square is close to 2. Consider $\langle x \rangle = \frac{1}{1}, \frac{14}{10}, \frac{141}{100}, \frac{1414}{1000}, \frac{14142}{10000}, \frac{141423}{100000} \cdots$. The decimal expansion 1.41423... provides the same information in an abbreviated form. Each term in $\langle x \rangle$ is rational. The further we go, the closer the squares of the terms come to 2. The sequence is nondecreasing, and each term is smaller than $15/10$. Thus, $3/2$ is an upper bound for the set of terms, as are $142/100$ and $1415/1000$. We believe that the smallest upper bound should be a number whose square is 2, but how do we know that such a number exists?

We will use the Completeness Axiom to prove the existence of a real number whose square is 2. We recall some terminology about bounds and then restate the axiom.

13.3. Definition. Let S be a set of real numbers. A number $\alpha \in \mathbb{R}$ is an **upper bound** for S if $x \leq \alpha$ for all $x \in S$. An upper bound α for S is

256

the **least upper bound** or **supremum** of S if S has no upper bound less than α. Similarly, α is a **lower bound** for S if $x \geq \alpha$ for all $x \in S$, and a lower bound α is the **greatest lower bound** or **infimum** for S if S has no lower bound greater than α. We use $\sup(S)$ and $\inf(S)$ to denote the supremum and infimum of S, when they exist.

13.4. Axiom. (The Completeness Axiom for $\mathbb{R}$) Every nonempty subset of $\mathbb{R}$ that has an upper bound has a least upper bound. ∎

The Completeness Axiom has many equivalent versions. Axiom 13.4 as stated is the **Least Upper Bound Property**. It is equivalent to the **Greatest Lower Bound Property**, which states that every nonempty subset of $\mathbb{R}$ that has a lower bound has a greatest lower bound (Exercise 21). The Greatest Lower Bound Property is an analogue for $\mathbb{R}$ of the Well-Ordering Property of $\mathbb{N}$.

13.5. Remark. Since $\alpha < \beta$ or $\alpha > \beta$ whenever α, β are distinct real numbers, a set cannot have more than one supremum or infimum. Thus we speak of *the* least upper bound. ∎

The least upper bound of a set need not be an element of the set. The word "has" in Axiom 13.4 does *not* mean "contains".

13.6. Example. *Sups and infs.* Let $S = \{x \in \mathbb{R}: 0 < x < 1\}$. The infimum and supremum of S are 0 and 1; these lie in $\mathbb{R}$ but not in S. The set S has no maximum. A set $T \subseteq \mathbb{R}$ has a maximum if and only if $\sup(T)$ exists and belongs to T, in which case $\sup(T)$ is the maximum. The analogous statement holds for the minimum.

As another example, consider $S = \{x: x^3 - 3x^2 + 2x < 0\}$. By factoring or by graphing $y = x^3 - 3x^2 + 2x$, we find that $S = (-\infty, 0) \cup (1, 2)$. In this case $\sup(S) = 2$, but $\inf(S)$ does not exist. ∎

13.7. Solution. *Existence of $\sqrt{2}$.* Let $S = \{x \in \mathbb{R}: x^2 < 2\}$. Because the squaring function preserves order on the positive real numbers, every positive number whose square is at least 2 is an upper bound for S. Thus S has an upper bound, and the Completeness Axiom yields a least upper bound α for S. We will prove that $\alpha^2 = 2$.

When $x^2 < 2$, we show that x is not an upper bound by finding a number greater than x whose square is less than 2. Thus $\alpha^2 \geq 2$. When $x^2 > 2$, we show that x is not a least upper bound by finding a number less than x whose square is greater than 2. Thus $\alpha^2 \leq 2$. Together, these yield $\alpha^2 = 2$.

When $x^2 \neq 2$, consider the numbers x and $2/x$. Their product is 2, and their squares are on opposite sides of 2. Let $y = \frac{1}{2}(x + 2/x)$ be their arithmetic mean; y is between x and $2/x$. Furthermore, since $x \neq 2/x$, the AGM Inequality (Proposition 1.4) yields $y^2 > x(2/x) = 2$ (see Exercise 13 for another proof). From $y^2 > 2$, we have $2 > 4/y^2 = (2/y)^2$. Thus

$$(\min\{x, \tfrac{2}{x}\})^2 < (\tfrac{2}{y})^2 < 2 < y^2 < (\max\{x, \tfrac{2}{x}\})^2.$$

Thus we choose y when $x^2 > 2$ and choose $2/y$ when $x^2 < 2$ to obtain a number with the desired properties.   ∎

In the next section, we will give another proof of the existence of a real number whose square is 2.

13.8. Example. *The rationals are "incomplete".* Let $S = \{x \in \mathbb{Q}: x^2 < 2\}$. By the argument in Solution 13.7, the square of the real number sup S is 2. We proved in Chapter 8 that there is no such rational number. Although S is bounded in $\mathbb{Q}$, it has no supremum in $\mathbb{Q}$.   ∎

In Appendix A, we begin with the natural numbers as known, and we construct the real numbers from them. We define addition, multiplication, and order for the real numbers, and we prove that the result is a complete ordered field. All results about $\mathbb{R}$ follow from the axioms of a complete ordered field; in the text, we assume its existence. Motivated by the construction of $\mathbb{R}$, the mathematician Leopold Kronecker (1823–1891) said, "God created the integers; all the rest is the work of man."

Archimedes (287?–212 BC) asked whether placing segments of unit length end to end would produce arbitrarily long segments. That is, do the natural numbers form an unbounded set of real numbers? This may seem obvious: since each natural number is exceeded by the next, there is no largest natural number, but this proves only that $\mathbb{N}$ has no bound in $\mathbb{N}$. Is it possible that some real number is an upper bound for $\mathbb{N}$? The Completeness Axiom says no. Without the Completeness Axiom, there are ordered fields in which $\mathbb{N}$ is a bounded set (Exercise 40).

13.9. Theorem. (The Archimedean Property) Given positive real numbers a, b, there exists a natural number n such that $na > b$. Equivalently, no real number is an upper bound for the set $\mathbb{N}$.

Proof: We prove first that $\mathbb{N}$ has no upper bound in $\mathbb{R}$. If $\mathbb{N}$ has an upper bound, then by the Completeness Axiom $\mathbb{N}$ has a least upper bound α. Since α is the least upper bound, $\alpha - 1$ is not an upper bound, and hence there is a natural number n such that $n > \alpha - 1$. The properties of arithmetic now yield $n + 1 > \alpha$, contradicting the choice of α as an upper bound. Thus $\mathbb{N}$ has no upper bound. In particular, b/a is not an upper bound, so there exists $n \in \mathbb{N}$ such that $n > b/a$, and hence $na > b$.   ∎

We use the Archimedean Property often in the following way: given a real number α, we can choose a natural number N such that $N > \alpha$.

LIMITS AND MONOTONE CONVERGENCE

The absolute value function is vital for our discussions in this chapter, because $|x - L|$ is the **distance** between x and L. For a positive number ϵ, we read "$|x - L| < \epsilon$" as "the distance between x and L is less than epsilon". The inequality $|x - L| < \epsilon$ is equivalent to the two inequalities

$$L - \epsilon < x < L + \epsilon.$$

This double inequality determines an interval for x; the picture helps us understand the definition of "limit".

13.10. Definition. A sequence $\langle a \rangle$ of real numbers has **limit** $L \in \mathbb{R}$ if, for every $\epsilon > 0$, there exists $N \in \mathbb{N}$ (specified in terms of ϵ) such that $n \geq N$ implies $|a_n - L| < \epsilon$. A sequence **converges** if it has a limit. We write $a_n \to L$ to mean "a_n converges to (the limit) L".

We also write $a_n \to L$ as $\lim a_n = L$ or $\lim_{n \to \infty} a_n = L$. We read the latter as "the limit of a_n as n goes to infinity is L". The phrase "as n goes to infinity" (written "$n \to \infty$") means that we study the behavior when n is arbitrarily large; there is no element of $\mathbb{R}$ called "infinity".

The Greek letter ϵ denotes an arbitrary positive number, usually thought of as small. The order of the quantifiers for N and ϵ in the definition is crucial: the choice of N generally depends on the value of ϵ.

13.11. Example. *Convergent sequences.* When using the definition to prove convergence, we choose N appropriately in terms of ϵ.

1) Let $a_n = 3 + \frac{2}{n}$. We claim that $\langle a \rangle$ converges to 3. Given $\epsilon > 0$, we need $N \in \mathbb{N}$ such that $n \geq N$ implies $\frac{2}{n} < \epsilon$. The Archimedean property yields $N \in \mathbb{N}$ so that $N > \frac{2}{\epsilon}$. Now $n \geq N$ yields $\frac{2}{n} \leq \frac{2}{N} < \epsilon$, and hence $|a_n - 3| < \epsilon$. This proves that $a_n \to 3$. No one value of N works for all ϵ.

2) Let $a_n = c$. Every constant sequence converges; here $a_n \to c$. For each $\epsilon > 0$, we choose $N = 1$. Now $n \geq N$ implies $|a_n - c| = 0 < \epsilon$.   ∎

A convergent sequence can serve as a basis of comparison for proving that other sequences converge. For example, when $\langle a \rangle$ converges to 0, we may be able to prove that $b_n \to L$ by comparing $|b_n - L|$ with a_n.

13.12. Proposition. If $|b_n - L| \leq a_n$ for all n and $a_n \to 0$, then $b_n \to L$.

Proof: We use the definition of convergence. Given $\epsilon > 0$, the convergence of $\langle a \rangle$ to 0 guarantees the existence of $N \in \mathbb{N}$ such that $n \geq N$ implies $|a_n| < \epsilon$. Thus $n \geq N$ implies $|b_n - L| \leq a_n < \epsilon$, and this choice of N in terms of ϵ shows that $\langle b \rangle$ converges to L.   ∎

We abbreviate "sequence of real numbers" to "sequence". We often think of a sequence as the set of its terms (in order), although more precisely a sequence is a function from $\mathbb{N}$ to $\mathbb{R}$. Thus we may speak of the supremum of a sequence $\langle a \rangle$ and write $\sup \langle a \rangle$ (similarly for infimum).

A sequence converges to L when, for every $\epsilon > 0$, the graph of the sequence "eventually" remains within a band of width 2ϵ centered on the horizontal line defined by $y = L$. We make the notion of "eventually" precise in the next definition.

13.13. Definition. For each $n \in \mathbb{N}$, let $P(n)$ denote a mathematical statement. We say that $P(n)$ **holds for sufficiently large** n if there exists $N \in \mathbb{N}$ such that $P(n)$ is true whenever $n \geq N$.

In this language, the notation $a_n \to L$ means "for $\epsilon > 0$, the value a_n is within ϵ of L for sufficiently large n." The threshold N that marks the beginning of "sufficiently large n" depends on ϵ.

We speak of *the* limit of a sequence; the next lemma justifies this. If a sequence converged to both L and M, then eventually its elements would be arbitrarily close to both L and M. This cannot happen if L and M differ.

13.14. Lemma. Every convergent sequence has a unique limit.

Proof: Suppose that $a_n \to L$ and $a_n \to M$, but $L \neq M$; by symmetry, we may assume $M > L$. Using $\epsilon = (M - L)/2$ in the definition of convergence, we are given the existence of $N_1, N_2 \in \mathbb{N}$ such that $n \geq N_1$ implies $|a_n - L| < \epsilon$ and $n \geq N_2$ implies $|a_n - M| < \epsilon$. Let $N = \max\{N_1, N_2\}$. If $n \geq N$, then $a_n < L + \epsilon = M - \epsilon < a_n$, which is a contradiction. ∎

We can characterize infs and sups using sequences. We use "sequence in S" to mean a sequence of elements of S.

13.15. Proposition. If $S \subseteq \mathbb{R}$, then $\alpha = \sup(S)$ if and only if α is an upper bound for S and there is a sequence in S converging to α.

Proof: Suppose that $\alpha = \sup(S)$. Since α is the least upper bound for S, the number $\alpha - 1/n$ is not an upper bound, and hence for each n there exists $a_n \in S$ such that $a_n > \alpha - 1/n$. Given $\epsilon > 0$, by Theorem 13.9 we can choose $N \in \mathbb{N}$ such that $N > 1/\epsilon$. For $n \geq N$, we then have

$$\alpha - \epsilon < \alpha - 1/n < a_n \leq \alpha < \alpha + \epsilon,$$

and hence $|a_n - \alpha| < \epsilon$. Therefore, $\langle a \rangle \to \alpha$.

Conversely, let α be an upper bound for S, and let $\langle a \rangle$ be a sequence in S converging to α. To prove that α is the least upper bound, we show that each number less than α is not an upper bound for S. Let β be a number less than α, and set $\epsilon = \alpha - \beta$. Since $\langle a \rangle \to \alpha$, we have $|a_n - \alpha| < \epsilon$ for sufficiently large n. Since $a_n > \alpha - \epsilon = \beta$ and $a_n \in S$, we conclude that β is not an upper bound for S. Thus α is the least upper bound. ∎

A set has a maximum when it contains its supremum α. In this case, the constant sequence $\alpha, \alpha, \alpha, \cdots$ is a sequence in S converging to $\sup(S)$.

13.16. Theorem. (Monotone Convergence Theorem) Every bounded monotone sequence of real numbers has a limit: a bounded nondecreasing sequence converges to its supremum, and a bounded nonincreasing sequence converges to its infimum.

Proof: By symmetry, we need only consider the nondecreasing case. Since $\langle a \rangle$ is bounded, it has a supremum; call it L. Given $\epsilon > 0$, it suffices to show that a_n is within ϵ of L for sufficiently large n.

Since L is the least upper bound for $\langle a \rangle$, we know that $L - \epsilon$ is not an upper bound for $\langle a \rangle$. Hence there exists $N \in \mathbb{N}$ such that $a_N > L - \epsilon$. Since $\langle a \rangle$ is nondecreasing and L is an upper bound, $n \geq N$ implies

$$L - \epsilon < a_N \leq a_n \leq L < L + \epsilon.$$

Hence $n \geq N$ implies $|L - a_n| < \epsilon$, and this choice of N in terms of ϵ shows that $\langle a \rangle$ converges to its supremum. ∎

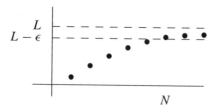

The Monotone Convergence Theorem yields convergence without directly verifying the definition. It reduces finding a limit to finding an infimum or supremum. In most cases, this is little help. For example, the sequence with $a_n = 1/n$ is decreasing and bounded. The limit exists, but proving that 0 is the infimum is no easier than proving that 0 is the limit.

Theorem 13.16 nevertheless has theoretical uses and applies in problems where knowing that the limit exists makes it easier to find it (see Chapter 14). Here we use it to give another proof of the existence of $\sqrt{2}$. The same ideas will yield decimal expansions for all real numbers.

We seek a positive number α whose square is 2. Since $1^2 < 2$ and $2^2 > 2$, we begin with the interval $[1, 2]$. We partition this interval into 10 subintervals of equal length. Numbers below 1.4 are too small and numbers above 1.5 are too large, so we restrict our attention to the interval $[1.4, 1.5]$. At the next step, we restrict our attention to $[1.41, 1.42]$.

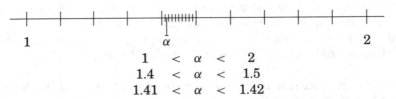

$$1 \quad < \quad \alpha \quad < \quad 2$$
$$1.4 \quad < \quad \alpha \quad < \quad 1.5$$
$$1.41 \quad < \quad \alpha \quad < \quad 1.42$$

Continuing this process, we generate a nondecreasing sequence of lower bounds and a nonincreasing sequence of upper bounds. We prove that they converge to the same real number and that the square of this real number is 2. Making these steps precise requires several general results about limits.

13.17. Lemma. If $a_n \leq M$ for $n \in \mathbb{N}$ and $a_n \to L$, then $L \leq M$.

Proof: If $L > M$, then we set $\epsilon = L - M$. Since $a_n \to L$, we have $|a_n - L| < \epsilon$ for sufficiently large n. For such n, we have $M = L - \epsilon < a_n$, which contradicts $a_n \leq M$. ∎

13.18. Proposition. If $\langle l \rangle$ is nondecreasing, $\langle r \rangle$ is nonincreasing, and $r_n - l_n \to 0$, then $\langle l \rangle$ and $\langle r \rangle$ converge and have the same limit.

Proof: We first claim that $l_m \leq r_m$ for all m, which we prove by contradiction. If not, then there exist $m \in \mathbb{N}$ and $\epsilon > 0$ such that $l_m - r_m \geq \epsilon$. The hypotheses of monotonicity then imply that $l_n - r_n \geq l_m - r_m \geq \epsilon$ for $n \geq m$, which contradicts $r_n - l_n \to 0$. Thus $l_m \leq r_m$ for all m.

This implies the stronger statement that $l_m \leq r_n$ for all $m, n \in \mathbb{N}$. If $m < n$, then $l_m \leq l_n \leq r_n$. If $m > n$, then $l_m \leq r_m \leq r_n$.

Now $l_m \leq r_1$ and $r_m \geq l_1$ tell us that $\langle l \rangle$ and $\langle r \rangle$ are bounded. The Monotone Convergence Theorem implies that $\langle l \rangle$ converges to its supremum and that $\langle r \rangle$ converges to its infimum. Let $L = \lim l_n = \sup\{l_n\}$ and $R = \lim r_n = \inf\{r_n\}$. From $l_m \leq r_n$, Lemma 13.17 yields $L \leq r_n$. Applying the analogue of Lemma 13.17 for lower bounds yields $L \leq R$.

To prove equality, suppose that $L < R$. Since $L = \sup\{l_n\}$ and $R = \inf\{r_n\}$, we have $l_n \leq L$ and $r_n \geq R$ for all n. This yields $r_n - l_n \geq R - L$ for all n, which contradicts $r_n - l_n \to 0$. ∎

13.19. Example. Without monotonicity, Proposition 13.18 fails. For example, when $l_n = (-1)^n - 1/n$ and $r_n = (-1)^n + 1/n$, we have $l_n - r_n \to 0$, but neither sequence converges. ∎

When we want to prove $b_n \to M$ using $a_n \to L$, we express $|b_n - M|$ in terms of quantities that we can make small using the knowledge that $|a_n - L|$ is small for sufficiently large n. In the next result, we introduce the quantity $|a_n - L|$ by subtracting and adding the number L. This technique often enables us to introduce quantities we can make small; we then group terms appropriately and apply the Triangle Inequality.

13.20. Lemma. If $a_n \to L$, then $a_n^2 \to L^2$.

Proof: For $\epsilon > 0$, we prove that $\left|a_n^2 - L^2\right| < \epsilon$ for sufficiently large n. We first write $\left|a_n^2 - L^2\right| = |(a_n - L)(a_n + L)| = |a_n - L|\,|a_n + L|$. We can make $|a_n - L|$ small for sufficiently large n, and $a_n \to L$ suggests that we can make $|a_n + L|$ close to $|2L|$.

In particular, by $a_n \to L$ we have $N_1 \in \mathbb{N}$ such that $n \geq N_1$ implies $|a_n - L| < 1$. Using the Triangle Inequality (Proposition 1.3) after introducing the quantity $a_n - L$ yields

$$|a_n + L| = |a_n - L + 2L| \leq |a_n - L| + |2L| \leq |2L| + 1.$$

Also, there exists N_2 such that $n \geq N_2$ implies $|a_n - L| < \epsilon/(|2L| + 1)$. Let $N = \max\{N_1, N_2\}$. For $n \geq N$, we have

$$\left|a_n^2 - L^2\right| = |a_n - L|\,|a_n + L| < \frac{\epsilon}{|2L| + 1}(|2L| + 1) = \epsilon. \qquad \blacksquare$$

13.21. Lemma. If $k \geq 2$, then $\frac{1}{k^n} \to 0$.

Proof: Since $k \geq 2$ and $n < 2^n$ (Proposition 3.16), we have $1/k^n \leq 1/2^n < 1/n$. Thus $n > 1/\epsilon$ implies $|1/k^n - 0| = 1/k^n < 1/n < \epsilon$. $\qquad \blacksquare$

We combine these lemmas to prove the existence of square roots of all positive real numbers.

13.22. Solution. *Existence of $\sqrt{x}$ for $x \geq 0$, by sequences.* Let l_n be the largest multiple of $1/10^n$ whose square is at most x ("multiple" means that $10^n l_n$ is an integer). By the Archimedean property, l_n exists. Since every multiple of $1/10^{n-1}$ is also a multiple of $1/10^n$, $\langle l \rangle$ is nondecreasing.

Similarly, let r_n be the smallest multiple of $1/10^n$ whose square exceeds x; $\langle r \rangle$ is nonincreasing. In addition, we have $r_n - l_n = 1/10^n$, which by Lemma 13.21 converges to 0. Thus Proposition 13.18 implies that $\langle l \rangle$ and $\langle r \rangle$ converge and have the same limit L.

We want to prove that $L^2 = x$. Since $l_n \to L$, Lemma 13.20 yields $l_n^2 \to L^2$. Since $l_n^2 \leq x$, Lemma 13.17 yields $L^2 \leq x$. Similarly, we obtain $r_n^2 \to L^2$ and $L^2 \geq x$. Thus $L^2 = x$. $\qquad \blacksquare$

DECIMAL EXPANSION AND UNCOUNTABILITY

The canonical representation of a rational number writes it as a fraction in lowest terms; how do we represent real numbers? The most familiar description is the decimal expansion. As a bonus, decimal expansions allow us to prove that $\mathbb{R}$ is uncountable.

The procedure for generating a canonical decimal expansion generalizes without additional effort to generate a canonical representation of a real number with respect to any base k. Thus we write the definitions and proofs for the general situation; the special case of decimal expansions is obtained by setting $k = 10$.

In Chapter 5, we obtained the base k representation of each positive integer. This expresses the positive integer a as $\sum_{j=0}^{m} c_j k^j$ using integers $c_0, \ldots, c_m$ with $0 \le c_j \le k - 1$ for $0 \le j \le m$ (and $c_m \ne 0$). Furthermore, the list $c_0, \ldots, c_m$ is uniquely determined.

When $\alpha < 0$, we use the expansion of $-\alpha$ and prepend a minus sign, so we may restrict our attention to positive numbers. A k-ary expansion of a positive real number α adds to the base k representation of the **integer part** $\lfloor \alpha \rfloor$ a k-ary expansion of the **fractional part** $\alpha - \lfloor \alpha \rfloor$. (Here $\lfloor \alpha \rfloor$ denotes the greatest integer less than or equal to α.)

Since we deal with the integer part separately, we may assume that $0 \le \alpha < 1$. The k-ary expansion of α expresses α as a sum using powers of $1/k$ (negative powers of k), such as $\sum_{j \ge 1} c_j / k^j$. Since we cannot yet sum infinitely many terms, we treat the truncations ($l_n = \sum_{j=1}^{n} c_j / k^j$) as a sequence of approximations to α. We have a k-ary expansion when this sequence converges to α.

We obtain the canonical k-ary expansion for α by essentially the method used for the sequence $\langle l \rangle$ in Solution 13.22 that converges to $\sqrt{x}$. We partition $[0, 1]$ into k equal intervals. The lower endpoint of the interval containing α is the first approximation. When we further subdivide the interval containing α into k subintervals, the lower endpoint of the subinterval containing α in the second approximation. Iterating this procedure generates as much of the expansion as we want.

Because the mathematical issue is convergence, we focus on the sequence of approximations to α. We could simultaneously generate the sequence of coefficients. In the decimal case, these are the digits c_n in the familiar abbreviation $.c_1 c_2 c_3 c_4 \cdots$ for the expansion of α. If α belongs to the ith interval in the subdivision at the nth step, then we set $c_n = i - 1$. Because we always subdivide into k intervals, each c_n is an integer between 0 and $k - 1$.

13.23. Example. *Decimal and binary expansions of 2/7.* The largest multiples of the first six powers of $1/10$ that are less than $2/7$ are $\frac{2}{10}$, $\frac{28}{100}$, $\frac{285}{1000}$, $\frac{2857}{10000}$, $\frac{28571}{100000}$, and $\frac{285714}{1000000}$, and the sequence of coefficients in the expansion begins $.285714$. This list repeats; the decimal expansion of $2/7$ is periodic with period 6. Exercise 36 discusses the length of the period.

The largest multiples of the first six powers of $1/2$ that are less than $2/7$ are $\frac{0}{2}$, $\frac{1}{4}$, $\frac{2}{8}$, $\frac{4}{16}$, $\frac{9}{32}$, and $\frac{18}{64}$. Since $\frac{1}{4} > \frac{0}{2}$ and $\frac{9}{32} > \frac{4}{16}$, we obtain 1s in the corresponding positions. Thus the binary expansion of $2/7$ begins $.010010$. The expansion repeats with period 3. ∎

In order to generate a decimal or k-ary expansion of a real number α, we must be able to compare α with rational numbers to determine which is larger. For example, we tested whether a rational number was less than $\sqrt{2}$ by testing whether its square was less than 2.

13.24. Definition. Let α be a real number with $0 \le \alpha < 1$, and let $k > 1$ be an integer. The **canonical k-ary expansion** is the sequence $\langle l \rangle$ defined by letting l_n be the largest multiple of k^{-n} such that $l_n \le \alpha$. For $k = 2, 3, 10$, the k-ary expansion is the **binary, ternary**, or **decimal** expansion, respectively.

Comment. The term "k-ary expansion" is also used for the sequence $\langle c \rangle$ defined by $l_n - l_{n-1} = c_n/k^n$ (by convention, $l_0 = 0$). Since l_{n-1} is a multiple of $1/k^{n-1}$, it is also a multiple of $1/k^n$, and therefore c_n is an integer. Since $l_n = \sum_{j=1}^n c_j/k^j$, knowing $\langle l \rangle$ is equivalent to knowing $\langle c \rangle$.

We next prove that real numbers have k-ary expansions. The proof provides an iterative algorithm that generates the expansion when we are able to compare the real number with powers of $1/k$. The approximations $\langle l \rangle$ are precisely the lower endpoints of the intervals chosen in the subdivision algorithm.

13.25. Theorem. Let $k \ge 2$ be a natural number.
a) Each real number in $[0, 1)$ has a canonical k-ary expansion.
b) Each k-ary expansion represents a real number in $[0, 1)$.

Proof: (a) Let $\langle l \rangle$ be the sequence of approximations defined in Definition 13.24. The existence of l_n follows from the Archimedean property; a sufficiently large multiple of $1/k^n$ exceeds α.

The definition of l_n yields $l_n \le \alpha < l_n + 1/k^n$. Let $r_n = l_n + 1/k^n$; thus $|l_n - r_n| = 1/k^n$. By Lemma 13.21, we have $l_n - r_n \to 0$. Since the multiples of $1/k^n$ include the multiples of $1/k^{n-1}$, we have $l_n \ge l_{n-1}$ and $r_n \le r_{n-1}$. Now Proposition 13.18 implies that $\langle l \rangle$ and $\langle r \rangle$ converge and have the same limit L. Since $l_n \le \alpha < r_n$ for all n, Lemma 13.17 (on both sides) yields $L \le \alpha \le L$, and thus $\alpha = L$.

(b) Given a sequence $\langle c \rangle$ of integers with $0 \le c_n \le k - 1$, let $l_n = \sum_{i=1}^n c^i/k^i$. Since $l_n = l_{n-1} + c_n/k^n$, the sequence is nondecreasing. Next we show that $l_n \le 1$ for all n. Using the geometric sum (Corollary 3.14), we obtain

$$l_n \le \sum_{j=1}^n \frac{k-1}{k^j} = \frac{k-1}{k} \sum_{j=1}^n \left(\frac{1}{k}\right)^{j-1} = \left(\frac{k-1}{k}\right)\left(\frac{(1/k)^n - 1}{(1/k) - 1}\right) = 1 - \frac{1}{k^n} < 1.$$

By the Monotone Convergence Theorem, $L = \lim L_n$ exists, and $L \le 1$. $\blacksquare$

We accept the k-ary expansion of α as a representation of α since $l_n \to \alpha$. Nevertheless, a philosophical question remains: What does it mean to "know" a number? We feel comfortable with integers and perhaps with fractions. In fact, the decimal expansion of a rational number terminates or eventually repeats, and this happens only for rational numbers (see Exercise 14.38). To what extent do we understand irrational numbers such as $\sqrt{2}$ or π? We can compute their decimal expansions with arbitrarily high accuracy, but no one will ever know the full expansions. (Indeed, knowledge of particular coefficients in the expansions is more a matter of curiosity than of mathematical substance.) The precise definitions of numbers such as $\sqrt{2}$ and π involve properties that characterize them and do not rely on their decimal expansions.

13.26. Remark. Rational numbers representable as fractions with a power of k in the denominator have two k-ary expansions. For example, $1/2$ equals both $0.100000\cdots$ and $0.011111\cdots$ in binary, just as $0.99999\cdots$ equals 1 in decimal. These are the only ways to obtain multiple k-ary representations for a single number (Exercise 35). The algorithm used in Definition 13.24 always chooses the expansion with infinitely many 0s when there is an alternative expansion ending in repeating $(k-1)$s. ∎

We have proved that $\mathbb{Q}$ is countable. We will show that there is no bijection from $\mathbb{N}$ to $\mathbb{R}$, and hence $\mathbb{R}$ is uncountable (Definition 4.43). The proof, due to Georg Cantor (1845–1918), is one instance of "Cantor's Diagonalization Argument".

13.27. Theorem. (Cantor) The set of real numbers is uncountable.

Proof: It suffices to show that $[0, 1]$ is uncountable (see Exercise 7). If not, then we have a bijection from $\mathbb{N}$ to $[0, 1]$. This is a sequence $\langle x \rangle$ that lists all numbers in $[0, 1]$, in some order. By considering the canonical decimal expansions, we will construct a number not on the list.

$$\begin{aligned}
x_1 &= c_{1,1}c_{1,2}c_{1,3} \\
x_2 &= c_{2,1}c_{2,2}c_{2,3} \\
x_3 &= c_{3,1}c_{3,2}c_{3,3} \\
&\vdots
\end{aligned}$$

Suppose that the expansions appear in order as indicated above. We build a canonical decimal expansion that disagrees with every expansion in our list. Let $a_n = 1$ if $c_{n,n} = 0$, and $a_n = 0$ if $c_{n,n} > 0$. Now $\langle a \rangle$ disagrees in position n with the expansion of x_n. Furthermore, since $\langle a \rangle$ has no 9, $\langle a \rangle$ cannot be the alternative expansion of any number in our list. Therefore, the expansion $\langle a \rangle$ does not represent a number in our list. By Theorem 13.25, $\langle a \rangle$ is the canonical expansion of some real number. Thus our list does not contain expansions for all real numbers in $[0, 1]$. ∎

There is nothing special about the use of 0 and 1 to generate the sequence $\langle a \rangle$ in this proof. We could as well have changed 9s to 4s, 6s to

2s, and everything else to 6s. All we needed was to change every value and avoid repeating 9s.

The set of real numbers whose k-ary expansions have only finitely many nonzero terms is countable; in fact, this is a subset of $\mathbb{Q}$. The number of nonzero terms can be "arbitrarily large" but not infinite. The possibility that more than one digit in the expansion may appear infinitely often makes the set of real numbers uncountable.

HOW TO APPROACH PROBLEMS

The technical aspects of real numbers require care; it took mathematicians centuries to understand and develop the foundations of analysis. It will help to keep a few guidelines in mind.

1) Remember that definitions and hypotheses are your friends. They often indicate how the proof should proceed.

2) Obvious-sounding statements involving inequalities can often be proved by contradiction. (See Lemma 13.17, for example.)

3) The Monotone Convergence Theorem can sometimes prove that a limit exists even when finding the limit may be difficult.

4) Think of the decimal expansion of a real number as a sum in terms of powers of ten.

The Monotone Convergence Theorem can make it easy to prove that a limit exists, even though it may be hard to find its value. The theorem has two hypotheses: the sequence must be monotone, and it must be bounded. In Exercise 30, for example, x_1, x_2, x_3 are $\frac{1}{2}, \frac{7}{12}, \frac{74}{120}$, respectively. The sequence appears to be increasing. We compare x_n and x_{n+1} to prove that $x_n < x_{n+1}$. To obtain an upper bound, it is not necessary to prove the best upper bound (that would be the limit!); any upper bound will do. The expression for x_n is a sum of n terms; showing that each is small enough will prove a bound.

Expansions in a general base k are easily understood by analogy with decimal expansions. What does the decimal number 198.32 really mean? It is an abbreviation for

$$1 * 10^2 + 9 * 10^1 + 8 * 10^0 + 3 * 10^{-1} + 2 * 10^{-2}.$$

Writing $x = 198.32$ is more efficient, but we must remember the meaning of the notation. Exercises 14–18 reinforce this understanding by using different bases; keep the definition of k-ary expansion in mind.

EXERCISES

13.1. (−) For $n \in \mathbb{N}$, let $x_n = n$ and $y_n = 1/n$. For each of $\langle x \rangle$, $\langle y \rangle$, determine whether the sequence is monotone and whether it is bounded.

13.2. (−) Consider the proverb "A lot of a little makes a lot". Which result in Chapter 13 does this describe?

13.3. (−) Give a counterexample to the following false statement, and add one word to correct it (other than "not").
    "Every bounded sequence of real numbers converges."

13.4. (−) The statement below is false. Correct it by changing two symbols.
    "The interval (a, b) contains its infimum and its supremum."

13.5. (−) Suppose that the sequence $\langle x \rangle$ does not converge to zero. The statement below is false. Correct it by changing the quantification of "n".
    "There exists $\epsilon > 0$ so that for all n, $|x_n| > \epsilon$."

13.6. Find the flaw in the following argument that claims to prove that the set of real numbers between 0 and 1 is countable: "Using decimal expansions, we can list the numbers in the interval $(0, 1)$ as follows: .1, .2, .3, ..., .9, .01, ..., .09, .11, ..., .19, ..., .99, .001, ..., .009, .011, ..., .019, .021, ..., etc." In other words, we first list the numbers whose last nonzero digit is in the tenths place, then those whose last nonzero digit is in the hundredths place, and so on.

13.7. (−) Prove that every infinite subset of a countable set is countable. Prove that every set that contains an uncountable set is uncountable. Conclude that $\mathbb{R}$ is uncountable if $[0, 1]$ is uncountable.

•　　•　　•　　•　　•

For Exercises 8–12, determine whether the statement is true or false. If true, provide a proof; if false, provide a counterexample.

13.8. If S is a bounded set of real numbers, and S contains sup(S) and inf(S), then S is a closed interval.

13.9. If $f \colon \mathbb{R} \to \mathbb{R}$ is defined by $f(x) = \frac{2x-8}{x^2-8x+17}$, then the supremum of the image of f is 1.

13.10. Every positive irrational number is the limit of a nondecreasing sequence of rational numbers.

13.11. Suppose that $\langle a \rangle$ and $\langle b \rangle$ converge.
    a) If $\lim a_n < \lim b_n$, then there exists $N \in \mathbb{N}$ such that $n \geq N$ implies $a_n < b_n$.
    b) If $\lim a_n \leq \lim b_n$, then there exists $N \in \mathbb{N}$ such that $n \geq N$ implies $a_n \leq b_n$.

13.12. If S is a bounded set of real numbers, and $x_n \to \sup(S)$ and $y_n \to \inf(S)$, then $\lim \frac{x_n + y_n}{2} \in S$.

•　　•　　•　　•　　•

13.13. Suppose that $x > 0$ and $x^2 \neq 2$. Let $y = \frac{1}{2}(x + 2/x)$. Show directly that $y^2 > 2$. (Hint: Express $y^2 - 2$ as a square. Solution 13.7 gave a different proof.)

13.14. (−) Compute the first six places of the canonical 3-ary expansion of 1/10.

13.15. (−) In base 10, the fractions $\frac{1}{2} = .5$, $\frac{1}{5} = .2$, and $\frac{1}{10} = .1$ are the only reciprocals of positive integers whose decimal expansion consists of one digit followed by all zeros. Determine all such fractions in base 8 and in base 9.

13.16. Suppose that the 26 symbols used in base 26 have the values $A = 0$, $B = 1$, $C = 2$, etc. What decimal number does BAD in base 26 represent? How about $.MMMMMMMMMMMMM \cdots$?

13.17. Let $q = 2n + 1$ for some natural number n. Compute the base q expansion of 1/2. (Assume the existence of symbols for the numbers $0, 1, \ldots, q - 1$.)

13.18. Let $q = 3n + 1$ for some natural number n. Compute the base q expansion of 1/3. (Assume the existence of symbols for the numbers $0, 1, \ldots, q - 1$.)

13.19. Let f be a bounded function on an interval I. Prove that
$$\sup(\{-f(x): x \in I\}) = -\inf(\{f(x): x \in I\}).$$

13.20. For each set S below, obtain a sequence in S converging to sup S and a sequence in S converging to inf S.
    a) $S = \{x \in \mathbb{R}: 0 \leq x < 1\}$.   b) $S = \{\frac{2+(-1)^n}{n}: n \in \mathbb{N}\}$.

13.21. (−) Prove that the Least Upper Bound Property holds for $\mathbb{R}$ if and only if the Greatest Lower Bound Property holds for $\mathbb{R}$.

13.22. For each set S below, determine whether S is bounded, and determine sup(S) and inf(S), if they exist.
    a) $S = \{x: x^2 < 5x\}$.
    b) $S = \{x : 2x^2 < x^3 + x\}$.
    c) $S = \{x : 4x^2 > x^3 + x\}$.

13.23. Let A and B be nonempty subsets of $\mathbb{R}$. Let $C = \{x + y : x \in A, y \in B\}$. Prove that if A and B have upper bounds, then C has a least upper bound, and $\sup C = \sup A + \sup B$. (Hint: Use Proposition 13.15.)

13.24. Let $f, g: \mathbb{R} \to \mathbb{R}$ be bounded functions such that $f(x) \leq g(x)$ for all x. Let F denote the image of f, and let G denote the image of g. Give examples of such pairs of functions, with pictures, such that
    a) $\sup(F) < \inf(G)$.
    b) $\sup(F) = \inf(G)$.
    c) $\sup(F) > \inf(G)$.

13.25. Use the definition of limit to prove that $\lim \sqrt{1 + n^{-1}} = 1$.

13.26. Use the definition of limit to prove that $\lim[(1 + a_n)^{-1}] = \frac{1}{2}$ if $\lim a_n = 1$.

13.27. (!) Let $a_n = \sqrt{n^2 + n} - n$. Compute $\lim a_n$. (Hint: Multiply and divide a_n by $\sqrt{n^2 + n} + n$, simplify the result, and use Exercises 13.25–13.26.)

13.28. Suppose that $x_n \to 0$ and that $|y_n| \leq 1$ for $n \in \mathbb{N}$. Find the flaw in the following computation for $\lim(x_n y_n)$, and give a valid proof that $\lim(x_n y_n) = 0$:
$$\lim(x_n y_n) = \lim(x_n) \lim(y_n) = 0 \cdot \lim(y_n) = 0.$$

13.29. Let $x_n = (1 + n)/(1 + 2n)$. Prove that $\lim_{n \to \infty} x_n$ exists by using Monotone Convergence. Prove that $\lim_{n \to \infty} x_n = 1/2$ by using the definition of limit.

13.30. (!) Let $x_n = \frac{1}{n+1} + \frac{1}{n+2} + \cdots + \frac{1}{2n}$. Prove that $\lim_{n \to \infty} x_n$ exists. (Comment: In fact, the limit equals $\ln 2$, but this information is not needed for this exercise.)

13.31. (+) Prove that $x_n = (1 + (1/n)^n$ defines a bounded monotone sequence. (Hint: Simplify the ratio x_{n+1}/x_n and use $(1 - a)^n \geq 1 - na$ (Corollary 3.20).

13.32. *The Nested Interval Property.* Let $\{I_n\}$ be a sequence of closed intervals, with I_n of length d_n, such that $I_{n+1} \subseteq I_n$ for all n and $d_n \to 0$. The Nested Interval Property states that for such a sequence, there is exactly one point that belongs to each I_n. Prove the following statements:
   a) The Completeness Axiom implies the Nested Interval Property.
   b) The Nested Interval Property implies the Completeness Axiom.

13.33. For each k greater than 1, compute the k-ary expansion of $1/2$.

13.34. (!) Prove that there is a rational number between any two irrational real numbers, and an irrational number between any two rational numbers.

13.35. (!) Prove that a real number has more than one k-ary expansion if and only if it is expressible as a fraction using a denominator that is a power of k.

13.36. (!) Let a and b be natural numbers, expressed in decimal expansion.
   a) Prove that long division of a by b yields the decimal expansion of a/b.
   b) Use the Pigeonhole Principle and long division to prove that the decimal expansion of a/b has a period of length less than b.

13.37. Explain why the technique of Theorem 13.27 does not prove that $\mathbb{Q}$ is uncountable. "Proceeding as in Theorem 13.27, we list the expansions of numbers in $\mathbb{Q}$ and create an expansion $\langle a \rangle$ for a number y that is not on our list. This contradicts the hypothesis that $\mathbb{Q}$ is countable."

13.38. (!) Let S be the set of subsets of $\mathbb{N}$. Let $T = \{x \in \mathbb{R} \colon 0 \leq x < 1\}$. Prove that S and T have the same cardinality.

13.39. (+) Prove that $\mathbb{R} \times \mathbb{R}$ has the same cardinality as $\mathbb{R}$.

13.40. (+) *An ordered field in which $\mathbb{N}$ is bounded.* Let F be the set of expressions of the form $a = \sum_{i \in \mathbb{Z}} a_i x^i$, where each $a_i \in \mathbb{R}$ and $\{i < 0 \colon a_i \neq 0\}$ is finite. (Here x is a formal symbol, not a number.) An element $a \in F$ is *positive* if the least-indexed nonzero coefficient a_k in the expression for a is positive. The *sum* of $a \in F$ and $b \in F$ is the element $c \in F$ defined by $c_i = a_i + b_i$ for $i \in \mathbb{Z}$. The *product* of $a \in F$ and $b \in F$ is the element $c \in F$ defined by $c_j = \sum_{i \in \mathbb{Z}} a_i b_{j-i}$ for $j \in \mathbb{Z}$.
   a) Prove that the sum and product of two elements of F is an element of F.
   b) We have defined addition, multiplication, and order on F. Prove that with these operations, F is an ordered field.
   c) Interpret each real number α as the element $a \in F$ with $a_i = 0$ for all i, except $a_0 = \alpha$; this interprets $\mathbb{R}$ as a subset of F. Prove that $\mathbb{N}$ is a bounded set in F. Conclude that F does not satisfy the Archimedean property.

Chapter 14

Sequences and Series

In Chapter 13 we defined convergent sequences and used them to develop decimal expansions for real numbers. A decimal expansion expresses the real number as an infinite sum or "series". The usefulness of this approach suggests developing the theory of sequences and series.

A sequence converges when its values cluster around a limiting value; hence the values of a convergent sequence must eventually be close together. The central result of this chapter is the converse: when the terms are eventually close together, the sequence must have a limit. This leads to a deep understanding of the Completeness Axiom.

In this chapter we also prove related results that we apply to the theory of calculus in later chapters. We obtain several criteria for the convergence of infinite series, and we solve several problems where understanding an infinite series is the main issue.

14.1. Problem. *"Rationalization" of repeating decimals.* Find a simple expression as a rational number for the number with decimal expansion $.abcdedede\cdots$. Which decimal expansions yield rational numbers? ∎

14.2. Problem. *The Tennis Problem.* Suppose that the points in a tennis game are independent and that the server wins each point with probability p. The first player to have at least four points and at least two more points than the opponent wins the game. What is the probability that the server wins the game? ∎

PROPERTIES OF CONVERGENT SEQUENCES

Many proofs about limits use a type of argument suggested by a discussion about errors in measurements. In a laboratory, we may measure two quantities L, M as $L \pm 2$ and $M \pm 3$ due to experimental error. Because the errors may have the same sign, we write $(L+M) \pm 5$ for the sum. In order to ensure that the error in measuring the sum is at most ϵ, we determine each of L, M within accuracy $\epsilon/2$. Bounding an error in terms of two given errors is a standard technique called an $\epsilon/2$-**argument**.

We begin with a good example of the $\epsilon/2$-argument.

14.3. Lemma. If $a_n \to L$ and $b_n - a_n \to 0$, then $b_n \to L$.

Proof: Given $\epsilon > 0$, we show that $|b_n - L| < \epsilon$ when n is sufficiently large. To do this, we express $|b_n - L|$ as a sum of terms that are small when n is sufficiently large. The hypotheses imply the existence of N_1 and N_2 such that $n \geq N_1$ implies $|a_n - L| < \epsilon/2$ and $n \geq N_2$ implies $|b_n - a_n| < \epsilon/2$.

Let $N = \max\{N_1, N_2\}$. For $n \geq N$, (using the Triangle Inequality—Proposition 1.3) we have

$$|b_n - L| = |b_n - a_n + a_n - L| \leq |b_n - a_n| + |a_n - L| < \frac{\epsilon}{2} + \frac{\epsilon}{2} = \epsilon. \quad \blacksquare$$

In proving Lemma 14.3, we did not assume that $\langle b \rangle$ converges. When proving $b_n \to L$, it does not suffice to assume $b_n \to M$ and obtain a contradiction, because this neglects the possibility that $\langle b \rangle$ does not converge.

14.4. Remark. *The form of the $\epsilon/2$-argument.* In these arguments we prove statements about the convergence of some sequence, often using a hypothesis about convergence of other sequences. Our treatment of the "ϵ" in the definition of convergence depends on whether we are *proving* convergence or *using* convergence. When proving that $b_n \to L$, we are given ϵ and must find $N \in \mathbb{N}$ such that $n \geq N$ implies $|b_n - L| < \epsilon$.

To do this, we express $|b_n - L|$ in terms of other quantities that we can make arbitrarily small by choosing n sufficiently large. This may use known convergence of another sequence $\langle a \rangle$ (as in Proposition 13.12 or Lemma 14.3). When we know that $a_n \to M$, we are guaranteed for every positive number ϵ' the existence of some $N' \in \mathbb{N}$ such that $|a_n - M| < \epsilon'$. We may choose ϵ' as we wish. Since we want to draw a conclusion about the number ϵ we were given, we will choose ϵ' in terms of ϵ. We choose ϵ' for each contributing error so that the contributions to $|b_n - L|$ will sum to less than ϵ for sufficiently large n. In the proof of Lemma 14.3, we have two contributions to $|b_n - L|$, and for each we choose $\epsilon' = \epsilon/2$. In some arguments, the contributions to the error are more complicated.

We then construct N in terms of all resulting N'. We must guarantee several conditions that each occurs when the index is sufficiently large. When N is the maximum of the thresholds, all the conditions occur. $\quad \blacksquare$

14.5. Theorem. (Limits and Arithmetic) If $a_n \to L$ and $b_n \to M$, then
a) $a_n + b_n \to L + M$.

b) $a_n b_n \to LM$ (special case: $c a_n \to cL$).

c) $a_n / b_n \to L/M$ (if $M \neq 0$ and b_n is never 0).

Proof: (a) Given $\epsilon > 0$, we want to find an N so that $n \geq N$ implies $|a_n + b_n - (L + M)| < \epsilon$. Since $a_n \to L$ and $b_n \to M$, we can make $|a_n - L|$ and $|b_n - M|$ small by making n large; we choose $N_1, N_2 \in \mathbb{N}$ so that $n \geq N_1$ implies $|a_n - L| < \epsilon/2$ and $n \geq N_2$ implies $|b_n - M| < \epsilon/2$. If $N = \max\{N_1, N_2\}$, then $n \geq N$ implies

$$|a_n + b_n - (L + M)| = |a_n - L + b_n - M| \leq |a_n - L| + |b_n - M| < \frac{\epsilon}{2} + \frac{\epsilon}{2} = \epsilon.$$

By the definition of convergence, we have proved that $a_n + b_n \to L + M$.

(b) We seek N in terms of ϵ so that $n \geq N$ implies $|a_n b_n - LM| < \epsilon$. We rewrite $a_n b_n - LM$ to express it in terms of quantities that we can make small by making n large: $a_n b_n - LM = a_n(b_n - M) + (a_n - L)M$. Since $a_n \to L$ and $b_n \to M$, we can choose N_1 and N_2 so that $n \geq N_1$ implies $|a_n - L| < \frac{\epsilon}{2(1+|M|)}$ (and $|a_n| < \epsilon + |L|$) and $n \geq N_2$ implies $|b_n - M| < \frac{\epsilon}{2(\epsilon+|L|)}$. If $N = \max\{N_1, N_2\}$, then $n \geq N$ implies

$$|a_n b_n - LM| = |a_n b_n - a_n M + a_n M - LM| = |a_n(b_n - M) + (a_n - L)M|$$
$$\leq |a_n|\,|b_n - M| + |a_n - L|\,|M|$$
$$< \frac{(\epsilon + |L|) \cdot \epsilon}{2(\epsilon + |L|)} + \frac{\epsilon \cdot |M|}{2(1 + |M|)} < \frac{\epsilon}{2} + \frac{\epsilon}{2} = \epsilon.$$

We leave the proof of (c) as Exercise 14.   ∎

Our next property of limits is geometrically intuitive. It rests on the observation that if two numbers are each within ϵ of L, then every number between them is also within ϵ of L.

14.6. Theorem. (The Squeeze Theorem) Suppose that $a_n \leq b_n \leq c_n$ for all n. If $a_n \to L$ and $c_n \to L$, then also $b_n \to L$.

Proof: Given $\epsilon > 0$, we need N so that $n \geq N$ implies $|b_n - L| < \epsilon$. Since $a_n \to L$ and $c_n \to L$, we can choose N_1, N_2 so that $n \geq N_1$ implies $|a_n - L| < \epsilon$ and $n \geq N_2$ implies $|c_n - L| < \epsilon$. When $N = \max\{N_1, N_2\}$, $n \geq N$ implies

$$L - \epsilon < a_n \leq b_n \leq c_n < L + \epsilon.$$

We have satisfied the definition of convergence to obtain $b_n \to L$.   ∎

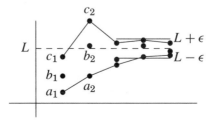

Setting $L = M = 0$ in Theorem 14.5a tells us that the sum of two sequences converging to 0 converges to 0. Making the product converge to 0 requires that one sequence converges to 0, and this suffices when the other is bounded.

14.7. Proposition. If $a_n \to 0$ and $\langle b \rangle$ is bounded, then $a_n b_n \to 0$.

Proof: Consider $\epsilon > 0$. Since $\langle b \rangle$ is bounded, there exists a positive M such that $|b_n| \le M$ for all N. Since $a_n \to 0$, there exists N such that $n \ge N$ implies $|a_n| < \epsilon/M$. Thus $n \ge N$ implies $|a_n b_n| < (\epsilon/M)M = \epsilon$. We now have $a_n b_n \to 0$ by the definition of convergence. ∎

Although the basic method of proving limits involves epsilons and inequalities, there are many situations when clever algebraic manipulation simplifies the proof of the inequalities. One class of examples arises when a sequence x_n is defined by $x_n = a_{n+1} - a_n$. When $\lim a_n$ exists, also $\lim x_n$ exists and must be 0, by Theorem 14.5a. It is possible for $\lim x_n$ to exist even when $\lim a_n$ does not.

14.8. Example. Let $x_n = \sqrt{n+1} - \sqrt{n}$; we prove that $x_n \to 0$. Multiplying and dividing x_n by $\sqrt{n+1} + \sqrt{n}$ yields $x_n = 1/(\sqrt{n+1} + \sqrt{n}) < n^{-1/2}$. Thus $|x_n - 0| < n^{-1/2}$. Since $n^{-1/2} \to 0$, also $x_n \to 0$, by Proposition 13.12. ∎

Our next technique applies in many situations. Given an equation involving terms of a convergent sequence, letting the index tend to infinity on both sides of the equation may yield an equation for the limit.

14.9. Example. *Convergence to $\sqrt{2}$* (compare with Solution 13.7). We construct a sequence of rational numbers that converges to $\sqrt{2}$. We know that $x^2 = 2$ if and only if $x = 2/x$. Given a positive number x_1 other than $\sqrt{2}$, one of $\{x_1, 2/x_1\}$ is larger than $\sqrt{2}$ and the other is smaller. We hope that the average of x_1 and $2/x_1$ is closer to $\sqrt{2}$ than x_1 is. Given $x_1 > 0$, this suggests that we define $\langle x \rangle$ by setting $x_{n+1} = \frac{1}{2}(x_n + 2/x_n)$ for $n \ge 1$.

The two sides of the recurrence are different names for the same sequence. If $\langle x \rangle$ converges, we therefore have $\lim x_{n+1} = \lim(x_n/2 + 1/x_n)$. The arithmetic properties of limits then tell us that the limit L must satisfy $L = L/2 + 1/L$, which requires $L^2 = 2$. If $x_n > 0$, then x_{n+1} is the average of two positive numbers and is also positive, so when $x_1 > 0$ the only possible limit is the positive square root of 2. We still must show that the sequence does have a limit, for each positive initial guess.

Rewriting the recurrence as $x_{n+1} - x_n = \frac{1}{2}(-x_n + 2/x_n)$ shows that $x_{n+1} < x_n$ if and only if $x_n > \sqrt{2}$. Furthermore, since x_{n+1} is the average of x_n and $2/x_n$, the AGM Inequality (Proposition 1.4) implies that $x_{n+1} > \sqrt{2}$. Thus after the initial term the terms of $\langle x \rangle$ are greater than $\sqrt{2}$ and form a decreasing sequence. The sequence starting at x_2 satisfies the hypotheses of the Monotone Convergence Theorem and thus converges. ∎

Our next proof employs a technique similar to that in Example 14.9. First we establish that $\langle b \rangle$ converges. Once we know that $\langle b \rangle$ converges to some value L, we find an equation for L.

14.10. Proposition. If $\langle b \rangle$ is a sequence such that $|b_{n+1}| / |b_n| \to x$ with $0 \le x < 1$, then $b_n \to 0$. In particular, $\lim_{n \to \infty} t^n = 0$ when $|t| < 1$.

Proof: It suffices to show that $|b_n| \to 0$, so we may assume that $b_n > 0$ for all n. Since $b_{n+1}/b_n \to x$, we can use $\epsilon = 1 - x$ in the definition of convergence to obtain N such that $n \ge N$ implies $b_{n+1}/b_n < 1$. Hence $\langle b \rangle$ is positive and decreasing after b_N, and the Monotone Convergence Theorem guarantees that it has a limit L.

Since 0 is a lower bound for $\langle b \rangle$, we have $L \ge 0$. If $L \ne 0$, then we have $x = \lim \frac{b_{n+1}}{b_n} = \frac{\lim b_{n+1}}{\lim b_n} = \frac{L}{L} = 1$. This contradicts our hypothesis that $x < 1$, and therefore $b_n \to 0$. ∎

When a sequence converges, it is natural to expect that averaging its initial terms produces a sequence that converges to the same limit. Proving this illustrates another way of bounding a quantity we *want* to make small by a sum of quantities we *can* make small.

14.11. Proposition. If $\langle a \rangle$ converges to L, and $\langle b \rangle$ is defined by $b_n = \frac{1}{n} \sum_{k=1}^{n} a_k$, then $\langle b \rangle$ also converges to L.

Proof: Consider $\epsilon > 0$. The Triangle Inequality for n terms (Exercise 3.22) yields

$$|b_n - L| = \frac{1}{n} \left| \sum_{k=1}^{n} (a_k - L) \right| \le \frac{1}{n} \sum_{k=1}^{n} |a_k - L|.$$

Since $a_n \to L$, we can choose N_1 large enough that $k \ge N_1$ implies $|a_k - L| < \epsilon/2$. Furthermore since $a_k - L \to 0$, the set $\{|a_k - L| : k \in \mathbb{N}\}$ is bounded, and we can choose M such that $|a_k - L| < M$ for all k. Choose $N_2 > N_1 \epsilon/(2M)$. For $n \ge \max\{N_1, N_2\}$, we break the sum bounding $|b_n - L|$ into two pieces. We have

$$|b_n - L| \le \frac{1}{n} \sum_{k=1}^{N_1} |a_k - L| + \frac{1}{n} \sum_{k=N_1+1}^{n} |a_k - L|$$

$$< \frac{1}{N_2} N_1 M + \frac{1}{n}(n - N_1)\frac{\epsilon}{2} < \frac{\epsilon}{2} + \frac{\epsilon}{2} = \epsilon$$

Setting $N = \max\{N_1, N_2\}$, we have shown that $n \ge N$ implies $|b_n - L| < \epsilon$, and thus $b_n \to L$. ∎

CAUCHY SEQUENCES

We may want to know whether a sequence converges without needing to know the limit. Our criteria so far are the definition of convergence (which requires that we know the limit) and the Monotone Convergence Theorem (which applies only to monotone sequences). We prove in this section that it suffices to show that for each $\epsilon > 0$, all terms with sufficiently large indices are within ϵ of each other.

14.12. Definition. A sequence $\langle a \rangle$ is a **Cauchy sequence** if for every $\epsilon > 0$, there exists $N \in \mathbb{N}$ (depending on ϵ) such that $n, m \geq N$ implies $|a_n - a_m| < \epsilon$.

This property is named for Augustin Cauchy (1789–1857). The Triangle Inequality tells us that if each of two numbers is within $\epsilon/2$ of some number L, then the distance between them is at most ϵ. This is the basis of the next proof, a particularly clean example of the $\epsilon/2$-argument.

14.13. Proposition. Every convergent sequence is a Cauchy sequence.

Proof: Let $\langle a \rangle$ be a convergent sequence. To prove that $\langle a \rangle$ is a Cauchy sequence, for each $\epsilon > 0$ we must choose $N \in \mathbb{N}$ so that every two terms after a_N differ by at most ϵ. Let $L = \lim a_n$. Given $\epsilon > 0$, we can apply the definition of convergence of $\langle a \rangle$ for the number $\epsilon/2$. This yields $N \in \mathbb{N}$ such that $n \geq N$ implies $|a_n - L| < \epsilon/2$. If we choose $n, m \geq N$, then we have the desired inequality

$$|a_m - a_n| = |a_m - L + L - a_n| \leq |a_m - L| + |L - a_n| < \epsilon/2 + \epsilon/2 = \epsilon. \qquad \blacksquare$$

The converse of Proposition 14.13 is the fundamental result about convergence, called the **Cauchy Convergence Criterion**. It is equivalent to the Completeness Axiom in the following sense. If we had taken the Cauchy Convergence Criterion as an axiom, then we could have derived the Completeness Axiom from it as a theorem. Each of these is a precise mathematical formulation of the intuitive notion that the real numbers have no gaps. We need to develop several tools in order to prove the Cauchy Convergence Criterion.

14.14. Lemma. Every Cauchy sequence is bounded.

Proof: Let $\langle a \rangle$ be a Cauchy sequence. Using $\epsilon = 1$ in the definition, we obtain $N \in \mathbb{N}$ such that $m, n \geq N$ implies $|a_n - a_m| < 1$. Now $n \geq N$ implies $|a_n - a_{N+1}| < 1$, which yields $|a_n| < |a_{N+1}| + 1$. Letting $M = \max\{|a_{N+1}| + 1, |a_1|, |a_2|, \ldots, |a_N|\}$, we have $|a_n| < M$ for all $n \in \mathbb{N}$. $\qquad \blacksquare$

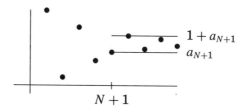

Every bounded monotone sequence is convergent, but boundedness alone does not guarantee convergence. When $a_n = (-1)^n + 1/n$, the sequence does not converge, but the terms with n even form a convergent subsequence. We prove next that every bounded sequence has a convergent subsequence.

14.15. Definition. A **subsequence** of $\langle a \rangle$ is a sequence $\langle b \rangle$ defined by $b_k = a_{n_k}$, where $n_1 < n_2 < \cdots$ is an increasing sequence of indices.

14.16. Example. *Subsequences.* If $a_n = 2n - 1$ and $n_k = k^2$, then $b_k = 2k^2 - 1$. ∎

n	1	2	3	4	5	6	7	8	9
a_n	1	3	5	7	9	11	13	15	17
k	1		2						3
b_k	1		7						17

We can interpret subsequences as composite functions. When $\langle b \rangle$ is a subsequence of $\langle a \rangle$, the function $b \colon \mathbb{N} \to \mathbb{R}$ is the composition of the function $a \colon \mathbb{N} \to \mathbb{R}$ with the increasing function $n \colon \mathbb{N} \to \mathbb{N}$; using the notation of functions instead of subscripts, we obtain $b(k) = a(n(k))$.

The next theorem reveals the importance of subsequences. It uses a version of the Pigeonhole Principle for infinite sets. When the union of two sets is infinite, at least one of them must be infinite. We apply this to sets of real numbers obtained by repeatedly bisecting intervals.

14.17. Theorem.   (Bolzano-Weierstrass Theorem) Every bounded sequence of real numbers has a convergent subsequence.

Proof: Let $\langle x \rangle$ be a sequence with $L < x_n < M$ for all $n \in \mathbb{N}$. We construct a convergent subsequence $\langle b \rangle$, where $b_k = x_{n_k}$. We will choose b_k so that $a_k \le b_k \le c_k$, where $\langle a \rangle$ and $\langle c \rangle$ are sequences converging to the same limit K. The squeeze theorem then implies $b_k \to K$.

Set $a_1 = L$ and $c_1 = M$. We construct $\langle a \rangle$ and $\langle c \rangle$ iteratively. Having specified a_k and c_k, let $z_k = (a_k + c_k)/2$ be the midpoint of the interval between them. If there are infinitely many terms of $\langle x \rangle$ in the lower half $[a_k, z_k]$, then we set $a_{k+1} = a_k$ and $c_{k+1} = z_k$. Otherwise, we set $a_{k+1} = z_k$ and $c_{k+1} = c_k$.

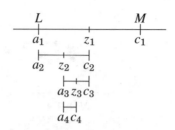

For each $k \in \mathbb{N}$, we claim that $[a_k, c_k]$ contains infinitely many terms of $\langle x \rangle$ and has length $(M - L)/2^{k-1}$. We prove this by induction on k. For $k = 1$, the interval $[a_1, c_1]$ contains all of $\langle x \rangle$ and has length $M - L$. For the induction step, suppose that the claim holds for some $k \geq 1$. Each of $[a_k, z_k]$ and $[z_k, c_k]$ is half the length of $[a_k, c_k]$, so we have $c_{k+1} - a_{k+1} = (c_k - a_k)/2 = (M - L)/2^k$. Also, since $[a_k, c_k]$ contains infinitely many terms of $\langle x \rangle$ (by the induction hypothesis), there must be infinitely many elements of $\langle x \rangle$ in $[a_k, z_k]$ or in $[z_k, c_k]$.

By construction, $\langle a \rangle$ is increasing and $\langle c \rangle$ is decreasing, and we have proved that $c_k - a_k \to 0$. Proposition 13.18 implies that these sequences converge and have the same limit, which we call K. It remains only to select a subsequence $\langle b \rangle$ of $\langle x \rangle$ such that $b_k \in [a_k, c_k]$; we do this iteratively. We must have $b_k = x_{n_k}$ with $\{n_k\}$ being an increasing sequence.

We choose $b_1 = x_1$; that is, $n_1 = 1$. Suppose we have chosen $n_1, \ldots, n_{k-1}$ as the indices in $\langle x \rangle$ for the first $k - 1$ terms of $\langle b \rangle$. Because $[a_k, c_k]$ contains infinitely many terms of $\langle x \rangle$, we can choose such a term whose index is larger than all of those previously chosen. Let this index be n_k. With $b_k = x_{n_k}$ for all k, we have constructed a subsequence $\langle b \rangle$ with $a_k \leq b_k \leq c_k$ for all k, and hence $b_k \to K$. ∎

The method of proof used here, in which we successively select the upper or lower half of the current interval to continue the search, is called the **method of bisection**. Applied to a convergent sequence, this method develops the binary expansion of the limit.

14.18. Example. *Binary expansion by the method of bisection.* Suppose we want the binary expansion of a real number α in the interval $[0, 1]$. We can produce the sequence of zeroes and ones iteratively by the method of bisecting intervals. The original (first) interval is $[0, 1]$. If α is less than $1/2$, the first digit is 0, otherwise it is 1. In general, we let the nth digit be 0 when α is in the first half of the nth interval and 1 when α is in the second half. The next interval is obtained by bisecting the current interval and selecting the half containing α. ∎

Now we are ready to prove the main result about Cauchy sequences.

14.19. Theorem. (Cauchy Convergence Criterion) A sequence of real numbers converges if and only if it is a Cauchy sequence.

Proof: We have already proved that a convergent sequence is a Cauchy sequence, using an $\epsilon/2$-argument. For the converse, suppose that $\langle a \rangle$ is a Cauchy sequence. By Lemma 14.14, $\langle a \rangle$ is bounded. By the Bolzano-Weierstrass Theorem, $\langle a \rangle$ has a convergent subsequence $\langle b \rangle$ with $b_k = a_{n_k}$.

Let $L = \lim b_k$; we prove that also $a_n \to L$. Consider $\epsilon > 0$. Since $\langle a \rangle$ is a Cauchy sequence, we can choose N_1 such that $n, m \geq N_1$ implies $|a_n - a_m| < \epsilon/2$. Since $b_k \to L$, we can choose N_2 such that $k \geq N_2$ implies $|b_k - L| < \epsilon/2$. Set $N = \max\{N_1, N_2\}$. Since $\{n_k\}$ is an increasing sequence of indices, we have $n_k \geq k$. Hence $k \geq N$ implies $n_k \geq N$, and we compute

$$|a_k - L| = |a_k - b_k + b_k - L| \leq |a_k - a_{n_k}| + |b_k - L| < \epsilon. \qquad \blacksquare$$

In the next section we apply Theorem 14.19 to study conditions for convergence of series. In Chapter 16 we apply it to study convergence for series of functions.

INFINITE SERIES

We have discussed how to evaluate various finite sums. Now we consider summing the terms in an infinite sequence. This is not generally possible. Even when it is possible, the value may depend on the order in which the terms are summed (Exercise 54). Therefore, we need a precise definition for the sum of an infinite series.

14.20. Definition. Let $\langle a \rangle$ be a sequence of real numbers. The formal expression $\sum_{k=1}^{\infty} a_k$ is an **infinite series**. The number $s_n = \sum_{k=1}^{n} a_k$ is the nth **partial sum** of the series. The infinite series $\sum_{k=1}^{\infty} a_k$ **converges** if $\lim_{n \to \infty} s_n$ exists; otherwise the series **diverges**. When $\sum_{k=1}^{\infty} a_k$ converges, we write $L = \lim s_n = \sum_{k=1}^{\infty} a_k$ and say that the sum of the series equals L.

14.21. Example. *Sequence versus partial sums.* Let $a_n = 1/2^n$, and let $s_n = \sum_{k=1}^{n} a_k = 1 - 1/2^n$. Thus $s_n \to 1$, and hence $\sum_{k=1}^{\infty} 1/2^n = 1$.

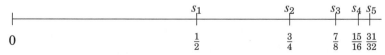

As another example, we present a table of the terms and the partial sums when $a_n = (-1)^{n+1}/n$. This is an alternating series with terms of decreasing absolute value tending to zero. Every such series converges, as shown in Exercise 52. The sum of this series is .693147 to six decimal places; the table below suggests that it converges slowly. ∎

n	1	2	3	4	5	6	7	$\cdots$	99
a_n	1	$-.5$	.333	$-.25$	.2	$-.166$	.143	$\cdots$	.010101
$\sum_{k=1}^{n} a_k$	1	.5	.833	.583	.783	.617	.760	$\cdots$	.698172

14.22. Proposition. (The Distributive Law) If $\sum_{k=1}^{\infty} a_k$ converges and $c \in \mathbb{R}$, then $\sum_{k=1}^{\infty} ca_k$ also converges and equals $c \sum_{k=1}^{\infty} a_k$.

Proof: The nth partial sum for $\sum_{k=1}^{\infty} ca_k$ equals cs_n, where s_n is the nth partial sum for $\sum_{k=1}^{\infty} a_k$. By Theorem 14.5, $\lim cs_n = c \lim s_n$, and the result follows. ∎

14.23. Solution. *"Rationalization" of repeating decimals.* Let x equal the repeating decimal $.abcabcabcabc \cdots$. By the definition of decimal expansion, $x = (100a + 10b + c)/1000 + (100a + 10b + c)/1000000 + \cdots$. Therefore, $1000x = 100a + 10b + c + x$; here we have used Proposition 14.22. Solving for x, we obtain $x = (100a + 10b + c)/999$. (Or, $x = abc/999$.) ∎

This procedure generalizes to rationalize any decimal expansion that eventually repeats (Exercise 38). Alternatively, we could use Proposition 14.22 to write $x = (100a + 10b + c) \sum_{k=1}^{\infty} (1/1000)^k$ and obtain the same result by summing the geometric series.

14.24. Theorem. (The Geometric Series) Given $x \in \mathbb{R}$, the **geometric series** $\sum_{k=0}^{\infty} x^k$ converges to $\frac{1}{1-x}$ if $|x| < 1$ and diverges otherwise.

Proof: When $x \neq 1$, the partial sum $s_n = \sum_{k=0}^{n} x^k$ equals $(1 - x^{n+1})/(1 - x)$ (Corollary 3.14). Now we apply the properties of limits. Because $x^{n+1} \to 0$ if $|x| < 1$ and x^{n+1} does not converge if $x > 1$, the sequence of partial sums (and hence the series $\sum_{k=0}^{\infty} x^k$) converges to $1/(1-x)$ if $|x| < 1$ and diverges if $|x| > 1$. When $x = 1$, the nth partial sum is $n + 1$; when $x = -1$, the partial sums alternate between 1 and 0. Hence the series also diverges when $|x| = 1$. ∎

We have indexed the geometric series beginning at 0 instead of 1. When $|x| < 1$, the series $\sum_{k=1}^{\infty} x^k$ differs from $\sum_{k=0}^{\infty} x^k$ by 1; we have $\sum_{k=1}^{\infty} x^k = \frac{1}{1-x} - 1 = \frac{x}{1-x}$.

14.25. Example. *The Multiplier Effect.* Suppose that the typical individual in a society spends a fraction t of all new or extra income, where $0 < t < 1$. Economists call t the *marginal propensity to consume.* Summing the geometric series explains the "multiplier effect" in the economy. When a typical individual receives an extra dollar in wages, he or she spends $\$t$. The recipient of that $\$t$ then spends $\$t^2$, and so on. The total increase in economic activity is $\sum_{k=0}^{\infty} t^k = 1/(1-t)$ dollars. The higher the marginal propensity to consume, the greater the multiplier effect. ∎

14.26. Solution. *The Tennis Problem.* The server wins each particular point with probability p; let $q = 1 - p$. First we consider the ways the server can win with exactly four points. The other player may score zero, one, or two points before the server scores four, and the total probability for these mutually exclusive possibilities is $p^4 + \binom{4}{1}p^4 q + \binom{5}{2}p^4 q^2$. The server may also win after the game reaches a 3-3 tie; a 3-3 tie happens with probability $\binom{6}{3}p^3 q^3$. In this situation, the probability that the server wins after a total of exactly $2k+2$ more points is $(2pq)^k p^2$. Summing this over all $k \geq 0$ yields the geometric series $p^2 \sum_{k=0}^{\infty}(2pq)^k = p^2/(1 - 2pq)$. Hence the total probability is

$$p^4(1 + 4q + 10q^2) + \frac{20p^5 q^3}{1 - 2pq}.$$

This equals .736 when $p = .6$ and .901 when $p = .7$; this suggests the difficulty of "breaking serve".

The probability x of winning after the score is tied with at least three points each can be computed in another way. The server can win the next two points or can win after splitting the next two points, which repeats the tied situation. Hence $x = p^2 + 2pqx$, or $x = p^2/(1 - 2pq)$. This is essentially the same computation as summing a geometric series by using the distributive law. ∎

Many students confuse the words "sequence" and "series" in mathematics, perhaps because the use of the word "series" in English is similar to the use of the word "sequence" in mathematics. In English, we generally use "series" to mean a **finite sequence** (*list*) of events, such as a "series" of baseball games.

A series converges if its sequence of partial sums has a limit; otherwise it diverges. We say that a series "diverges to positive infinity" if, for every $M \in \mathbb{R}$, there is an $N \in \mathbb{N}$ such that $n \geq N$ implies $s_n > M$. By the Monotone Convergence Theorem, a divergent series of positive numbers must diverge to positive infinity. The definition of divergence to negative infinity is similar, with $s_n < M$ in place of $s_n > M$. Series may diverge in other ways. The series $1 - 2 + 3 - 4 = \sum_{k=1}^{\infty} k(-1)^{k+1}$ has arbitrarily large partial sums of both signs. The series $\sum_{k=1}^{\infty}(-1)^k$ has bounded partial sums, but the sequence of partial sums does not converge. We begin with a necessary condition for convergence.

14.27. Lemma. If the series $\sum_{k \geq 0} a_k$ converges, then $a_n \to 0$.

Proof: Let $s_n = \sum_{k=1}^{n} a_k$; thus $a_n = s_n - s_{n-1}$. Because the series converges, the sequence $\langle s \rangle$ has a limit L. Therefore, $a_n \to L - L = 0$.

Second proof. If $\langle s \rangle$ converges, then $\langle s \rangle$ is a Cauchy sequence, so for every $\epsilon > 0$ we have $|s_n - s_{n-1}| < \epsilon$ for sufficiently large n. We again have $a_n = s_n - s_{n-1}$, and hence $a_n \to 0$ by the definition of convergence. ∎

The converse of Lemma 14.27 is false. In the next example, the sequence of terms converges to 0, but so slowly that the series diverges. Convergence of a series of positive numbers requires that the sequence of terms converges to 0 sufficiently rapidly.

14.28. Example. *The harmonic series.* Consider $\sum_{k=1}^{\infty} 1/k$. To see that $\sum_{k=1}^{\infty} 1/k$ diverges even though $1/k \to 0$, we compare this with another divergent series whose terms approach 0. Let $\langle c \rangle = \frac{1}{2}, \frac{1}{4}, \frac{1}{4}, \frac{1}{8}, \frac{1}{8}, \frac{1}{8}, \frac{1}{8}, \frac{1}{16}, \cdots$; here there are 2^{j-1} copies of $1/2^j$ for each $j \geq 1$. Since the copies of $1/2^j$ for each fixed j sum to $1/2$, for each $M \in \mathbb{N}$ the partial sum $\sum_{k=1}^{n} c_k$ exceeds M for large enough n, and $\sum_{k=1}^{\infty} c_k$ diverges. The last copy of $1/2^j$ in $\langle c \rangle$ is the $2^j - 1$th term. Thus we have $1/k > c_k$ for every k. For each n, summing n of these inequalities yields $\sum_{k=1}^{n} 1/k > \sum_{k=1}^{n} c_k$. Hence $\sum_{k=1}^{\infty} 1/k$ also diverges. ∎

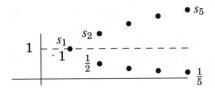

At first, we can only prove convergence of series by applying the definition. Applying the definition requires finding a formula for the partial sum $s_k = \sum_{n=1}^{k} a_n$ in terms of k, and then determining whether the sequence defined by this function converges. Seldom can we compute the limit of the partial sums directly, so we need other methods to test for convergence of a series. Example 14.29 illustrates the "comparison test". When the test applies, it settles the question of convergence.

14.29. Proposition. (Comparison test) Suppose that $c_n \geq 0$ for all n. If $\sum_{n=1}^{\infty} c_n$ converges and $|a_n| \leq c_n$ for all $n \geq N$, then $\sum_{n=1}^{\infty} a_n$ converges. If $\sum_{n=1}^{\infty} c_n$ diverges to ∞ and $a_n \geq c_n$ for all n, then also $\sum_{n=1}^{\infty} a_n$ diverges to ∞.

Proof: Let $s_k = \sum_{n=1}^{k} a_n$, and let $S_k = \sum_{n=1}^{k} c_n$. Since $\sum_{n=1}^{\infty} c_n$ converges, $\langle S \rangle$ is a Cauchy sequence; we show that $\langle s \rangle$ is also a Cauchy sequence. Given $\epsilon > 0$, choose N such that $m, n \geq N$ implies that $|S_m - S_n| < \epsilon$. Given $m > n \geq N$, we have

$$|s_m - s_n| = \left| \sum_{i=n+1}^{m} a_i \right| \leq \sum_{i=n+1}^{m} |a_i| \leq \sum_{i=n+1}^{m} c_i = |S_m - S_n| < \epsilon.$$

Hence $\langle s \rangle$ is a Cauchy sequence and converges. We leave the second statement to Exercise 49. ∎

14.30. Corollary. If $\sum |a_n|$ converges, then $\sum a_n$ converges.

Proof: Apply the comparison test with $c_n = |a_n|$. ∎

Applying the comparison test to prove convergence requires known convergent series for comparisons, such as the geometric series. The behavior of the geometric series suggests another general test for convergence. Consider a series $\sum_{k=1}^{\infty} a_k$ with positive terms, and let $c_k = a_{k+1}/a_k$. If $\langle a \rangle$ is a geometric series, then $\langle c \rangle$ is constant. If $\langle c \rangle$ is not constant but has a limit ρ, then the convergence criterion is the same: the series converges if $\rho < 1$ and diverges if $\rho > 1$. The letter "ρ" suggests "ratio". The test is inconclusive when $\rho = 1$, but Exercise 65 develops a refinement (Raabe's test) that sometimes applies even when $\rho = 1$.

14.31. Theorem. (Ratio test) Let $\langle a \rangle$ be a sequence with $|a_{k+1}/a_k| \to \rho$. If $\rho < 1$, then $\sum_{k=1}^{\infty} a_k$ converges. If $\rho > 1$, then $\sum_{k=1}^{\infty} a_k$ diverges.

Proof: We leave the proof for $\rho > 1$ to Exercise 56; assume $\rho < 1$. By Corollary 14.30, it suffices to show that $\sum_{k=1}^{\infty} |a_k|$ converges. Since the hypothesis involves only the absolute values of the terms, we only need to consider the case where each a_k is positive. Let $\langle s \rangle$ be the sequence of partial sums: $s_n = \sum_{k=1}^{n} a_k$. We prove that $\langle s \rangle$ is a Cauchy sequence.

Let ϵ be a positive number, and choose β between ρ and 1. Since $a_{k+1}/a_k \to \rho$, we can choose N_1 such that $k \geq N_1$ implies $a_{k+1}/a_k < \beta$. In particular, this implies $a_{k+j} < a_k \beta^j$ for $k \geq N_1$ and $j \geq 1$. We have already proved that $a_{k+1}/a_k \to \rho < 1$ implies $a_k \to 0$ (Proposition 14.10). Hence there also exists N_2 such that $k \geq N_2$ implies $a_k < (1 - \beta)\epsilon$. Choose $N = \max\{N_1, N_2\}$. Given any $k, l \geq N$ with $l \geq k$ we have

$$0 < s_l - s_k = \sum_{n=k+1}^{l} a_n < \sum_{j=1}^{l-k} (1-\beta)\epsilon\beta^j < (1-\beta)\epsilon \sum_{j=0}^{\infty} \beta^j = \epsilon.$$

Hence the sequence of partial sums is a Cauchy sequence; by Theorem 14.19, it converges. ∎

The ratio test enables us to define the exponential function via series. We discuss its crucial properties in Chapter 17.

14.32. Example. *Exponential series.* Given $x \in \mathbb{R}$, the **exponential function** is defined by $\exp(x) = \sum_{n=0}^{\infty} x^n/n!$. For each $x \in \mathbb{R}$, this series converges by the ratio test, since the ratio of successive terms is $\frac{|x|}{n+1}$, which converges to 0. ∎

Our last convergence test will be applied in Chapter 17.

14.33. Theorem. (Root test) Let $\langle a \rangle$ be a sequence such that $|a_n|^{1/n} \to \rho$. If $\rho < 1$, then $\sum_{k=1}^{\infty} a_k$ converges. If $\rho > 1$, then $\sum_{k=1}^{\infty} a_k$ diverges.

Proof: (Exercise 68).   ∎

We can extend the root test to a larger class of series by refining the concept of limit. Given a bounded sequence $\langle b \rangle$, let $L_m = \sup_{n>m} b_n$. By the Completeness Axiom, each L_m is defined. Each successive supremum is taken over a more restricted set, so $\{L_m\}$ is a nonincreasing sequence bounded below by $\inf\{b_n\}$. Thus the Monotone Convergence Theorem implies that $\{L_m\}$ converges.

14.34. Definition. The lim sup of a bounded sequence $\langle b \rangle$ is the number $L = \lim_{m \to \infty} \sup_{n>m} b_n$. We write this as $L = \limsup b_n$. When $\langle b \rangle$ has no upper bound, we say that $\limsup b_n$ is infinite.

14.35. Example. If $a_n = (-1)^n + 1/n$, then $\lim a_n$ does not exist. Nevertheless, $\sup_{n \geq 1} a_n = 3/2$, $\sup_{n \geq 3} a_n = 5/4$, $\sup_{n \geq 5} a_n = 7/6$, etc. In general, $\sup_{n \geq 2k} = \sup_{n \geq 2k-1} = \frac{2k+1}{2k}$. Thus $\limsup a_n = 1$.   ∎

Observe that if $\lim b_n = \rho$, then also $\limsup b_n = \rho$. More generally, $\limsup b_n$ is the supremum of the limits of all convergent subsequences of $\langle b \rangle$. We use lim sup to extend the root test for convergence of a series. This extension will be useful in Chapter 17.

14.36. Theorem. (Root test) Let $\langle a \rangle$ be a sequence with $L = \limsup |a_n|^{1/n}$. If $L < 1$, then $\sum_{k=1}^{\infty} a_k$ converges. If $L > 1$ (or L is infinite), then $\sum_{k=1}^{\infty} a_k$ diverges.

Proof: The proof is like that of Theorem 14.33 (see Exercise 69).   ∎

HOW TO APPROACH PROBLEMS

Chapters 13–18 concern *elementary analysis*, which is based on limits and inequalities. The notion of a limit ultimately rests on the ability to approximate something arbitrarily well, and approximation is done via inequalities. Due to the standard use of the Greek letter ϵ, we call this subject *epsilonics*. Nearly all of the theorems and exercises in Chapter 14 follow from epsilonics. We present some tricks of the trade.

1) Interpret $|x - a|$ as the distance between x and a.

2) Learn the fundamental principles of epsilonics (stated below).

3) Sometimes proving that a given limit equals L should be done in two steps. First prove that the limit exists. Then, assuming that it exists, take limits of both sides of the defining relation to obtain an equation for the limit.

4) Understand the relationship between a infinite series and its sequence of partial sums.

The mathematical inequality $|y - a| < \epsilon$ has a precise geometrical meaning. We think of a as a given number, y as a variable to be measured, and ϵ as a level of accuracy. The inequality says that we can approximate y by a to within epsilon accuracy.

14.37. Example. Consider the sentence "The temperature T is about 80 degrees". One interpretation of this might be $|T - 80| < 5$, or equivalently, $75 < T < 85$. Perhaps the estimate is more accurate, yielding $|T - 80| < 2$. The right side describes the accuracy of the measurement.   ∎

The definition of $\lim y_n = L$ requires an inequality for each positive epsilon. Given $\epsilon > 0$, we must be able to approximate y_n (for all sufficiently large n) by L within an accuracy of ϵ. We expect that n might have to be larger when ϵ is smaller. There are various techniques for proving such inequalities.

Suppose that we want to show that some quantity w is small (close to zero in some sense). We write $w = w - v + v$ and hence $|w| \le |w - v| + |v|$. Thus to show that w is small it suffices to show that both $|w - v|$ and $|v|$ are small. This epitomizes proofs in analysis.

It helps to interpret $|x - a|$ as the distance between x and a. We often use this with the inequality

$$|x - z| \le |x - y| + |y - z|,$$

which is a version of the Triangle Inequality. The inequality states that the distance between x and z is no more than the the distance between x and y plus the distance between y and z. If these two distances can be made small, then so can the distance between x and z.

The fundamental principles of epsilonics.

To prove that a sequence has a given limit, we must show that a certain expression can be made sufficiently small when the index is sufficiently large. In general, we want to express the desired quantity in terms of various contributions, all of which can be made as small as desired.

One general approach is "to show that b is small, it suffices to show that b is close to a and also that a is small". This uses $|b| \le |b - a| + |a|$ and the information that both terms on the right can be made small. See Theorem 14.5 for examples.

A second approach is "to show that a nonnegative quantity is small, it suffices to show that a larger quantity is small". This uses $|a| \le C |b|$ and the information that $|b|$ can be made small. When our goal is to show that a given quantity is small, we often show that a *larger* quantity is small. This might seem strange, but it is useful when the larger quantity is easier to understand and hence easier to bound. The sine function provides a nice example here; its formal definition appears in Chapter 17.

14.38. Example. $\lim_{n \to \infty} \frac{\sin(n)}{n} = 0$. Given $\epsilon > 0$, we must find N such that $n \geq N$ implies $|\frac{\sin(n)}{n}| < \epsilon$. Since $|\sin(n)| \leq 1$ holds for all n, it suffices to bound $1/n$. We choose N so that $N > 1/\epsilon$. Now $n \geq N$ implies

$$\left| \frac{\sin(n)}{n} \right| \leq \frac{1}{n} < \epsilon.$$

The crucial point was to replace $|\sin(n)|$ by 1, making the quantity larger, but simpler, and still sufficiently small. This powerful technique applies many times in the remaining chapters of this book.   ∎

Finding limits via equations.

Another technique in this chapter is useful for evaluating limits. We first prove that a limit exists, perhaps by the Monotone Convergence Theorem. Then, we find the possible values that the limit can have under the assumption that it exists. Consider Exercises 19–26 in this light. For example, suppose that in Exercise 20 we have determined somehow that the limit exists. We give it a name, say L. The recursive definition $x_{n+1} = \sqrt{1 + x_n}$ implies by the arithmetic properties of limits that $L^2 = 1 + L$. From here it is easy to solve for L.

This technique can also be used to show that a sequence or series fails to converge. Assume that it converges to a number L, and then find an equation or inequality for L. If this has no solutions, then the limit cannot exist. Below we give an example using series.

Infinite Series.

An infinite series is nothing more than the sequence of its partial sums. To appreciate this, think of the decimal expansion $.a_1 a_2 a_3 a_4 \cdots$ of a number in the interval $[0, 1]$. The number is determined by the sequence $.a_1, .a_1 a_2, .a_1 a_2 a_3, .a_1 a_2 a_3 a_4, \ldots$ of partial sums of the series $\sum_{n=1}^{\infty} a_n 10^{-n}$. The decimal expansion is nothing but an infinite series; a finite part of this expansion is a partial sum.

Many problems on infinite series involve determining whether a given series converges or diverges. This can be a delicate issue, but most problems in textbooks are fairly easy to solve by standard tests. The most useful test is the comparison test; it expresses a simple intuition. A series of positive terms converges when its terms decay to zero fast enough. If $|c_n|$ decays to zero rapidly enough for $\sum |c_n|$ to converge, and $|a_n| \leq |c_n|$, then $\sum a_n$ converges also, as shown in Proposition 14.29.

We illustrate the use of contradiction to prove that a series diverges.

14.39. Example. Suppose that $0 < a_n \leq a_{2n} + a_{2n+1}$ for $n \geq 1$. We show that $\sum_{n=1}^{\infty} a_n$ diverges. To get started, we suppose that the series converges to some number L, and see what happens. We have

$$L = a_1 + a_2 + a_3 + a_4 + a_5 + \cdots$$

We are given that $a_2 + a_3 \geq a_1$, that $a_4 + a_5 \geq a_2$, that $a_6 + a_7 \geq a_3$, and so on. Substituting these into the first expression yields

$$L \geq a_1 + a_1 + a_2 + a_3 + \cdots$$

In general, we have replaced each pair of terms $a_{2n} + a_{2n+1}$ with a_n without increasing the sum. Now we have obtained $L \geq a_1 + L$, contradicting the assertion that $a_1 > 0$.

We write the proof more formally, using proof by contradiction. Assume that the series converges to some real number L. Since all the terms are positive, we may group them in any order when we sum them. Since $a_1 > 0$, we obtain the contradiction

$$L > \sum_{n=2}^{\infty} a_j = \sum_{k=1}^{\infty}(a_{2k} + a_{2k+1}) \geq \sum_{k=1}^{\infty} a_k = L.$$

This contradiction shows that the series diverges. ∎

We close by discussing another test for convergence of series; this test can be used to solve most such problems in calculus books. When it is difficult to verify the inequalities required to invoke the comparison test, one often can invoke the *limit comparison test* of Exercise 58.

This test generalizes the comparison test as follows. Let $\langle a \rangle$ and $\langle b \rangle$ be sequences of positive numbers, and suppose that b_k/a_k converges to a nonzero real number L. Then $\sum_{k=1}^{\infty} b_k$ converges if and only if $\sum_{k=1}^{\infty} a_k$ converges. We sketch one possible proof, assuming that $\sum_{k=1}^{\infty} a_k$ converges. First write $b_k = La_k + e_k$ for some "error" e_k, and observe that $e_k/a_k \to 0$, and hence that e_k/a_k is bounded. Apply the ordinary comparison test to $\sum_{k=1}^{\infty} e_k$, and observe that $\sum_{k=1}^{\infty} b_k$ is now the sum of two convergent series.

EXERCISES

14.1. (−) Find an unbounded sequence that has no convergent subsequence and an unbounded sequence that has a convergent subsequence.

14.2. (−) For each condition below, give an example of an unbounded sequence $\langle a \rangle$ such that $a_{n+1} - a_n > 0$ for all n and the specified condition holds.
 a) $\lim(a_{n+1} - a_n) = 0$.
 b) $\lim(a_{n+1} - a_n)$ does not exist.
 c) $\lim(a_{n+1} - a_n) = L$, where $L > 0$.

14.3. (−) For each condition below, give examples of sequences $\langle a \rangle$ and $\langle b \rangle$ such that $\lim a_n = 0$, $\lim b_n$ does not exist, and the specified condition holds.
 a) $\lim(a_n b_n) = 0$.

b) $\lim(a_n b_n) = 1$.

c) $\lim(a_n b_n)$ does not exist.

14.4. (−) Suppose that $x_{n+1} = \sqrt{1 + x_n^2}$ for all $n \in \mathbb{N}$. Show that $\langle x \rangle$ does not converge.

14.5. (−) Find a counterexample to the following false statement.
"If $a_n < b_n$ for all n and $\sum b_n$ converges, then $\sum a_n$ converges."

14.6. (−) What number has $.111\ldots$ as its k-ary expansion?

14.7. (−) Compute the binary expansion of $\sqrt{2}$ to six places.

•   •   •   •   •

For Exercises 8–12, determine whether the statement is true or false. If true, provide a proof; if false, provide a counterexample.

14.8. Let $\langle x \rangle$ be a sequence of real numbers.
a) If $\langle x \rangle$ is unbounded, then $\langle x \rangle$ has no limit.
b) If $\langle x \rangle$ is not monotone, then $\langle x \rangle$ has no limit.

14.9. Suppose that $x_n \to L$.
a) For all $\epsilon > 0$, there exists $n \in \mathbb{N}$ such that $|x_{n+1} - x_n| < \epsilon$.
b) There exists $n \in \mathbb{N}$ such that for all $\epsilon > 0$, $|x_{n+1} - x_n| < \epsilon$.
c) There exists $\epsilon > 0$ such that for all $n \in \mathbb{N}$, $|x_{n+1} - x_n| < \epsilon$.
d) For all $n \in \mathbb{N}$, there exists $\epsilon > 0$ such that $|x_{n+1} - x_n| < \epsilon$.

14.10. Let $\langle x \rangle$ be a sequence of real numbers.
a) If $\langle x \rangle$ converges, then there exists $n \in \mathbb{N}$ such that $|x_{n+1} - x_n| < 1/2^n$.
b) If $|x_{n+1} - x_n| < 1/2^n$ for all $n \in \mathbb{N}$, then $\langle x \rangle$ converges.

14.11. a) If $x_1 = 1$ and $x_{n+1} = x_n + 1/n$ for $n \geq 1$, then $\langle x \rangle$ is bounded.
b) If $y_1 = 1$ and $y_{n+1} = y_n + 1/n^2$ for $n \geq 1$, then $\langle y \rangle$ is bounded.

14.12. If $a_n \to 0$ and $b_n \to 0$, then $\sum a_n b_n$ converges.

•   •   •   •   •

14.13. Prove that if $\langle a \rangle$ converges, then every subsequence of $\langle a \rangle$ converges and has the same limit as a.

14.14. Let $\langle a \rangle$ and $\langle b \rangle$ be sequences, with $b_n \neq 0$ for all n. Prove that if $a_n \to L$ and $b_n \to M \neq 0$, then $a_n/b_n \to L/M$. (Hint: First prove it when $a_n = 1$ for all n.)

14.15. Suppose that $b \leq L + \epsilon$ for all $\epsilon > 0$. Prove that $b \leq L$.

14.16. Let $a_n = p(n)/q(n)$, where p and q are polynomials and the degree of q is greater than the degree of p. Use properties of limits to prove that $a_n \to 0$.

14.17. (!) Let $a_n = p(n)x^n$, where p is a polynomial in n and $|x| < 1$. Prove that $a_n \to 0$. (Hint: Consider the ratio a_{n+1}/a_n. Comment: When $|x| < 1$, this exercise shows that x^n tends to zero so fast that multiplication by a polynomial in n does not affect the limit. Thus exponential decay dominates polynomial growth.)

14.18. If $a_1 = 1$ and $a_n = \sqrt{3a_{n-1} + 4}$ for $n > 1$, prove that $a_n < 4$ for all $n \in \mathbb{N}$.

14.19. Suppose that $x_1 = 1$ and that $2x_{n+1} = x_n + 3/x_n$ for $n \geq 1$. Prove that $\lim_{n \to \infty} x_n$ exists, and find the limit.

14.20. Suppose that $x_1 > -1$ and that $x_{n+1} = \sqrt{1 + x_n}$ for $n \geq 1$. Prove that $\lim_{n \to \infty} x_n$ exists, and find the limit.

14.21. Let c be a real number greater than 1. Let $\langle x \rangle$ be the sequence defined by $x_1 = c$ and $x_{n+1} = x_n^2$ for $n \geq 1$. Prove that $\langle x \rangle$ is unbounded.

14.22. (!) Prove that $c^{1/n} \to 1$ when c is a positive real number.

14.23. Suppose that $f_1(x) = x$ for $x \in \mathbb{R}$ and that $f_{n+1}(x) = (f_n(x))^2/2$ for $n \geq 1$. If $\lim_{n \to \infty} f_n(x)$ exists, what can the limit equal? For which x is the sequence $\{f_n(x)\}$, strictly increasing, constant, or strictly decreasing? Use this information to determine how $\lim_{n \to \infty} f_n(x)$ depends on x.

14.24. Let $\langle x \rangle$ be a sequence satisfying the recurrence $x_{n+1} = x_n^2 - 4x_n + 6$.
　　a) If $\lim_{n \to \infty} x_n$ exists and equals L, what possible values can L have?
　　b) The behavior of x_n as $n \to \infty$ depends on the initial value x_0. For each $x_0 \in \mathbb{R}$, describe this behavior. (Hint: Graph the functions defined by x and $x_n^2 - 4x_n + 6$ and interpret the graphs, or obtain a recurrence for the sequence $\{y_n\}$ defined by $y_n = x_n - 2$ and study its behavior.)

14.25. (+) *Generalization of Exercise 14.24.* Suppose that $\langle x \rangle$ satisfies $x_n = f(x_{n-1})$ for $n \geq 1$, where $f(x) = x^2 + Ax + B$. Determine the possible values of $\lim_{n \to \infty} x_n$. Determine the limiting behavior of x_n in terms of x_0, A, and B.

14.26. Suppose that $a_{n+2} = (\alpha + \beta)a_{n+1} - \alpha\beta a_n$ with $\beta \neq \alpha$, and suppose that $a_0 = a_1 = 1$. Find the limit as $n \to \infty$ of a_{n+1}/a_n.

14.27. (!) For $c > 0$, let $x_n = (c^n + 1)^{1/n}$. Determine $\lim_{n \to \infty} x_n$. More generally, find $\lim_{n \to \infty} (a^n + b^n)^{1/n}$. (Hint: Consider $c < 1$ first and use the Squeeze Theorem.)

14.28. (!) *Alternative proof of the Bolzano-Weierstrass Theorem* (Theorem 14.17).
　　a) Use the Monotone Convergence Theorem to show that every bounded sequence with a monotone subsequence has a convergent subsequence.
　　b) Prove that every bounded sequence has a monotone subsequence. (Hint: In a sequence $\langle a \rangle$, call an index n a *peak* if $a_m < a_n$ for $m > n$. Consider two cases, accordingly to whether $\langle a \rangle$ has infinitely many peaks.)

14.29. A **limit point** of a sequence $\langle a \rangle$ is a number L to which some subsequence of $\langle a \rangle$ converges. Construct a sequence with infinitely many limit points.

14.30. Let $\langle x \rangle$ be the sequence given by $x_1 = 1$ and $x_{n+1} = 1/(x_1 + \cdots + x_n)$ for $n \geq 1$. Prove that $\langle x \rangle$ converges, and obtain the limit.

14.31. (!) Given $x_1 \geq 0$, define $x_{n+1} = \frac{x_n + 2}{x_n + 1}$ for $n \geq 0$. Prove that $x_n \to \sqrt{2}$. (Hint: The sequence is not monotone, but it is possible to show that $\left| x_{n+1} - \sqrt{2} \right| < \left| x_n - \sqrt{2} \right|$ for all n.

14.32. (!) A runaway train is hurtling toward a brick wall at the speed of 100 miles per hour. When it is two miles from the wall, a fly begins to fly repeatedly between the train and the wall at the speed of 200 miles per hour. Determine how far the fly travels before it is smashed.

14.33. (−) Suppose that $\sum_{k=1}^{\infty} a_k$ and $\sum_{k=1}^{\infty} b_k$ converge to A and B, respectively. Prove that $\sum_{k=1}^{\infty}(a_k + b_k)$ converges and equals $A + B$.

14.34. Find the expansion of $1/2$ in base 3, with proof. Determine the rational number that has ternary expansion $.121212\cdots$.

14.35. a) What number x has expansion $.141414\cdots$ in base 10?
   b) What number y has expansion $.141414\cdots$ in base 5?

14.36. (!) Consider $.247247247\cdots$, expressed as a decimal expansion. Write this as a rational number with numerator and denominator in base 10. Now consider the expression $.247247247\cdots$ as an expansion in base 8. Write this as a rational number with numerator and denominator in base 10.

14.37. Describe geometrically the set of numbers in the interval $[0, 1]$ whose ternary expansions contain no 1's. Prove that the set is uncountable.

14.38. (!) A k-ary expansion is **eventually periodic** if after some initial portion, the remainder is a repeating list of some finite length (this includes terminating expansions, where the repeating list is "0").
   a) Prove that the every k-ary expansion of a rational number is eventually periodic. (Hint: First prove this for rational numbers of the form j/s with $0 \le j < s$. Then use this and k-ary expansions of integers to prove the claim in the general case.)
   b) Prove the converse of part (a): if the k-ary expansion of x is eventually periodic, then x is rational.

14.39. (−) Determine whether $\sum_{n=1}^{\infty} \frac{1}{10^{n!}}$ is rational.

14.40. *Alternative approach to the geometric series.* Suppose that $y = 1/(1 - x)$, which is equivalent to $y = 1 + xy$. Suppose that $|x| < 1$ and that we have an initial guess y_0 for y. The equation $y = 1 + xy$ suggests two algorithms.
   a) Given y_0, define the sequence $\langle y \rangle$ by $y_{n+1} = 1 + xy_n$ for $n \ge 0$. Prove that $\langle y \rangle$ converges to $1/(1 - x)$.
   b) Given y_0, define $\langle y \rangle$ instead by $y_n = 1 + xy_{n+1}$, so $y_{n+1} = (y_n - 1)/x$. Why does this algorithm fail, even when $x \ne 0$?

14.41. Suppose that a sequence of measurements is to be made. Each measurement involves some error but can be made to any specified accuracy. How can it be guaranteed that the total error is at most 1?

14.42. (+) *Measure zero.* A set $S \subset \mathbb{R}$ has **measure zero** if, for every $\epsilon > 0$, there is a countable collection of intervals whose union contains S, such that the sum of the lengths of the intervals is less than ϵ. Prove that a union of countably many sets of measure zero also has measure zero. Conclude that the set of rational numbers has measure zero. (Hint: This uses what might be called the "ultimate" $\epsilon/2$ argument; consider $\epsilon/2^n$ for each n.)

14.43. Compute $\sum_{n=1}^{\infty}(\frac{x}{x+1})^n$. What assumptions must be made about x?

14.44. Compute $\sum_{n=1}^{\infty} \frac{1}{n(n+1)}$. Use this to obtain upper and lower bounds on $\sum_{n=1}^{\infty} \frac{1}{n^2}$. (Comment: The exact value of $\sum_{n=1}^{\infty} \frac{1}{n^2}$ is $\pi^2/6$.)

14.45. (!) Suppose that the nth partial sum of a series equals $1/n$, for $n \ge 1$. Determine the nth term in the series.

14.46. Suppose that $b_k = c_k - c_{k-1}$, where $\langle c \rangle$ is a sequence such that $c_0 = 1$ and $\lim_{k \to \infty} c_k = 0$. Use the definition of series to determine $\sum_{k=1}^{\infty} b_k$.

14.47. (!) Suppose that $\sum a_n^2$ and $\sum b_n^2$ both converge. Prove that $\sum a_n b_n$ converges. (Hint: Use the AGM Inequality and the comparison test.)

14.48. Change the Tennis Problem (Problem 14.2) so that the winner is the first one to reach four points. What is the server's probability of winning the game?

14.49. *Comparison test for divergence.* Suppose that $\sum_{k=1}^{\infty} c_k$ diverges to ∞, and suppose $a_k \geq c_k$ for all k. Prove that $\sum_{k=1}^{\infty} a_k$ diverges to ∞.

14.50. Determine whether $1 + \frac{1}{3} + \frac{1}{5} + \frac{1}{7} + \cdots$ converges.

14.51. (+) Prove that e is irrational. (Hint: If e were rational, then there would be a natural number n such that $n!e = n! \sum_{k=0}^{\infty} 1/k!$ would be an integer. This would imply that $n! \sum_{k=n+1}^{\infty} 1/k!$ is an integer. Use this to obtain a contradiction.)

14.52. (!) *Convergence of alternating series.*
   "If $\langle a \rangle$ is a sequence whose terms alternate in sign, converge to 0, and satisfy $|a_{k+1}| \leq |a_k|$ for all n, then the series $\sum_{k=0}^{\infty} a_k$ converges."
Give proofs of the statement above by the two methods below:
   a) Show that the partial sums form a Cauchy sequence.
   b) Use Proposition 13.18 and the Squeeze Theorem (Theorem 14.6).

14.53. Consider $\sum_{k=1}^{\infty} \frac{(-1)^{k+1}}{k} = 1 - \frac{1}{2} + \frac{1}{3} - \frac{1}{4} + \cdots$. By Exercise 14.52, this series converges. (It converges to $\ln 2$, but that is not needed here.) Prove that the sum of the series is less than $5/6$. The terms in the series can be summed in other orders. Prove that $1 + \frac{1}{3} - \frac{1}{2} + \frac{1}{5} + \frac{1}{7} - \frac{1}{4} + \frac{1}{9} + \frac{1}{11} - \frac{1}{6} + \cdots$ has sum greater than $5/6$ (in fact, the sum exceeds 1). Find a reordering of the terms to obtain (with proof!) a convergent series whose sum exceeds $3/2$.

14.54. Suppose that $\sum a_k$ converges, that $\sum |a_k|$ diverges, and that L is a real number. Prove that the terms of $\langle a \rangle$ can be reordered to obtain a series that converges to L. (Riemann)

14.55. (!) True or false: If $a_k \to 0$ and the sequence of partial sums is bounded, then $\sum_{k=1}^{\infty} a_k$ converges. Provide a proof or a counterexample.

14.56. *Ratio test for divergence.* Let $\langle a \rangle$ be a sequence such that $|a_{k+1}/a_k| \to \rho$ for some $\rho > 1$. Prove that $\sum_{k=1}^{\infty} a_k$ diverges.

14.57. In Example 12.33, we found the formula $f(x) = 1/(1 - x - x^2)$ for the generating function of the Fibonacci numbers. This implies that the series $\sum_{n=0}^{\infty} F_n x^n$ converges for all x such that $|x|$ is less than the smallest magnitude of a root of $1 - x - x^2$. Assuming this, use the ratio test to find $\lim F_{n+1}/F_n$. Compare this limit with the formula for the Fibonacci numbers in Solution 12.25.

14.58. (!) *Limit comparison test.* Let $\langle a \rangle$ and $\langle b \rangle$ be sequences of positive numbers, and suppose that b_k/a_k converges to a nonzero real number L. Prove that $\sum_{k=1}^{\infty} b_k$ converges if and only if $\sum_{k=1}^{\infty} a_k$ converges.

14.59. (−) Let $\langle a \rangle$ be a convergent sequence of positive numbers. Prove that $\sum_{k=1}^{\infty} \frac{1}{ka_k}$ diverges.

14.60. (−) For each series below, use Exercise 14.58 to test convergence.

a) $\sum_{n=1}^{\infty} \frac{2n^2+15n+2}{n^4+3n+1}$.   b) $\sum_{n=1}^{\infty} \frac{2n^2+15n+2}{n^3+3n+1}$.   c) $\sum_{n=1}^{\infty} \frac{3+5n+n^2}{2^n}$.

14.61. (!) Let p be a polynomial of degree d, and let q be a polynomial of degree at least $d+2$. Suppose that $q(x) \neq 0$ for $x > 0$. Prove that $\sum_{n=1}^{\infty} \frac{p(n)}{q(n)}$ converges.

14.62. (!) Use the limit comparison test and the geometric series to prove the convergence part of the ratio test.

14.63. (!) *Condensation test.*

a) Let $\langle a \rangle$ be a decreasing sequence of positive numbers. Prove that $\sum_{k=1}^{\infty} a_k$ converges if and only if $\sum_{j=0}^{\infty} 2^j a_{2^j}$ converges. (Hint: Compare the series below.)

$$a_1 + a_2 + a_3 + a_4 + a_5 + a_6 + a_7 + a_8 + \cdots$$
$$a_1 + a_2 + a_2 + a_4 + a_4 + a_4 + a_4 + a_8 + \cdots$$
$$a_1 + a_1 + a_2 + a_2 + a_3 + a_3 + a_4 + a_4 + \cdots$$

b) For $p \in \mathbb{R}$, prove by part (a) that $\sum_{k=1}^{\infty} k^{-p}$ converges if and only if $p > 1$.

14.64. Let $\langle a \rangle$ and $\langle b \rangle$ be sequences of positive numbers such that $\frac{b_{k+1}}{b_k} \leq \frac{a_{k+1}}{a_k}$ for sufficiently large k. Prove that if $\sum_{k=1}^{\infty} a_k$ converges, then $\sum_{k=1}^{\infty} b_k$ converges.

14.65. *Raabe's test.* The ratio test for convergence of series is inconclusive when the ratio of successive terms converges to 1. This can be overcome if the convergence is slow enough. Let p be a real number greater than 1.

a) (+) Prove that if $0 < x < 1$, then $(1 - px) < (1 - x)^p$.

b) Use part (a) and Exercise 14.64 with $a_k = 1/k^p$ to prove that $\sum_{k=1}^{\infty} b_k$ converges if $b_{k+1}/b_k \leq 1 - p/k$ for sufficiently large k (assume $b_k > 0$ for all k).

14.66. (+) Use the divergence of $\sum_{k=1}^{\infty} 1/k$ to prove that every nonzero rational number is a finite sum of reciprocals of distinct integers. (Such an expression is known as an **Egyptian fraction**.)

14.67. Use the Binomial Theorem to prove that $\exp(x + y) = \exp(x)\exp(y)$.

14.68. *Root test.* Let $\langle a \rangle$ be a sequence such that $|a_n|^{1/n} \to \rho$.

a) Prove that if $\rho < 1$, then $\sum_{k=1}^{\infty} a_k$ converges.

b) Prove that if $\rho > 1$, then $\sum_{k=1}^{\infty} a_k$ diverges.

c) Show by example that if $\rho = 1$, then $\sum_{k=1}^{\infty} a_k$ may converge or diverge.

14.69. *Root test with* lim sup. Let $\langle a \rangle$ be a sequence such that $L = \limsup |a_n|^{1/n}$.

a) Prove that if $L < 1$, then $\sum_{k=1}^{\infty} a_k$ converges.

b) Prove that if $L > 1$, then $\sum_{k=1}^{\infty} a_k$ diverges.

Chapter 15

Continuous Functions

Continuity is the precise mathematical formulation of an intuitive idea. For many phenomena, a small change in the input results in a small change in the output. Before defining continuous functions, we describe several problems where the idea arises.

15.1. Problem. *The Antipodal Point Problem.* Consider a circular wire. If the temperature does not change abruptly from point to point, then some pair of opposite points on the circle have the same temperature. Why?

∎

15.2. Problem. *The Jewel Thieves Problem.* Two jewel thieves have stolen a circular necklace. The necklace has an even number of diamonds and an even number of rubies. The thieves want to split the necklace so that each keeps half of each type of jewel. Because the links are made of gold, the thieves do not want to cut many links. Is it always possible, no matter how the jewels are arranged, to make two cuts so that each thief gets a segment containing half the jewels of each type?

∎

15.3. Example. *The Butterfly Effect.* If a butterfly flaps its wings in Moscow, does the resulting wind current affect weather patterns in New York? For years scientists believed the answer to be no, but recent studies suggest otherwise. Physical phenomena depend on many variables, resulting in possible "chaotic" behavior.[†] The flapping of the butterfly's wings produces a small change in one variable. Depending on the role of other variables, this small change may produce a significant change far away. See Exercise 17 for a related example.

∎

[†]J. Gleick, *Chaos: Making a New Science*, Viking Press (New York, 1987), Chapter 1.

LIMITS AND CONTINUITY

We discuss limits and continuity for a function $f: I \to \mathbb{R}$, where the domain I is a subset of $\mathbb{R}$. It may be an open interval, a closed interval, or an interval with one point removed. We first introduce the required notion of limit for values $f(x)$ as x approaches a.

When computing such a limit we do not consider what happens at a, but only at elements of the domain close to a. A **neighborhood** of a is an open interval containing a. We define a **deleted neighborhood** of a to be the set obtained by removing a from a neighborhood of a. For example, $\{x \in \mathbb{R}: |x - a| < \delta\}$ is a neighborhood of a and $\{x \in \mathbb{R}: 0 < |x - a| < \delta\}$ is a deleted neighborhood of a.

15.4. Definition. Consider f defined on a deleted neighborhood of a. We say that the **limit** of $f(x)$ as x approaches a is L if for every $\epsilon > 0$ there exists $\delta > 0$ such that $0 < |x - a| < \delta$ implies $|f(x) - L| < \epsilon$. We write this as $\lim_{x \to a} f(x) = L$ or as "$f(x) \to L$ as $x \to a$", which we read as "$f(x)$ approaches L as x approaches a".

15.5. Example. *The role of epsilon and delta.* If $f(x) \to L$ as $x \to a$, then we can make $f(x)$ as close as we like to L by making x sufficiently close to a. For example, $\lim_{x \to 10} x^2 = 100$. If we want to guarantee that x^2 is within $\epsilon = 1$ of 100, then we may choose $\delta = .04$. Choosing $\delta = .05$ does not suffice, because $10.05^2 = 101.0025$. ∎

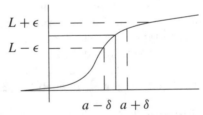

The definition of $\lim_{x \to a} f(x) = L$ says that given a desired "tolerance" ϵ, there is a real number $\delta > 0$ such that each input x within δ of a (except a itself) produces an output $f(x)$ within ϵ of L. By comparing this definition with that of convergence for sequences, we see that δ plays the same role here that N plays there.

$$(\forall \epsilon > 0)(\exists \delta > 0)\big[(0 < |x - a| < \delta) \Rightarrow (|f(x) - L| < \epsilon)\big]$$
$$(\forall \epsilon > 0)(\exists N \in \mathbb{N})\big[(n > N) \Rightarrow (|a_n - L| < \epsilon)\big]$$

Let us also consider what it means for $\lim_{x \to a} f(x) = L$ to be false. In this situation there exists some $\epsilon > 0$, say $\epsilon = \epsilon^*$, such that for every $\delta > 0$ (no matter how small), there is some x within δ of a such that $|f(x) - L| \geq \epsilon^*$. In particular, taking $\delta = 1/n$, we find a number x_n such that $|x_n - a| < 1/n$ but $|f(x_n) - L| \geq \epsilon^*$. The ability to construct such a sequence will be helpful in writing proofs by contradiction.

15.6. Example. Let $f(x) = cx\sin(1/x)$ for $x \neq 0$, where c is a positive constant. The values of the sine function are bounded by ± 1. Hence $|f(x)| \leq c\,|x|$. In proving that $\lim_{x \to 0} f(x) = 0$, we do not consider $x = 0$, where f is not defined. Given $\epsilon > 0$, we can choose δ (in terms of ϵ) to be ϵ/c. Then when $0 < |x - 0| < \delta$, we have $|x| < \epsilon/c$, and hence $|f(x) - 0| \leq c\,|x| < \epsilon$.

On the other hand, suppose $f(x)$ is defined to be the "sign" of x, meaning $+1$ when $x > 0$ and -1 when $x < 0$. We show that f has no limit at 0. No matter how we choose L, there is no deleted neighborhood of 0 on which $|f(x) - L| < 1$ for all x. If $L \geq 0$, the negative values of x are bad; if $L \leq 0$, the positive values of x are bad. ∎

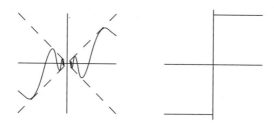

The definition of limit for function values parallels the definition of limit for sequences. Indeed, we can define limits for function values using sequences. We will prove that the two definitions are equivalent. The mapping that takes n to $f(x_n)$ is the composition of f with the sequence $\langle x \rangle$, and hence it defines a sequence $\langle y \rangle$. We write $f(x_n) \to L$ to mean that this sequence $\langle y \rangle$ converges to L.

15.7. Definition. A function f defined on a deleted neighborhood S of a **has sequential limit** L as x approaches a if $f(x_n) \to L$ for every sequence $\langle x \rangle$ converging to a in S.

15.8. Theorem. If f is defined on a deleted neighborhood of a, then the following two statements are equivalent:
A) $\lim_{x \to a} f(x) = L$.
B) f has sequential limit L as x approaches a.

Proof: We first prove A$\Rightarrow$B. Suppose that $\lim_{x \to a} f(x) = L$. To prove that f has sequential limit L, we consider an arbitrary sequence $\langle x \rangle$ in S converging to a and prove that $f(x_n)$ converges to L. Given $\epsilon > 0$, we need to find $N \in \mathbb{N}$ such that $n > N$ implies $|f(x_n) - L| < \epsilon$. Using the definition of $\lim_{x \to a} f(x) = L$, we know for this particular ϵ that there exists a number $\delta > 0$ for which $0 < |x - a| < \delta$ implies $|f(x) - L| < \epsilon$.

Now we use the definition of $x_n \to a$; given $\delta > 0$, we can choose $N' \in \mathbb{N}$ such that $n > N'$ implies $0 < |x_n - a| < \delta$. If we set $N = N'$, then $n > N$ implies $|f(x_n) - L| < \epsilon$.

We prove B$\Rightarrow$A by proving $\neg A \Rightarrow \neg B$. If $\lim_{x \to a} f(x) = L$ does not hold, then there is some $\epsilon^* > 0$ such that for every $\delta > 0$, some x satisfies $0 < |x - a| < \delta$ and $|f(x) - L| \geq \epsilon^*$. We consider ϵ^* and the sequence of choices for δ given by $\delta_n = 1/n$. Thus for each n we can find x_n such that $0 < |x_n - a| < 1/n$ and $|f(x_n) - L| \geq \epsilon^*$. This constructs a sequence $\langle x \rangle$ such that $x_n \to a$, but the sequence of values $f(x_n)$ does not converge to L. Hence f does not have sequential limit L as x approaches a. ∎

When we prove that some hypothesis H about limits of functions implies a conclusion C about limits of functions, we use "ϵ" as we did for limits of sequences. The statements H and C have the same form; suppose C is $(\forall \epsilon)(\exists \delta)(\forall x)P(x)$ and H is $(\forall \epsilon')(\exists \delta')(\forall x)Q(x)$. Proving C requires proving a statement for every positive ϵ. When we invoke H, we may use the existence of a suitable δ' for any desired ϵ', since we have assumed that H is true. We make an appropriate choice of ϵ' in terms of ϵ. We did this twice in the proof of A$\Rightarrow$B above, using two different statements in the role of H. The ϵ' yields a δ', from which we construct the desired δ.

The equivalence of the two notions of limit implies that limits of function values have essentially the same properties as limits of sequences. Let $\square$ denote a binary operation on $\mathbb{R}$ (Definition 7.21); examples include addition, subtraction, multiplication, and division (excluding division by zero). Given functions f and g, we define $f \square g$ by setting $(f \square g)(x) = f(x) \square g(x)$ (as in Definition 1.25).

15.9. Lemma. Let $\square$ be a binary operation such that $b_n \square c_n \to L \square M$ whenever $b_n \to L$ and $c_n \to M$. If f, g are functions such that $\lim_{x \to a} f(x) = L$ and $\lim_{x \to a} g(x) = M$, then $\lim_{x \to a}(f \square g)(x) = L \square M$.

Proof: By definition, $(f \square g)(x_n) = f(x_n) \square g(x_n)$. If $x_n \to a$, then Theorem 15.8 implies that $f(x_n) \to L$ and $g(x_n) = M$. The hypothesis on $\square$ then yields $(f \square g)(x_n) \to L \square M$. By Theorem 15.8, $\lim_{x \to a}(f \square g)(x) = L \square M$. ∎

The definition of "$\lim_{x \to a} f(x)$" omits the situation $|x - a| = 0$. If the value of f at a is the same as the limit, then the graph of f has no "gap" at $x = a$, and we think of f as being "continuous" at a:

15.10. Definition. A function f defined on an interval containing a is **continuous** at a if $\lim_{x \to a} f(x) = f(a)$. Equivalently, f is continuous at a if for every $\epsilon > 0$ there is a $\delta > 0$ such that $|x - a| < \delta$ implies $|f(x) - f(a)| < \epsilon$.

A function is **continuous on an open interval** (c, d) if it is continuous at every point of (c, d). A function is **continuous on a**

closed interval $[c, d]$ if it is continuous on (c, d) and for every sequence $\langle x \rangle$ of numbers in $[c, d]$, $x_n \to c$ implies $f(x_n) \to f(c)$ and $x_n \to d$ implies $f(x_n) \to f(d)$.

15.11. Remark. *Interpretations of the definition of continuity.*

1) In the language of neighborhoods, Definition 15.10 says that f is continuous at a if and only if for every neighborhood T of $f(a)$ there is some neighborhood S of a such that $f(S) \subseteq T$.

2) By Theorem 15.8, f is continuous at a if and only if $f(x_n) \to f(a)$ whenever $x_n \to a$.

3) To prove that f is continuous at a, sometimes we check two statements separately: $\lim_{x \to a} f(x)$ exists, and $f(a)$ equals this limit.

4) The definition of continuity on a closed interval looks complicated, but it is easy to understand. The requirement for points in the open interval is unchanged. The required condition at each endpoint considers only sequences converging from within the interval.

5) We say "f is continuous" or "f is a continuous function" when the domain of f is $\mathbb{R}$ or an unspecified interval, and f is continuous on it. ∎

15.12. Corollary. If f and g are continuous at x, then $f + g$ and fg are continuous at x. If f is continuous and $f(x) \neq 0$, then $1/f$ is continuous at x. Every polynomial is continuous on $\mathbb{R}$. The ratio of two polynomials is continuous wherever the denominator is nonzero.

Proof: The first two statements follow directly from Lemma 15.9. We leave the proofs for polynomials to Exercise 20. ∎

We use the squeeze theorem for sequences (Theorem 14.6) to prove a sufficient condition for continuity at a point.

15.13. Proposition. (Squeeze Theorem for Continuity) Suppose that $A(x) \leq f(x) \leq C(x)$ for all x in an interval I containing α. If A and C are continuous at α and $A(\alpha) = C(\alpha)$, then f is continuous at α.

Proof: Let $\langle x \rangle$ be a sequence in I converging to α. By the sequential version of continuity, the sequences $A(x_n)$ and $C(x_n)$ converge to $A(\alpha) = L = C(\alpha)$. By the squeeze theorem, $f(x_n)$ also converges to L. This holds for every such sequence $\langle x \rangle$, so f has sequential limit L at α. Furthermore, $A(\alpha) = C(\alpha)$ implies $f(\alpha) = L$, and hence f is continuous at α. ∎

15.14. Example. If $|f(x)| \leq m\,|x|$ for some positive constant m and all x, then f is continuous at 0. Here $A(x) = -mx$ and $C(x) = mx$. ∎

When a function f is defined in a neighborhood of a, there are two ways for continuity to fail. One is that $\lim_{x \to a} f(x)$ exists but does not equal $f(a)$. This is called a "removable singularity"; by changing the definition of $f(a)$, we could make f continuous at a. The second type of failure is that $\lim_{x \to a} f(x)$ does not exist at all.

15.15. Example. *Failure of continuity.* Consider the functions defined by $f(x) = 1/x$, $g(x) = \sin(1/x)$, and $h(x) = \text{sign}(x)$ for $x \neq 0$. As x tends to 0, the first is unbounded, the second oscillates wildly, and the third has a "jump discontinuity". All three are discontinuous at 0.   ∎

The sequential version of limit yields a simple proof that the composition of continuous functions is continuous.

15.16. Theorem. (Continuity of composite functions) If f is continuous at x and g is continuous at $f(x)$, then the composite function $h = g \circ f$ is continuous at x.

Proof: It suffices to prove that if $\langle x \rangle$ is an arbitrary sequence converging to x, then the sequence defined by $z_n = h(x_n)$ converges to $h(x)$. Because f is continuous at x, the sequence defined by $y_n = f(x_n)$ converges to $f(x)$. Because $y_n \to f(x)$ and g is continuous at $f(x)$, we conclude $z_n = h(x_n) = g(y_n) \to g(f(x))$.   ∎

APPLICATIONS OF CONTINUITY

The epsilons and deltas in the definition of continuity hold more than theoretical interest. The question of how small δ must be to keep error from exceeding ϵ can be a question of engineering.

15.17. Example. *Constructing a rectangle.* Suppose we want to enclose a rectangular area of 32 square feet. We are given a wooden board 12 feet long and 1 foot wide. We cut the board into lengths of roughly 4 feet and 8 feet, split each piece lengthwise into two pieces 6 inches wide, and assemble the rectangle. How close must we make the board lengths to 4 and 8 in order to make the area within ϵ of 32?

Suppose that the long pieces have length $8 + x$. Ignoring losses due to sawdust, the short pieces have length $4 - x$. The resulting area is $(4 - x)(8 + x) = 32 - 4x - x^2$, and we want to choose x so that $\left| 4x + x^2 \right| < \epsilon$. For $0 < \epsilon < 4$, this requires $-2 + 2\sqrt{1 - \epsilon/4} < x < -2 + 2\sqrt{1 + \epsilon/4}$ (Exercise 15). When $\epsilon = 1$, the requirement is $-.268 < x < .236$ (asymmetric!), so we chose $\delta \leq .236$. To keep the area within one square foot of 32 square feet, we must cut the boards to within .236 feet (about 3 inches) of the desired lengths. Note that the desired inequalities on the error also hold in an interval around $x = -4$, when our cut is so bad that the short and long sides have switched roles.   ∎

Careful use of the definition of continuity enables us to prove several important statements suggested by geometric intuition. One such statement is the Intermediate Value Theorem (Theorem 15.19), suggested by the illustration below. When f is continuous on $[a, b]$ and $f(a) < 0 < f(b)$, the graph of f must cross the horizontal axis between a and b, and this provides a solution to the equation $f(x) = 0$.

The validity of conclusions drawn from geometric reasoning may depend on aspects of the Completeness Axiom for $\mathbb{R}$ that are not evident in the picture. Consider what happens when the domain is $\mathbb{Q}$. Let $f: \mathbb{Q} \to \mathbb{R}$ be defined by $f(x) = x^2 - 2$, as illustrated below. There is no x in the domain of f such that $f(x) = 0$. The graph looks the same whether the domain is $\mathbb{Q}$ or $\mathbb{R}$, but the Intermediate Value Theorem fails even for polynomials defined on $\mathbb{Q}$.

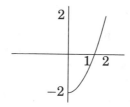

15.18. Lemma. If f is continuous on a neighborhood of a and $f(a) \neq 0$, then there exists some $\delta > 0$ such that $|x - a| < \delta$ implies that $f(x)$ is nonzero and has the same sign as $f(a)$.

Proof: Suppose that $f(a) \neq 0$. Setting $\epsilon = |f(a)|$ in the definition of continuity, we know there exists $\delta > 0$ such that $|x - a| < \delta$ implies $|f(x) - f(a)| < |f(a)|$. Hence for every x between $a - \delta$ and $a + \delta$, the distance from $f(a)$ to $f(x)$ is less than the distance from $f(a)$ to 0. This implies that $f(x)$ has the same sign as $f(a)$. ∎

We could also present this proof using the contrapositive. If $f(x)$ and $f(a)$ have opposite signs, then the vertical distance $|f(x) - f(a)|$ between them is $|f(x)| + |f(a)|$, which exceeds $|f(a)|$.

15.19. Theorem. (Intermediate Value Theorem) If f is continuous on $[a, b]$, and $f(a) < y < f(b)$, then there exists an $x \in (a, b)$ such that $f(x) = y$. The conclusion also holds when $f(a) > y > f(b)$.

Proof: Let $S = \{x \in [a, b]: f(t) < y \text{ for all } t \in [a, x]\}$. Since $a \in S$, we know that S is non-empty. Also, b is an upper bound for S. By the Completeness Axiom, we know that S has a least upper bound α. We claim that $f(\alpha) = y$.

If $f(\alpha) \neq y$, then by Lemma 15.18 there exists δ such that $f(x) - y$ has the same sign as $f(\alpha) - y$ when x is within δ of α. If $f(\alpha) < y$, then $f(x) - y$ is negative when $\alpha \leq x < \alpha + \delta$, and α is not an upper bound for S. If $f(\alpha) > y$, then $f(x) - y$ is positive when $\alpha - \delta < x \leq \alpha$, and α is not the least upper bound for S. Both cases are impossible, and we conclude that $f(\alpha) = y$.

For the case $f(a) > y > f(b)$, see Exercise 11.   ∎

15.20. Example. Let $f(x) = x^5 - 12x - 13$. Being a polynomial, f is continuous. Since $f(2) = -5$ and $f(2.8) = 125.5$, the Intermediate Value Theorem implies that there exists $x \in (2, 2.8)$ such that $f(x) = 0$.

To obtain a better approximation, consider $f(2.4) = 37.8$. This implies that there is a solution between 2 and 2.4. Continuing this process, $f(2.2) = 12.1 > 0$, $f(2.1) = 2.6 > 0$, but $f(2.05) = -1.39 < 0$. Therefore, there is a solution between 2.05 and 2.1. We can continue this *bisection* process to approximate the solution as accurately as desired. Additional steps lead to 2.067916, accurate to six decimal places.   ∎

The method of bisection provides an alternative proof of the Intermediate Value Theorem. Suppose that f is continuous and that $f(a_0)$ and $f(c_0)$ have opposite signs. By successively replacing one of $\{a_n, c_n\}$ with their average (unless we find an exact solution), we create two bounded monotone sequences $\langle a \rangle$ and $\langle c \rangle$ such that $f(a_n) < 0$, $f(c_n) > 0$, and $\lim a_n = L = \lim c_n$. Since f is sequentially continuous, we have $\lim f(a_n) = f(L) = \lim f(c_n)$. Since $f(a_n) < 0$, we have $f(L) = \lim f(a_n) \leq 0$. Since $f(c_n) > 0$, we have $f(L) = \lim f(c_n) \geq 0$. Therefore $f(L) = 0$. The method converges slowly; in Chapter 16 we will discuss a faster method that sometimes works.

We next consider a continuous function $f : [0, 1] \to [0, 1]$. Graphing f suggests that f has a fixed point (a point where $f(x) = x$). The Intermediate Value Theorem provides the proof.

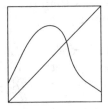

15.21. Corollary. If $f : [0, 1] \to [0, 1]$ is continuous, then f has a fixed point.

Proof: The function f has a fixed point at x^* if and only if the function g defined by $g(x) = x - f(x)$ is 0 at x^*. If f is continuous, then g is continuous, since it is the difference of continuous functions (see Corollary 15.12). Since $0 \leq f(x) \leq 1$ for all x, we have $g(0) \leq 0$ and $g(1) \geq 0$. If neither 0 nor 1 is a fixed point, then we can apply the Intermediate Value Theorem to conclude that there is a number $x^* \in (0, 1)$ such that $g(x^*) = 0$. Now x^* is the desired fixed point for f.   ∎

15.22. Solution. *The Antipodal Point Problem.* We want to prove that some pair of opposite points on a circular wire have the same temperature. When the wire has circumference c, we can represent the temperature as a continuous function f on the interval $[0, c]$ with $f(0) = f(c)$. We extend the domain of f to $\mathbb{R}$ by setting $f(b) = f(a)$ whenever $b - a$ is an integer multiple of c. The temperature at the point opposite x is then $f(x + c/2)$.

We seek an x^* such that $f(x^*) = f(x^* + c/2)$. Consider the function g defined by $g(x) = f(x) - f(x + c/2)$. If g is always zero, then the temperature is constant, and our conclusion holds. Otherwise, since $g(x + c/2) = -g(x)$, the function g attains both positive and negative values. Since g is a difference of continuous functions, g is continuous. Applying the Intermediate Value Theorem to the interval between x and $x + c/2$ yields a number x^* such that $g(x^*) = 0$, and hence $f(x^*) = f(x^* + c/2)$. ∎

15.23. Solution. *The Jewel Thieves Problem.* Consider a circular necklace with $2k$ diamonds and $2l$ rubies in some order. We must find $k + l$ consecutive jewels consisting of k diamonds and l rubies. This is a discrete problem whose solution parallels that of the Antipodal Point Problem.

We consider cuts that capture $k + l$ jewels for the first thief. When we shift the position of the cut counterclockwise by one, we drop one jewel and gain another. This could leave the number of diamonds unchanged or change it by one. Since we always capture $k + l$ jewels, getting k diamonds ensures getting l rubies; thus we focus only on the number of diamonds. Let $f(i)$ be the number of diamonds among the $k + l$ jewels starting with the ith jewel. We can extend this so that $f(i + 2k + 2l) = f(i)$.

Moving the starting point from i to $i + k + l$ transforms the set of beads captured for the first thief into the complementary set. If the first set has too many diamonds, then the second has too few. Thus $f(i) - k$ and $f(i + k + l) - k$ have opposite signs. Since f is integer-valued and its value changes by at most one when its argument changes by one, $f(i) - k$ cannot change sign without being 0 somewhere along the way. This is a discrete version of the Intermediate Value Theorem. ∎

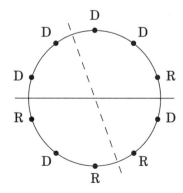

CONTINUITY AND CLOSED INTERVALS

We next show that a continuous function defined on a closed and bounded interval is bounded (that is, its image is a bounded set).

15.24. Theorem. If f is continuous on $[a, b]$, then f is bounded.

Proof: Let f be continuous on $[a, b]$. If f is not bounded on $[a, b]$, then for every n there is an x_n in $[a, b]$ such that $|f(x_n)| > n$. The sequence $\langle x \rangle$ is bounded, since $a \le x_n \le b$. By the Bolzano-Weierstrass Theorem, $\langle x \rangle$ has a convergent subsequence $\{x_{n_k}: k \in \mathbb{N}\}$. Let c be the limit of this subsequence; note that $a \le x_{n_k} \le b$ implies $a \le c \le b$.

Since f is continuous at c and $x_{n_k} \to c$, the sequential version of continuity implies $\lim_{k \to \infty} f(x_{n_k}) = f(c)$. Hence $\{f(x_{n_k}): k \in \mathbb{N}\}$ is a convergent sequence. On the other hand, our choice of $\langle x \rangle$ implies $|f(x_{n_k})| > n_k$, which means $\{|f(x_{n_k})|\}$ is unbounded and $f(x_{n_k})$ does not converge. The contradiction implies that f is bounded on $[a, b]$. ∎

15.25. Example. *The importance of closed intervals.* The function defined by $f(x) = 1/x$ is continuous on the open interval $(0, 1)$, but it is not bounded on this interval. Where does the proof fail? ∎

Suppose that f is continuous on $[a, b]$. We now know that f takes on a bounded set S of values, which by the Completeness Axiom has a supremum and infimum. The next theorem asserts that f attains its supremum and its infimum; thus f has a maximum and a minimum value.

15.26. Theorem. (The Maximum-Minimum Theorem) If f is a continuous function on $[a, b]$, then there exist $c_1, c_2 \in [a, b]$ such that $f(c_1) \le f(x) \le f(c_2)$ for all $x \in [a, b]$.

Proof: By Theorem 15.24, the set $S = \{f(x): a \le x \le b\}$ is bounded, so we can set $\alpha = \inf(S)$ and $\beta = \sup(S)$. We prove that $\beta = f(x)$ for some $x \in [a, b]$. The definition of supremum guarantees a sequence $\langle y \rangle$ in S such that $y_n \to \beta$ (see Proposition 13.15). Since y_n belongs to the set S of images of f on the interval $[a, b]$, there is a number $x_n \in [a, b]$ such that $f(x_n) = y_n$. Since $a \le x_n \le b$ for each n, the sequence $\langle x \rangle$ is bounded, and the Bolzano-Weierstrass Theorem guarantees a convergent subsequence $\{x_{n_k}: k \in \mathbb{N}\}$. Let $c = \lim_{k \to \infty} x_{n_k}$; since $a \le x_n \le b$, we also have $a \le c \le b$.

We claim that $f(c) = \beta$. Since $\lim_{k \to \infty} x_{n_k} = c$ and f is continuous at c, we have $\lim_{k \to \infty} f(x_{n_k}) = f(c)$. This sequence of values is a subsequence of $\langle y \rangle$, and $y_n \to \beta$. Every subsequence of a convergent sequence converges to the same limit, so the value $f(c)$ to which $f(x_{n_k})$ converges must also be β.

A similar argument establishes the result for the minimum, or we can apply the statement about the maximum to the function $-f$. ∎

15.27. Example. *The importance of bounded intervals.* The function defined by $f(x) = 1/(1 + x^2)$ is bounded for $x \geq 0$, since its values are between 0 and 1, but it does not attain its infimum, which is 0. ∎

We next introduce uniform continuity, a property stronger than continuity. For functions on closed and bounded intervals, the properties are equivalent, a fact we apply in Chapter 17.

15.28. Definition. A function f is **uniformly continuous** on an interval I if for every $\epsilon > 0$, there exists a δ such that $y, x \in I$ and $|y - x| < \delta$ together imply $|f(y) - f(x)| < \epsilon$.

This property is harder to satisfy than continuity at each point because we require more of δ. Instead of merely guaranteeing that $f(y)$ is within ϵ of $f(x)$ whenever y is within some δ of a specific x, we want the same δ to work for every $x \in I$. When f is merely continuous at each x in I, the δ that we choose to make $|y - x| < \delta$ imply $|f(y) - f(x)| < \epsilon$ can depend on *both* ϵ and x. The strengthening of the definition is in the order of the quantifiers (compare with Example 2.11):

pointwise: $(\forall \epsilon > 0)(\forall x \in I)(\exists \delta > 0)(|y - x| < \delta \Rightarrow |f(y) - f(x)| < \epsilon)$

uniform: $(\forall \epsilon > 0)(\exists \delta > 0)(\forall x \in I)(|y - x| < \delta \Rightarrow |f(y) - f(x)| < \epsilon)$

15.29. Example. *Uniform continuity versus continuity at each point.* The function $f(x) = 1/x$ is continuous on the open interval $(0, 1)$, but it is not uniformly continuous on $(0, 1)$. If $x < y$, then

$$|f(x) - f(y)| = (1/x) - (1/y) = (y - x)/xy < (y - x)/x^2.$$

If $y - x$ is very small, then $|f(x) - f(y)|$ is very close to $(y - x)/x^2$ (by the continuity of f). To make $|f(x) - f(y)| < \epsilon$, we need δ to be smaller than $x^2 \epsilon$; we choose $\delta = \epsilon x^2/2$. We must choose a smaller δ as x becomes smaller. No single choice of δ works throughout the interval. ∎

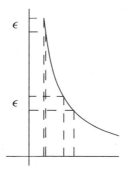

The next result helps illuminate uniform continuity.

15.30. Proposition. If f is uniformly continuous on (a, b) and $\langle x \rangle$ is a sequence in (a, b) converging to L, then $f(x_n)$ converges.

Proof: We show $f(x_n)$ is a Cauchy sequence. Given $\epsilon > 0$, uniform continuity of f yields $\delta > 0$ such that $|x_n - x_m| < \delta$ implies $|f(x_n) - f(x_m)| < \epsilon$. Since convergence makes $\langle x \rangle$ a Cauchy sequence, given $\delta > 0$ there exists $N \in \mathbb{N}$ such that $n, m \geq N$ implies $|x_n - x_m| < \delta$. With these choices, $n, m \geq N$ implies $|f(x_n) - f(x_m)| < \epsilon$. Hence $f(x_n)$ converges. ∎

15.31. Example. Continuity of f is not enough to yield the conclusion in Proposition 15.30. Let $x_n = 1/n$ and $f(x) = 1/x$. Now f is continuous on $(0, 1)$ and $x_n \to 0$. However, $f(x_n) = n$, which does not converge. ∎

15.32. Theorem. If f is continuous on $[a, b]$, then f is uniformly continuous on $[a, b]$.

Proof: We prove the contrapositive of the claim. Thus we assume that f is not uniformly continuous. We use the Bolzano-Weierstrass Theorem and the sequential version of continuity to conclude that f is not continuous.

Negating the definition of uniform continuity of f on $[a, b]$ yields

(∗): there exists a number $\epsilon^* > 0$ such that for every $\delta > 0$, there is a pair $x, y \in [a, b]$ satisfying $|y - x| < \delta$ and $|f(y) - f(x)| \geq \epsilon^*$.

For this ϵ^* in (∗), let (y_n, x_n) be the pair that results when $\delta = 1/n$. Thus $|y_n - x_n| < 1/n$ and $|f(y_n) - f(x_n)| \geq \epsilon^*$. Note that $a \leq x_n \leq b$ for all n, and hence $\langle x \rangle$ is bounded.

By the Bolzano-Weierstrass Theorem, $\langle x \rangle$ has a convergent subsequence $\{x_{n_k}: k \in \mathbb{N}\}$ converging to some $c \in [a, b]$. Since $y_n - x_n \to 0$, we also have $y_{n_k} \to c$. We now have sequences $\{x_{n_k}\}$ and $\{y_{n_k}\}$ converging to c whose images remain far apart (since $\left|f(x_{n_k}) - f(y_{n_k})\right| \geq \epsilon^*$). Thus f is not (sequentially) continuous. ∎

EXERCISES

For Exercises 1–10, determine whether the statement is true or false. If true, provide a proof; if false, provide a counterexample.

15.1. (−) There is a continuous $f: \mathbb{R} \to \mathbb{R}$ such that $f(x) = (-1)^k$ for $k \in \mathbb{Z}$.

15.2. (−) There is a continuous $f: \mathbb{R} \to \mathbb{R}$ such that $f(x) = 0$ if and only if $x \in \mathbb{Z}$.

15.3. (−) If f is continuous on $\mathbb{R}$, and $f(x) = 0$ for $x \in \mathbb{Q}$, then f is constant.

15.4. (−) There exists $x > 1$ such that $\frac{x^2+5}{3+x^7} = 1$.

15.5. (−) The function f defined by $f(x) = |x|^3$ is continuous at all $x \in \mathbb{R}$.

15.6. If $f + g$ and fg are continuous, then f and g are continuous.

15.7. Let f, g, h be continuous on the interval $[a, b]$. If $f(a) < g(a) < h(a)$ and $f(b) > g(b) > h(b)$, then there exists $c \in [a, b]$ such that $f(c) = g(c) = h(c)$.

15.8. If $|f|$ is continuous, then f is continuous.

15.9. (!) Suppose that f and g are continuous on $\mathbb{R}$.
a) If $f(x) > g(x)$ for all $x > 0$, then $f(0) > g(0)$.
b) f/g is continuous at all $x \in \mathbb{R}$.
c) If $0 < f(x) < g(x)$ for all x, then there is some $x \in \mathbb{R}$ such that $f(x)/g(x)$ is the maximum value of f/g.
d) If $f(x) \le g(x)$ for all x, and g is never 0, then f/g is bounded.
e) If $f(x)$ is rational for each x, then f is constant.

15.10. a) If f is continuous on $\mathbb{R}$, then f is bounded.
b) If f is continuous on $[0, 1]$, then f is bounded.
c) There is a function from $\mathbb{R}$ to $\mathbb{R}$ that is continuous at exactly one point.
d) If f is continuous on $\mathbb{R}$ and is bounded, then f attains its supremum.

$\bullet \qquad \bullet \qquad \bullet \qquad \bullet \qquad \bullet$

15.11. ($-$) Prove that the Intermediate Value Theorem remains true when the hypothesis $f(a) < y < f(b)$ is replaced with $f(a) > y > f(b)$.

15.12. Construct a function f such that there are sequences $\langle a \rangle$ and $\langle b \rangle$ converging to 0 such that $f(a_n)$ converges but $f(b_n)$ is unbounded. Does there exist such a function f that is continuous at 0?

15.13. Prove that the absolute value function is continuous (use "ϵ-δ").

15.14. Let f be defined by $f(x) = 1/x$, and let $a = .5$. How large can δ be if it is required that $f(x)$ is within $.1$ of $f(a)$ when x is within δ of a?

15.15. (!) When $f(x) = x^2 + 4x$, $\lim_{x \to 0} f(x) = 0$. How small must δ be so that $|x| < \delta$ implies that $|f(x)| < \epsilon$? Express δ as a function of ϵ. Assume that $\epsilon < 4$.

15.16. ($-$) Suppose that $\lim_{x \to 0} f(x) = 0$. Prove that for all $n \in \mathbb{N}$, there exists x_n such that $|f(x_n)| < 1/n$.

15.17. Let $f(a, n) = (1 + a)^n$, where a and n are positive.
a) For constant a, how does $f(a, n)$ behave as $n \to \infty$? For constant n, how does $f(a, n)$ behave as $a \to 0$?
b) Let L be a real number with $L \ge 1$. Prove that there exists a sequence $\langle a \rangle$ such that $a_n \to 0$ and $f(a_n, n) \to L$ as $n \to \infty$. In other words, depending on the rate at which a_n approaches 0, f may approach any value.

15.18. (!) *Often discontinuous functions.*
a) Let $f : \mathbb{R} \to \mathbb{R}$ be the function defined by $f(x) = 0$ if $x \in \mathbb{Q}$ and $f(x) = 1$ if $x \notin \mathbb{Q}$. Prove that f is discontinuous at every real number.
b) Let $g : \mathbb{R} \to \mathbb{R}$ be the function defined by $g(x) = 0$ if $x \in \mathbb{Q}$ and $g(x) = cx$ if $x \notin \mathbb{Q}$, where c is a nonzero real number. Prove that g is continuous at 0 and discontinuous at every other real number.

15.19. (!) Give two proofs that if $|f(x) - f(a)| \le c|x - a|$ for some positive constant c and all x, then f is continuous at a. One proof should use the ϵ, δ definition, and the other should apply general results about continuity.

15.20. Prove that every polynomial is continuous on $\mathbb{R}$. Prove that the ratio of two polynomials is continuous at every point where the denominator is nonzero.

15.21. ($-$) Prove that there exists $x \in [1, 2]$ such that $x^5 + 2x + 5 = x^4 + 10$.

15.22. Let f and g be continuous on $[a, b]$. Suppose that $f(a) > g(a)$ and that $f(b) < g(b)$. Prove that there exists $c \in [a, b]$ such that $f(c) = g(c)$.

15.23. (!) Let f and g be continuous on $[a, b]$. Suppose that $f(a) = (1/2)g(a)$ and that $f(b) = 2g(b)$. Show by example that there need not exist $c \in [a, b]$ such that $f(c) = g(c)$. Prove that such a c must exist if $g(x) \geq 0$ for $x \in [a, b]$.

15.24. (!) Prove that every polynomial of odd degree has at least one real zero.

15.25. Given a positive real number ϵ, prove that there is a positive real number c (depending on ϵ, but not on x or y) such that $|xy| \leq \epsilon x^2 + cy^2$ for all $x, y \in \mathbb{R}$.

15.26. Write out the proof that a continuous function on $[a, b]$ has a lower bound, using the Bolzano-Weierstrass Theorem and the definition of continuity.

15.27. (−) Prove that f is continuous if and only if $-f$ is continuous. Use this to show that if f is continuous on $[a, b]$ and $f(a) > y > f(b)$, then there exists $c \in (a, b)$ such that $f(c) = y$.

15.28. Let $P = \{x \in \mathbb{R} : x > 0\}$. Let $f : P \to P$ be continuous and injective.
a) Prove that the inverse of f, defined on the image of f, is continuous.
b) Let $x_1 = c$ for some $c \in P$, and let $x_{n+1} = f(\sum_{j=1}^{n} x_j)$ for $n \geq 1$. Prove that if $\langle x \rangle$ converges, then its limit is 0. (Hint: Prove that $\sum_{j=1}^{n} x_j$ converges.)

15.29. (!) Let $f_n(x) = (x^n + 1)^{1/n}$, defined on $\{x \in \mathbb{R} : x \geq 0\}$. Sketch the graphs of f_1 and f_2. For $x > 0$, compute $g(x) = \lim_{n \to \infty} f_n(x)$. Graph g.

15.30. Use the method of bisection to compute $\sqrt{10}$ correct to four decimal places. Use this method to solve $x^7 - 5x^3 + 10 = 0$ correct to two decimal places.

15.31. Find a counterexample to the following statement: If f is a real-valued function of two variables and all the limits described below exist, then

$$\lim_{y \to 0} \lim_{x \to 0} f(x, y) = \lim_{x \to 0} \lim_{y \to 0} f(x, y).$$

(When taking a limit in one variable, other variables are treated as constants.)

15.32. There are containers of gas along a circular track. The total amount of gas is exactly enough to fuel a car once around the track. Prove that there is some starting place from which the car can complete the trip without running out of gas.

15.33. (+) Let n be a positive integer, and suppose f is continuous on $[0, 1]$ and $f(0) = f(1)$. Prove that the graph of f has a horizontal chord of length $1/n$. In other words, prove that there exists $x \in [0, (n-1)/n]$ such that $f(x + 1/n) = f(x)$. (Comment: Surprisingly, for any α that is not the reciprocal of an integer, we can construct such a function f that has no horizontal chord of length α.)

15.34. (!) Let f be continuous on an interval I. For each $a \in I$ and $\epsilon > 0$, let $m(a, \epsilon) = \sup(\{\delta : |x - a| < \delta$ implies $|f(x) - f(a)| < \epsilon\})$. What property must $m(a, \epsilon)$ satisfy for f to be uniformly continuous on I?

15.35. *Continuous functions with constant multiplicity.*
a) Construct a continuous function $f : \mathbb{R} \to \mathbb{R}$ such that every real number occurs as the image of exactly three numbers.
b) (+) Let $f : \mathbb{R} \to \mathbb{R}$ be continuous. Suppose that each $z \in \mathbb{R}$ occurs as the image of exactly k numbers. Prove that k must be odd. (Hint: Try to draw the graph of such a function with k even. For k even, use the Intermediate Value Theorem and the Maximum-Minimum Theorem to complete a proof by contradiction.)

Chapter 16

Differentiation

The famous calculus text by George Thomas begins, "Calculus is the mathematics of change and motion."[†] Studying how physical quantities change with time leads to the notions of continuity and differentiability. Our preparatory work with limits enables us to prove the basic theorems of differential calculus and to fully appreciate them.

16.1. Problem. *Linear approximations.* A square with area 64 has sides of length 8. By continuity, the sides of a square with area 65 have length near 8. Is there an easy way to improve this estimate for $\sqrt{65}$? ∎

16.2. Problem. *Solving equations iteratively.* In Example 15.20, we used the bisection method to approximate a zero of a function. Is there a faster algorithm? Using the tangent to the graph, we can aim toward a better guess. When does this process converge to a solution? ∎

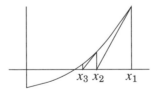

16.3. Problem. *Circle of Curvature.* Suppose a particle moves along a smooth curve in the plane. At each instant, its motion is closely approximated by motion along a circle. The center and radius of the best approximating circle change as the particle moves. The reciprocal of the radius measures the "curvature" of the motion. How can it be found? ∎

16.4. Problem. Does $\lim_{n\to\infty} \lim_{x\to 1} f_n(x) = \lim_{x\to 1} \lim_{n\to\infty} f_n(x)$? ∎

[†]G. Thomas, *Calculus and Analytic Geometry*, Addison-Wesley (Reading, 1968), 1.

THE DERIVATIVE

The derivative is the precise formulation of the notion of rate of change. The ratio $(f(b) - f(a))/(b - a)$ is the average rate of change in $f(x)$ as x changes from a to b. As b approaches a, this ratio approaches the instantaneous rate of change at a. A familiar example is the relationship between an odometer and a speedometer. The odometer measures distance traveled; the speedometer measures the speed, which is the instantaneous rate of change of distance traveled. Thus if we have traveled $f(a)$ miles at time a, and $f(b)$ miles at time b, then $(f(b) - f(a))/(b - a)$ represents our average speed in the time interval from a to b. The limit of this ratio as b approaches a equals the speed at time a.

The derivative also has a simple geometric interpretation. Given f, the slope $m_{a,b}$ of the line through $(a, f(a))$ and $(b, f(b))$ is $\frac{f(b)-f(a)}{b-a}$. As $b \to a$, this slope $m_{a,b}$ approaches the slope of the graph of f at $(a, f(a))$.

16.5. Definition. A function f is **differentiable** at x if $\lim_{h\to 0} \frac{f(x+h)-f(x)}{h}$ exists. When this limit exists, its value is the **derivative** of f at x. The derivative at x is written as $f'(x)$ or $\frac{df}{dx}(x)$. The ratio $\frac{f(x+h)-f(x)}{h}$ is the **difference quotient**.

16.6. Definition. A **linear approximation** to f at x is a linear function whose graph passes through $(x, f(x))$. An **error function** is a function e defined in a neighborhood of 0 such that $\lim_{h\to 0} e(h)/h = 0$.

The derivative of f at x equals the slope of the line tangent to the graph of f at $(x, f(x))$. If we move from $(x, f(x))$ along this line to $(x + h, y)$ for small h, then we expect y to be close to $f(x + h)$. This motivates the alternative definition given below. The derivative of a function f at x is the slope of the unique linear approximation to f at x such that the difference $f(x + h) - f(x) - f'(x)h$ defines an error function.

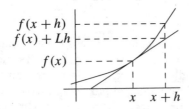

16.7. Definition. The function f is **differentiable** at $x \in \mathbb{R}$ if f is defined in a neighborhood of x and there exists $L \in \mathbb{R}$ such that the function e_x defined by $f(x + h) = f(x) + Lh + e_x(h)$ is an error function. If so, L is called the **derivative** of f at x, and we write $L = f'(x)$.

16.8. Example. If $f(x) = mx + b$, then $f(x + h) = f(x) + mh$. The error function e_x is identically zero, and $f'(x) = m$ for all x. ∎

16.9. Lemma. The two definitions of the derivative are equivalent.
Proof: For $h \neq 0$, the equation $f(x + h) - f(x) - Lh = e(h)$ is equivalent to $\frac{f(x+h)-f(x)}{h} = L + e(h)/h$. Thus $\lim_{h \to 0} e(h)/h = 0$ if and only if $\lim_{h \to 0} \frac{f(x+h)-f(x)}{h} = L$. ∎

In Definition 16.7, the error function e_x and value L depend on x. We write the derivative at x as $f'(x)$ to emphasize its dependence on x.

16.10. Example. *If $f(x) = x^n$ for $n \in \mathbb{N}$, then $f'(x) = nx^{n-1}$.* Using the Binomial Theorem, $(x + h)^n = \sum_{k=0}^{n} \binom{n}{k} x^{n-k} h^k$. We compute

$$\frac{f(x + h) - f(x)}{h} = \sum_{k=1}^{n} \binom{n}{k} x^{n-k} h^{k-1} = nx^{n-1} + hg_x(h),$$

where $g_x(h)$ is a polynomial in h. As $h \to 0$, $\frac{f(x+h)-f(x)}{h} \to nx^{n-1}$.

We can perform the same computation using linear approximations. We expand $f(x + h) = (x + h)^n$ using the Binomial Theorem as before. We obtain $(x + h)^n = x^n + nx^{n-1}h + e_x(h)$, where $e_x(h)$ is a polynomial in h that is divisible by h^2. Hence $e_x(h)/h \to 0$. Since $e_x(h)$ equals $f(x + h) - f(x) - nx^{n-1}h$, we obtain $f'(x) = nx^{n-1}$. ∎

16.11. Example. *If $f(x) = \sqrt{x}$ for $x > 0$, then $f'(x) = \frac{1}{2\sqrt{x}}$.* We compute

$$\frac{f(x+h)-f(x)}{h} = \frac{\sqrt{x+h}-\sqrt{x}}{h} = \frac{\sqrt{x+h}-\sqrt{x}}{h} \frac{\sqrt{x+h}+\sqrt{x}}{\sqrt{x+h}+\sqrt{x}} = \frac{x+h-x}{h \cdot (\sqrt{x+h}+\sqrt{x})}.$$

Simplifying and letting $h \to 0$ yields $1/(2\sqrt{x})$ for the derivative. ∎

16.12. Solution. *Linear approximation of square roots* (Problem 16.1). By Example 16.11 and Definition 16.7, $\sqrt{x + h} = \sqrt{x} + \frac{1}{2\sqrt{x}}h + e(h)$. With $x = 64$ and $h = 1$, we have $\sqrt{65} \approx 8 + \frac{1}{16}1 = 8.0625$. The actual value is 8.06226 to five decimal places. ∎

16.13. Remark. Many standard approximations use the linear approximation definition of the derivative. For example, when h is small, we approximate $1/(1 - h)$ by $1 + h$, $(1 + h)^\alpha$ by $1 + \alpha h$, $\sin h$ by h, and e^h by $1 + h$. This requires knowing the values of the derivatives at $x = 0$ of $1/(1 - x)$, $(1 + x)^\alpha$, $\sin x$, and e^x, which are $1, \alpha, 1, 1$, respectively. (We cannot compute the last two until we give definitions for $\sin x$ and e^x.) ∎

We discuss error functions in order to prove formulas for derivatives.

16.14. Lemma. Suppose that e, e_1, e_2 are error functions.

    a) $e(h) \to 0$ as $h \to 0$.

    b) The sum $e_1 + e_2$ is an error function.

    c) If $c \in \mathbb{R}$ and u is a bounded function defined on a neighborhood of
       0, then the products ce and ue are error functions.

Proof: We use the elementary properties of limits for sums and products
(Lemma 15.9). (a) We have $e(h) = h \cdot [e(h)/h] \to 0 \cdot 0 = 0$. (b) We have
$(e_1 + e_2)(h)/h = e_1(h)/h + e_2(h)/h \to 0 + 0 = 0$. (c) When $c \in \mathbb{R}$, we
have $ce(h)/h \to c \cdot 0 = 0$. For the statement about the product ue, let c be
an upper bound on $|u|$ in a neighborhood of 0. Since $0 \le |u(h)e(h)/h| \le
c\,|e(h)/h|$, the result for ue follows from the Squeeze Theorem.   ■

16.15. Lemma. Suppose that e is an error function.

    a) If $c \in \mathbb{R}$, then $\frac{1}{1+ch+e(h)}$ equals $1 - ch$ plus an error function.

    b) If $s(h) \to 0$, then the composition $e \circ s$ is an error function.

Proof: (a) Computing the difference yields

$$\frac{1}{1+ch+e(h)} - (1-ch) = \frac{1-(1-ch)(1+ch+e(h))}{1+ch+e(h)} = \frac{e(h)(ch-1)+c^2h^2}{1+ch+e(h)}.$$

We now divide by h and let $h \to 0$. Since $e(h)/h \to 0$, $1-ch \to 1$, $c^2h \to 0$,
and $1 - ch + e(h) \to 1$, the displayed espression tends to 0. This yields
the desired conclusion.

    (b) Since e is an error function, for each $\epsilon > 0$ there is a $\delta > 0$ such that
$|t| < \delta$ implies $|e(t)| \le |t|\,\epsilon$. Therefore, $|e(s(h))| \le |s(h)|\,\epsilon$ for $|s(h)| < \delta$.
Since $s(h) \to 0$, we can choose δ' such that $|h| < \delta'$ implies $|s(h)| < \delta$, and
hence $|h| < \delta'$ implies $|e(s(h))/h| < \epsilon$. This proves that $(e \circ s)(h)/h \to 0$. ■

16.16. Theorem. If f and g are differentiable at x and c is a constant,
then $f + g$, cf, $f \cdot g$, and f/g (if $g(x) \ne 0$) are differentiable at x. The
derivatives are

    a) $(f+g)'(x) = f'(x) + g'(x)$,

    b) $(cf)'(x) = c \cdot f'(x)$,

    c) $(fg)'(x) = f(x)g'(x) + f'(x)g(x)$ (product rule),

    d) $(f/g)'(x) = \frac{g(x)f'(x)-f(x)g'(x)}{[g(x)]^2}$ (quotient rule).

Proof: We obtain the derivatives as linear approximations. In these computations, x, $f(x)$, $g(x)$, $f'(x)$, $g'(x)$ do not change as h changes. In each
case the derivative is the coefficient of h.

    (a) By definition, $(f+g)(x+h) = f(x+h) + g(x+h)$. Using this and
the linear approximations for f and g, we compute

$$(f+g)(x+h) = f(x) + f'(x)h + e_1(h) + g(x) + g'(x)h + e_2(h)$$
$$= (f+g)(x) + (f'(x)+g'(x))h + (e_1+e_2)(h).$$

By Lemma 16.14b, $e_1 + e_2$ is an error function, and hence extracting the coefficient of h yields $(f + g)'(x) = f'(x) + g'(x)$.

(b) Recall that $(cf)(x + h) = c \cdot f(x + h)$. We compute

$$(cf)(x + h) = c[f(x) + f'(x)h + e(h)] = cf(x) + cf'(x)h + ce(h).$$

By Lemma 16.14c, ce is an error function, and hence $(cf)'(x) = c \cdot f'(x)$.

(c) Recall that $(fg)(x + h) = f(x + h) \cdot g(x + h)$. Again we compute

$$
\begin{aligned}
(fg)(x + h) &= [f(x) + f'(x)h + e_1(h)] \cdot [g(x) + g'(x)h + e_2(h)] \\
&= f(x)g(x) + [f'(x)g(x) + f(x)g'(x)]h + f'(x)g'(x)h^2 \\
&\quad + e_1(h)[g(x) + g'(x)h] + e_2(h)[f(x) + f'(x)h] + e_1(h)e_2(h).
\end{aligned}
$$

By Lemma 16.14c, each of the last four terms defines an error function. By Lemma 16.14b, their sum is an error function, and hence $(fg)'(x) = f'(x)g(x) + f(x)g'(x)$.

(d) The formula for $(f/g)'(x)$ follows from (c) and the case where f is identically 1 (Exercise 13). To differentiate $1/g$, we compute

$$\frac{1}{g(x + h)} = \frac{1}{g(x) + g'(x)h + e(h)} = \frac{1}{g(x)} \frac{1}{1 + \frac{g'(x)h + e(h)}{g(x)}}.$$

By Lemma 16.14c with $c = 1/g(x)$, $e(h)/g(x)$ is an error function. By Lemma 16.15a with $c = g'(x)/g(x)$, we can rewrite the second factor as $1 - g'(x)h/g(x) + e_3(x)$, where e_3 is an error function. Hence

$$\frac{1}{g(x + h)} = \frac{1}{g(x)}\left[1 - \frac{g'(x)}{g(x)}h + e_3(h)\right] = \frac{1}{g(x)} - \frac{g'(x)}{[g(x)]^2}h + e_4(h),$$

where e_4 is an error function. Therefore, $(1/g)'(x) = g'(x)/[g(x)]^2$. ∎

16.17. Corollary. A polynomial is differentiable at every point. More generally, the ratio of two polynomials is differentiable at every point where the denominator is nonzero.

Proof: Since x^n is differentiable, this follows from Theorem 16.16. ∎

Differentiability is a stronger condition than continuity.

16.18. Theorem. If f is differentiable at x, then f is continuous at x.

Proof: Using linear approximations, $f(x + h) = f(x) + f'(x)h + e(h)$, where e is an error function. Since $\lim_{h \to 0} e(h) = 0$ (Lemma 16.14a), we have $\lim_{h \to 0} f(x) + f'(x)h + e(h) = f(x)$, and hence f is continuous at x.

Second proof. $f(x + h) - f(x) = \frac{f(x+h) - f(x)}{h} \cdot h \to f'(x) \cdot 0 = 0$. ∎

16.19. Example. *Continuous but not differentiable.* The absolute value function is continuous but not differentiable at 0. The difference quotient is $\frac{|h|}{h}$, which has no limit as $h \to 0$.

More generally, if g is bounded, but not continuous at 0, then the function f defined by $f(x) = xg(x)$ for $x \neq 0$ and $f(0) = 0$ is continuous at 0 but not differentiable at 0. The difference quotient is $g(h)$, which has no limit as $h \to 0$. In Examples 16.74–16.75, we present continuous functions that are not differentiable anywhere! ∎

We next give a sufficient condition for differentiability at a point that is an analogue of the Squeeze Theorem for continuous functions (Proposition 15.13). When a function f is squeezed between differentiable functions having equal values and equal derivatives at a point, f must also have that derivative there (see Example 16.47).

16.20. Theorem. (Squeeze Theorem for Differentiability) Let A and C be functions differentiable at x, with $A(x) = C(x)$ and $A'(x) = C'(x) = L$. If $A(t) \leq f(t) \leq C(t)$ for t in a neighborhood of x, then f is differentiable at x, and $f'(x) = L$.

Proof: The hypotheses imply that $A(x) = f(x) = C(x)$. Hence $A(x + h) - A(x) \leq f(x + h) - f(x) \leq C(x + h) - C(x)$. Using error functions, this becomes $A'(x)h + e_1(h) \leq f(x + h) - f(x) \leq C'(x)h + e_2(h)$. Subtracting Lh yields $e_1(h) \leq f(x + h) - f(x) - Lh \leq e_2(h)$. Since a function squeezed between two error functions is itself an error function (Exercise 17), f is differentiable at x with derivative L. ∎

16.21. Corollary. If $|g(t) - g(x)| \leq c\,|t - x|^{1+\alpha}$ for all t, where c and α are positive constants, then g is differentiable at x, and $g'(x) = 0$.

Proof: This is the special case of Theorem 16.20 with $f(t) = g(t) - g(x)$, $A(t) = -c\,|t - x|^{1+\alpha}$, and $C(t) = c\,|t - x|^{1+\alpha}$. ∎

We next study differentiation of composite functions, obtaining a useful formula called the "chain rule". The idea is simple: if f and g are differentiable, then $g \circ f$ is differentiable, and its linear approximation is the composition of the linear approximations to g and f.

Suppose that f and g are themselves linear functions. If $f(x) = ax + b$ and $g(x) = cx + d$, then $(g \circ f)(x) = c(ax + b) + d = acx + (bc + d)$. The composition is also linear, and its derivative is the product of the derivatives of f at x and g at $f(x)$.

16.22. Theorem. (Chain Rule) If f is differentiable at x and g is differentiable at $f(x)$, then the composite function $\phi = g \circ f$ is differentiable at x, and $\phi'(x) = g'(f(x))f'(x)$.

Proof: Let $y = f(x)$, and write $L = f'(x)$ and $M = g'(y)$. Differentiability of f and g allows us to write $f(x + h) = f(x) + Lh + e_1(h)$ and $g(y + k) = g(y) + Mk + e_2(k)$, where e_1 and e_2 are error functions. To prove that ϕ is differentiable at x, we seek a real number N such that $\phi(x + h) = \phi(x) + Nh + e(h)$, where $e(h)/h \to 0$.

To do this, we evaluate $\phi(x + h)$. Letting $k = Lh + e_1(h)$, we have

$$g(f(x + h)) = g(f(x) + Lh + e_1(h)) = g(f(x) + k)$$
$$= g(f(x)) + Mk + e_2(k)$$
$$= g(f(x)) + MLh + [Me_1(h) + e_2(Lh + e_1(h))].$$

Let $e(h) = Me_1(h) + e_2(Lh + e_1(h))$. If we can prove that $e(h)/h \to 0$, then $\phi'(x)$ exists and equals $ML = g'(f(x))f'(x)$.

By Lemma 16.14, $Me_1(h)$ is an error function. Because the sum of error functions is an error function, showing that $e_2(Lh + e_1(h))/h \to 0$ will imply that e is an error function and will complete the proof.

Let $s(h) = Lh + e_1(h)$. We have $e_2(s(h)) = 0$ when $s(h) = 0$, and otherwise we write $\left|\frac{e_2(s(h))}{h}\right| = \left|\frac{e_2(s(h))}{s(h)}\right|\left|\frac{s(h)}{h}\right|$. Since $e_1(h)/h \to 0$, we have $s(h)/h \to L$. Since $s(h) \to 0$ and e_2 is an error function, we have $e_2(s(h))/s(h) \to 0$, by Lemma 16.15b. ∎

16.23. Example. Given $m, n \in \mathbb{N}$, let $f(x) = (x^n + 1)^m$. Then $f'(x) = m(x^n + 1)^{m-1}nx^{n-1}$. ∎

16.24. Proposition. If f is differentiable and strictly monotone, then f^{-1} exists and is differentiable, and $\frac{df^{-1}(y)}{dy} = \frac{1}{f'(f^{-1}(y))}$.

Proof: (Sketch). After showing that f^{-1} is differentiable, we differentiate both sides of $y = f(f^{-1}(y))$ using the chain rule. See Exercise 36. ∎

The definition of the derivative via linear approximation and our proof of the chain rule extend to functions of several variables. All the formal rules of differentiation (such as the product rule, quotient rule, etc.) follow as corollaries of the general chain rule. This approach makes Theorem 16.16, for example, trivial.

APPLICATIONS OF THE DERIVATIVE

Differential calculus provides a method for finding the maximum and minimum values of a function on an interval. We know from the Maximum-Minimum Theorem that a continuous function on a closed and bounded interval attains its supremum and its infimum, but the proof tells us neither how to compute them nor where they occur.

16.25. Definition. A **local maximum** for the function f occurs at x if $f(t) \leq f(x)$ for all t in some neighborhood of x. Similarly, a **local minimum** occurs at x if $f(t) \geq f(x)$ for all t in some neighborhood of x. A **local extremum** occurs where a local maximum or a local minimum occurs.

The next theorem gives a necessary (but not sufficient) condition for a local extremum of a differentiable function to occur at x; the derivative must be 0 at x. To find the extreme values of f on the interval $[a, b]$, we thus check only the endpoints and the places where the derivative is 0.

16.26. Theorem. If f is differentiable at x and a local extremum of f occurs at x, then $f'(x) = 0$.

Proof: We first suppose that the local extremum is a local maximum. Thus there exists $\delta > 0$ such that $|h| < \delta$ implies $f(x + h) \leq f(x)$. Since f is differentiable at x, the limit of the difference quotient $[f(x + h) - f(x)]/h$ exists as $h \to 0$. This ratio is nonnegative when $-\delta < h < 0$ and nonpositive when $0 < h < \delta$. Hence the limit L must satisfy both $L \geq 0$ and $L \leq 0$, and we conclude that $L = 0$.

If a local minimum occurs at x, then we can apply the same argument with all inequalities reversed, or we can apply the result about local maximums to the differentiable function $-f$. ∎

16.27. Example. *Maximum area in a pen.* A farmer plans to use a feet of wire fencing to form three sides of a rectangular pen against the side of a barn. The pen will stand out x feet from the wall, for some x. How should x be chosen to maximize the area of the rectangle?

The dimensions of the rectangle are x by $a - 2x$; the area is $f(x) = x(a - 2x)$, with $0 \leq x \leq a/2$. The function f has the value 0 at both endpoints. Its derivative is $a - 4x$, which is 0 at $x = a/4$. The maximum area is $a^2/8$, achieved when $x = a/4$. We can also minimize any quadratic polynomial without using calculus (see Exercise 1.28). ∎

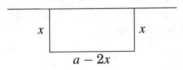

16.28. Example. *Necessary but not sufficient.* The condition $f'(x) = 0$ is necessary but not sufficient for a local extremum at x. If $f(x) = x^3$, then $f'(0) = 0$, but this f has no local extremum. ∎

A differentiable function must have a local extremum between two numbers where it has the same value. If the function has a larger value somewhere on this interval, then it has a local maximum, and otherwise it has a local minimum. In other words, "what goes up and comes down must turn around." The next theorem makes this statement precise.

16.29. Theorem. (Rolle's Theorem) If f is differentiable on (a, b), is continuous on $[a, b]$, and $f(a) = f(b)$, then there exists a $c \in (a, b)$ such that $f'(c) = 0$.

Proof: If f is constant on $[a, b]$, then, $f'(x) = 0$ for all $x \in (a, b)$; hence the conclusion holds in this case. If $f(x) \leq f(a)$ for all $x \in (a, b)$, then we can consider $-f$ instead. Hence we may assume that there exists $x \in (a, b)$ with $f(x) > f(a)$.

Since f is continuous on $[a, b]$, we know from the Maximum-Minimum Theorem that f achieves its maximum value on the interval $[a, b]$. The maximum cannot occur at a or b, since $f(a) = f(b) < f(x)$. Hence it occurs at some $c \in (a, b)$, and thus $f(c)$ must be a local maximum. The necessary condition for local extrema (Theorem 16.26) now yields $f'(c) = 0$. ∎

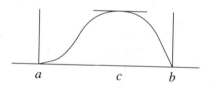

$$a \qquad\qquad c \qquad b$$

16.30. Example. *Necessity of continuity.* The hypothesis of continuity on $[a, b]$ is necessary in Rolle's Theorem. Consider the function f defined by $f(x) = x$ for $0 \leq x < 1$, but $f(1) = 0$. Then $f(0) = f(1) = 0$ and f is differentiable on $(0, 1)$, but $f'(x) = 1$ for all $x \in (0, 1)$. ∎

Rolle's Theorem leads to the Mean Value Theorem. When $a \neq b$, the equation of the line through (a, A) and (b, B) is $y = \frac{b-x}{b-a}A + \frac{x-a}{b-a}B$. The slope of this line is $m_{a,b} = \frac{f(b)-f(a)}{b-a}$ when $A = f(a)$ and $B = f(b)$.

16.31. Theorem. (Mean Value Theorem) If f is differentiable on (a, b) and continuous on $[a, b]$, then there exists $c \in (a, b)$ such that $f'(c) = \frac{f(b)-f(a)}{b-a}$.

Proof: By subtracting a linear function from f, we obtain a function to which Rolle's Theorem applies. The linear function defined by $g(x) = \frac{b-x}{b-a}f(a) + \frac{x-a}{b-a}f(b)$ satisfies $g(a) = f(a)$ and $g(b) = f(b)$. Letting $h(x) = f(x) - g(x)$, we have $h(a) = h(b) = 0$. Also h is differentiable on (a, b) and continuous on $[a, b]$, since it is the difference of two functions that have those properties. Hence Rolle's Theorem applies to h and yields some $c \in (a, b)$ such that $h'(c) = 0$. But $h'(x) = f'(x) - g'(x) = f'(x) - \frac{f(b)-f(a)}{b-a}$. Hence $f'(c) = \frac{f(b)-f(a)}{b-a}$, as desired. ∎

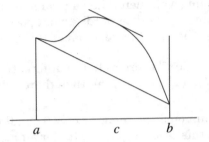

$$a \qquad c \qquad b$$

16.32. Corollary. If f is differentiable on an open interval I, and f' is 0 on I, then f is constant on I.

Proof: If there are two numbers a, b in the interval at which f has different values, then the Mean Value Theorem says that the derivative is nonzero somewhere between them. ∎

We illustrate the Mean Value Theorem with two examples.

16.33. Example. Each driver entering the Pennsylvania Turnpike receives a card noting the entry point and time and returns it upon exiting. A driver entered at Pittsburgh and exited four hours later at Philadelphia, 300 miles away. He received a speeding ticket! When a driver travels 300 miles in four hours, the Mean Value Theorem implies that the speed was 75 miles per hour at some point along the way. ∎

16.34. Example. Consider the equation $(x + y)^n = x^n + y^n$, where n is an integer greater than 1. The equation holds whenever x or y is zero; are there other solutions? The Binomial Theorem forbids solutions with x and y having the same sign; the Mean Value Theorem facilitates the analysis of the more difficult case.

For $y \neq 0$, the solutions are the zeros of $f(x) = (x + y)^n - x^n - y^n$. Obviously, $f(0) = 0$. Suppose that $f(x^*) = 0$ for some $x^* \neq 0$. By the Mean Value Theorem, $f'(c) = 0$ for some c between 0 and x^*. However, $f'(x) = n(x + y)^{n-1} - nx^{n-1}$, so $f'(c) = 0$ requires $(c + y)^{n-1} = c^{n-1}$.

This is impossible when n is even, since exponentiation by an odd power is injective. When n is odd, $f'(c) = 0$ requires $c + y = \pm c$ (Example 4.26). This holds if and only if $c = -y/2$. Indeed, f has a zero at $x = -y$ when n is odd. This is the only other zero for f; if f had a third zero, then the Mean Value Theorem would require a second zero for f'. ∎

A generalization of the Mean Value Theorem due to Cauchy allows us to compute limits of ratios.

16.35. Theorem. (Cauchy Mean Value Theorem) If f and g are differentiable on (a, b) and continuous on $[a, b]$, then there exists $c \in (a, b)$ such that $[f(b) - f(a)]g'(c) = [g(b) - g(a)]f'(c)$.

Proof: Define F on $[a, b]$ by

$$F(x) = [f(b) - f(a)] \cdot [g(x) - g(a)] - [g(b) - g(a)] \cdot [f(x) - f(a)].$$

Since F satisfies the hypotheses of Rolle's Theorem, there exists $c \in (a, b)$ such that $F'(c) = 0$. By differentiating F with respect to x and evaluating at c, we obtain the desired formula. ∎

The method for computing limits of ratios, called l'Hôpital's Rule, is named for Marquis de l'Hôpital (1661–1704). It was discovered by John Bernoulli (1667–1748) and given to l'Hôpital in return for salary.

The limit of a quotient f/g is the quotient of the limits of f and g when both limits exist and $\lim g$ is nonzero. When both limits equal 0, the limit of the quotient may still exist, and we may be able to compute it using derivatives. See Exercise 38 for a version that applies whenever $g'(a) \neq 0$.

16.36. Theorem. (l'Hôpital's Rule) Suppose that $\lim_{x \to a} f(x) = 0$ and $\lim_{x \to a} g(x) = 0$. If f and g are differentiable in an interval containing a and $\lim_{x \to a} \frac{f'(x)}{g'(x)}$ exists, then $\lim_{x \to a} \frac{f(x)}{g(x)} = \lim_{x \to a} \frac{f'(x)}{g'(x)}$.

Proof: Since f and g are differentiable at a, they are continuous at a, and hence $f(a) = g(a) = 0$. The existence of $L = \lim_{x \to a} \frac{f'(x)}{g'(x)}$ presumes that $g'(x) \neq 0$ for x in some deleted neighborhood of a. We prove that $\lim_{x \to a} \frac{f(x)}{g(x)} = L$ by using the ϵ-δ definition of limit.

Given $\epsilon > 0$, there exists $\delta > 0$ such that $0 < |x - a| < \delta$ implies $\left| (f'/g')(x) - L \right| < \epsilon$. Consider such an x. Applying the Cauchy Mean Value Theorem yields a number c between a and x such that

$$\frac{f(x)}{g(x)} = \frac{f(x) - f(a)}{g(x) - g(a)} = \frac{f'(c)}{g'(c)}$$

Since $|c - a| < |x - a|$, we have $|c - a| < \delta$. Hence our choice of δ yields

$$\left| \frac{f(x)}{g(x)} - L \right| = \left| \frac{f'(c)}{g'(c)} - L \right| < \epsilon$$

Thus $0 < |x - a| < \delta$ implies $|f/g(x) - L| < \epsilon$, and so the limit is L. ∎

16.37. Example. Because the numerator and denominator are both 0 when $x = 2$, we have $\lim_{x \to 2} \frac{x^3 - 2x^2 + x - 2}{x^2 - 7x + 10} = \lim_{x \to 2} \frac{3x^2 - 4x + 1}{2x - 7} = -\frac{5}{3}$. ∎

Since ∞ is not a real number, we cannot evaluate functions at "infinity", but we *can* study the behavior of a function "near infinity".

16.38. Definition. For $L \in \mathbb{R}$, we say that $\lim_{x \to \infty} f(x) = L$ if for every $\epsilon > 0$ there is an $M > 0$ such that $x \geq M$ implies $|f(x) - L| < \epsilon$.

16.39. Example. $\lim_{x \to \infty} \frac{x}{1 + x^2} = 0$. Given $\epsilon > 0$, we let $M = 1/\epsilon$. Now $x \geq M$ implies $|f(x) - 0| = \frac{x}{1 + x^2} < \frac{1}{x} \leq \frac{1}{M} = \epsilon$. ∎

16.40. Definition. When $a \in \mathbb{R} \cup \{\infty\}$, we write $\lim_{x \to a} f(x) = \infty$ if $\lim_{x \to a} \frac{1}{f(x)} = 0$. Similarly, $\lim x_n = \infty$ if $\lim \frac{1}{x_n} = 0$.

16.41. Example. $\lim_{x \to \infty} \frac{1+x^2}{x} = \infty$ (compare with Example 16.39). ∎

The theory of $\lim_{x \to \infty}$ is parallel to that of $\lim_{x \to a}$. To keep the same intuition and language, we define a **neighborhood of** ∞ to be an interval of the form $(M, \infty) = \{x \in \mathbb{R} : x > M\}$.

Another version of l'Hôpital's Rule applies when the functions in the numerator and denominator tend to infinity. This version can be proved in a manner similar to Theorem 16.36 (see Exercise 40).

16.42. Theorem. (l'Hôpital's Rule, second form) Given $a \in \mathbb{R} \cup \{\infty\}$, suppose that $\lim_{x \to a} f(x) = \infty$ and $\lim_{x \to a} g(x) = \infty$. If f and g are differentiable in a neighborhood of a and $\lim_{x \to a} \frac{f'(x)}{g'(x)}$ exists, then $\lim_{x \to a} \frac{f(x)}{g(x)} = \lim_{x \to a} \frac{f'(x)}{g'(x)}$. ∎

NEWTON'S METHOD

We return to the study of equations. A **zero** of the function f is a solution to $f(x) = 0$. Sir Isaac Newton (1642–1727) invented an algorithm that often produces a sequence $\langle x \rangle$ converging to a zero of a differentiable function f. Given a guess x_n for a solution, Problem 16.2 suggests using the linear approximation to f at x_n to make a better guess. Let ℓ be the line tangent to the graph of f at $(x_n, f(x_n))$. We move to where ℓ intersects the horizontal axis. This yields the next guess $x_{n+1} = x_n - \frac{f(x_n)}{f'(x_n)}$.

16.43. Algorithm. (Newton's Method.) Given an initial guess x_0 for a zero of a differentiable function f, Newton's Method generates a sequence of guesses via the recurrence $x_{n+1} = x_n - \frac{f(x_n)}{f'(x_n)}$ for $n \geq 0$. ∎

16.44. Proposition. Suppose that f is differentiable and that f' is continuous. If $\langle x \rangle$ is a sequence defined by $x_{n+1} = x_n - \frac{f(x_n)}{f'(x_n)}$ for $n \geq 0$, and $x_n \to L$, then $f(L) = 0$.

Proof: Since $\langle x \rangle$ converges, $x_{n+1} - x_n \to 0$. Hence $f(x_n)/f'(x_n) \to 0$. Since f is continuously differentiable, $f'(x_n) \to f'(L)$. Hence $\lim f(x_n) = \lim \frac{f(x_n)}{f'(x_n)} f'(x_n) = 0 \cdot f'(L) = 0$. Being differentiable, f is continuous, and thus $f(L) = \lim f(x_n) = 0$. ∎

Newton's Method certainly fails when $f'(x_n) = 0$ for some n. Even when $f'(x_n)$ is never zero, the process may not converge. It may be hard to tell which initial guesses yield convergence, and it may be impossible to converge to some solutions without guessing them exactly as x_0. We now explore favorable cases.

16.45. Example. *Newton's Method for pth roots of real numbers.* For the recurrence $x_{n+1} = \frac{1}{2}(x_n + 2/x_n)$, we proved in Example 14.9 that $\langle x \rangle$ converges to $\sqrt{2}$ if $x_0 > 0$. This recurrence also arises when applying Newton's Method to the function f defined by $f(x) = x^2 - 2$.

The pth root of a positive real number a is a zero of the function defined by $f(x) = x^p - a$. Because $f'(x) = px^{p-1}$, Newton's Method yields the recurrence $x_{n+1} = (1 - 1/p)x_n + (1/p)(a/x_n^{p-1})$. If the resulting sequence converges, then by Proposition 16.44 the limit is a pth root of a.

The limit does exist, by generalizing Example 14.9 (see Exercise 56) or by applying Theorem 16.54. Furthermore, convergence to the limit is much faster than by the method of bisection; see Exercise 48. ∎

The conclusion of Proposition 16.44 need not hold when f' is not continuous (Exercise 57). In the study of differentiable functions, we often need the stronger hypothesis that f' is continuous. In the next section, we need the yet stronger hypothesis that f' is differentiable.

16.46. Definition. A function f is **continuously differentiable** on an open interval if f' exists and is continuous there. When f' is differentiable, we write its derivative as f'' and say that f is **twice differentiable**. For $k \geq 2$, the **kth derivative** $f^{(k)}$ of f is the derivative of $f^{(k-1)}$, when it exists. A function f is **smooth** or **infinitely differentiable** if for each $k \in \mathbb{N}$, the kth derivative exists.

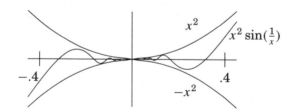

16.47. Example. *Differentiable but not continuously differentiable.* Consider the function f defined by $f(0) = 0$ and $f(x) = x^2 \sin(1/x)$ if $x \neq 0$. Since $|\sin y| \leq 1$ for all y, we have $|f(x)| \leq x^2$ for all $x \neq 0$. This inequality also holds at $x = 0$, since $f(0) = 0$. Hence Corollary 16.21 implies that f is differentiable at 0 and that $f'(0) = 0$. To compute $f'(x)$ when $x \neq 0$, we use the chain rule and the product rule. By the chain rule, the derivative of $\sin(1/x)$ is $-x^{-2} \cos(1/x)$. By the product rule, this yields $f'(x) = 2x \sin(1/x) - \cos(1/x)$ for $x \neq 0$. Since $\lim_{x \to 0} f'(x)$ does not exist, f' is not continuous at 0. ∎

CONVEXITY AND CURVATURE

Geometric considerations yield an important class of functions where Newton's Method works, including those in Example 16.45.

16.48. Definition. A function f is **convex** on an interval I if, for all x, z, y in I with $x < z < y$, the point $(z, f(z))$ lies at or below the line segment joining $(x, f(x))$ and $(y, f(y))$. Equivalently, for all $t \in [0, 1]$, we have the **convexity inequality**

$$f((1 - t)x + ty) \le (1 - t)f(x) + tf(y).$$

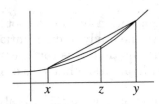

The equivalence of the two statements in the definition follows from setting $z = (1 - t)x + ty$. The expression $(1 - t)x + ty$ is called a "convex combination" of x and y. A convex combination is a weighted average. We can interpret the convexity inequality as saying that every weighted average of two function values is larger than the function value at the corresponding weighted average of the arguments. To show that a continuous function is convex, it suffices to know that the convexity inequality holds for each pair x, y when $t = 1/2$ (Exercise 55). The convexity inequality has a simple interpretation in terms of slopes.

16.49. Lemma. Suppose $f : [x, y] \to \mathbb{R}$. If $0 < t < 1$ and $z = (1-t)x + ty$, then the following are equivalent:
  A) $f(z) \le (1 - t)f(x) + tf(y)$ (the convexity inequality).
  B) $m_{x,z} \le m_{x,y}$.
  C) $m_{x,y} \le m_{z,y}$.

Proof: (A)$\Leftrightarrow$(B). We rewrite (A) as $f(z) - f(x) \le t[f(y) - f(x)]$. Since $t = (z-x)/(y-x)$, we have $f(z)-f(x) \le \frac{z-x}{y-x}[f(y)-f(x)]$. Since $z - x > 0$, we divide by it to obtain $m_{x,z} \le m_{x,y}$. These steps are reversible.

(A)$\Leftrightarrow$(C). We rewrite (A) as $f(z)-f(y) \le (1-t)[f(x)-f(y)]$. Since $1 - t = (y - z)/(y - x)$, we obtain $f(z) - f(y) \le \frac{y-z}{y-x}[f(x) - f(y)]$. Multiplying this inequality by -1 and dividing by $y - z$ yields $m_{z,y} \ge m_{x,y}$. These steps are reversible. ∎

The geometric definition of convexity does not mention differentiation, but Lemma 16.49 and the Mean Value Theorem combine to characterize convex differentiable functions.

16.50. Theorem. A differentiable function f is convex on an interval I if and only if f' is nondecreasing on I.

Proof: Suppose first that f is convex, and choose $x, y \in I$ with $x < y$. For all z between x and y, Lemma 16.49 yields $m_{x,z} \le m_{x,y} \le m_{z,y}$. Letting z decrease to x in the first inequality yields $f'(x) \le m_{x,y}$. Letting z increase to y in the second inequality yields $f'(y) \ge m_{x,y}$. Therefore, $f'(x) \le f'(y)$ and f' is nondecreasing on I.

If f is not convex on I, then Lemma 16.49 yields points x, z, y in I with $x < z < y$ such that $m_{x,z} > m_{x,y}$ and $m_{x,y} > m_{z,y}$. Since f is differentiable, the Mean Value Theorem yields points c, d with $x < c < z$ and $z < d < y$ such that $f'(c) = m_{x,z}$ and $f'(d) = m_{z,y}$. Now $f'(c) > f'(d)$, and f' is not nondecreasing. ∎

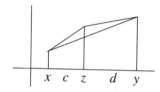

$x \quad c \quad z \quad d \quad y$

16.51. Corollary. If f is twice differentiable, then f is convex if and only if f'' is nonnegative.

Proof: When f is twice differentiable, each condition is equivalent to f' being nondecreasing. This uses Theorem 16.50 and the application of Exercise 32 to f' and f''. ∎

16.52. Example. *pth roots.* Suppose $f(x) = x^p - a$, so $f'(x) = px^{p-1}$ (the case where p is not an integer is covered in Exercise 17.29b). If $p > 0$ and $x > 0$, then $f'(x) > 0$, and hence f is injective for $x > 0$. If $p \ge 1$, then $f''(x) = p(p-1)x^{p-2} > 0$, so f is convex for $x > 0$. We will prove that Newton's Method converges for convex differentiable functions; hence we can use it to compute the unique positive pth root of a to any desired accuracy. ∎

16.53. Lemma. If differentiable functions f and g are equal at a and satisfy $f'(x) > g'(x) > 0$ for $x > a$, then $f(x) > g(x)$ for $x > a$.

Proof: By the Cauchy Mean Value Theorem, there exists $c \in (a, x)$ such that $[f(x) - f(a)]g'(c) = [g(x) - g(a)]f'(c)$. Since $f'(c) > g'(c) > 0$, we have $f(x) - f(a) > g(x) - g(a)$. Since $f(a) = g(a)$, we conclude that $f(x) > g(x)$. ∎

16.54. Theorem. Suppose f is convex and differentiable and has a zero. If $f'(x_0) \neq 0$, then Newton's Method starting at x_0 converges to a zero of f. Furthermore, all zeros of f arise in this way.

Proof: Consider four sets of real numbers:

$$S = \{x : f(x) > 0, f'(x) < 0\} \qquad U = \{x : f(x) < 0, f'(x) > 0\}$$
$$T = \{x : f(x) < 0, f'(x) < 0\} \qquad V = \{x : f(x) > 0, f'(x) > 0\}$$

By Theorem 16.50, f' is nondecreasing. Hence every element of $S \cup T$ is less than every element of $U \cup V$. Because a function with positive derivative is increasing, every element of U is less than every element of V, and similarly every element of S is less than every element of T. Thus the four sets appear in the order illustrated.

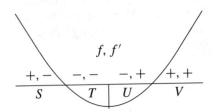

We claim that Newton's Method converges to a zero of f between S and T if $x_0 \in S \cup T$ and to a zero of f between U and V if $x_0 \in U \cup V$. By symmetry, we may assume $x_0 \in U \cup V$. The picture suggests that $x_0 \in U$ implies $x_1 \in V$. To see this, we let g be the linear approximation to f at x_0 and use Lemma 16.53 to conclude that $f(x_1) > g(x_1) = 0$.

Now suppose $x_n \in V$. The assumption that f has a zero guarantees that V has a lower bound. Since f and f' are both positive on V, we have $x_{n+1} < x_n$. Since $(x_{n+1}, 0)$ is on the tangent to the graph at $(x, f(x))$, we again obtain $f(x_{n+1}) > g(x_{n+1}) = 0$, and thus $x_{n+1} \in V$.

Since the sequence decreases and remains in V, the Monotone Convergence Theorem implies that it converges to a number at least inf V. As in the proof of Proposition 16.44, the sequence can only converge to a zero of f. If $T \cup U$ is non-empty and f has two zeros, then we can find them by starting the sequence in $U \cup V$ and in $S \cup T$. ∎

16.55. Example. For the function defined by $f(x) = x^2 - 2$, Newton's Method converges to $\sqrt{2}$ when $x_0 > 0$ and to $-\sqrt{2}$ when $x_0 < 0$. When $x_0 = 0$, the sequence is undefined. ∎

We close this section with a discussion of curvature.

16.56. Solution. *Circle of curvature (optional).* Suppose that the function g defined on an open interval I containing x_0 is twice differentiable. The subset $\{(x, g(x)): x \in I\}$ of the plane is a curve γ. We determine the radius of curvature at the point (x_0, y_0) on γ.

The circle of curvature C is the set $\{(x, y): (x - a)^2 + (y - b)^2 = r^2\}$, where the pair (a, b) is the center of the circle and r is its radius. To determine the three unknowns a, b, r, we need to specify three pieces of information. We want the function y describing C to agree with g, g', g'' at x_0, so we require

$$y(x_0) = g(x_0) \quad \text{(the circle intersects } \gamma \text{ at } (x_0, y_0))$$
$$y'(x_0) = g'(x_0) \quad \text{(the circle is tangent to } \gamma \text{ at } (x_0, y_0))$$
$$y''(x_0) = g''(x_0) \quad \text{(the circle has the same curvature as } \gamma \text{ at } (x_0, y_0))$$

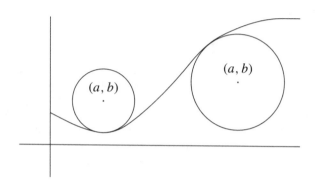

The equation for the circle determines (two choices of) y implicitly as a function of x. Our calculations do not require making this choice at the start. We differentiate both sides of $(x - a)^2 + (y - b)^2 = r^2$ with respect to x to obtain $2(x - a) + 2(y - b)y'(x) = 0$. This yields $y'(x) = -\frac{x-a}{y-b}$. Differentiating this by using the quotient rule and the chain rule yields

$$y''(x) = -\frac{(y-b)-(x-a)y'(x)}{(y-b)^2} = -\frac{(y-b)^2+(x-a)^2}{(y-b)^3}.$$

We evaluate these expressions at x_0 to write the required equations in terms of the parameters a, b, r.

$$1) \quad (x_0 - a)^2 + (y_0 - b)^2 = r^2$$
$$2) \quad -\frac{x_0-a}{y_0-b} = g'(x_0)$$
$$3) \quad -\frac{(y_0-b)^2+(x_0-a)^2}{(y_0-b)^3} = g''(x_0).$$

From (2) and (1), we obtain $1 + (g'(x_0))^2 = \frac{(y_0-b)^2+(x_0-a)^2}{(y_0-b)^2} = \frac{r^2}{(y_0-b)^2}$. We rewrite (3) as $g''(x_0) = -\frac{r^2}{(y_0-b)^3}$. By eliminating $(y_0 - b)$, we obtain an expression for r^2:

$$r^2 = \left[\frac{r^2}{(y_0-b)^2}\right]^3 \left[-\frac{r^2}{(y_0-b)^3}\right]^{-2} = \frac{[1+g'(x_0)^2]^3}{g''(x_0)^2}.$$

When $g''(x_0) \neq 0$, the radius of curvature is $r = \frac{[1+g'(x_0)^2]^{3/2}}{|g''(x_0)|}$.

When $g''(x_0) = 0$, the circle degenerates to a straight line (the radius becomes infinite). The **curvature** of γ at the point $(x, g(x))$ is defined to be the reciprocal of the radius of curvature. This reflects the intuition that curvature should be larger when the graph is more curved. (Some authors define curvature by measuring the rate at which the line tangent to the curve is changing.) Because the numerator in the formula for r is never zero, curvature is well-defined, and it equals 0 when $g''(x_0) = 0$.

When $g''(x_0)$ is positive, the circle lies above the curve γ, and $y - b$ is the negative square root of $r^2 - (x_0 - a)^2$. When $g''(x_0)$ is negative, $y - b$ is the positive square root. Computing with y^2 instead of y enabled us to consider both cases simultaneously.   ∎

SERIES OF FUNCTIONS

In the remainder of this chapter, we study sequences and series in which each term is a function. After proving the basic theorems about convergence of series of functions, we use such series to construct examples of continuous but nowhere differentiable functions. Constructing functions via series (as described in the next example) also has many scientific applications.

16.57. Example. *Power series and Fourier series.* A **power series** is a series of the form $\sum_{n=0}^{\infty} a_n (x - p)^n$. We view this as the sum of the functions f_n defined by $f_n(x) = a_n (x - p)^n$. Perhaps the most important power series is given by the exponential function $\exp(x) = \sum_{n=0}^{\infty} \frac{x^n}{n!}$. In Chapter 14, we thought of x as fixed when writing this series. Now we treat the summands as functions of x. Convergence of power series is fairly easy to understand.

Fourier series, named for Joseph Fourier (1768–1830), are series of the form $\sum_{n=0}^{\infty} (a_n \sin(nx) + b_n \cos(nx))$. Convergence is a delicate matter for Fourier series. Physicists and engineers often use them, because they can represent rather general functions by superposition (summation) of waves. Convergence of Fourier series is too delicate a matter to be discussed in this book, but we mention a simple example the reader might enjoy. Consider $\sum_{n=1}^{\infty} \frac{\sin(nx)}{n}$. Graph the first few partial sums to see what is happening; to what function does the series seem to converge? Both Fourier series and power series are best understood using complex numbers, which we introduce in Chapter 18.   ∎

We begin by defining two notions of convergence for sequences of functions. The limits of such sequences will themselves be functions. We use the notation $\{f_n\}$ rather than $\langle f \rangle$ to name the sequence in order to use f to name the limit function. *Pointwise* convergence means that for each x,

$\lim_{n\to\infty} f_n(x) = f(x)$. *Uniform* convergence is a stronger notion. We have written the definitions so that the only difference is the interchange of the order of two quantifiers (compare with Example 2.11).

16.58. Definition. Let $\{f_n\}$ be a sequence of functions defined on an interval I. The sequence $\{f_n\}$ **converges pointwise** to f on I if for every $\epsilon > 0$ and $x \in I$, there exists $N \in \mathbb{N}$ such that $n \geq N$ implies $|f_n(x) - f(x)| < \epsilon$. The sequence $\{f_n\}$ **converges uniformly** to f on I if for every $\epsilon > 0$, there exists $N \in \mathbb{N}$ such that, for $x \in I$, we have that $n \geq N$ implies $|f_n(x) - f(x)| < \epsilon$. The sequence $\{f_n\}$ is **uniformly Cauchy** on I if for every $\epsilon > 0$, there exists $N \in \mathbb{N}$ such that, for all choices of $n, m \geq N$ and $x \in I$, we have $|f_n(x) - f_m(x)| < \epsilon$.

pointwise: $(\forall \epsilon > 0)(\forall x \in I)(\exists N \in \mathbb{N})(n \geq N \Rightarrow |f_n(x) - f(x)| < \epsilon)$

uniform: $(\forall \epsilon > 0)(\exists N \in \mathbb{N})(\forall x \in I)(n \geq N \Rightarrow |f_n(x) - f(x)| < \epsilon)$

16.59. Remark. *Testing uniform convergence.* Uniform convergence asks us to prove that many sequences (one for each x) converge "at the same rate". To do so, we must control the worst case. In particular, $\{f_n\}$ converges uniformly to f on I if and only if $\sup_{x \in I} |f_n(x) - f(x)|$ converges to 0 as n tends to infinity. ∎

This remark is particularly easy to apply when we can bound $\sup_{x \in I} |f_n(x) - f(x)|$ easily, as in the next example and in Exercise 71.

16.60. Example. We contrast two sequences of functions defined on the interval $[0, 1]$. Let $f_n(x) = x$ for $0 \leq x \leq 1/n$ and $f_n(x) = 0$ for $1/n < x \leq 1$. Let $g_n(x) = nx$ for $0 \leq x \leq 1/n$ and $g_n(x) = 0$ for $1/n < x \leq 1$. Both $\{f_n\}$ and $\{g_n\}$ converge pointwise to the zero function. Since $\max_{x \in [0,1]} |f_n(x)| = 1/n$, the sequence $\{f_n\}$ converges uniformly. On the other hand, $\max_{x \in [0,1]} |g_n(x)| = 1$, so $\{g_n\}$ does not converge uniformly. ∎

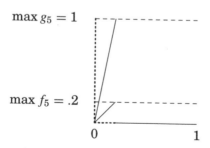

Uniform convergence requires more than pointwise convergence, because a single natural number N chosen in terms of ϵ must work simultaneously at every $x \in I$. Similarly, uniformly Cauchy requires more than each sequence $f_n(x)$ being a Cauchy sequence, because the same natural number must work simultaneously at all x.

16.61. Lemma. Suppose $\{f_n\}$ is a sequence of bounded functions on an interval I. Then $\{f_n\}$ converges uniformly to some function on I if and only if $\{f_n\}$ is uniformly Cauchy on I.

Proof: Suppose $\{f_n\}$ converges uniformly to f. Consider $\epsilon > 0$. From the definition of uniform convergence, we obtain a natural number N such that $n \geq N$ and $x \in I$ imply $|f_n(x) - f(x)| < \epsilon/2$. Choose $n, m \geq N$. Now $\{f_n\}$ is uniformly Cauchy, because for each $x \in I$ we have

$$|f_n(x) - f_m(x)| \leq |f_n(x) - f(x)| + |f(x) - f_m(x)| < \epsilon/2 + \epsilon/2 = \epsilon.$$

Conversely, suppose $\{f_n\}$ is uniformly Cauchy. For each x, the numbers $f_n(x)$ form a Cauchy sequence indexed by n. By the Cauchy Convergence Criterion, this sequence has a limit; call it $f(x)$. Doing this for each $x \in I$ defines a function f on I; we claim that $\{f_n\}$ converges uniformly to f. Consider $\epsilon > 0$, and choose ϵ' such that $0 < \epsilon' < \epsilon$. From the definition of uniformly Cauchy, we obtain a natural number N such that $n, m \geq N$ and $x \in I$ imply $|f_n(x) - f_m(x)| < \epsilon'$. Keeping n fixed, let $m \to \infty$. Using the definition of f, the continuity of the absolute value function, and the preservation of inequalities under limits (Lemma 13.17), we have

$$|f_n(x) - f(x)| = \left| f_n(x) - \lim_{m \to \infty} f_m(x) \right| = \lim_{m \to \infty} |f_n(x) - f_m(x)| \leq \epsilon' < \epsilon.$$

This inequality proves that $\{f_n\}$ converges uniformly to f on I. ∎

16.62. Definition. Let $\{g_n\}$ be a sequence of bounded functions on an interval I. If for each $x \in I$, the series $\sum_{n=1}^{\infty} g_n(x)$ converges, then setting $g(x) = \sum_{n=1}^{\infty} g_n(x)$ defines a function on I; we say that the **series** of functions $\sum_{n=1}^{\infty} g_n$ **converges pointwise** to g. Given such a sequence $\{g_n\}$, let $f_n = \sum_{k=1}^{n} g_k$. The series $\sum_{n=1}^{\infty} g_n$ **converges uniformly** to g on I if $\{f_n\}$ converges uniformly to g on I.

16.63. Corollary. (Weierstrass M-test) Let $\{g_n\}$ be a sequence of bounded functions on an interval I, with $|g_n(x)| \leq M_n$ for $x \in I$. If $\sum_{n=1}^{\infty} M_n$ converges, then $\sum_{n=1}^{\infty} g_n$ converges uniformly on I.

Proof: By Lemma 16.61, it suffices to show that the sequence $\{f_n\}$ of partial sums is uniformly Cauchy. Consider $\epsilon > 0$. We have $f_n - f_m = \sum_{k=m+1}^{n} g_k$. Hence $x \in I$ implies that

$$|f_n(x) - f_m(x)| = \left| \sum_{k=m+1}^{n} g_k(x) \right| \leq \sum_{k=m+1}^{n} |g_k(x)| \leq \sum_{k=m+1}^{n} M_k.$$

The last term equals $s_n - s_m$, where $\langle s \rangle$ is the sequence of partial sums. Since $\sum_{k=1}^{\infty} M_k$ converges, $\langle s \rangle$ is a Cauchy sequence. Hence we can choose $N \in \mathbb{N}$ such that $n, m \geq N$ implies $|s_n - s_m| < \epsilon$. Together with the displayed inequality, this yields $|f_n(x) - f_m(x)| < \epsilon$ whenever $n, m \geq N$ and $x \in I$. Thus $\{f_n\}$ is uniformly convergent. ∎

We use this notation because M_n "majorizes" g_n. The corollary provides an easy way to prove uniform convergence of series of functions.

16.64. Example. *Applying the M-test.* Let $g_n(x) = x^n / n^2$, and let $M_n = 1/n^2$. Because $\sum_{n=1}^{\infty} 1/n^2$ converges (Exercise 14.44 or Exercise 14.44) and $|g_n(x)| \leq M_n$, we conclude that $\sum_{n=1}^{\infty} g_n$ converges uniformly on $[-1, 1]$. ∎

We next obtain a sufficient condition for continuity of the limit of a convergent sequence of functions.

16.65. Theorem. If $\{f_n\}$ are continuous on an interval I, and $\{f_n\}$ converges uniformly to f on I, then f is continuous on I.

Proof: Given $\epsilon > 0$, uniform convergence allows us to choose a natural number N such that $n \geq N$ and $x \in I$ imply $|f_n(x) - f(x)| < \epsilon/3$. For each $a \in I$, we prove that f is continuous at a. For $x \in I$ and any fixed $n \geq N$,

$$|f(x) - f(a)| \leq |f(x) - f_n(x)| + |f_n(x) - f_n(a)| + |f_n(a) - f(a)|$$
$$\leq \epsilon/3 + |f_n(x) - f_n(a)| + \epsilon/3.$$

Since f_n is continuous, we can choose $\delta > 0$ such that $|x - a| < \delta$ implies $|f_n(x) - f_n(a)| < \epsilon/3$. With this choice of δ in terms of ϵ, we have $|f(x) - f(a)| < \epsilon$ when $|x - a| < \delta$. Hence f is continuous at a. ∎

16.66. Example. The hypothesis of uniform convergence is needed in Theorem 16.65. Suppose f_n is defined on $[0, 1]$ by $f_n(x) = x^n$. The sequence $\{f_n\}$ converges pointwise to the function f defined by $f(x) = 0$ for $0 \leq x < 1$ and $f(1) = 1$. The limit function f is not continuous. ∎

16.67. Corollary. If $\{g_n\}$ are continuous functions on an interval I, and $\sum_{n=1}^{\infty} g_n$ converges uniformly to g on I, then g is continuous on I.

Proof: The sum of n continuous functions is a continuous function. Hence $f_n = \sum_{k=1}^{n} g_k$ is continuous, and the theorem applies. ∎

16.68. Corollary. If $\{g_n\}$ is a sequence of bounded continuous functions, with bounds $|g_n(x)| \leq a_n$ such that $\sum_{n=1}^{\infty} a_n$ converges, then $g = \sum_{n=1}^{\infty} g_n$ is a continuous function.

Proof: Combine Corollary 16.63 and Corollary 16.67. ∎

16.69. Example. Define $g: \mathbb{R} \to \mathbb{R}$ by $g(x) = \sum_{n=1}^{\infty} \exp(-nx^2)/n^2$. Here each g_n is continuous (see Exercise 62 for continuity of the exponential function). Also $\exp(-nx^2) \leq 1$ for all n and x, and $\sum_{n=1}^{\infty} 1/n^2$ converges. We conclude that g is continuous on $\mathbb{R}$. ∎

These results about uniform convergence allow us to interchange certain limits involving infinite series.

16.70. Corollary. If $\{f_n\}$ is a uniformly convergent sequence of continuous functions on I and $a \in I$, then

$$\lim_{n \to \infty} \lim_{x \to a} f_n(x) = \lim_{x \to a} \lim_{n \to \infty} f_n(x).$$

Proof: Suppose $\{f_n\}$ converges to f. By Theorem 16.65, f is continuous. Therefore, $\lim_{x \to a} f(x) = f(a) = \lim_{n \to \infty} f_n(a)$. Because each f_n is continuous, the last expression equals $\lim_{n \to \infty} \lim_{x \to a} f_n(x)$. Because $f_n \to f$, the first expression equals $\lim_{x \to a} \lim_{n \to \infty} f_n(x)$. ∎

16.71. Solution. *Failure of interchange of limits* (Problem 16.4). Corollary 16.70 is false without uniform convergence. Define a sequence of functions on $[0, 2]$ by $f_n(x) = x^n$ for $0 \leq x \leq 1$ and $f_n(x) = (2 - x)^n$ for $1 \leq x \leq 2$. Each f_n is continuous, but (see Exercise 70)

$$\lim_{x \to 1} \lim_{n \to \infty} f_n(x) = 0 \neq 1 = \lim_{n \to \infty} \lim_{x \to 1} f_n(x).$$ ∎

16.72. Corollary. (Interchange of limit and uniformly convergent sum) Let each g_n be continuous on I. If $\sum_{n=1}^{\infty} g_n$ converges uniformly on I, and $a \in I$, then

$$\lim_{x \to a} \sum_{n=1}^{\infty} g_n(x) = \sum_{n=1}^{\infty} g_n(a) = \sum_{n=1}^{\infty} \lim_{x \to a} g_n(x).$$

Proof: By Corollary 16.67, $\{g_n\}$ converges to a function g that is continuous on I. Continuity of g implies $\lim_{x \to a} g(x) = g(a)$, as desired. ∎

16.73. Example. *Failure of interchange of limits.* Consider summing the entries in this infinite matrix:

$$\begin{pmatrix} 0 & 1 & 0 & 0 & \cdots \\ -1 & 0 & 1 & 0 & \cdots \\ 0 & -1 & 0 & 1 & \cdots \\ 0 & 0 & -1 & 0 & \cdots \\ \vdots & \vdots & \vdots & \vdots & \ddots \end{pmatrix} \qquad a_{i,j} = \begin{cases} +1 & \text{if } j = i+1 \\ -1 & \text{if } j = i-1 \\ 0 & \text{otherwise} \end{cases}$$

One way to do this is to evaluate $\sum_{i=1}^{k} \sum_{j=1}^{k} a_{i,j}$ and let k tend to infinity. This corresponds to considering k by k matrices in the upper left corner. For each k, the entries sum to 0, so the limit is 0.

Alternatively, we can think of the infinite sum as taking the limit of $\sum_{i=1}^{m} \sum_{j=1}^{n} a_{i,j}$ as n and m approach infinity. The sum is 1 when $n > m$ (more columns than rows), and the sum is -1 when $m > n$ (more rows than columns). Therefore, $\lim_{m \to \infty} \lim_{n \to \infty} \sum_{i=1}^{m} \sum_{j=1}^{n} a_{i,j} = 1$ and $\lim_{n \to \infty} \lim_{m \to \infty} \sum_{i=1}^{m} \sum_{j=1}^{n} a_{i,j} = -1$. Thus the value depends on the order in which we perform the two limit operations. ∎

Uniform convergence helps in constructing functions that are continuous on an interval but not differentiable anywhere. We present two classical examples; the first we only sketch.

16.74. Example. *Everywhere continuous and nowhere differentiable.* We define a sequence of functions $\{f_n\}$, each mapping $[0, 1]$ to $[0, 1]$. We start with the function $f_0(x) = x$. It is easiest to define each successive function by considering their graphs. To obtain f_1, break the graph of f_0 into thirds, reflect the middle third of the graph through its average height, and connect the ends of this new segment to the ends of the interval. To obtain f_{n+1} from f_n, do this to each segment in the graph of f_n; we illustrate the first two steps.

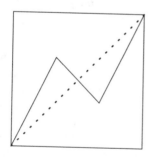 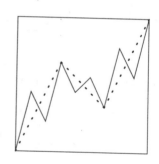

It is fairly easy to show that $\{f_n\}$ is uniformly Cauchy and converges pointwise to a continuous function f (Exercise 73). It is much more difficult to prove that f is nowhere differentiable. ∎

16.75. Example. *Everywhere continuous and nowhere differentiable.* Given $x \in \mathbb{R}$, let $d(x)$ denote the distance from x to the nearest integer. Define g by $g(x) = \sum_{n=0}^{\infty} d(10^n x)/10^n$. We show that g is continuous on all of $\mathbb{R}$ but is not differentiable anywhere. This construction is due to Bartel Leendert van der Waerden (1903–1996).

The continuity of g is easy to prove. Since $d(x) = \inf_{n \in \mathbb{Z}} |x - n|$, and since the absolute value function is continuous, the function d is continuous as well. Hence the function g_n defined by $g_n(x) = d(10^n x)/10^n$ is continuous. Since $0 \leq d(x) \leq \frac{1}{2}$ for all x, we have $|g_n(x)| \leq \frac{1}{2} \cdot 10^{-n}$ for all x. Since $\sum_{n=0}^{\infty} \frac{1}{2} \cdot 10^{-n} = 10/18$ (geometric series), we know by the Weierstrass M-test that the series $\sum_{n=0}^{\infty} g_n$ converges uniformly on $\mathbb{R}$. Corollary 16.67 then implies that g is continuous everywhere.

Next we prove that g is nowhere differentiable. Consider $x \in \mathbb{R}$. Let $.a_1 a_2 a_3 \cdots$ be the decimal expansion of its fractional part. We define a sequence $\langle h \rangle$ as follows: When a_m is not 4 or 9, we put $h_m = 10^{-m}$. When a_m equals 4 or 9, we put $h_m = -10^{-m}$. This choice makes $g(x + h_m) - g(x)$ easy to compute. If a_m does not equal 4 or 9, then from each of $10^n x + 10^{n-m}$ and $10^n x$ we move up to the nearest integer, or from each we move down. If a_m equals 4 or 9, then the same phenomenon occurs for $10^n x - 10^{n-m}$ and $10^n x$. In each case, we are comparing $d(10^n(x + h_m))$ with $d(10^n x)$ (see Exercise 74 for an example).

To obtain a formula for the difference quotient $[g(x + h_m) - g(x)]/h_m$, we let $\alpha_n = +1$ if $a_n \in \{0, 1, 2, 3, 4\}$ and $\alpha_n = -1$ if $a_n \in \{5, 6, 7, 8, 9\}$. We now compute

$$\frac{d(10^n(x + h_m)) - d(10^n x)}{h_m} = \begin{cases} \alpha_n 10^n & \text{if } n < m \\ 0 & \text{if } n \geq m. \end{cases}$$

From this, we have

$$\frac{g(x + h_m) - g(x)}{h_m} = \sum_{n=0}^{\infty} \frac{d(10^n x + 10^n h_m) - d(10^n x)}{10^n h_m} = \sum_{n=0}^{m-1} \alpha_n.$$

The infinite sum in the expression for g collapses to a finite sum in the difference quotient. Furthermore, the difference quotient for h_m is the sum of the first m terms of some sequence of $+1$s and -1s. The existence of a limit for the difference quotients $[g(x + h_m) - g(x)]/h_m$ is equivalent to the convergence of the series $\sum_{n=0}^{\infty} \alpha_n$. A necessary condition for convergence of a series is that the terms approach 0; this fails here, since each α_n is 1 or -1. Hence this sequence of difference quotients does not converge, and therefore g is not differentiable at x. ∎

Although it is difficult to explicitly describe functions that are continuous but nowhere differentiable, more advanced considerations imply that "almost all" continuous functions are nowhere differentiable.

EXERCISES

16.1. For $x \neq 0$, determine $\lim_{h \to 0} \frac{1}{h} \left(\frac{1}{(x+h)^2} - \frac{1}{x^2} \right)$. Think before computing!

16.2. What does the chain rule say when f and g are linear functions?

16.3. How does the material of this chapter justify the method of interpolation used when reading tables of values?

16.4. Is the temperature changing slowly or rapidly near the time when it reaches its high for the day? How is this related to the material of this chapter?

16.5. Construct a function f such that f^2 is differentiable at every point while f is differentiable at no point.

For Exercises 6–9, determine whether the statement is true or false. If true, provide a proof; if false, provide a counterexample. Unless otherwise stated, all functions have domain $\mathbb{R}$ and target $\mathbb{R}$.

16.6. There is a function f such that $f(x+h) = f(x) + h$ for all $x, h \in \mathbb{R}$.

16.7. There is a function f such that $f(x+h) = f(x) + h^2$ for all $x, h \in \mathbb{R}$.

16.8. There is a differentiable function $f \colon \mathbb{R} \to \mathbb{R}$ such that $f'(x) = -1$ for $x < 0$ and $f'(x) = 1$ for $x > 0$.

16.9. If both $f + g$ and fg are differentiable, then f and g are differentiable. Compare with Exercise 1.49 and Exercise 15.6.

•   •   •   •   •

16.10. Let $f(x) = \prod_{j=1}^{n}(x + a_j)$, where $a_1, \ldots, a_n \in \mathbb{R}$. Compute $f'(x)$.

16.11. Derive the product rule for differentiation using difference quotients. (Hint: Add and subtract an appropriate quantity to the numerator.)

16.12. Suppose that g is differentiable at x and that $g(x) \neq 0$. The following argument fails to prove that $\frac{d}{dx}\frac{1}{g(x)} = \frac{-g'(x)}{(g(x))^2}$. Find the flaw.

Since $1 = g(x) \cdot \frac{1}{g(x)}$, the product rule yields $0 = \frac{g'(x)}{g(x)} + g(x)\frac{d}{dx}\frac{1}{g(x)}$. Solving for $\frac{d}{dx}\frac{1}{g(x)}$ yields $\frac{d}{dx}\frac{1}{g(x)} = \frac{-g'(x)}{(g(x))^2}$.

16.13. Let f and g be differentiable at x, and suppose that $g(x) \neq 0$. Using the product rule and the formula for $(1/g)'(x)$, prove the quotient rule

$$(f/g)'(x) = \frac{g(x)f'(x) - f(x)g'(x)}{[g(x)]^2}.$$

16.14. (!) Compute the derivative of the cube root function using either definition. (Hint: Use the factorization $a^3 - b^3 = (a-b)(a^2+ab+b^2)$ to simplify the difference of cube roots.)

16.15. The following inductive argument fails to prove that $(d/dx)x^n = nx^{n-1}$ for nonnegative integers n. Explain the error and correct the proof.
Basis step ($n = 0$): $\lim_{h\to 0}(1-1)/h = 0$. Inductive step ($n > 0$): Using the induction hypothesis for $n - 1$ and the product rule for differentiation,

$$\frac{d}{dx}x^n = \frac{d}{dx}xx^{n-1} = x \cdot (n-1)x^{n-2} + 1 \cdot x^{n-1} = nx^{n-1}.$$

16.16. Let $r = p/q$, where $p \in \mathbb{Z}$ and $q \in \mathbb{N}$. We define x^r to be $(x^p)^{1/q}$. Determine $f'(x)$, where $f(x) = x^r$. (Hint: We have determined this already for $r \in \mathbb{N}$. Derive the formula first for $p = 1$ and then for $r \in \mathbb{Q}$. Comment: When $r \in \mathbb{R}$, the same formula holds for f'. The proof uses properties of the exponential function and appears in Exercise 17.29.)

16.17. Suppose that e_1 and e_2 are error functions and that $e_1(h) \le e(h) \le e_2(h)$ for all h in a neighborhood of 0. Prove that e is an error function.

16.18. (!) Let $f(x) = x + x^2$ if x is rational, and let $f(x) = x$ if x is irrational. Prove that f is differentiable at $x = 0$.

16.19. (−) *Sufficient conditions for differentiability at a point.*
   a) Suppose that $|f(x)| \le x^2 + x^4$ for all x. Prove that $f'(0)$ exists.
   b) Suppose that $|f(x)| \le g(x)$, where $g(x) \ge 0$ for all x and $g'(0) = g(0) = 0$. Prove that $f'(0)$ exists.
   c) Suppose that g is a bounded function and that $f(x) = (x - a)^2 g(x)$ for all x. Prove that $f'(a)$ exists.

16.20. Suppose that $|f(x) - f(y)| \le |g(x) - g(y)|$ for all $x, y \in \mathbb{R}$, and suppose g is differentiable at a with $g'(a) = 0$. Prove using difference quotients that f is differentiable at a and that $f'(a) = 0$.

16.21. Let f be differentiable, with $f(0) = 0$. Let $g(x) = f(x)/x$ for $x \ne 0$.
   a) How should $g(0)$ be defined to make g continuous at 0?
   b) If $g(0)$ is defined so that g is continuous at 0, does it follow that g is differentiable at 0? Give a proof or a counterexample.

16.22. (−) The volume of a ball of radius r is $\frac{4}{3}\pi r^3$. Suppose that air is escaping from a ball at the rate of 36 cubic inches per second. How fast is the radius of the ball decreasing at the moment when the radius is 6 inches?

16.23. (−) What real number most exceeds its square?

16.24. Let $f(x) = ax^2 + bx + c$ with $a > 0$. Determine the minimum value of f on $\mathbb{R}$. What condition on a, b, c is necessary and sufficient for the minimum value to be positive?

16.25. A company wishes to set the price of its new liquid to maximize profit. A marketing analysis indicates that if the price is set at x dollars per gallon, then the number of gallons sold per day will be $g(x) = 1000/(5 + x)$. The government, wishing to stimulate production, will also pay the company (per day) \$50 times $\sqrt{g(x)}$. Determine the maximum and minimum values of the company's daily profit and the prices that yield these values. (Assume zero production cost.)

16.26. (!) Suppose that $m_1, \dots, m_k$ are nonnegative real numbers with sum n.
   a) Using calculus and induction, prove that $\sum_{i<j} m_i m_j \le (1 - \frac{1}{k})\frac{n^2}{2}$, with equality only when $m_1 = \cdots = m_k$.
   b) In the case where $m_1, \dots, m_k$ are integers, give a combinatorial proof that $\sum_{i<j} m_i m_j$ is maximized when each m_i is $\lfloor n/k \rfloor$ or $\lceil n/k \rceil$.

16.27. (!) Prove that two differentiable functions on an interval (a, b) have the same derivative if and only if they differ by a constant.

16.28. Derive the Mean Value Theorem from the Cauchy Mean Value Theorem

16.29. (−) Suppose that $f(x) = x^3$, $g(x) = x^2$, $a = 0$, and $b = 1$. Find the value c guaranteed by the Cauchy Mean Value Theorem.

16.30. Let f be differentiable on $[a, b]$, with $f'(a) < y < f'(b)$. Prove that there exists $c \in (a, b)$ such that $f'(c) = y$. (Comment: This is the Intermediate Value Property for the function f'. It does not require that f' be continuous.)

16.31. (!) Let f be differentiable, with $f'(x) < 1$ for all x. Prove that f has at most one fixed point. (Recall that x is a fixed point of f if $f(x) = x$.)

16.32. (−) Let f be differentiable. Prove that f' is nonnegative everywhere if and only if f is nondecreasing.

16.33. Let f be differentiable, with $f'(0) > 0$. Suppose also that f is not monotone in any neighborhood of 0. Explain why f' must be discontinuous at 0. Construct an example of such a function f. (Hint: Modify Example 16.19).

16.34. Let f be differentiable, and suppose that f and f' are positive on $\mathbb{R}$. Prove that the function $g = f/(1 + f)$ is bounded and increasing.

16.35. Let f be differentiable on $[a, b]$, with $f(a) = f(b) = 0$. Determine f under the condition that $f'(x) \geq 0$ for all $x \in [a, b]$.

16.36. Let f be a function on an interval S that is monotone and differentiable.
   a) Explain why f has an inverse.
   b) Prove that f^{-1} is differentiable, with $\frac{d(f^{-1}(y))}{dy} = \frac{1}{f'(f^{-1}(y))}$.

16.37. An **operator** is a function whose domain and target are themselves sets of functions. For example, differentiation is an operator on the set of differentiable functions. Here we define another operator A on this set. The image of the function f under the operator A is the function A_f whose value at x is $\lim_{t \to 1} \frac{f(tx) - f(x)}{tf(x) - f(x)}$. (If $f(x) = 0$, then A_f is not defined at x.)
   a) Use l'Hôpital's Rule to compute A_f when f is continuously differentiable.
   b) Use part (a) to compute $(Af)(x)$ when $f(x) = x^n$ and when $f(x) = e^x$.
   c) When f' is not continuous, the computation using l'Hôpital's Rule is not valid. Give a direct proof of the formula in part (a) that is valid even when f' is not continuous. (Hint: Replace t by $1 + h$. When $x \neq 0$, replace h_x by u.)

16.38. (!) *l'Hôpital's Rule, weak form.* Let f and g be differentiable in a neighborhood of a. Suppose that $f(a) = g(a) = 0$ and $g'(a) \neq 0$. Use the definition of the derivative as a linear approximation to prove that $\lim_{x \to a} \frac{f(x)}{g(x)} = \frac{f'(a)}{g'(a)}$.

16.39. (!) In this problem, all limits are taken as $x \to a$, where $a \in \mathbb{R} \cup \{\infty\}$. Let f and g be differentiable. Suppose that $\lim f(x) = \infty$ and $\lim g(x) = \infty$, and that $\lim f(x)/g(x) = L$ and $\lim f'(x)/g'(x) = M$. Assuming that $L \neq 0$, prove that $L = M$. (Hint: Apply l'Hôpital's Rule (Theorem 16.36) to compute $\lim_{x \to a} \frac{1/g(x)}{1/f(x)}$.)

16.40. (+) Prove Theorem 16.42, using the Cauchy Mean Value Theorem.

16.41. The **first forward difference** of a function $f: \mathbb{R} \to \mathbb{R}$ is the function Δf defined by $\Delta f(x) = f(x + 1) - f(x)$. The k**th forward difference** of f is defined by $\Delta^k f(x) = \Delta^{k-1} f(x + 1) - \Delta^{k-1} f(x)$.
   a) Prove that $\Delta^k f(x) = \sum_{j=0}^{k} (-1)^j \binom{k}{j} f(x + j)$.
   b) Prove that $f^{(k)}(x) = \lim_{h \to 0} \frac{1}{h} \sum_{j=0}^{k} (-1)^j \binom{k}{j} f(x + jh)]$ when the limit exists.

16.42. Suppose that f is smooth (Definition 16.46). Prove that f is a polynomial of degree at most k if and only if $f^{(k+1)}(x) = 0$ for all x.

16.43. (+) Suppose that f is smooth, that $f(0) = 0$, and that f has a local minimum at 0. If $f^{(j)}(0) \neq 0$ for some natural number j, let k be the smallest such number. Prove that k is even. Give an example of a smooth function f such that $f(x) = 0$ if and only if $x = 0$, and $f^{(j)}(0) = 0$ for all $j \in \mathbb{N}$.

16.44. Suppose that f and g are smooth. Compute the kth derivative of $f \circ g$, for $1 \leq k \leq 5$. Describe the form of the expression for general k. (Comment: The sum of the coefficients of terms where f is differentiated j times is known as the **Stirling number** $S(k, j)$. It equals the number of ways to partition a set of k elements into j nonempty subsets.)

16.45. (−) Using an initial guess of 1 for the solution, apply Newton's Method to seek a solution to the equation $x^5 = 33$ and compute the first four iterations. Repeat this with an initial guess of 2. (Use a calculator.)

16.46. Find a quadratic function f for which the recurrence generated by Newton's Method is $x_{n+1} = \frac{1}{2}(x_n - 1/x_n)$. Use the graph of the function to explain the behavior of the recurrence as $n \to \infty$.

16.47. Find a differentiable function f and a sequence $\langle x \rangle$ such that $x_n \to 0$, $f'(x_n) \to \infty$, and $f(x_n) = 1$ for every n. Determine $\lim[x_n - f(x_n)/f'(x_n)]$. What does this exercise say about Proposition 16.44?

16.48. (!) Given a differentiable function f, let $g(x) = x - \frac{f(x)}{f'(x)}$ when $f'(x) \neq 0$. The function g is the function that generates x_{n+1} from x_n in Newton's Method.
    a) Verify that $g(x) = x$ if and only if $f(x) = 0$.
    b) When $f(x) = x^2 - 2$, verify that $g(x) - \sqrt{2} = \frac{1}{2x}(x - \sqrt{2})^2$.
    c) Use (b) to show that when Newton's Method is applied to $x^2 - 2$ with $x_0 = 1$, the value of x_5 is within 2^{-31} of $\sqrt{2}$.
    d) What can be said about Newton's Method in general when a is a zero of f and $|g(x) - a| \leq c|x - a|^2$ for some constant c and for x near a?

16.49. (!) Suppose that f and g are convex and $c \in \mathbb{R}$. Which of the three functions $f + g$, $c \cdot f$, and $f \cdot g$ must be convex? (Give proofs or counterexamples.)

16.50. Let f be convex on the interval $[a, b]$. Prove that the maximum of f on $[a, b]$ is $f(a)$ or $f(b)$. (Comment: Convex functions need not be differentiable.)

16.51. Suppose that f is twice differentiable and that f'' is nonnegative everywhere. Given that $f(a) = A$ and $f(b) = B$, what is the maximum possible value of $f((a + b)/2)$? For what function is this bound attained?

16.52. Which polynomials of odd degree are convex on $\mathbb{R}$?

16.53. Characterize the fourth-degree polynomials that are convex on $\mathbb{R}$ by giving a necessary and sufficient condition on the coefficients.

16.54. Let Y be a random variable that takes only values in $\{y_1, \ldots, y_n\}$, with corresponding probabilities $p_1, \ldots, p_n$. Suppose that $-1 \leq y_i \leq 1$ for all i. Suppose also that the expectation of Y is 0 and that f is convex. Prove that the expectation of $f(Y)$ is at most $[f(1) + f(-1)]/2$.

16.55. (+) Suppose that f is continuous and that $f(\frac{x+y}{2}) \le \frac{f(x)+f(y)}{2}$ for all $x, y \in \mathbb{R}$. Prove that f is convex. (Hint: First prove that the inequality of Definition 16.48 holds when t is a fraction whose denominator is a power of 2, then apply the continuity of f.)

16.56. (+) Starting with $x_0 = a$, define a sequence $\langle x \rangle$ by $x_{n+1} = (1 - 1/p)x_n + (1/p)(a/x_n^{p-1})$ for $n \ge 0$. By using the convexity of x^p as a function of x, but without differentiability or Newton's Method, prove that $x_n \to a^{1/p}$.

16.57. (+) Consider the polynomial f defined by $f(x) = (x-a)(x-b)(x-c)(x-d)$ with $a < b < c < d$. Describe the set of starting points x_0 such that Newton's Method converges to a zero of f. (Hint: Draw very careful pictures. The set of starting points x_0 that fail is an uncountable set.)

16.58. (−) Suppose that $f_n \to f$ uniformly and $g_n \to g$ uniformly on an interval I. Prove that $f_n + g_n \to f + g$ uniformly on I. Give an example of sequences f_n and g_n such that each converges pointwise but not uniformly, and yet $f_n + g_n$ does converge uniformly.

16.59. Prove or disprove each statement below.
 a) $\sum_{n=0}^{\infty} e^{-nx}/2^n$ converges uniformly for $x \in \mathbb{R}$.
 b) $\sum_{n=0}^{\infty} e^{-nx}/2^n$ converges uniformly for $x \ge 0$.

16.60. Define $f_n \colon \mathbb{R} \to \mathbb{R}$ by $f_n(x) = n^2/(x^2 + n^2)$, and define $f = \lim f_n$. Determine f. Does $\{f_n\}$ converge uniformly to f?

16.61. Let $f_n(x) = x^2/(x^2 + n^2)$.
 a) Prove that f_n converges pointwise to 0 everywhere on $\mathbb{R}$.
 b) Prove that f_n does not converge uniformly to 0 on $\mathbb{R}$.

16.62. Recall that $\exp(x) = \sum_{n=0}^{\infty} x^n/n!$. Define g_n by $g_n(x) = x^n/n!$.
 a) Prove that $\sum_{n=0}^{\infty} g_n$ converges uniformly to $\exp(x)$ on any bounded interval I (and hence $\exp(x)$ is continuous).
 b) Prove that $\exp(x + y) = \exp(x)\exp(y)$.
 c) Determine $\lim_{h \to 0}(\exp(h) - 1)/h$. (Comment: l'Hôpital's rule cannot be applied here, since we do not yet know that $\exp(x)$ defines a differentiable function. The series definition of $\exp(h)$ must be used.)
 d) Use (b) and (c) to prove that $(d/dx)(\exp(x)) = \exp(x)$.

16.63. Given $a > 0$, and define f by $f(x) = \exp(-ax^2)$. Determine where f is convex. Sketch the graph of f.

16.64. (!) Find an explicit formula for $\sum_{n=0}^{\infty} n^2 x^n$ when $|x| < 1$. Using derivatives of the geometric series or Application 12.37, describe $\sum_{n=0}^{\infty} q(n)x^n$, where q is a polynomial. (Hint: The answer is a polynomial in $1/(1-x)$.)

16.65. (+) Consider two kinds of baseball players. One hits singles with probability p; the other hits home runs with probability $p/4$. The players otherwise strike out. Assume that singles advance each runner by two bases. Compare a team composed of the home run hitters with a team composed of the singles hitters. How many runs does each team expect to score per inning?

16.66. (!) Express $\sum_{k=0}^{n} kx^k$ as a ratio of two polynomials in x.

16.67. Let q be a polynomial. Prove that $\sum_{k=0}^{n} q(k)x^k$ is the ratio of two polynomials in x.

16.68. Suppose that $0 < p < 1$. Let X be a random variable such that Prob $(X = n) = p(1 - p)^n$, for each nonnegative integer n.
    a) The probability generating function for X is $\phi(t) = \sum_{n=0}^{\infty}$ Prob $(X = n)t^n$. Find an explicit formula for $\phi(t)$ by evaluating the sum. Use this to verify that these probabilities sum to 1.
    b) Compute $E(X)$.
    c) Obtain a simple formula for Prob $(X \le 20)$.

16.69. Let $y(x) = x^n$, where $n \ge 2$.
    a) Find the curvature at the point $(x, y(x))$.
    b) Find an equation for the value of x where the curvature is maximized.
    c) Solve the equation in (b) to find the value of x where the graph of $y(x) = x^3$ has greatest curvature.

16.70. $(-)$ Check the computations in Solution 16.71.

16.71. (!) *Critical exponent for uniform convergence.*
    a) Suppose that $f_n(x) = x^n(1 - x)$. Prove that $f_n(x) \to 0$ uniformly on $[0, 1]$.
    b) Suppose that $f_n(x) = n^2 x^n(1-x)$. Prove that $f_n(x)$ converges to 0 pointwise but not uniformly on $[0, 1]$.
    c) Suppose that $f_n(x) = n^{\alpha} x^n(1 - x)$, where $\alpha \ge 0$. Prove that $f_n(x) \to 0$ uniformly on $[0, 1]$ if and only if $\alpha < 1$. (Assume that $(1 - 1/n)^n \to e^{-1}$.)

16.72. Let g be a bounded and differentiable function on $\mathbb{R}$ such that $\lim_{x \to \infty} g'(x)$ does not exist. Let $f_n(x) = \frac{1}{n}g(nx)$. Prove that f_n converges uniformly on $\mathbb{R}$ but that f_n' does not even converge.

16.73. Prove that the sequence defined in Example 16.74 is uniformly Cauchy.

16.74. Consider the proof that the function g in Example 16.75 is nowhere differentiable. Compute the difference quotient $[g(x + h_m) - g(x)]/h_m$ for all m in the following two cases: $x = 0$ and $x = .1496$.

16.75. (+) Define $f: \mathbb{R} \to \mathbb{R}$ by $f(x) = \sum_{n=0}^{\infty} \frac{\sin(3^n x)}{2^n}$. Prove that f is continuous on $\mathbb{R}$ and is not differentiable at 0. (Comment: In fact, f is nowhere differentiable.)

16.76. Given a continuous and nowhere differentiable function on $\mathbb{R}$, construct from it a continuous function that is differentiable at exactly one point.

Chapter 17

Integration

Integration is the mathematical process that enables us to calculate areas and volumes; it is the continuous analogue of summation. In this chapter, we present the theory of integration and its relationship to differentiation and infinite series. The Fundamental Theorem of Calculus shows that integration is an inverse process to differentiation.

We use integration to define the logarithm function and define the exponential function to be its inverse. This definition yields the same function as the series definition in Example 14.32. We also define the trigonometric functions by infinite series. This leads to a definition of π and a proof, using integration, that the area inside the unit circle is π. Integration also arises in discussion of probability and expectation and in physical considerations such as work and center of mass, since the average value of a function over a set can be expressed as an integral.

17.1. Problem. *Area and Limits.* Let $T_1, T_2, \ldots$ be a sequence of triangles in the plane. Suppose that the sequence of triangles converges to a region T. Can we conclude that $\text{Area}(T) = \lim \text{Area}(T_n)$? ∎

17.2. Problem. *The Rainfall Problem.* Suppose that rain is falling uniformly over a square region. What fraction of the raindrops do we expect to fall within an inscribed circle? ∎

17.3. Problem. *Continuous Compounding.* A person puts p dollars in a savings account paying $x\%$ interest per year. If the interest is compounded once per year, then after m years the total amount is $p(1+x)^m$. If the interest is compounded n times per year, then the total amount after m years is $p(1 + x/n)^{nm}$. The effective interest rate or *yield* depends on how often interest is compounded. For example, 5% interest ($x = .05$) compounded daily has a yield of 5.13%. The yield under continuous compounding is $\lim_{n \to \infty}(1 + x/n)^n - 1$. What is the value of this limit? ∎

DEFINITION OF THE INTEGRAL

What do we mean by the "area" of a region in the plane? This question leads to deep mathematical issues, involving subtle properties of geometry and limits. The area of a square is defined to be the square of the length of each side. Using squares, we can obtain upper and lower bounds for what we believe to be the area of a bounded region R.

We lay a fine grid of squares over the plane. Let S_1 be the union of the squares that are completely contained in R, and let S_2 be the union of the squares that contain points of R. Because R is bounded, S_1 and S_2 consist of finitely many squares of equal size; we count the squares and multiply by the area of each square to find the areas of S_1 and S_2. We obtain Area(S_1) ≤ Area(S_2), and we believe that the area of R is between them. By using a finer grid, we make Area(S_1) and Area(S_2) closer together.

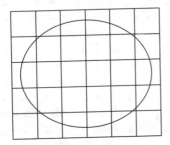 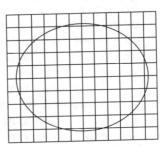

The first picture above, using unit squares, suggests that the area inside the ellipse satisfies $8 \leq$ Area(R) ≤ 26. The finer grid suggests that $13 \leq$ Area(R) ≤ 23. Let **U** be the set of upper bounds we can obtain using grids, and let **L** be the set of lower bounds. If sup **L** = inf **U** = a, then we believe that Area(R) = a.

We use this idea of upper and lower approximation to define integrals. Let f be a continuous (hence bounded) positive-valued function on the interval $[a, b]$. Let R be the region defined by $\{(x, y) \in \mathbb{R}^2: a \leq x \leq b$ and $0 \leq y \leq f(x)\}$.

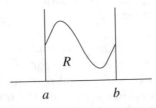

It is efficient to use rectangles instead of squares to obtain upper and lower bounds on Area(R). We break the interval $[a, b]$ into subintervals to obtain the bases of the rectangles, and we use the values of f on these subintervals to determine the heights of the rectangles.

17.4. Definition. A **partition** of $[a, b]$ is a collection of n subintervals $\{[x_{i-1}, x_i]\}$ such that $a = x_0 \leq \cdots \leq x_n = b$. We specify a partition P by its **breakpoints** $x_0, \ldots, x_n$. A partition Q is a **refinement** of a partition P if each breakpoint for P is also a breakpoint for Q. The **least common refinement** of two partitions is the partition whose set of breakpoints is the union of their sets of breakpoints.

17.5. Definition. Let $f : [a, b] \to \mathbb{R}$ be a bounded function. Let P be a partition of $[a, b]$ specified by $x_0, \ldots, x_n$. Let

$$l_i = \inf\{f(x): x \in [x_{i-1}, x_i]\} \quad \text{and} \quad u_i = \sup\{f(x): x \in [x_{i-1}, x_i]\}.$$

The **lower sum** of f corresponding to P is the sum $L(f, P) = \sum_{i=1}^n (x_i - x_{i-1})l_i$. The **upper sum** of f on P corresponding to P is the sum $U(f, P) = \sum_{i=1}^n (x_i - x_{i-1})u_i$.

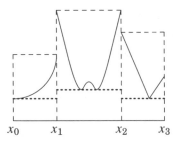

$$x_0 \qquad x_1 \qquad\qquad x_2 \qquad x_3$$

17.6. Example. Always $L(f, P) \leq U(f, P)$. The figure shows an interval partitioned into three subintervals. The dashed horizontal line above the interval $[x_{i-1}, x_i]$ is at height u_i above the axis, and the dotted horizontal line is at height l_i. The number $(x_i - x_{i-1})u_i$ is the area of the taller rectangle above $[x_{i-1}, x_i]$. Thus the upper sum $U(f, P)$ is an upper bound on the area under the graph of f. Similarly, $L(f, P)$ is a lower bound. ∎

For the infimum and supremum of the values of a function f over a set S, we write $\inf_S f$ and $\sup_S f$.

17.7. Remark. If $S \subseteq T$, then $\inf_T f \leq \inf_S f \leq \sup_S f \leq \sup_T f$. ∎

17.8. Lemma. Let $f: [a, b] \to \mathbb{R}$, and let P, Q, R be partitions of $[a, b]$.
   a) If R is a refinement of P, then $L(f, P) \leq L(f, R) \leq U(f, R) \leq U(f, P)$.
   b) $L(f, P) \leq U(f, Q)$.

Proof: In (a), the case where R is obtained from P by adding one more breakpoint is illustrated below and follows from Remark 17.7. The general case follows by induction on the number of breakpoints added.

   Statement (b) is obtained from (a) by letting R be the least common refinement of P and Q. Exercise 8 requests the details for both parts. ∎

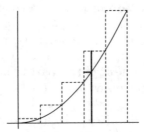

If f is a constant function, then $L(f, P) = U(f, P)$ for every partition P. Otherwise, $L(f, P) < U(f, P)$ for some P. We hope to make these numbers closer by refining the partition. The quantity squeezed between them is what we will call the integral of f on $[a, b]$. There are three equivalent ways to make this notion precise.

17.9. Proposition. Given a bounded function f defined on the interval $[a, b]$, the following three statements are equivalent.
   a) For every $\epsilon > 0$, there exists a partition R of $[a, b]$ such that
      $$U(f, R) - L(f, R) < \epsilon.$$
   b) $\sup_P L(f, P) = \inf_Q U(f, Q)$.
   c) There is a sequence $\{R_n\}$ of partitions such that
      $$\lim_{n \to \infty} L(f, R_n) = \lim_{n \to \infty} U(f, R_n).$$

Proof: By Lemma 17.8, $L(f, P) \le U(f, Q)$ for partitions P, Q of $[a, b]$. Hence the set of lower sums is bounded above by each upper sum, and the set of upper sums is bounded below by each lower sum. By the Completeness Axiom, $l = \sup_P L(f, P)$ and $u = \inf_Q U(f, Q)$ exist, and $l \le u$.

If (a) holds, then for each n we have a partition R_n such that $U(f, R_n) - L(f, R_n) < 1/n$. Since $L(f, R_n) \le l \le u \le U(f, R_n)$, we have $|u - l| < 1/n$. Since n is arbitrary, l must equal u. Thus (a) implies (b).

Assume (b). By the elementary properties of infs and sups (Proposition 13.15), we have partitions P_n and Q_n such that $l - L(f, P_n) < 1/(2n)$ and $U(f, Q_n) - u < 1/(2n)$. Since $l = u$, adding the inequalities yields $U(f, Q_n) - L(f, P_n) < 1/n$. Let R_n be the least common refinement of P_n and Q_n. Lemma 17.8 now implies that

$$U(f, R_n) - L(f, R_n) \le U(f, Q_n) - L(f, P_n) < 1/n.$$

Thus (b) implies (c).

If (c) holds, then for each $\epsilon > 0$ there exists n such that $U(f, R_n) - L(f, R_n) < \epsilon$. Thus R_n is the desired partition R for (a).   ∎

17.10. Definition. The (bounded) function f is **integrable on** $[a, b]$ if the equivalent conditions in Proposition 17.9 hold. When f is integrable, the **integral of** f from a to b, written as $\int_a^b f(x)\,dx$ or $\int_a^b f$, is the common value of $\sup_P L(f, P)$ and $\inf_Q U(f, Q)$.

Comments. The full meaning of the notation "dx" is beyond the scope of this book; we view it as specifying the variable of integration. Informally, we also think of dx as the base of an infinitesimal rectangle at x with height $f(x)$. The integral sign is a stylized "S" representing the sum of the infinitesimal areas $f(x)\,dx$. By analogy with *summand*, we call f the **integrand**. When the integrand f is nonnegative on $[a, b]$, we define the **area under the graph** of f to be $\int_a^b f(x)\,dx$.

17.11. Remark. Let f be integrable on $[a, b]$.

a) Using the partition consisting of just one interval, we obtain

$$(b - a)\,\inf_{x \in [a,b]} f(x) \le \int_a^b f \le (b - a)\,\sup_{x \in [a,b]} f(x).$$

b) When P is a partition of $[a, b]$, the definitions of lower sum and upper sum imply that $L(f, P) \le \int_a^b f \le U(f, P)$. Each of $U(f, P)$ and $L(f, P)$ approximates the integral, and they squeeze $\int_a^b f$.

c) If f is nonnegative, then $\int_a^b f$ is nonnegative, since $L(f, P) \ge 0$ for each partition P.   ∎

To integrate continuous or monotone functions, we can use partitions whose intervals have equal size. Similar ideas apply also to functions that are piecewise continuous or piecewise monotone (continuous or monotone on each subinterval in some partition of $[a, b]$).

17.12. Example. For $p \in \mathbb{N}$, we obtain $\int_0^1 x^p\,dx = 1/(p + 1)$. Let P_n be the partition of $[0, 1]$ into n equal parts. The upper and lower sums are

$$U(f, P_n) = \sum_{i=1}^{n} \frac{1}{n}\left(\frac{i}{n}\right)^p = n^{-(p+1)} \sum_{i=1}^{n} i^p$$

$$L(f, P_n) = \sum_{i=1}^{n} \frac{1}{n}\left(\frac{i-1}{n}\right)^p = n^{-(p+1)} \sum_{i=0}^{n-1} i^p.$$

By Theorem 5.31, $\sum_{i=1}^{n} i^p$ is a polynomial in n with leading term $\frac{1}{p+1} n^{p+1}$. Substituting this into the formula for $U(f, P_n)$ and taking the limit as $n \to \infty$ yields $\lim U(f, P_n) = \frac{1}{p+1}$. Since $L(f, P_n) = U(f, P_n) - \frac{1}{n}$, also $\lim L(f, P_n) = \frac{1}{p+1}$. Since the limits are equal, the integral is $\frac{1}{p+1}$.   ∎

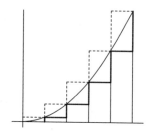

Definition 17.10 for $\int_a^b f(x)\,dx$ assumes that $a \le b$ and implies that $\int_a^a f(x)\,dx = 0$. When $a > b$, we define $\int_a^b f(x)\,dx = -\int_b^a f(x)\,dx$, if the latter exists. This suggests that when integrating from right to left we are "erasing" area under the curve.

We can now develop the theory of integration. First we relate infimum and supremum to sums.

17.13. Lemma. When f and g are bounded real-valued functions on S, $\inf_S(f + g) \ge \inf_S f + \inf_S g$ and $\sup_S(f + g) \le \sup_S f + \sup_S g$.

Proof: Let $B = S \times S$, and let $A = \{(x, x): x \in S\}$. Since $A \subseteq B$ yields $\inf_A h \ge \inf_B h$, we have

$$\inf_S(f + g) = \inf_{(x,x)\in A} (f(x) + g(x)) \ge \inf_{(x,y)\in B} (f(x) + g(y)) = \inf_S f + \inf_S g.$$

This proves the first inequality; the second is similar (Exercise 6). ∎

17.14. Proposition. (Linearity of Integration) If f, g are integrable on $[a, b]$, and $c \in \mathbb{R}$, then $f + g$ and cf are integrable on $[a, b]$, and the following formulas hold:
a) $\int_a^b (f + g) = \int_a^b f + \int_a^b g$,
b) $\int_a^b cf = c \int_a^b f$.

Proof: We first consider (a). Since f and g are integrable, Proposition 17.9 yields sequences $\{P_n\}$ and $\{Q_n\}$ of partitions of $[a, b]$ such that $\lim L(f, P_n) = \lim U(f, P_n)$ and $\lim L(g, Q_n) = \lim U(g, Q_n)$. Let R_n be the least common refinement of P_n and Q_n. By Lemma 17.8,

$$L(f, P_n) \le L(f, R_n) \le U(f, R_n) \le U(f, P_n).$$

Hence $\lim L(f, R_n) = \lim U(f, R_n)$; also, $\lim L(g, R_n) = \lim U(g, R_n)$.

The crucial point is to squeeze the lower and upper sums for $f + g$ between $L(f, R_n) + L(g, R_n)$ and $U(f, R_n) + U(g, R_n)$. For each subinterval J_i of R_n, Lemma 17.13 yields $\inf_{J_i} f + \inf_{J_i} g \le \inf_{J_i}(f + g)$. Multiplying by the length of J_i, applying the distributive law, and summing over i yields the first inequality below. The third inequality follows similarly. The middle inequality is always true (Example 17.6). Together,

$$L(f, R_n) + L(g, R_n) \le L(f + g, R_n) \le U(f + g, R_n) \le U(f, R_n) + U(g, R_n).$$

Here the leftmost and rightmost expressions converge to $\int_a^b f + \int_a^b g$. Hence the middle terms, squeezed between them, have the same limit. By Proposition 17.9, $f + g$ is integrable, and $\int_a^b (f + g) = \int_a^b f + \int_a^b g$.

We prove (b) for $c \ge 0$ and leave the case $c = -1$ to Exercise 9; the case $c < 0$ follows from these. Since f is integrable, there is a sequence

$\{P_n\}$ such that $L(f, P_n) \to \int_a^b f$ and $U(f, P_n) \to \int_a^b f$. Since $c \geq 0$, we have $\inf_J cf = c \inf_J f$ and $\sup_J cf = c \sup_J f$. Hence $L(cf, P_n) = cL(f, P_n) \to c\int_a^b f$ and $U(cf, P_n) = cU(f, P_n) \to c\int_a^b f$. By Proposition 17.9, cf is thus integrable, and $\int_a^b cf = c\int_a^b f$. ∎

17.15. Corollary. Let f and g be integrable on $[a, b]$. If $f \leq g$, then

$$\int_a^b f \leq \int_a^b g.$$

Proof: Apply Remark 17.11c to $g - f$ and use Proposition 17.14. ∎

17.16. Proposition. If f is integrable on $[a, b]$, and $c \in [a, b]$, then

$$\int_a^b f = \int_a^c f + \int_c^b f.$$

Proof: Since f is integrable on $[a, b]$, f is integrable on both $[a, c]$ and $[c, b]$ (Exercise 10). This yields sequences of partitions of $[a, c]$ and $[c, b]$ such that

$$\lim L(f, P_n) = \lim U(f, P_n) = \int_a^c f$$

$$\lim L(f, Q_n) = \lim U(f, Q_n) = \int_c^b f.$$

Let R_n be the partition of $[a, b]$ whose set of breakpoints is the union of the sets of breakpoints of P_n and Q_n. We have

$$L(f, R_n) = L(f, P_n) + L(f, Q_n)$$
$$U(f, R_n) = U(f, P_n) + U(f, Q_n).$$

Hence both $L(f, R_n)$ and $U(f, R_n)$ converge to $\int_a^c f + \int_c^b f$. ∎

17.17. Proposition. If f is integrable on $[a, b]$, then $|f|$ is integrable on $[a, b]$, and $\left|\int_a^b f\right| \leq \int_a^b |f| \leq (b - a) \sup_{[a,b]} |f|$.

Proof: Given $\epsilon > 0$, we need to find a partition P of $[a, b]$ such that $U(|f|, P) - L(|f|, P) < \epsilon$. Since f is integrable, there is a partition P such that $U(f, P) - L(f, P) < \epsilon$. We claim that the same partition P produces the desired inequality for $|f|$.

We first prove the following inequality for an interval I.

$$\sup_I(|f|) - \inf_I(|f|) \leq \sup_I(f) - \inf_I(f) \tag{$*$}$$

If $\inf_I(f) \geq 0$, then $|f| = f$ on I, and the two sides of $(*)$ are the same. If $\sup_I(f) \leq 0$, then $|f| = -f$ on I. The identity $\sup_I(-f) = -\inf_I(f)$ (Exercise 13.19) now yields equality in $(*)$ on I via

$$\sup_I(|f|) - \inf_I(|f|) = \sup_I(-f) - \inf_I(-f) = -\inf_I(f) + \sup_I(f).$$

In the remaining case, $\inf_I(f) < 0 < \sup_I(f)$. Now

$$\sup_I(|f|) - \inf_I(|f|) < \sup_I(|f|)$$

$$= \max\{\sup_I(f), -\inf_I(f)\} < \sup_I(f) - \inf_I(f).$$

The partition P determines intervals I_j of lengths c_j. Since $(*)$ holds,

$$U(|f|, P) - L(|f|, p) = \sum c_j(\sup_{I_j}|f| - \inf_{I_j}|f|)$$

$$\leq \sum c_j(\sup_{I_j}(f) - \inf_{I_j}(f))$$

$$= U(f, P) - L(f, P) < \epsilon$$

We conclude that $|f|$ is integrable on I.

The bound $\int_a^b |f| \leq (b - a)\sup(|f|)$ is an instance of Remark 17.11b. Since both $f \leq |f|$ and $-f \leq |f|$, the inequality $\left|\int_a^b f\right| \leq \int_a^b |f|$ follows from Corollary 17.15. ∎

These results justify *defining* area using integrals; they yield some of the properties we believe about area. Proposition 17.14a and Proposition 17.16 express the area of a region as the sum of the areas of two regions composing it. Proposition 17.14b explains how area behaves under a vertical change of scale (along with a change of orientation if $c < 0$). The statement of how area behaves under a horizontal change of scale is a special case of the change of variables formula (Theorem 17.24). Proposition 17.17 generalizes the discrete Triangle Inequality $|\sum x_i| \leq \sum |x_i|$ to integrals ("continuous sums").

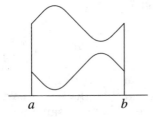

Proposition 17.14a

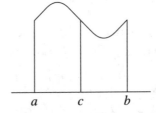

Proposition 17.16

Not all bounded functions are integrable. For example, the function that is 1 on the rationals and 0 on the irrationals is not integrable on any interval (Exercise 11). We next prove that every continuous function is integrable. Every bounded monotone function is integrable, regardless of whether it is continuous (Exercise 14).

17.18. Theorem. If f is continuous on the interval $[a, b]$, then f is integrable on $[a, b]$.

Proof: By Theorem 15.24 and Theorem 15.32, every function that is continuous on $[a, b]$ is bounded and uniformly continuous on $[a, b]$. Given $\epsilon > 0$, we seek a partition P such that $U(f, P) - L(f, P) < \epsilon$. Uniform continuity on $[a, b]$ yields a number $\delta > 0$ such that $t', t \in [a, b]$ with $|t' - t| < \delta$ implies $|f(t') - f(t)| < \frac{\epsilon}{b-a}$. For $n > (b - a)/\delta$, we let P be the partition of $[a, b]$ into n subintervals $J_1, \ldots, J_n$ of equal length. Since each subinterval has length $(b - a)/n$, the elements of J_i differ by less than δ. Hence $\sup_{J_i} f - \inf_{J_i} f < \frac{\epsilon}{b-a}$. By summing the contributions from each J_i, we have $U(f, P) - L(f, P) < \frac{b-a}{n} \sum_{i=1}^{n} \frac{\epsilon}{b-a} = \epsilon$. Hence P has the desired property, which proves that f is integrable on $[a, b]$. ∎

More advanced analysis considers other types of integrals. Our definition is the **Riemann integral**, due to G. F. B. Riemann (1826–1866). This definition applies only to bounded intervals and to bounded functions. "Improper integrals" sometimes overcome these limitations.

17.19. Definition. We define $\int_a^\infty f(x)\, dx$ to be $\lim_{b \to \infty} \int_a^b f(x)\, dx$, if this limit exists. When f is unbounded at a, we define $\int_a^b f(x)\, dx$ to be $\lim_{\epsilon \to 0} \int_{a+\epsilon}^b f(x)\, dx$, if this limit (through positive values of ϵ) exists. Integrals of these two types are **improper integrals**.

The definition of improper integral for unbounded intervals is analogous to the definition of infinite series using partial sums. Many important functions in mathematics can be expressed as improper integrals. For example, after we have defined the exponential function, we can show that $n! = \int_0^\infty e^{-x} x^n\, dx$ (Exercise 49).

Consider also the integral $\int_0^1 x^\alpha\, dx$ for $\alpha \in (-1, 0)$. Here the integrand is unbounded at 0. Using elementary calculus, we compute $\int_\epsilon^1 x^\alpha\, dx = \frac{1}{\alpha+1} - \frac{\epsilon^{\alpha+1}}{\alpha+1}$. As $\epsilon \to 0$, this approaches $1/(\alpha + 1)$. The justification for evaluating $\int_\epsilon^1 x^\alpha\, dx$ in this way is the Fundamental Theorem of Calculus, which we develop next.

THE FUNDAMENTAL THEOREM OF CALCULUS

The Fundamental Theorem of Calculus states precisely the sense in which differentiation and integration are inverse operations. This is the basis for the method of antiderivatives used to find indefinite integrals. From the Fundamental Theorem we obtain several techniques of integration, such as change of variables and integration by parts.

We treat $\int_a^x f(t)\, dt$ as a function of x, calling it $F(x)$. Thus $F(x)$ is the area under the graph of f from a to x. The first form of the Fundamental Theorem of Calculus states that at x, this area is changing at the rate $f(x)$. Thus, a continuous function is the derivative of its integral. In Theorem 17.22 we prove a second form, stating that a continuously differentiable function is the integral of its derivative.

17.20. Theorem. (Fundamental Theorem of Calculus) Let f be integrable on $[a, b]$, and let $F(x) = \int_a^x f(t)\, dt$ for $a < x < b$. If f is continuous at x, then F is differentiable at x, and $F'(x) = f(x)$.

Proof: Because f is integrable on $[a, b]$, it is also integrable on $[a, x]$ (Exercise 10), so $F(x)$ is defined for x in (a, b). To prove that $F'(x) = f(x)$, we show that $F(x + h) = F(x) + hf(x) + e(h)$, where e is an error function. Using Proposition 17.14a, we see that

$$e(h) = F(x + h) - F(x) - hf(x) = \int_x^{x+h} f(t)\, dt - hf(x).$$

Since $f(x)$ is a constant in terms of t, we have $hf(x) = \int_x^{x+h} f(x)\, dt$. Thus $e(h) = \int_x^{x+h} [f(t) - f(x)]\, dt$.

To show that e is an error function, we prove that $e(h)/h \to 0$. Let J be the interval $[x, x + h]$ if $h > 0$ and $[x + h, x]$ if $h < 0$; observe that the length of J is $|h|$. Using Proposition 17.17, we see that

$$|e(h)/h| = \tfrac{1}{|h|} \left| \int_x^{x+h} (f(t) - f(x))\, dt \right| \le \sup_{t \in J} |f(t) - f(x)|.$$

Since f is continuous at x, this converges to 0 as $h \to 0$. ∎

17.21. Example. *Necessity of continuity.* In the Fundamental Theorem of Calculus, we cannot drop the hypothesis that f is continuous at x. Suppose that $f(x) = 1$ for $0 \le x \le 1$ and that $f(x) = -1$ for $-1 \le x < 0$. It follows that f is integrable on $[-1, 1]$. Let $F(x) = \int_{-1}^x f(t)\, dt = |x| - 1$. Thus f is not continuous at 0, and F is not differentiable at 0. ∎

We define the **average value** of a function f integrable on $[a, b]$ to be $\frac{1}{b-a} \int_a^b f$. Theorem 17.20 states that the slope $m_{x,x+h}$ for the function F converges to $f(x)$ as $h \to 0$. The proof parallels that of Proposition 14.11, where we studied the limit of a sequence of averages.

The second form of the Fundamental Theorem states that the slope $m_{a,x}$ for F equals the average value of F' on the interval $[a, x]$.

17.22. Theorem. (Fundamental Theorem of Calculus, second form) If F is continuously differentiable on an open interval containing $[a, b]$, then $\int_a^x F'(t)\, dt = F(x) - F(a)$ for all $x \in [a, b]$.

Proof: Let $G(x) = \int_a^x F'(t)\, dt - (F(x) - F(a))$; this is well-defined since F' is integrable. By the first version of the Fundamental Theorem, G is differentiable, and $G'(x) = F'(x) - F'(x) = 0$ for all $x \in [a, b]$. By Corollary 16.32, G is constant. Since $G(a) = 0$, we have $G(x) = 0$ for all x. Thus the integral equals $F(x) - F(a)$. ∎

The Fundamental Theorem of Calculus enables us to evaluate $\int_a^x f$ by finding a function F whose derivative is f, thus avoiding the use of upper and lower sums. It is convenient to write $F(b) - F(a)$ as $F(x)|_a^b$, which we read as "$F(x)$ evaluated from $x = a$ to $x = b$." Since functions with the same derivative differ by a constant, we sometimes introduce an additive constant C and informally say "the **indefinite integral** of $f(x)$ is $F(x) + C$." The notation $F(x) + C$ represents an equivalence class of functions; here two functions are equivalent when they differ by a constant. The constant C disappears when we evaluate $F(b) - F(a)$.

17.23. Example. *Example 17.12 revisited.* When $p \neq -1$, we have $(d/dx)\frac{x^{p+1}}{p+1} = x^p$. The Fundamental Theorem of Calculus yields

$$\int_0^1 x^p = \frac{x^{p+1}}{p+1}\bigg|_0^1 = \frac{1}{p+1} - 0 = \frac{1}{p+1}.$$ ∎

Since Example 17.23 does not use Theorem 5.31, it makes precise the approach suggested in Example 5.46 for computing the leading term of $\sum_{i=1}^n i^k$. By estimating the area given by the integral using trapezoids instead of rectangles, once can prove Theorem 5.31: see Exercise 32.

The Fundamental Theorem of Calculus implies the change of variables formula for a definite integral.

17.24. Theorem. (Change of Variables) Let f be continuous on $[a, b]$, and let g be a continuously differentiable bijection from the interval $[a, b]$ to the interval $[g(a), g(b)]$. For every $x \in [a, b]$,

$$\int_a^x f(g(z))g'(z)dz = \int_{g(a)}^{g(x)} f(t)\, dt.$$

Proof: Both sides depend on a and x. Fixing a, we view them as functions of x. By the Fundamental Theorem of Calculus, both these functions are differentiable (using the continuity of $(f \circ g) \cdot g'$). We show they are equal by showing that they have the same derivative and that they agree at $x = a$ (equality then follows from Corollary 16.32).

Both functions are 0 at $x = a$. The derivative of the function on the left is $f(g(x))g'(x)$, by the Fundamental Theorem. The derivative of the function on the right is also $f(g(x))g'(x)$, by the Fundamental Theorem and the chain rule. ∎

The product rule for differentiation combines with the Fundamental Theorem of Calculus to yield an important technique of integration.

17.25. Theorem. (Integration by Parts) If u and v are continuously differentiable, then

$$\int_a^b u \cdot v' = (uv)\Big|_a^b - \int_a^b v \cdot u',$$

where $(uv)\Big|_a^b = u(b)v(b) - u(a)v(a)$.

Proof: Let $F = uv$. By the product rule, $F' = u'v + uv'$, which is continuous. By the Fundamental Theorem of Calculus, $\int_a^b [u'v+uv'] = F(b)-F(a)$. Subtracting $\int_a^b v \cdot u'$ from both sides completes the proof. ∎

Integration by parts has a particularly nice geometric interpretation for monotone functions. A monotone continuous function is a bijection from its domain to its image, and therefore it has an inverse.

17.26. Theorem. If f is a monotone increasing function, with inverse function f^{-1}, then $\int_a^b f(x)\,dx = yf^{-1}(y)\Big|_s^t - \int_s^t f^{-1}(y)\,dy$, where $s = f(a)$ and $t = f(b)$.

Proof: By translating the origin if necessary, we may reduce the computation to the case where a and s are positive; similarly we may assume $b \geq a$ and $t \geq s$ (Exercise 34). It suffices to prove that $\int_a^b f(x)\,dx + \int_s^t f^{-1}(y)\,dy = yf^{-1}(y)\Big|_s^t$. Both sides compute the same area, as suggested by the picture below. The right side is the area of the difference between the two rectangles. The left side computes this as the sum of two pieces, separated by the graph of f.

If $xf'(x)$ is integrable as a function of x, then we can also prove this using integration by parts and Theorem 17.24. We make the change of variables $x = f^{-1}(y)$.

$$\int_a^b f(x)\,dx = xf(x)\Big|_a^b - \int_a^b xf'(x)\,dx = f^{-1}(y)y\Big|_s^t - \int_s^t f^{-1}(y)\,dy. \quad \blacksquare$$

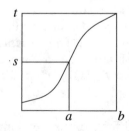

EXPONENTIALS AND LOGARITHMS

We are now prepared to define the logarithm and with it the exponential function. We use the Fundamental Theorem of Calculus to derive their basic properties.

17.27. Definition. For $x > 0$, the **natural logarithm** of x (written $\ln x$) equals $\int_1^x \frac{1}{t}\,dt$.

17.28. Theorem. The natural logarithm is a strictly increasing function having the property that $\ln(xy) = \ln x + \ln y$.

Proof: Since $1/t$ is continuous for $t > 0$, the integral $\int_1^x \frac{1}{t}\,dt$ exists as a function of x. By the Fundamental Theorem of Calculus, its derivative is $1/x$, which is positive, and hence the natural logarithm is a strictly increasing function.

To prove that $\ln(xy) = \ln x + \ln y$, we let $g(x) = \ln(xy) - \ln x$ for fixed y and show that g has the constant value $\ln y$. To show first that g is constant, we show that $g'(x) = 0$ for all x. Using the Fundamental Theorem of Calculus and the chain rule, we have

$$g'(x) = \frac{1}{xy}\frac{d}{dx}(xy) - \frac{1}{x} = \frac{1}{xy}y - \frac{1}{x} = 0.$$

Thus g is constant. Since $\ln 1 = 0$, setting $x = 1$ yields the result $g(x) = \ln(xy) - \ln x = \ln y$. ∎

The property $\ln(xy) = \ln x + \ln y$ implies that $\ln x$ is unbounded above as x increases and unbounded below as x approaches 0 (Exercise 27). Since the natural logarithm also is strictly increasing, it is a bijection from the set of positive real numbers to $\mathbb{R}$. We give a clean definition of exponentiation by using the inverse of the logarithm function.

17.29. Definition. The **exponential function** is the bijection from $\mathbb{R}$ to the set of positive real numbers that is the inverse of the logarithm function. The value of the exponential function at x, written e^x, is the number $y > 0$ such that $x = \ln y$. For $a > 0$, we define a^x to be $e^{x \ln a}$.

The number we call "e" is the unique y such that $\int_1^y (1/t)\,dt = 1$. The notation for the exponential function is motivated by the formula $e^{a+b} = e^a e^b$, which follows from Theorem 17.28 (see Exercise 29).

Many applied problems involve a mechanism in which the growth rate of a time-dependent function is proportional to its value. These include compounding of interest, analysis of current in electrical circuits, and problems of radioactive decay. The exponential function arises in such problems because it is its own derivative.

17.30. Corollary. If $g(y) = e^y$, then $g'(y) = g(y)$. Furthermore, if $a > 0$ and $h(y) = a^y$, then $h'(y) = a^y \ln a = h(y) \ln(a)$.

Proof: The function g is defined to be the inverse of $\ln$, which is strictly increasing. By the Fundamental Theorem of Calculus, $(d/dx) \ln x = 1/x$. Letting $f(x) = \ln x$, we have $g = f^{-1}$. By Proposition 16.24, g is differentiable, and $g'(y) = 1/f'(g(y)) = e^y$. The second statement follows from the first statement and the chain rule, using $a^y = e^{y \ln(a)}$. ∎

    The differential equation $g'(x) = cg(x)$ is analogous to the recurrence $a_{n+1} - a_n = ca_n$. In each case, the change is proportional to the function value. The recurrence has solution $a_n = A(1+c)^n$, where $A = a_0$ (Theorem 12.16), and the differential equation has solution $g(x) = Ae^{cx} = A(e^c)^x$, where $A = g(0)$. The analogy extends to higher-order constant-coefficient differential equations. The constant-coefficient second-order differential equation $g''(x) - (\alpha + \beta)g'(x) + \alpha\beta g(x) = 0$ has solution $g(x) = A_1 e^{\alpha x} + A_2 e^{\beta x}$ when $\alpha \neq \beta$. Here again we need two initial conditions to determine A_1 and A_2. We will not explore the vast subject of differential equations.

    One simple application of the exponential function is its role in the compounding of interest. This application requires the evaluation of a certain limit, which is sometimes given as the definition of e^x.

17.31. Theorem. If $x \in \mathbb{R}$ and $a_n = (1 + x/n)^n$, then $a_n \to e^x$.

Proof: Letting $t = 1/n$, we consider $(1 + xt)^{1/t}$. For $t \neq 0$, this is a continuous function of t. By the sequential definition of convergence, $\lim_{n \to \infty} a_n$ must be the limit of this function as $t \to 0$, if it exists. Let $f(t) = \ln((1 + tx)^{1/t}) = \frac{\ln(1+tx)}{t}$. We can use l'Hôpital's rule (Theorem 16.36) to evaluate $\lim_{t \to 0} f(t) = \lim_{t \to 0} \frac{x}{1+tx} = x$. Thus $\ln(a_n) \to x$. Since the exponential function is continuous, $a_n \to e^x$. ∎

17.32. Solution. *Continuous Compounding* (Problem 17.3). Under continuous compounding, an amount p in a savings account at interest rate $x\%$ grows to $p \lim_{n \to \infty} (1 + x/n)^n = pe^x$ after one year. Hence the yield is $e^x - 1$. For small x, the yield is approximately $x + x^2/2$. ∎

    The alert reader will recall that we defined $\exp(x)$ in Example 14.32 using series. We prove in Application 17.52 that this is the same function as the exponential function defined in Definition 17.29. To do so, we develop machinery allowing us to differentiate a power series term by term. The same technique yields the definitions and properties of the trigonometric functions sine and cosine, which we discuss before presenting the technical details of differentiating term by term.

TRIGONOMETRIC FUNCTIONS AND π

The fundamental constant π and the sine and cosine functions arise throughout mathematics and science; π is the area inside the unit circle, and the trigonometric functions describe the relationship between the sides and the angles of a right triangle with hypotenuse 1. Because geometric reasoning is difficult to make precise, we instead define π and these functions using series. This leads to a proof that π is the area of the unit circle. It is possible to define the sine and cosine functions rigorously using geometric reasoning; this also would allow us to prove Proposition 17.34 and from it the series expansions that we use as the definitions. Our approach is direct but not geometric.

17.33. Definition. The **sine** and **cosine** functions are defined on $\mathbb{R}$ by

$$\sin x = \sum_{n=0}^{\infty} \frac{(-1)^n}{(2n+1)!} x^{2n+1} \qquad \text{and} \qquad \cos x = \sum_{n=0}^{\infty} \frac{(-1)^n}{(2n)!} x^{2n}.$$

17.34. Proposition. The sine and cosine functions are defined and differentiable on all of $\mathbb{R}$. The derivatives are $(d/dx)\sin x = \cos x$ and $(d/dx)\cos x = -\sin x$.

Proof: These functions are defined for all x because the ratio test gives an infinite radius of convergence for the series. For a fixed x, the absolute value of ratio of successive terms in the series is $\frac{x^2}{(2n+3)(2n+2)}$ for $\sin x$ and $\frac{x^2}{(2n+2)(2n+1)}$ for $\cos x$; both converge to 0. To show that sine and cosine are differentiable, we would like to interchange the order of differentiation and summation so that we can differentiate the series term by term. Theorem 17.51 allows us to do this. When we differentiate term by term, the series for $\sin x$ becomes the series for $\cos x$, and the series for $\cos x$ becomes the series for $-\sin x$. ∎

17.35. Proposition. $\sin^2 x + \cos^2 x = 1$ for all $x \in \mathbb{R}$.

Proof: From the series definition, we obtain $\sin 0 = 0$ and $\cos 0 = 1$. Now let $f(x) = \sin^2 x + \cos^2 x$. Using the chain rule, we compute $f'(x) = 2 \sin x \cos x - 2 \cos x \sin x = 0$. By Corollary 16.32, f is constant, so $f(x) = f(0) = 1$ for all x. ∎

17.36. Corollary. The sine and cosine functions are bounded, with $|\sin x| \leq 1$ and $|\cos x| \leq 1$ for all x.

Proof: Immediate from Proposition 17.35. ∎

17.37. Proposition. There is a point $x_0 > 0$ such that $\cos x_0 = 0$.

Proof: Since $\cos 0 = 1 > 0$ and differentiability implies continuity, it suffices by the Intermediate Value Theorem to show that there exists a positive x such that $\cos x < 0$. We show that $\cos 2$ is negative. From the definition, $\cos 2 = 1 - \frac{2^2}{2} + \frac{2^4}{24} + \sum_{n=3}^{\infty} \frac{(-1)^n}{(2n)!} x^{2n}$. The first three terms sum to $-1/3$. We consider the remaining terms in pairs. The two terms with $n = 2k - 1$ and $n = 2k$ sum to

$$\frac{(-1)^{2k-1}}{(4k-2)!} 2^{4k-2} + \frac{(-1)^{2k}}{(4k)!} 2^{4k} = -\frac{2^{4k-2}}{(4k-2)!} \left(1 - \frac{2^2}{4k(4k-1)} \right).$$

This is negative when $k(4k - 1) > 1$. Hence each successive pair of the remaining terms is negative, and $\cos 2 < -1/3$. ∎

Thus the set $S = \{x > 0 : \cos x = 0\}$ is nonempty. Since S is bounded below by 0, it has an nonnegative infimum α. The set S contains a sequence $\langle x \rangle$ converging to α. Since the cosine function is continuous, $\cos \alpha = \cos(\lim x_n) = \lim(\cos x_n) = 0$. Since $\cos 0 = 1 \neq 0$, we have $\alpha > 0$.

17.38. Definition. The number π is defined to be 2α, where α is the smallest positive zero of $\cos x$.

Since $\cos 2 < 0$, we have shown that $\pi < 4$. We will soon obtain more accurate estimates for π. First we relate π to the area of a circle.

17.39. Lemma. $\int_0^{\pi/2} \sin^2 x \, dx = \int_0^{\pi/2} \cos^2 x \, dx = \pi/4$.

Proof: By Proposition 17.35, we have

$$\int_0^{\pi/2} \sin^2 x \, dx = \int_0^{\pi/2} (1 - \cos^2 x) \, dx = \pi/2 - \int_0^{\pi/2} \cos^2 x \, dx.$$

Hence the two desired integrals sum to $\pi/2$.

Using integration by parts (Theorem 17.25), we have

$$\int_0^{\pi/2} \cos x (\cos x \, dx) = \sin x \cos x \Big|_0^{\pi/2} + \int_0^{\pi/2} \sin^2 x \, dx.$$

Since $\sin 0 = 0$ and $\cos(\pi/2) = 0$, the first term is 0. Hence the two integrals in the statement of the lemma are equal. Since they sum to $\pi/2$, each equals $\pi/4$. ∎

17.40. Proposition. The functions mapping x to $\sin x$ and x to $\cos x$ are bijections from $[0, \pi/2]$ to $[0, 1]$.

Proof: By the definition of π, the sine function is increasing and the cosine function is decreasing on the interval $[0, \pi/2]$; hence they are injective. Since the values are 0 and 1 at the endpoints of $[0, \pi/2]$, the Intermediate Value Theorem implies that they are surjective. ∎

17.41. Theorem. The area inside a circle of radius 1 is π.

Proof: Consider a circle with center at the origin. The circle is defined to be $\{(u, v) \in \mathbb{R}^2 : u^2 + v^2 = 1\}$. By symmetry, the area is four times the area inside the quarter circle in the first quadrant. This is the area enclosed by the axes and by the curve $v = \sqrt{1 - u^2}$. Because $\sin x$ defines a bijection from $[0, \pi/2]$ to $[0, 1]$, we can let $u = \sin x$ and use the change of variables formula (Theorem 17.24) and Lemma 17.39 to compute $\int_0^1 \sqrt{1 - u^2}\, du = \int_0^{\pi/2} \cos^2 x\, dx = \pi/4$. ∎

17.42. Solution. *The Rainfall Problem* (Problem 17.2). If the rain falls uniformly, then the fraction of the rain that falls inside the circle is the ratio of the area of the circle to the area of the square. The sides of the square are twice as long as the radius of the circle. By our understanding of scale factors in area, the ratio of the two areas remains the same no matter what the radius is. When the radius is 1, the area of the circle is π and the area of the square is 4. Hence the answer is $\pi/4$.

We interpret this to mean that $\pi/4$ is the probability that a randomly chosen raindrop falls within the circle. This suggests the role that integrals and area play in probability. ∎

The value of π has been computed to millions of decimal places. To ten places it is 3.1415926535. The methods we have developed yield some crude estimates for π. First consider the formula we have using the integral: $\pi = 4 \int_0^1 \sqrt{1 - x^2}\, dx$. We can use the definition of integration to approximate π. Let P be the partition of $[0, 1]$ into 100 equal pieces. For $f(x) = \sqrt{1 - x^2}$, we have $L(f, P) = (1/100^2) \sum_{k=1}^{100}(100^2 - k^2)^{1/2}$ and $U(f, P) = (1/100^2) \sum_{k=0}^{99}(100^2 - k^2)^{1/2}$. Using a calculator, we obtain $4L(f, P) = 3.12042$ and $4U(f, P) = 3.16042$. The convergence is slow.

We can also obtain a series that converges to π. Using the change of variables $x = \sin y / \cos y$, we have $\int_0^1 (1 + x^2)^{-1} dx = \int_0^{\pi/4} dy = \pi/4$. Since $\left|1/(1 + x^2)\right| < 1$ for $x > 0$, we can expand $\left|1/(1 + x^2)\right|$ using a geometric series to obtain $\sum_{n=0}^{\infty}(-x^2)^n$ as the integrand. Theorem 17.43 justifies the interchange of the summation and integration operations (applied with f_n equal to the nth partial sum of the series). Integrating term by term, we obtain $\pi/4 = \sum_{n=0}^{\infty}(-1)^n \int_0^1 x^{2n} dx = 1 - \frac{1}{3} + \frac{1}{5} - \frac{1}{7} + \cdots$. Taking 100 terms in the sum and multiplying by 4 yields the approximation 3.15149, which differs from π by about .01. Again convergence is slow.

A more geometric method is to inscribe regular n-gons in a circle of radius one and find their areas. Taking the limit as $n \to \infty$ yields the area of the circle, π. This approach is due to Pythagoras and also converges slowly. There are many methods for computing the decimal expansion of π, some of which converge much more rapidly than these.

A RETURN TO INFINITE SERIES

We would like to be able to differentiate or integrate a power series term by term. By Theorem 16.16 or Proposition 17.14a, respectively, we may interchange differentiation or integration with finite sums. These interchanges are not generally valid with infinite sums. Our work with uniform convergence leads us to general theorems (Theorem 17.43 for integrals and Theorem 17.45 for derivatives) giving circumstances under which we can interchange summation with integration and differentiation. The results apply in particular to convergent power series.

17.43. Theorem. Suppose that $\{f_n\}$ is a sequence of continuous functions on an interval $[a, b]$ and that $x \in [a, b]$. If $\{f_n\}$ converges uniformly to f on $[a, b]$, then $\int_a^x f_n(t)\, dt \to \int_a^x f(t)\, dt$.

Proof: By Theorem 16.65, f is continuous, and thus by Theorem 17.18 it is integrable. We prove that $\int_a^x f_n(t)\, dt - \int_a^x f(t)\, dt$ converges to 0. Consider $\epsilon > 0$. Applying Proposition 17.14 and Proposition 17.17, we have

$$\left| \int_a^x f_n(t)\, dt - \int_a^x f(t)\, dt \right| \le \int_a^x |f_n(t) - f(t)|\, dt.$$

Using uniform convergence, we can choose N such that $n \ge N$ and $t \in [a, b]$ implies $|f_n(t) - f(t)| < \epsilon/(b - a)$. Thus $n \ge N$ implies $\int_a^x |f_n(t) - f(t)|\, dt < \epsilon$, as desired. ∎

17.44. Solution. *Area and Limits.* The conclusion of Theorem 17.43 is not valid without the hypothesis of uniform convergence. This leads to a counterexample for Problem 17.1. Define f_n on the interval $[0, 2]$ by

$$f_n(x) = \begin{cases} n^2 x & \text{if } 0 \le x \le 1/n \\ 2n - n^2 x & \text{if } 1/n \le x \le 2/n \\ 0 & \text{if } 2/n \le x \le 1. \end{cases}$$

Below we graph f_2. The region under the graph of f_n is an isosceles triangle with base $2/n$, height n, and hence area 1. Thus $\lim_{n \to \infty} \int_0^2 f_n(x)\, dx = 1$. On the other hand, f_n converges pointwise to the zero function on $[0, 2]$. Thus $\int_0^2 \lim f_n(x)\, dx = 0 \ne 1 = \lim_{n \to \infty} \int_0^2 f_n(x)\, dx$. ∎

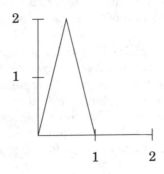

17.45. Theorem. Let $\{F_n\}$ be a sequence of continuously differentiable functions on an interval $[a, b]$, and suppose that $F_n(a)$ converges. If the sequence $\{F_n'\}$ converges uniformly to f on $[a, b]$, then the sequence $\{F_n\}$ converges to a continuously differentiable function F whose derivative is f.

Proof: Using the Fundamental Theorem of Calculus, we may write $F_n(x) - F_n(a) = \int_a^x F_n'(t)\,dt$. Since F_n' converges uniformly, Theorem 16.65 implies that the limit function f is continuous. Also, Theorem 17.43 implies that $\int_a^x F_n'(t)\,dt$ converges to $\int_a^x f(t)\,dt$. By hypothesis, $F_n(a)$ converges to some number c. Define F by $F(x) = c + \int_a^x f(t)\,dt$; we have shown that $\lim_{n\to\infty} F_n(x) = F(x)$.

Since the integrand f is continuous, the Fundamental Theorem guarantees that F is differentiable and that $F' = f$. Hence F is continuously differentiable. ∎

17.46. Theorem. For every power series $\sum_{n=0}^{\infty} c_n x^n$, there is an $R \geq 0$ such that the series converges for $|x| < R$ and diverges for $|x| > R$.

Proof: Let $L = \limsup(|c_n|^{1/n})$; note that $\limsup(|c_n x^n|^{1/n}) = L\,|x|$. By the root test (Theorem 14.36), the series converges when $L\,|x| < 1$ and diverges when $L\,|x| > 1$. Therefore the claim follows using $R = 1/L$. (When $\limsup |a_n|^{1/n} = \infty$, we set $R = 0$). ∎

17.47. Definition. The **radius of convergence** of a power series is the number R associated with it by Theorem 17.46.

17.48. Example. For a power series in x with radius of convergence R, anything can happen when $|x| = R$. Consider the three series below.

$$\text{a) } \sum \frac{x^n}{n^2} \qquad \text{b) } \sum \frac{x^n}{n} \qquad \text{c) } \sum \frac{x^{2n}}{2n}$$

In each case, $R = 1$. In case (a), we have convergence at $x = \pm 1$ In case (b), we have convergence at $x = -1$ and divergence at $x = 1$. In case (c), we have divergence at $x = \pm 1$. Exercise 46 requests the details. ∎

17.49. Proposition. Let $\sum_{n=0}^{\infty} a_n x^n$ be a power series that converges when $x = R$. If $0 < r < R$, then the series converges uniformly on the interval $[-r, r]$.

Proof: By Lemma 16.61, it suffices to show that the sequence of partial sums is uniformly Cauchy. Observe first that since the series converges at R, there is a constant C such that $|a_n R^n| \leq C$ for all n. Comparing partial sums $s_N(x)$ and $s_M(x)$ for $M < N$ yields

$$\left| \sum_{n=0}^{N} a_n x^n - \sum_{n=0}^{M} a_n x^n \right| = \left| \sum_{n=M+1}^{N} a_n x^n \right| \le \sum_{n=M+1}^{N} |a_n x^n|$$

$$= \sum_{n=M+1}^{N} |a_n R^n| \left| \frac{x}{R} \right|^n$$

$$\le C \sum_{n=M+1}^{N} \left| \frac{x}{R} \right|^n \le C \sum_{n=M+1}^{N} \left(\frac{r}{R} \right)^n.$$

The last expression is bounded by ϵ when M is sufficiently large, since it is $(r/R)^{M+1}$ times a convergent geometric series. Thus the sequence of partial sums is uniformly Cauchy. ∎

17.50. Lemma. $\limsup |a_n|^{1/n} = \limsup |na_n|^{1/(n-1)}$.

Proof: This follows from $\lim n^{1/(n-1)} = 1$ (see Exercise 48). ∎

17.51. Theorem. Let $\sum_{n=0}^{\infty} a_n x^n$ be a power series with radius of convergence R. If $0 < r < R$, then the function F defined on $[-r, r]$ by $F(x) = \sum_{n=0}^{\infty} a_n x^n$ is differentiable on $(-r, r)$, with $F'(x) = \sum_{n=1}^{\infty} a_n n x^{n-1}$.

Proof: By Lemma 17.50 and Theorem 14.36, the radius of convergence of the series $\sum_{n=1}^{\infty} a_n n x^{n-1}$ is at least R. By Proposition 17.49, the series converges uniformly on $[-r, r]$. We know that $F(x)$ converges for $x = r$. The conclusion now follows from Theorem 17.45. ∎

We now prove that our definitions of the exponential function agree.

17.52. Application. The function defined by $\exp(x) = \sum_{k=0}^{\infty} x^k/k!$ is the inverse of the natural logarithm.

Proof: In Example 14.32, we proved that $f(x) = \sum_{k=0}^{\infty} x^k/k!$ converges for all x. Since $(1/k!)kx^{k-1} = x^{k-1}/(k-1)!$, the series resulting from term-by-term differentiation is the same as the original series. Theorem 17.51 then tells us that f is differentiable and that $f' = f$.

We next show that $f(x) > 0$ for all x. This holds for $x > 0$ because then the coefficients of the power series are positive. Because $f(0) = 1$ and $\exp(a+b) = \exp(a)\exp(b)$ (Exercise 14.67), we have $1 = f(0) = f(x - x) = f(x)f(-x)$. Hence $f(x) = 1/f(-x) > 0$ if $x < 0$.

To prove that f is the inverse of the natural logarithm, we prove that the composition $g = \ln \circ f$ is the identity function. Since f is positive, g is well-defined. Because f is differentiable, g is also differentiable. Using the chain rule, we have $g'(x) = (1/f(x))f'(x) = 1$. Also, $f(0) = 1$ implies $g(0) = 0$. By the Fundamental Theorem of Calculus,

$$g(x) = \int_0^x g'(t) \, dt = \int_0^x 1 \, dt = x$$

for all x. Hence g is the identity function, as desired. ∎

EXERCISES

17.1. $(-)$ Let $f(x) = \min\{x, 2-x\}$ and $g(x) = \max\{x, 2-x\}$. Evaluate $\int_0^2 f(x)\,dx$ and $\int_0^2 g(x)\,dx$.

17.2. $(-)$ Determine the yields on bank accounts paying 6% simple interest, 6% interest compounded daily, and 6% compounded continuously.

17.3. $(-)$ How many years does it take to double the value of a bank account paying 4% simple interest? How many years if the interest rate is $p\%$?

17.4. Give a proof or a counterexample: "If f is bounded and nonconstant on $[0, 1]$, then for each partition P of the interval, $L(f, P) < U(f, P)$."

17.5. Give a proof or a counterexample: "If f is continuous and nonconstant on $[0, 1]$, then for each partition P of the interval, $L(f, P) < U(f, P)$."

17.6. $(-)$ Let f and g be bounded real-valued functions on a set S. Prove that $\sup_S(f + g) \le \sup_S f + \sup_S g$. Give an example where the two sides differ.

17.7. $(-)$ Let $f(x) = x^2$, and let P_n be the partition of $[1, 3]$ into n subintervals of equal length. Compute formulas for $L(f, P_n)$ and $U(f, P_n)$ in terms of n. Verify that they have the same limit. Determine how large n must be to ensure that $U(f, P_n)$ is within .01 of $\int_1^3 f(x)\,dx$.

17.8. Consider $f\colon [a, b] \to \mathbb{R}$. For partitions P, Q, R of $[a, b]$, prove that
   a) $L(f, P) \le L(f, R) \le U(f, R) \le U(f, P)$ when R is a refinement of P. (Hint: Consider the case where R has one more breakpoint than P and apply induction.)
   b) $L(f, P) \le U(f, Q)$. (Hint: Consider their least common refinement.)

17.9. Prove that if f is integrable on $[a, b]$, then $-f$ is integrable on $[a, b]$, with $\int_a^b(-f) = -\int_a^b f$. What is the geometric interpretation of $\int_a^b(f - g) = \int_a^b f - \int_a^b g$?

17.10. Let f be integrable on $[a, b]$. For $a < c < b$, prove that f is integrable on $[a, c]$ and on $[c, b]$.

17.11. Define $f\colon [0, 1] \to [0, 1]$ by $f(x) = 1$ if x is rational and $f(x) = 0$ if x is irrational. Prove that f is not integrable.

17.12. Give an example of a function f such that $|f|$ is integrable on $[0, 1]$ but f is not integrable on $[0, 1]$.

17.13. (!) *Mean Value Theorem for integrals.* Let f be continuous on $[a, b]$. Prove that there exists $c \in [a, b]$ such that $f(c) = \frac{1}{b-a}\int_a^b f$. (Hint: First prove the special case when $\int_a^b f = 0$. Consider the function $f - \frac{1}{b-a}\int_a^b f$ to reduce the general statement to this case.)

17.14. (!) *Integration of monotone functions.* Let f be increasing on the interval $[a, b]$, and let P_n be the partition of $[a, b]$ into n intervals of equal length. Obtain a formula for $U(f, P_n) - L(f, P_n)$. Use this to show that f is integrable on $[a, b]$.

17.15. Let f be continuous on the interval $[a, b]$.
   a) Prove that if $f(x) \ge 0$ for $x \in [a, b]$ and f is not everywhere zero on $[a, b]$, then $\int_a^b f(x)\,dx > 0$.

b) Prove that if $\int_a^b f(t)g(t)\,dt = 0$ for every continuous function g on $[a, b]$, then $f(x) = 0$ for $a \le x \le b$.

17.16. Let $g(x) = \int_0^x (1 + t^2)^{-1}\,dt + \int_0^{1/x} (1 + t^2)^{-1}\,dt$. Prove that g is constant. (Comment: We have shown that $\int_0^1 (1 + t^2)^{-1}\,dt = \frac{\pi}{4}$, so the constant must be $\frac{\pi}{2}$.)

17.17. Suppose that $f : [0, 1] \to [0, 1]$ is a bijection with $f(0) = 1$ and $f(1) = 0$. Prove that $\int_0^1 f(x)\,dx = \int_0^1 f^{-1}(y)\,dy$.

17.18. Use Exercise 17.17 to show that $\int_0^1 (1 - x^a)^{1/b}\,dx = \int_0^1 (1 - x^b)^{1/a}\,dx$. Evaluate the integral when a and b are positive integers

17.19. (!) For $x > 0$, determine $\lim_{h \to 0} \left(\frac{1}{h} \ln(\frac{x+h}{x}) \right)$.

17.20. (!) Evaluate $\frac{1}{n} \sum_{k=1}^n \ln(k/n)$ as a function of n. Interpret the sum as a lower sum of an improper integral, and evaluate its limit as $n \to \infty$.

17.21. (!) Let N be a positive integer, and let $a_n = \sum_{j=n+1}^{(N+1)n} (1/j)$.
   a) Consider the lower sum $L(f, P)$ where $f(x) = 1/x$ and P is the partition of $[1, N + 1]$ into Nn pieces. Change the index of summation to obtain $L(f, P) = a_n$.
   b) Evaluate $\lim a_n$.

17.22. Use the definition of $\ln$ as an integral to prove that $\ln(\frac{x+1}{x}) > \frac{1}{x+1}$ when $x > 0$. (Hint: Use a lower sum for an appropriate integral.)

17.23. Let $f(x) = (1 + 1/x)^x$ for $x > 0$. Prove that f is increasing. (Hint: Find $f'(x)$ and use Exercise 17.22.)

17.24. Prove the product rule $(fg)' = f'g + fg'$ by considering $(d/dx)(\ln(fg))$.

17.25. (+) For $n \in \mathbb{N}$ and $b \ge 1$, prove that $\ln(b) \ge n(1 - b^{-1/n})$. (Hint: Use a lower sum for the integral.)

17.26. Use Theorem 17.26 to find indefinite integrals of $\ln(x)$ and $\tan^{-1}(x)$.

17.27. Use the property $\ln(xy) = \ln x + \ln y$ to prove that the logarithm function is unbounded above and below.

17.28. Let f be continuous, and suppose that $f(x) = \int_0^x f(t)\,dt + c$. Determine f.

17.29. *Properties of exponentiation.*
   a) Use the properties of the logarithm to prove that $e^{x+y} = e^x e^y$.
   b) For $\alpha > 0$, compute $(d/dx)x^\alpha$. (Hint: Use Definition 17.29.)
   c) Complete the details of proving that $(d/dx)a^x = a^x \ln a$ (Corollary 17.30).

17.30. (+) For $x, a > 0$, find all solutions to the equation $x^a = a^x$.

17.31. Compute the geometric sum $\sum_{k=0}^n e^{kx}$. Differentiate both sides p times to prove that $\sum_{k=0}^n k^p$ is the value at $x = 0$ of $(\frac{d}{dx})^p \frac{1 - e^{(n+1)x}}{1 - e^x}$. (Comment: Compare with Theorem 5.31.)

17.32. (+) Use trapezoids to obtain upper and lower bounds on $\int_0^n x^k\,dx$. Use this to prove Theorem 5.31.

17.33. (!) Define a sequence $\{f_n\}$ of functions by $f_n(x) = ae^{-anx} - be^{-bnx}$, where a, b are real constants with $0 < a < b$. Compute $\sum_{n=1}^\infty \int_0^\infty f_n(x)\,dx$, and compute $\int_0^\infty \sum_{n=1}^\infty f_n(x)\,dx$. (Hint: These are not equal!)

17.34. Let $f: \mathbb{R} \to \mathbb{R}$ be monotone increasing. Suppose that $0 \le a \le b$, $s = f(a)$, $t = f(b)$, and $0 \le s \le t$. The proof of Theorem 17.26 shows that

$$\int_a^b f(x)\, dx = yf^{-1}(y)\Big|_s^t - \int_s^t f^{-1}(y)\, dy.$$

Prove that this formula still holds when the requirements $0 \le a \le b$ and $0 \le s \le t$ are weakened to $a \le b$ and $s \le t$. This completes the proof of Theorem 17.26. (Hint: Use substitution to reduce to the case where a and s are positive.)

17.35. By the Fundamental Theorem of Calculus, $\int_0^1 e^x\, dx = e - 1$. The steps below evaluate the integral as a limit of sums.

a) Write down the lower sum $L(f, P_n)$, where $f(x) = e^x$ and P_n is the partition of $[0, 1]$ into n equal parts.

b) Use a finite geometric sum to evaluate the sum in part (a).

c) Verify directly that $\lim_{n\to\infty} L(f, P_n) = e - 1$. (What properties of the exponential function does this use?)

17.36. (+) Evaluate $\lim_{n\to\infty} \sum_{k=1}^n (n^2 + nk)^{-1/2}$.

17.37. (!) Evaluate $\lim_{x\to 0} x \ln x$ and $\lim_{x\to\infty} \frac{\ln x}{x}$. (Hint: Use l'Hôpital's Rule.)

17.38. (+) Let $\langle x \rangle$ be the sequence defined by $x_1 = \sqrt{2}$ and $x_{n+1} = (\sqrt{2})^{x_n}$ for $n \ge 1$. Prove that $\langle x \rangle$ converges and determine the limit.

17.39. (!) Let $f(x) = x/\ln x$ for $x > 1$. Find the minimum value of f. Use this information to determine which is larger: π^e or e^π.

17.40. (+) Suppose that $f(x) = u(x) \prod_{i=1}^n (x - a_i)$, where $a_i \ne 0$ for all i, and u is differentiable and never zero. Obtain a formula for $\sum(1/a_i)$. (Comment: This generalizes Exercise 3.54.)

17.41. (+) Suppose that $h: \mathbb{R} \to \mathbb{R}$ and that $h(x^n) = h(x)$ for all $n \in \mathbb{N}$ and $x \in \mathbb{R}$.

a) Prove that if h is continuous at $x = 1$, then h is constant.

b) Show that without this assumption h need not be constant.

c) Suppose that $f(x^n) = nx^{n-1} f(x)$ for all $x > 0$ and all $n \in \mathbb{N}$. Suppose also that $\lim_{x\to 1} f(x)/\ln(x)$ exists. What does this imply about f?

17.42. Let f and g be differentiable. Compute the derivative of f^g.

17.43. *AGM Inequality.*

a) Prove that $y^a z^{1-a} \le ay + (1 - a)z$ for all positive y, z and $0 \le a \le 1$. Determine when equality can hold.

b) Let $x_1, \ldots, x_n$ be a list of n positive real numbers. Prove that $(\sum_{i=1}^n x_i)/n \ge (\prod_{i=1}^n x_i)^{1/n}$, with equality only when $x_1 = \cdots = x_n$. (Hint: Part (a) can be applied to give a proof by induction on n.)

c) Let $a_1, \ldots, a_n$ be nonnegative real numbers. Find the maximum of $\prod_{i=1}^n x_i^{a_i}$ subject to $\sum x_i = 1$.

d) Use part (c) to give a different proof of part (b).

17.44. Use the unboundedness of $\ln x$ to prove that $\sum_{n=1}^\infty 1/n$ diverges.

17.45. (!) For $0 < \epsilon < 1$, consider $\int_\epsilon^1 \ln(x)\, dx$.

a) Evaluate this integral using Theorem 17.26.

b) Take the limit as $\epsilon \to 0$ of the answer to part (a) to evaluate the improper integral $\int_0^1 \ln(x)\,dx$.

c) Use upper sums to justify the statement that $\lim_{n\to\infty} \frac{1}{n}\sum_{k=1}^n \ln(k/n)$ equals the answer to part (b).

d) Rewrite the expression in part (c) to prove that $\lim_{n\to\infty} \frac{(n!)^{1/n}}{n} = \frac{1}{e}$. (Comment: This is a weak form of Stirling's Formula, which is used to approximate $n!$. Stirling's Formula states that $n!$ is approximately $n^n e^{-n}\sqrt{2\pi n}$.)

17.46. For each series below, determine the radius of convergence R, and determine the behavior of the series when $|x| = R$.

a) $\displaystyle\sum \frac{x^n}{n^2}$ b) $\displaystyle\sum \frac{x^n}{n}$ c) $\displaystyle\sum \frac{x^{2n}}{2n}$ d) $\displaystyle\sum \frac{x^n n^n}{n!}$

17.47. (+) Let f be continous on $[a,b]$. Compute $\lim_{n\to\infty}(\int_a^b |f|^n)^{1/n}$. (Comment: Compare with Exercise 14.27.)

17.48. Let $\langle a \rangle$ be a bounded sequence, and suppose that $\lim b_n = 1$. Prove that $\limsup a_n b_n = \limsup a_n$.

17.49. (!) Let f be continuous and nonnegative on $[0, \infty)$.

a) Prove that $\int_0^\infty f(x)\,dx$ exists if $\lim_{x\to\infty} \frac{f(x+1)}{f(x)}$ exists and is less than 1.

b) Prove that $\int_0^\infty f(x)\,dx$ exists if $\lim_{x\to\infty}(f(x))^{1/x}$ exists and is less than 1.

c) In parts (a) and (b), prove that the integrals do not exist if the specified limits exist but exceed 1.

17.50. (!) Let x, y, t be positive real numbers.

a) Prove that $t^2 + t(x+y) + (\frac{x+y}{2})^2 \geq t^2 + t(x+y) + xy \geq t^2 + 2t\sqrt{xy} + xy$.

b) After taking reciprocals of the expressions in part (a), integrate from 0 to ∞ with respect to t to prove that

$$\frac{x+y}{2} \geq \frac{x-y}{\ln(x) - \ln(y)} \geq \sqrt{xy}.$$

c) For $u \in \mathbb{R}$, use part (b) to show that $\frac{1}{2}(e^u + e^{-u}) \geq \frac{1}{2u}(e^u - e^{-u}) \geq 1$.

d) Prove part (c) directly using power series.

17.51. For $n \in \mathbb{N}$, use integration by parts to prove that $n! = \int_0^\infty e^{-x}x^n\,dx$.

17.52. The function Γ defined by $\Gamma(y) = \int_0^\infty e^{-x}x^{y-1}\,dx$ for $y > 0$ extends the notion of factorial to real arguments, with $\Gamma(n+1) = n!$.

a) Prove that the improper integral defining $\Gamma(y)$ converges when $y \geq 1$. (Hint: Use Exercise 17.49a.)

b) (+) When $0 < y < 1$, the integral defining $\Gamma(y)$ is also improper at the endpoint 0. Prove that this improper integral also converges.

c) Prove that $\Gamma(y+1) = y\Gamma(y)$.

d) Given that $\Gamma(\frac{1}{2}) = \sqrt{\pi}$, evaluate $\int_0^\infty e^{-x^2}\,dx$.

e) (+ +) Prove that $\Gamma(\frac{1}{2}) = \int_0^\infty e^{-x}x^{-1/2}\,dx = \sqrt{\pi}$.

Chapter 18

Complex Numbers

The complex number system extends the real number system by allowing solutions to the equation $t^2 = -1$. The resulting number system has many unexpected and useful properties essential to modern science as well as to pure mathematics. One beautiful application is the Fundamental Theorem of Algebra, which states that every nonconstant polynomial with complex coefficients has a zero. We prove this by extending to the complex numbers the ideas we have developed about convergence.

PROPERTIES OF THE COMPLEX NUMBERS

The real Cartesian plane $\mathbb{R}^2$ becomes the Euclidean plane when we define the distance between points by the usual Euclidean distance formula. It is useful in geometry and physics to view the points as vectors, but we will not do this here. Instead we define operations of arithmetic on $\mathbb{R}^2$ to make it into a field called $\mathbb{C}$. We call the elements of $\mathbb{C}$ *complex numbers*, and the Euclidean distance from the origin to a complex number is its *magnitude*. We now give the definitions of addition and multiplication that make $\mathbb{R}^2$ into a field with an element i satisfying $i^2 = -1$.

18.1. Definition. A **complex number** z is an ordered pair of real numbers. We write $z = (x, y)$ or $z = x + iy$, treating i as a formal symbol. The **sum** and **product** of complex numbers $z = (x, y)$ and $w = (a, b)$ are $z + w = (x + a, y + b)$ and $zw = (xa - yb, xb + ya)$. We write $\mathbb{C}$ for the set of complex numbers with these operations.

This definition for multiplication, written in terms of ordered pairs, is what results when we write $z = x + iy$ and $w = a + ib$, formally expand the product using the distributive law, and finally set $i^2 = -1$.

18.2. Example. $(1 + i)^2 = (1 + i)(1 + i) = 1 + 2i + i^2 = 2i$. ∎

18.3. Proposition. Under these operations of sum and product, $\mathbb{C}$ is a field. The identity element for addition is $0 + 0i$, and the identity element for multiplication is $1 + 0i$. The multiplicative inverse of $z \neq 0 + 0i$ is $(x - iy)/(x^2 + y^2)$.

Proof: (Exercises 1–3).   ∎

The expressions $x - iy$ and $x^2 + y^2$ in the formula for z^{-1} play prominent roles in complex analysis.

18.4. Definition. Given $z = x + iy$, the **conjugate** of z is the complex number $\overline{z} = x - iy$. The **magnitude** or **absolute value** of $z = x + iy$ is $|z| = \sqrt{x^2 + y^2} = \sqrt{z\overline{z}}$, which is the distance from (x, y) to the origin. We call x the **real part** and y the **imaginary part** of the complex number $z = x + iy$, writing $x = \text{Re}(z)$ and $y = \text{Im}(z)$.

18.5. Remark. When $\text{Im}(z) = y = 0$, addition and multiplication reduce to addition and multiplication of real numbers. We therefore identify $x + i0$ with $x \in \mathbb{R}$. In this sense, the field $\mathbb{R}$ is contained in the field $\mathbb{C}$.

Observe furthermore that $|x + i0|$ equals $|x|$, the ordinary absolute value of the real number x. Hence magnitude extends the notion of absolute value from $\mathbb{R}$ to $\mathbb{C}$. We also have $|\overline{z}| = |z|$. Using the conjugate, we can write the multiplicative inverse of z as $z^{-1} = \overline{z}/|z|^2$.   ∎

The Triangle Inequality also holds for complex numbers.

18.6. Proposition. (Triangle Inequality). For $z, w \in \mathbb{C}$,

$$|z + w| \leq |z| + |w|.$$

Furthermore, equality holds if and only if one of these numbers is a nonnegative real multiple of the other.

Proof: If $w = 0$, the inequality is trivial. Otherwise, we will prove that $|z + w|^2 \leq (|z| + |w|)^2$, which yields the desired inequality by taking positive square roots. Expanding the squares and simplifying shows that the inequality is equivalent to $\text{Re}(z\overline{w}) \leq |z|\,|w|$ (Exercise 5).

For each real number t, we compute

$$0 \leq |z + tw|^2 = |z|^2 + 2t\,\text{Re}(z\overline{w}) + t^2\,|w|^2.$$

We now choose $t = -\text{Re}(z\overline{w})/|w|^2$. This yields

$$0 \leq |z|^2 - \frac{2(\text{Re}(z\overline{w}))^2}{|w|^2} + \frac{(\text{Re}(z\overline{w}))^2}{|w|^4}\,|w|^2 = |z|^2 - \frac{(\text{Re}(z\overline{w}))^2}{|w|^2}.$$

Multiplying by $|w|^2$ yields $0 \leq |z|^2\,|w|^2 - (\text{Re}(z\overline{w}))^2$, as desired. We leave the condition for equality to Exercise 11.   ∎

The Triangle Inequality will also extend to infinite sums, after we have defined convergence for sequences and series of complex numbers. The definitions of limit and convergence in terms of absolute value are the same in $\mathbb{C}$ as in $\mathbb{R}$.

18.7. Definition. Suppose $\langle z \rangle$ is a sequence of complex numbers. We say that $\langle z \rangle$ **converges to** L or **has limit** L (written as $z_n \to L$ or $\lim(z_n) = L$) if for every positive real number ϵ, there exists $N \in \mathbb{N}$ such that $n \geq N$ implies $|z_n - L| < \epsilon$. A **Cauchy sequence** of complex numbers is a sequence such that for every positive real number ϵ, there exists $N \in \mathbb{N}$ such that $n, m \geq N$ implies $|z_n - z_m| < \epsilon$.

The Cauchy Convergence Criterion (a sequence converges if and only if it is a Cauchy sequence) holds in $\mathbb{C}$ as well as in $\mathbb{R}$. This extension follows from the corresponding result in $\mathbb{R}$ and the observation that $z_n \to L$ if and only if $\operatorname{Re}(z_n) \to \operatorname{Re}(L)$ and $\operatorname{Im}(z_n) \to \operatorname{Im}(L)$ (Exercise 10). We apply this to study the convergence of series.

18.8. Definition. A series $\sum_{n=0}^{\infty} w_n$ of complex numbers **converges** if its sequence of partial sums converges. A series $\sum_{n=0}^{\infty} w_n$ **converges absolutely** if $\sum_{n=0}^{\infty} |w_n|$ converges.

Absolute convergence implies convergence (as in Corollary 14.30).

18.9. Proposition. If $\langle z \rangle$ is a sequence of complex numbers such that $\sum_{n=0}^{\infty} |z_n|$ converges, then $\sum_{n=0}^{\infty} z_n$ converges, and $\left| \sum_{n=0}^{\infty} z_n \right| \leq \sum_{n=0}^{\infty} |z_n|$.

Proof: Since $\left| \sum_{n=0}^{N} z_n \right| \leq \sum_{n=0}^{N} |z_n|$ for each N, the inequality follows if $\sum_{n=0}^{\infty} z_n$ converges. When $\sum_{n=0}^{\infty} |z_n|$ converges, its sequence of partial sums is a Cauchy sequence. Hence there exists N' such that $N > M \geq N'$ implies $\sum_{n=M+1}^{N} |z_n| < \epsilon$. This implies that

$$\left| \sum_{n=0}^{N} z_n - \sum_{n=0}^{M} z_n \right| = \left| \sum_{n=M+1}^{N} z_n \right| \leq \sum_{n=M+1}^{N} |z_n| < \epsilon.$$

Hence the partial sums of $\sum_{n=0}^{\infty} z_n$ form a Cauchy sequence, and the series converges. ∎

Convergence tests for power series apply also in the complex case.

18.10. Proposition. (Ratio Test) Suppose $f(z) = \sum_{n=0}^{\infty} a_n z^n$ is a complex power series and that $|a_{n+1}/a_n|$ converges to L. Then $f(z)$ converges absolutely for all z with $L |z| < 1$. When $L = 0$, it converges for all z.

Proof: The ratio test (Theorem 14.31) for the real series $\sum_{n=0}^{\infty} |a_n z^n|$ considers the ratio $\frac{|a_{n+1} z^{n+1}|}{|a_n z^n|} = \left|\frac{a_{n+1}}{a_n}\right| |z|$. By hypothesis, its limit ρ is $L|z|$. When ρ is less than 1, $f(z)$ converges absolutely. ∎

The ratio test can also be used to show divergence. The series $g(z)$ diverges when $|z| > 1/L$, but the test is inconclusive when $|z| = 1/L$ (compare with Example 17.48).

For real numbers, we defined the exponential function to be the inverse of the logarithm function. We also derived a formula for the exponential function as a convergent power series. We use the power series to extend the exponential function to $\mathbb{C}$.

18.11. Definition. For every $z \in \mathbb{C}$, the value of the exponential function is defined to be $e^z = \sum_{n=0}^{\infty} z^n/n!$.

To show that the exponential function is well-defined, we apply the ratio test. Since the ratio of successive terms is $|z|/(n+1)$, which tends to 0, the series converges for every z. The familiar property $e^z e^w = e^{z+w}$ also extends to $\mathbb{C}$ (Exercise 12a). These results allow us to define the sine and cosine functions. For $\theta \in \mathbb{R}$, the complex numbers of the form $e^{i\theta}$ have magnitude 1 and form the unit circle centered at $(0,0)$ (Exercise 14). We define $\cos\theta = \mathrm{Re}(e^{i\theta})$ and $\sin\theta = \mathrm{Im}(e^{i\theta})$, and thus $e^{i\theta} = \cos\theta + i\sin\theta$. The series expansions for sine and cosine follow from the series for e^z:

$$\sin\theta = \theta - \theta^3/3! + \theta^5/5! - \cdots$$

$$\cos\theta = 1 - \theta^2/2! + \theta^4/4! - \cdots$$

18.12. Example. The Fourier series $\sum_{n=0}^{\infty} (a_n \sin(nx) + b_n \cos(nx))$ can be expressed as the imaginary part of a complex power series, using $a_n \sin(nx) + b_n \cos(nx) = \mathrm{Im}((a_n + ib_n)e^{inx})$. ∎

When we view the complex number z as a vector, we can express (x, y) as $|z|$ times a unit vector in the direction of (x, y). This leads to the polar coordinate representation of z, which is $z = |z|e^{i\theta}$ for some real number θ, called an **argument** of z. Because $e^{i(\theta+2n\pi)} = e^{i\theta}$ for all $n \in \mathbb{Z}$, we may assume that $0 \le \theta < 2\pi$. In taking roots of complex numbers, we must consider all possible choices for the argument.

18.13. Lemma. If z is a nonzero complex number and m is a positive integer, then $w^m = z$ has the m solutions $w = |z|^{1/m} e^{i(\theta+2k\pi)/m}$ for $0 \le k \le m - 1$. In the geometric view of $\mathbb{C}$, these are equally spaced on a circle centered at the origin.

Proof: (Exercise 16). ∎

LIMITS AND CONVERGENCE

We discuss the "topology" of $\mathbb{C}$ by defining open and closed sets.

18.14. Definition. Given $w \in \mathbb{C}$, the **open ball** $B_\epsilon(w)$ of radius ϵ around w is $\{z \in \mathbb{C}: |z - w| < \epsilon\}$. A subset S of $\mathbb{C}$ is **open** if for every $w \in S$ there exists $\epsilon > 0$ such that $B_\epsilon(w) \subset S$. A subset S of $\mathbb{C}$ is **closed** if $\mathbb{C} - S$ is open.

An open ball is an open set. These definitions apply equally well for real numbers, where open intervals replace open balls in the definition. Closed sets can be characterized by convergent sequences.

18.15. Theorem. A subset S of $\mathbb{C}$ is closed if and only if for every convergent sequence in S, the limit of the sequence also belongs to S.

Proof: Suppose S is closed and $\langle z \rangle$ is a sequence in S converging to L. If $L \notin S$, then by the definition of closed set there is an open ball $B_\epsilon(L)$ that is entirely outside S. This implies $|z_n - L| > \epsilon$ for all z_n, which contradicts the definition of convergence to L. Hence $L \in S$.

Conversely, suppose the limit of every convergent sequence of elements in S also belongs to S. If S is not closed, then the complement of S is not open. This implies there is some $L \in \mathbb{C} - S$ such that no open ball around L is contained in $\mathbb{C} - S$. In particular, for every $n \in N$, the ball $B_{1/n}(L)$ contains a point of S. We define a sequence in S by choosing $z_n \in B_{1/n}(L) \cap S$. This sequence converges to L, but $L \notin S$. From this contradiction, we conclude that S must be closed. ∎

18.16. Definition. A subset S of $\mathbb{C}$ is **bounded** if there exists a positive real number M such that $|z| \leq M$ for all $z \in S$. A subset S of $\mathbb{C}$ is **compact** if every sequence $\langle z \rangle$ of elements of S has a subsequence $\{z_{n_k}\}$ converging to a limit that belongs to S. A **closed rectangle** in $\mathbb{C}$ is a set $\{z = x + iy : a \leq x \leq b, \ c \leq y \leq d\}$ for some $a, b, c, d \in \mathbb{R}$.

18.17. Theorem. Every closed rectangle in $\mathbb{C}$ is compact.

Proof: (Sketch) The proof parallels that of the Bolzano-Weierstrass Theorem (Theorem 14.17). Given $\langle z \rangle$, we extract a convergent subsequence by dividing the rectangle into four subrectangles at each stage (Exercise 22), and choosing a subrectangle that contains z_n for infinitely many n. ∎

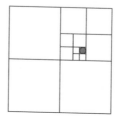

18.18. Theorem. A closed subset of a compact set in $\mathbb{C}$ is also compact.

Proof: Suppose $T \subseteq S$, with T closed and S compact. Let $\langle z \rangle$ be a sequence in T. Since $T \subseteq S$ and S is compact, $\langle z \rangle$ has a convergent subsequence $\{z_{n_k}\}$ whose limit belongs to S. Since T is closed, the limit is also in T. Hence T is compact. ∎

18.19. Theorem. A subset of $\mathbb{C}$ is compact if and only if it is closed and bounded.

Proof: Suppose first that S is closed and bounded. Because S is bounded, it is a subset of a closed rectangle. By the preceding two theorems, S is compact.

We prove the converse by contradiction. Suppose S is compact. If S is not closed, then S contains a convergent sequence $\langle z \rangle$ whose limit L is not in S. Every subsequence of $\langle z \rangle$ also converges to L. Since $L \notin S$, this violates the definition of compactness for S.

If S is not bounded, then we can define $\langle z \rangle$ by letting z_n be an element of S with magnitude greater than n. This has no convergent subsequence, again violating the definition of compactness. ∎

The definitions of limit and continuity for functions of a complex variable are much like those for functions of a real variable.

18.20. Definition. A complex-valued function f defined on an open ball around p is **continuous** at p if for each $\epsilon > 0$ there exists $\delta > 0$ so that $|z - p| < \delta$ implies $|f(z) - f(p)| < \epsilon$. In other words, for each $\epsilon > 0$ there exists $\delta > 0$ so that $f(B_\delta(p)) \subseteq B_\epsilon(f(p))$.

18.21. Proposition. Suppose $f: \mathbb{C} \to \mathbb{C}$. The following four statements are equivalent:
A) f is continuous.
B) for every open set T, $I_f(T)$ is open.
C) for every closed set T, $I_f(T)$ is closed.
D) for each sequence $\langle z \rangle$, $z_n \to w$ implies $f(z_n) \to f(w)$.

Proof: (Exercise 24). ∎

Since the interval $[0, r]$ is closed in $\mathbb{R}$, the set $\{z \in \mathbb{C}: |f(z)| \le r\}$ is closed whenever f is a continuous function from $\mathbb{C}$ to $\mathbb{R}$.

18.22. Theorem. Let f be a real-valued function that is defined and continuous on a compact subset S of $\mathbb{C}$. Then S contains elements at which f achieves its maximum and its minimum on S.

Proof: We first prove that f is bounded. If not, then for each $n \in \mathbb{N}$ we can find z_n such that $|f(z_n)| > n$. Since S is compact, $\langle z \rangle$ has a convergent subsequence $\{z_{n_k}\}$. By statement (D) of Proposition 18.21, $f(z_{n_k})$ converges. This contradicts $\left| f(z_{n_k}) \right| > n_k$.

We mimic the proof of the Minimum-Maximum Theorem (Theorem 15.26). Since f is bounded and real-valued, the set $f(S)$ has a supremum β. Let $\langle y \rangle$ be a sequence in $f(S)$ converging to β. Let $\langle z \rangle$ be a sequence such that $f(z_n) = y_n$. By compactness, $\langle z \rangle$ has a convergent subsequence $\{z_{n_k}\}$. Let $w = \lim z_{n_k}$. Since S is closed, $w \in S$. Since f is continuous, $f(w) = \beta$. The proof for the minimum is similar (Exercise 25).   ∎

THE FUNDAMENTAL THEOREM OF ALGEBRA

For the proof of the Fundamental Theorem of Algebra, we need to consider infinite limits.

18.23. Definition. We write $\lim_{z \to w} f(z) = L$ if for every $\epsilon > 0$ there exists $\delta > 0$ such that $0 < |z - w| < \delta$ implies $|f(z) - L| < \epsilon$. When $\langle z \rangle$ is a sequence of nonzero complex numbers, we write $z_n \to \infty$ if $|z_n|^{-1} \to 0$. We also write "$f(z) \to \infty$ as $z \to \infty$" if for every $\epsilon > 0$, there exists $\delta > 0$ such that $|z| > 1/\delta$ implies $|f(z)| > 1/\epsilon$.

A **nonconstant complex polynomial** is a function $p \colon \mathbb{C} \to \mathbb{C}$ defined by $p(z) = \sum_{j=0}^{k} a_j z^j$, where the coefficients $a_0, \ldots, a_k$ are complex numbers, $k \geq 1$, and $a_k \neq 0$. We call k the **degree** of p. If $p(z) = 0$, then z is a **zero** of p.

18.24. Lemma. If p is a nonconstant complex polynomial, then $p(z) \to \infty$ as $z \to \infty$.

Proof: A polynomial of degree k has at most k zeros, since a polynomial with zero α can be expressed as $(z - \alpha)$ times a polynomial of lower degree. (The proof of Lemma 3.23 remains valid in this setting.) Since the set of zeros is finite, we can choose M such that $p(z) \neq 0$ for $|z| > M$.

Let $w = 1/z$. It suffices to show that $\lim_{w \to 0} 1/[p(w^{-1})] = 0$. For $|w| < 1/M$, the function is well-defined. We compute

$$\frac{1}{p(w^{-1})} = \frac{1}{\sum_{j=0}^{k} a_j w^{-j}} = \frac{w^k}{\sum_{j=0}^{k} a_j w^{k-j}} = \frac{w^k}{\sum_{j=0}^{k} a_{k-j} w^j} = \frac{w^k}{a_k + \sum_{j=1}^{k} a_{k-j} w^j}.$$

As $w \to 0$, the numerator approaches 0 and the denominator approaches $a_k \neq 0$, so $1/[p(w^{-1})] \to 0$.   ∎

To prove the Fundamental Theorem of Algebra, we will find a z with $|p(z)|$ as small as possible. Compactness guarantees a minimum for $|p|$ on any closed ball, and Lemma 18.24 implies that $|p|$ is large outside a closed ball with large radius. These ideas enable us to reduce the problem of finding the zeros of an arbitrary polynomial to taking the rth root of a complex number.

18.25. Theorem. (Fundamental Theorem of Algebra) Every nonconstant complex polynomial has a zero in $\mathbb{C}$.

Proof: Suppose p is a nonconstant complex polynomial. By Lemma 18.24 there is an M such that $|z| > M$ implies $|p(z)| > |p(0)|$. Let $S_M = \{z \in \mathbb{C}: |z| \le M\}$. Since S_M is closed and bounded, it is compact (Theorem 18.19). Since p and the absolute value function are continuous, their composition $|p|$ is also continuous. By Theorem 18.22, $|p|$ achieves its minimum on S_M. Thus there is a z' with $\left|p(z')\right| \le |p(z)|$ for $|z| \le M$ and $\left|p(z')\right| \le |p(0)| < |p(z)|$ for $|z| > M$. Thus $\left|p(z')\right|$ is the minimum value for $|p|$ on all of $\mathbb{C}$.

We now use proof by contradiction to prove that $p(z') = 0$. If $p(z') \ne 0$, then we define

$$h(w) = \frac{p(z' + w)}{p(z')}.$$

Observe that $|h(w)| \ge 1$ for all $w \in \mathbb{C}$. Also, since z' is a constant, h is a polynomial in w with $h(0) = 1$. Thus $h(w) = 1 + \sum_{j=1}^{n} d_j w^j$. Let r be the smallest index for which $d_r \ne 0$.

We can now write $h(w) = 1 + d_r w^r + g(w)$, where $\lim_{w \to 0} \frac{g(w)}{w^r} = 0$. By the definition of limit, there is a $\delta > 0$ such that $0 < |w| < \delta$ implies $\frac{|g(w)|}{|w|^r} < \frac{1}{2}|d_r|$.

Choose a positive number α such that $\alpha < \delta^r |d_r|$. By Lemma 18.13, we can find $\zeta \in \mathbb{C}$ satisfying $d_r \zeta^r = -\alpha$. We have $|\zeta| < \delta$, since $|\zeta|^r = \left|\frac{\alpha}{d_r}\right| < \frac{\delta^r |d_r|}{|d_r|} = \delta^r$. From $|\zeta| < \delta$, we have $|g(\zeta)| \le \frac{1}{2}|d_r| \, |\zeta|^r = \frac{1}{2}\alpha$. From $d_r \zeta^r = -\alpha$, we have $h(\zeta) = 1 - \alpha + g(\zeta)$. By the Triangle Inequality,

$$|h(\zeta)| \le 1 - \alpha + |g(\zeta)| < 1 - \alpha + \tfrac{1}{2}\alpha < 1.$$

Since $|h(w)| \ge 1$ for all $w \in \mathbb{C}$, this is a contradiction. Thus $p(z') = 0$. ∎

18.26. Corollary. Every nonconstant complex polynomial of degree n can be expressed as a product of linear factors,

$$p(z) = c \prod_{j=1}^{n}(z - \alpha_i),$$

where c is a nonzero constant and each α_i is a zero of p in $\mathbb{C}$. ∎

One theme throughout this book has been the solution of equations. An equation that has no solution in a given number system may have a solution in a larger number system. If we think of mathematics as beginning with the natural numbers, then we introduce the integers to solve $x + n = 0$, the rational numbers to solve $mx + n = 0$, the real numbers to solve $x^2 - 2 = 0$, and finally the complex numbers to solve $x^2 + 1 = 0$. In $\mathbb{C}$ we can find the zeros of all polynomials. For this reason, the field of complex numbers is called the **algebraic closure** of $\mathbb{R}$.

EXERCISES

18.1. (−) Prove that $\mathbb{C}$ is a group under addition, with identity $(0, 0)$.

18.2. (−) *Multiplication of complex numbers.*
    a) Prove that $(1, 0)$ is an identity for multiplication.
    b) Prove that if $a^2 + b^2 \neq 0$, then $(\frac{a}{a^2+b^2}, \frac{-b}{a^2+b^2}) \cdot (a, b) = (1, 0)$. (Comment: This proves that $z^{-1} = \bar{z}/|z|^2$.)
    c) Prove that $\mathbb{C} - \{0\}$ is a group under multiplication.

18.3. (−) Prove that addition and multiplication of complex numbers are associative and commutative and satisfy the distributive law.

18.4. Determine all solutions to $x^2 + y^2 = 0$ with x and y real. Determine all solutions to $z^2 + w^2 = 0$ with z and w complex.

18.5. For complex numbers z, w, prove that $|zw| = |z|\,|w|$ and that $|z + w|^2 = |z|^2 + |w|^2 + 2\mathrm{Re}(z\bar{w})$.

18.6. Suppose w_1 and w_2 are distinct points in $\mathbb{C}$. Give a geometric description of the set $\{z: |z - w_1| = |z - w_2|\}$.

18.7. Prove the following properties of complex conjugation for all $z, w \in \mathbb{C}$:
    a) $\overline{zw} = \bar{z}\,\bar{w}$.   b) $\overline{z + w} = \bar{z} + \bar{w}$.   c) $|\bar{z}| = |z|$.

18.8. (−) Suppose $z = (x, y) \in \mathbb{C}$. Prove that $x = (z + \bar{z})/2$ and $y = (z - \bar{z})/2i$.

18.9. (−) Express the cube roots of 1 both in the form $re^{i\theta}$ and in the form $x + iy$.

18.10. Prove that $\langle z \rangle$ converges to A if and only if $\mathrm{Re}(z_n) \to \mathrm{Re}(A)$ and $\mathrm{Im}(z_n) \to \mathrm{Im}(A)$. Use this to prove that $\langle z \rangle$ converges if and only if it is a Cauchy sequence.

18.11. (!) For $z, w \in \mathbb{C}$, determine when $|z + w| = |z| + |w|$. Use this to solve Exercise 8.27.

18.12. (!) *Trigonometry and the exponential function.*
    a) Prove that $e^z e^w = e^{z+w}$ for all $z, w \in \mathbb{C}$. (Hint: Use the power series.)
    b) Use part (a) and the formula $e^{i\theta} = \cos\theta + i\sin\theta$ to prove that $\cos(n\theta)$ and $\sin(n\theta)$ are polynomials in the variables $\cos\theta$ and $\sin\theta$.
    c) Prove the trigonometric identity $\cos(3\theta) = 4\cos^3\theta - 3\cos\theta$. Obtain a similar formula for $\sin(3\theta)$.

18.13. Prove that conjugation is a continuous function. Apply the continuity of conjugation to prove that $\exp(\bar{z}) = \overline{\exp(z)}$.

18.14. Use Exercise 18.12 to conclude that $\left|e^{i\theta}\right| = 1$ for all $\theta \in \mathbb{R}$ and that $e^{2\pi i} = 1$.

18.15. By Exercise 18.12, $\cos\theta = \mathrm{Re}(e^{i\theta})$ and $\sin\theta = \mathrm{Im}(e^{i\theta})$.
    a) Write $\cos\theta$ and $\sin\theta$ in terms of $e^{i\theta}$ and $e^{-i\theta}$.
    b) (!) For $n \in \mathbb{N}$, use part (a) and the Binomial Theorem to evaluate
$$\int_0^{2\pi} (\cos\theta)^{2n} d\theta \qquad \text{and} \qquad \int_0^{2\pi} (\sin\theta)^{2n} d\theta.$$
    c) (+) Let n and m be nonnegative integers. Evaluate
$$\int_0^{2\pi} (\cos\theta)^{2n} (\sin\theta)^{2m} d\theta.$$

18.16. Suppose z is a nonzero complex number and m is a positive integer. Prove that $w^m = z$ has the m distinct solutions $w = |z|^{1/m} e^{i(\theta + 2k\pi)/m}$ for $0 \le k \le m - 1$. Plot these solutions for the case $m = 8$ and $z = 256i$.

18.17. Define $f: \mathbb{C} \to \mathbb{C}$ by $f(z) = iz$. Describe the functional digraph of f.

18.18. For $z \in \mathbb{C}$ and $z \ne 1$, prove that $\sum_{k=0}^{n-1} z^k = (1 - z^n)/(1 - z)$. Give a geometric interpretation of this result when z is an nth root of 1.

18.19. (!) Suppose $z^n = 1$. Obtain a simple formula for $\prod_{k=0}^{n-1} z^k$.

18.20. Prove that the set of nth roots of 1, under multiplication, form a group "isomorphic" to $\mathbb{Z}_n$.

18.21. Use the characteristic equation method (Theorem 12.22) to solve the recurrence $a_n = -a_{n-2}$ for $a_0 = 2$ and $a_1 = 4$, giving a single formula for a_n.

18.22. Fill in the details of the proof in Theorem 18.17 that every closed rectangle in $\mathbb{C}$ is compact.

18.23. Given $w \in \mathbb{C}$ and $r \in \mathbb{R}$, prove that the set $\{z \in \mathbb{C}: |z - w| \le r\}$ is closed.

18.24. Suppose $f: \mathbb{C} \to \mathbb{C}$. Prove that the following statements are equivalent:
   A) f is continuous.
   B) for every open set T, $I_f(T)$ is open.
   C) for every closed set T, $I_f(T)$ is closed.
   D) for each sequence $\langle z \rangle$, $z_n \to w$ implies $f(z_n) \to f(w)$.

18.25. Suppose f is a real-valued function that is defined and continuous on a compact subset S of $\mathbb{C}$. Prove that S contains an element at which f achieves its minimum on S. (Comment: This completes the proof of Theorem 18.22.)

18.26. Prove that on $\mathbb{C}$, every polynomial is continuous, the absolute value function is continuous, and the composition of continuous functions is continuous. Conclude that $|p|$ is continuous when p is a complex polynomial.

18.27. (Root test) Let $\sum_{n=0}^{\infty} a_n z^n$ be a power series. Let $L = \limsup |a_n|^{1/n}$. Prove that $\sum_{n=0}^{\infty} a_n z^n$ converges absolutely if $|z| < 1/L$ and diverges if $|z| > 1/L$.

18.28. Define $f: \mathbb{R} \to \mathbb{C}$ by $f(t) = (1 + it)^2/(1 + t^2)$. Prove that the image of f is the unit circle minus the point -1, and that $\lim_{t \to \pm\infty} f(t) = -1$. Describe the trigonometric relationship between t and θ if θ satisfies $f(t) = e^{i\theta}$. How does this problem relate to Pythagorean triples?

18.29. (+) Let $2r + 1$ be an odd positive integer, and let ω be a complex number such that $\omega^{2r+1} = 1$ but that $\omega^n \ne 1$ when n is a natural number less than $2r + 1$. Obtain explicit formulas (as rational numbers times binomial coefficients) for the nonzero coefficients in the polynomial g defined by

$$g(x, y) = 1 - \prod_{j=0}^{2r}(1 - \omega^j x - \omega^{2j} y).$$

Appendix A

From $\mathbb{N}$ to $\mathbb{R}$

It is possible to base mathematics on elementary set theory, but it may be more satisfying to begin with the natural numbers. From the set $\mathbb{N}$ of natural numbers, we construct first the integers ($\mathbb{Z}$), then the rational numbers ($\mathbb{Q}$), and finally the real numbers ($\mathbb{R}$). At each step, we define the arithmetic operations and sketch proofs of the desired arithmetic properties. More details appear on the World Wide Web, at http://www.math.uiuc.edu/~west/mt.

We begin with the natural numbers and with elementary concepts about sets and functions. This includes set operations, bijections, composition, and equivalence relations, as discussed in Chapters 1, 4, and 7.

We include the Well-Ordering Property of $\mathbb{N}$ among our assumptions so that we can use induction to define and study the arithmetic operations for $\mathbb{N}$. We define addition and multiplication of natural numbers, and then we sketch how to verify associativity, commutativity, the distributive law, and several other elementary properties.

To construct $\mathbb{Z}$, we consider pairs of natural numbers. We define a relation $\sim$ on $\mathbb{N} \times \mathbb{N}$ by $(a, b) \sim (c, d)$ if $a + d = b + c$. Each resulting equivalence class consists of pairs having the same "difference"; the integers are these equivalence classes. The class containing $(0, b)$ plays the role of the natural number b in a copy of $\mathbb{N}$ contained in $\mathbb{Z}$. We define arithmetic operations on these equivalence classes and show why they behave as desired.

We can add, subtract, and multiply within $\mathbb{Z}$, but we cannot generally divide. We construct the set $\mathbb{Q}$ of rational numbers to permit division, except by 0. To do this, we consider pairs of integers, in which the second is not zero. We define a relation $\sim$ on $\mathbb{Z} \times (\mathbb{Z} - \{0\})$ by $(a, b) \sim (c, d)$ if $ad = bc$; this is an equivalence relation. Each resulting equivalence class consists of pairs having the same "ratio"; the rational numbers are these equivalence classes. The class containing $(x, 1)$ plays the role of the integer x in a copy of $\mathbb{Z}$ contained in $\mathbb{Q}$. We define arithmetic operations and sketch how to prove that they behave as expected. In particular, given the arithmetic properties of $\mathbb{Z}$, we conclude that $\mathbb{Q}$ is an ordered field.

We have observed that $\mathbb{Q}$ does not contain quantities such as $\sqrt{2}$, which we believe exist. To remedy this defect, we "complete" $\mathbb{Q}$ into $\mathbb{R}$ by introducing limiting processes. We consider the set S of Cauchy sequences of elements of $\mathbb{Q}$. We define a relation $\sim$ on S by $\langle a \rangle \sim \langle b \rangle$ if $\langle a \rangle - \langle b \rangle$ converges to 0. We verify that this is an equivalence relation and define $\mathbb{R}$ to be the set of equivalence classes. The class containing the sequence with all terms equal to q plays the role of the rational number q in a copy of $\mathbb{Q}$ contained in $\mathbb{R}$. We define arithmetic on elements of $\mathbb{R}$ using arithmetic of sequences, and we sketch the proof that the resulting structure satisfies all the properties of a complete ordered field.

We have observed that in each construction ($\mathbb{Z}$ from $\mathbb{N}$, $\mathbb{Q}$ from $\mathbb{Z}$, $\mathbb{R}$ from $\mathbb{Q}$), the earlier system is included in the new system in a natural way. Thus the initial set $\mathbb{N}$ can be viewed as a subset of $\mathbb{R}$. In this injection of $\mathbb{N}$ into $\mathbb{R}$, the natural number 1 becomes the multiplicative identity in $\mathbb{R}$, and the successive natural numbers become the real numbers obtained by successively adding 1 to the multiplicative identity. From this it follows that the definition of $\mathbb{N}$ as a subset of $\mathbb{R}$ (Definition 3.5) produces the same subset of $\mathbb{R}$ that represents what we start with here in constructing $\mathbb{R}$.

Furthermore, we show at the end of this appendix that there is essentially only one complete ordered field. Thus our work is consistent. The mathematics we developed by beginning with a set called $\mathbb{R}$ satisfying Definitions 1.39–1.41 remains valid if we start from $\mathbb{N}$ and construct the set $\mathbb{R}$ with those properties.

THE NATURAL NUMBERS

We want to use the natural numbers as our starting point. We assume that $\mathbb{N}$ is a set with some additional but familiar structure. There is a relation $<$ on $\mathbb{N}$, along with the relations $>, \leq, \geq$ arising from it. We assume the following properties for the relation $<$ on $\mathbb{N}$.

A.1. Axiom.
a) (Trichotomy) For $m, n \in \mathbb{N}$, exactly one of $\{n = m, n < m, m < n\}$ holds.
b) (Transitivity) If $l, m, n \in \mathbb{N}$, then $l < m$ and $m < n$ imply $l < n$.   ∎

We next introduce the Well-Ordering Property and the successor function to allow us to use inductive arguments. The Well-Ordering Property guarantees that $\mathbb{N}$ itself has a least element, which we call 1.

A.2. Axiom. (Well-Ordering Property) Every nonempty subset of $\mathbb{N}$ has a least element.   ∎

A.3. Definition. The **successor function** $\sigma\colon \mathbb{N} \to \mathbb{N}$ is defined by letting $\sigma(n)$ be the least element in $\{k \in \mathbb{N}\colon k > n\}$.

Our final axiom for $\mathbb{N}$ is stated using the successor function. It completes our formal definition of $\mathbb{N}$ as a set with the properties we expect.

A.4. Axiom. The image of σ is $\mathbb{N} - \{1\}$. ∎

A.5. Lemma. The function σ is injective, and $\sigma(n) > n$ for all $n \in \mathbb{N}$.

Proof: Injectivity follows from the definition of σ and the Well-Ordering Property, and the second claim follows from the definition of σ. ∎

By Axiom A.1, σ is a bijection from $\mathbb{N}$ to $\mathbb{N} - \{1\}$. Together with Axiom A.2, this implies that $\mathbb{N} = \{1, \sigma(1), \sigma(\sigma(1)), \cdots\}$. We give the natural numbers the usual names $1, 2, 3, \cdots$ corresponding to this order.

Axioms A.2–A.4 yield the Principle of Induction (Exercise 3.64). Induction allows us to define addition and multiplication and to prove their fundamental properties. By these definitions, the sum and product of two natural numbers are natural numbers.

The successor function σ defines the operation of "adding one". Given any natural number n, we define "$n + 1$" to be (another name for) the natural number $\sigma(n)$. We can define addition of any pair natural numbers using iteration of σ.

A.6. Definition. The operation of adding k is a function $a_k\colon \mathbb{N} \to \mathbb{N}$. For $k = 1$, we put $a_1(n) = \sigma(n)$. Given a_k, we put $a_{k+1}(n) = \sigma(a_k(n))$. **Addition** is a binary operation on $\mathbb{N}$. For $n, k \in \mathbb{N}$, the **sum**, written $n + k$, is the natural number $a_k(n)$.

A.7. Proposition. Addition of natural numbers is associative and commutative.

Proof: Use induction on c to prove $a + (b + c) = (a + b) + c$. Prove $n + m = m + n$ in two steps. First, use induction on n to prove $n + 1 = 1 + n$. Then for each $n \in \mathbb{N}$ use induction on m to prove $n + m = m + n$. ∎

A.8. Definition. The operation of multiplying by k is a function $m_k\colon \mathbb{N} \to \mathbb{N}$. For $k = 1$, we put $m_1(n) = n$. Given the function m_k, we put $m_{k+1}(n) = m_k(n) + n$. **Multiplication** is a binary operation on $\mathbb{N}$. For $n, k \in \mathbb{N}$, the **product**, written $k \cdot n$ or kn, is defined to be the natural number $m_k(n)$.

By convention, multiplicative operations always take precedence over additive operations when both appear in an expression without parentheses: $ab + c$ means $(ab) + c$.

A.9. Proposition. The **distributive law** $a(b + c) = ab + ac$ holds in $\mathbb{N}$.

Proof: Use induction on a. ∎

A.10. Proposition. Multiplication of natural numbers is associative and commutative.

Proof: Use induction on a to prove $a(bc) = (ab)c$. Use induction on n to prove $n \cdot 1 = 1 \cdot n$. For each n, use induction on m to prove $nm = mn$. ∎

The properties of functions yield the cancellation properties of equalities in natural numbers.

A.11. Proposition. For every $k \in \mathbb{N}$, addition of k and multiplication by k are injective functions from $\mathbb{N}$ to $\mathbb{N}$. Furthermore, if $a, b, c \in \mathbb{N}$, then $a + c = b + c$ implies $a = b$, and $ca = cb$ implies $a = b$.

Proof: Since the composition of injective functions is injective (Proposition 4.30), the first statement follows inductively because both σ and the identity function are injective. The second statement then follows from the first. ∎

We close our discussion of the natural numbers by proving that the size of a finite set is a well-defined notion. Definition 4.36 states that a nonempty set S is finite if for some natural number n there is a bijection $f: S \to \{1, \dots, n\}$. The next result, stated also as Corollary 4.38, implies that the size of a set is a well-defined notion.

A.12. Proposition. If there is a bijection $f: [m] \to [n]$, then $m = n$.

Proof: Let $P(n)$ be the statement claimed. We use induction on n to prove $P(n)$ for all $n \in \mathbb{N} \cup \{0\}$. The main idea in the induction step is to remove n from $[n]$ and $f^{-1}(n)$ from $[m]$. Elementary properties of bijections then yield that $m - 1 = n - 1$. ∎

THE INTEGERS

Given the natural numbers, we can define the set $\mathbb{Z}$ of integers in several ways. We could define $\mathbb{Z}$ as a set of symbols. Let 0 be a symbol not in $\mathbb{N}$, and let $-\mathbb{N}$ denote the set of formal symbols $\{-m: m \in \mathbb{N}\}$. We can then define $\mathbb{Z} = \mathbb{N} \cup (-\mathbb{N}) \cup \{0\}$. This defines $\mathbb{Z}$ as a set, but it is difficult to verify that this set satisfies the usual properties of arithmetic.

Instead, we define $\mathbb{Z}$ using an equivalence relation. We define the relation $\sim$ on $\mathbb{N} \times \mathbb{N}$ by $(a, b) \sim (c, d)$ if $a + d = b + c$.

A.13. Proposition. The relation $\sim$ is an equivalence relation on $\mathbb{N} \times \mathbb{N}$.

Proof: We verify the definition of equivalence relation using arithmetic properties of $\mathbb{N}$, including Proposition A.11. ∎

We write $[(a, b)]$ for the equivalence class containing (a, b). We want negative numbers to represent differences, so we think geometrically of $(4, 0)$ as a "negative" distance from 4 to 0. Thus $\{(4, 0), (5, 1), (6, 2), \cdots\}$ is the class we wish to call "-4". This approach enables us to extend the arithmetic properties of $\mathbb{N}$ to $\mathbb{Z}$.

A.14. Definition. Addition and multiplication on $\mathbb{Z}$ are defined by
$$[(a, b)] + [(c, d)] = [(a + c, b + d)]$$
$$[(a, b)] \cdot [(c, d)] = [(ad + bc, ac + bd)].$$

In these expressions, the arithmetic operations on the left are being defined; those on the right involve arithmetic in $\mathbb{N}$.

A.15. Theorem. (Arithmetic Properties of $\mathbb{Z}$)
a) Addition and multiplication are well-defined.
b) Addition and multiplication have identity elements $[(n, n)]$ and $[(n, n + 1)]$, respectively.
c) Addition and multiplication are commutative and associative.
d) The distributive law holds.
e) Each element $[(a, b)]$ has an additive inverse $[(b, a)]$.

Proof: These are all simple verifications using Definition A.14 and the known arithmetic properties of $\mathbb{N}$. ∎

We define subtraction by $[(a, b)] - [(c, d)] = [(a, b)] + [(d, c)]$. We treat $\mathbb{N}$ as a subset of $\mathbb{Z}$ by identifying the number n with the class $[(0, n)]$. Since $[(0, a)] + [(0, b)] = [(0, a + b)]$ and $[(0, a)] \cdot [(0, b)] = [(0, ab)]$, these operations mirror the corresponding operations on $\mathbb{N}$.

Given $n \in \mathbb{N}$, we henceforth write $-n$ for $[(n, 0)]$, 0 for $[(n, n)]$, and 1 for $[(n, n + 1)]$. By Theorem A.15e, this defines $-n$ to be the additive inverse of n. We also use the minus sign for subtraction; it is natural to write $[(a, b)]$ as $b - a$. Now the arithmetic and order operations behave as we wish, and we have introduced subtraction and negative numbers.

A.16. Proposition. Suppose $a, b \in \mathbb{N}$. With $-b, a, b$ defined in $\mathbb{Z}$ as above, we have $a - b = a + (-b)$ and $-(-b) = b$.

Proof: First, $a + (-b) = [(0, a)] + [(b, 0)] = [(b, a)] = a - b$. Also, the additive inverse of the additive inverse of $[(0, b)]$ is $[(0, b)]$. ∎

It follows that $\{[(0, n)] : n \in \mathbb{N}\}$ is a positive set in $\mathbb{Z}$ (see Definition 1.40); membership is closed under addition and multiplication, and the trichotomy property holds.

THE RATIONAL NUMBERS

Next we discuss the construction of $\mathbb{Q}$ from $\mathbb{Z}$. First we note the similarity between the equivalence relation defined to construct $\mathbb{Z}$ from $\mathbb{N}$ and the equivalence relation defined to construct $\mathbb{Q}$ from $\mathbb{Z}$. In the first case, $(a, b) \sim (c, d)$ if $a + d = b + c$; in the second, $(a, b) \sim (c, d)$ if $ad = bc$. The similarity arises because both relations are designed to introduce inverses.

A.17. Lemma. Let $F = \mathbb{Z} \times (\mathbb{Z} - \{0\})$, and let $\sim$ be a relation on F defined by $(a, b) \sim (c, d)$ if and only if $ad = bc$. The relation $\sim$ is an equivalence relation on F.

Proof: We must verify the reflexive, symmetric, and transitive properties. *Reflexive property:* $(a, b) \sim (a, b)$, since $ab = ba$. *Symmetric property:* If $(a, b) \sim (c, d)$, then by definition $ad = bc$, which is equivalent to $cb = da$, which by definition means $(c, d) \sim (a, b)$. *Transitive property:* Suppose $(a, b) \sim (c, d)$ and $(c, d) \sim (e, f)$, which means $ad = bc$ and $cf = de$. To prove that $(a, b) \sim (e, f)$, we want to show that $af = be$. Multiplying the two known equations yields $adcf = bcde$. Since $d \neq 0$, we can cancel d to obtain $acf = bce$. If $c \neq 0$, then we can also cancel c to obtain $af = be$. If $c = 0$, then $ad = bc$ and $cf = de$ imply that a and e are 0 as well, and again we have $af = be$.   ∎

A.18. Definition. The set of **rational numbers**, written as $\mathbb{Q}$, is the set of equivalence classes of $F = \mathbb{Z} \times (\mathbb{Z} - \{0\})$ under the relation $\sim$ defined above. We write $\frac{m}{n}$ or m/n to denote the rational number that is the equivalence class containing the pair (m, n). We write $\frac{a}{b} = \frac{c}{d}$ to mean that (a, b) and (c, d) belong to the same equivalence class.

To use the rationals as numbers, we must first define addition, multiplication, the additive and multiplicative identities, and the positive set for $\mathbb{Q}$. We next specify these in terms of the known operations on integers. As in Chapter 1, we can then define subtraction by $x - y = x + (-y)$ and define order by $x < y$ when $y - x$ is positive.

A.19. Definition. We define the rational numbers 0 and 1 to be $\frac{0}{1}$ and $\frac{1}{1}$, respectively. The **sum** and **product** of $\frac{a}{b}, \frac{c}{d} \in \mathbb{Q}$ are

$$\frac{a}{b} + \frac{c}{d} = \frac{ad + bc}{bd} \quad \text{and} \quad \frac{a}{b} \cdot \frac{c}{d} = \frac{ac}{bd}.$$

The rational number $\frac{a}{b}$ is **positive** if $ab > 0$.

A.20. Theorem. With Definition A.19, the set $\mathbb{Q}$ of rational numbers forms an ordered field.

Proof: The proof has two main parts; both rely on integer arithmetic. We first verify that the operations in $\mathbb{Q}$ are well-defined. We must also verify that they make $\mathbb{Q}$ into an ordered field. We prove the order axioms here and leave the field axioms to Exercise 6. When proving that the operations are well-defined (independent of the representatives chosen from the classes), we select $\frac{a}{b} = \frac{a'}{b'}$ and $\frac{c}{d} = \frac{c'}{d'}$.

Addition is well-defined: $\frac{a}{b} + \frac{c}{d} = \frac{a'}{b'} + \frac{c'}{d'}$. It suffices to prove that $\frac{ad+bc}{bd} = \frac{a'd'+b'c'}{b'd'}$, by the definition of rational addition. By the definition of the equivalence relation, the condition for equality is $(ad + bc)b'd' = bd(a'd' + b'c')$. Using the properties of integer arithmetic, this condition becomes $(ab' - ba')dd' = bb'(cd' - dc')$. This equality follows from $ab' - ba' = cd' - dc' = 0$, which holds because $(a, b) \sim (a', b')$ and $(c, d) \sim (c', d')$.

Multiplication is well-defined: $\frac{a}{b} \cdot \frac{c}{d} = \frac{a'}{b'} \cdot \frac{c'}{d'}$. It suffices to prove that $\frac{ac}{bd} = \frac{a'c'}{b'd'}$, by the definition of rational multiplication. The equality is equivalent to $acb'd' = bda'c'$, which follows from $ab' = ba'$ and $cd' = dc'$.

The positive set is well-defined: $\frac{a}{b} > 0$ if and only if $\frac{a'}{b'} > 0$. This holds since $ab' = ba'$ requires that ab and $a'b'$ have the same sign.

The positive set is closed under addition. From $\frac{a}{b} > 0$ and $\frac{c}{d} > 0$, we have $ab > 0$ and $cd > 0$. This yields $(ad + bc)bd > 0$ for each choice of signs for a, b, c, d, and hence $\frac{a}{b} + \frac{c}{d} > 0$.

The positive set is closed under multiplication. Again $\frac{a}{b} > 0$ and $\frac{c}{d} > 0$ imply $ab > 0$ and $cd > 0$. Hence $(ac)(bd) = abcd > 0$, which yields $\frac{a}{b} \cdot \frac{c}{d} > 0$.

The trichotomy property holds. For each nonzero class $\frac{a}{b}$, we have $-\frac{a}{b} = \frac{-a}{b}$. The closure properties of the positive set imply that it contains exactly one of $\frac{a}{b}$ and $-\frac{a}{b}$. ∎

The function f defined by $f(m) = m/1$ is an injection from $\mathbb{Z}$ to $\mathbb{Q}$ that preserves all arithmetic properties of the integers. Thus we interpret the rational numbers having representatives of the form $m/1$ as the integers.

THE REAL NUMBERS

We construct $\mathbb{R}$ from $\mathbb{Q}$ by what mathematicians call "completion". We want the limits of Cauchy sequences to exist; if a sequence of rational numbers is Cauchy, but does not converge in $\mathbb{Q}$, then we decree that it has a limit in $\mathbb{R}$. Different sequences may approach the same value, so we will need to consider such pairs of sequences to be equivalent. Thus a real number will be an equivalence class of Cauchy sequences of rational numbers. We prove that the set of these equivalence classes forms a complete ordered field, leaving some of the details to the reader.

To complete this program, we must define objects $\alpha, \beta, \gamma, \cdots$ to be the elements of the set $\mathbb{R}$. We must designate elements **0** and **1** and define addition and multiplication on $\mathbb{R}$. We must prove that these operations satisfy the algebraic laws of Definition 1.39 (Steps 5, 6, 7 below). We must specify the subset of $\mathbb{R}$ that is positive and prove that it satisfies the axioms for a positive set in Definition 1.40 (Steps 4 and 8 below). Finally, we must prove that our system satisfies the Completeness Property.

A.21. What must be done.

Step 1) Define $\mathbb{R}$.

Step 2) Define **0** and **1**.

Step 3) Define addition and multiplication. Define for each α an additive inverse $-\alpha$ and for each $\beta \neq \mathbf{0}$ a multiplicative inverse β^{-1}.

Step 4) Define a subset P called the *positive set* and verify trichotomy.

Step 5) Verify the following laws for addition.

$$\alpha + \mathbf{0} = \alpha \qquad (\alpha + \beta) + \gamma = \alpha + (\beta + \gamma)$$
$$\alpha + -\alpha = \mathbf{0} \qquad \alpha + \beta = \beta + \alpha$$

Step 6) Verify the following laws for multiplication.

$$\alpha\mathbf{1} = \alpha \qquad (\alpha\beta)\gamma = \alpha(\beta\gamma)$$
$$\alpha\alpha^{-1} = \mathbf{1} \text{ if } \alpha \neq \mathbf{0} \qquad \alpha\beta = \beta\alpha$$

Step 7) Verify the distributive law $(\alpha + \beta)\gamma = \alpha\gamma + \beta\gamma$.

Step 8) Verify that addition and multiplication preserve order.

$$\alpha, \beta > \mathbf{0} \text{ implies both } \alpha + \beta > \mathbf{0} \text{ and } \alpha\beta > \mathbf{0}$$

Step 9) Prove that every nonempty subset $T \subset \mathbb{R}$ that is bounded above has a least upper bound. ∎

We assume that the rational numbers and their properties are known. Recall that $|x| = \max\{x, -x\}$. Note that reciprocals of large natural numbers are small positive rational numbers; these take the place of the real number ϵ in the definition of convergence.

We recall that a sequence of rational numbers is a function $a \colon \mathbb{N} \to \mathbb{Q}$; we write $\langle a \rangle$ to denote a sequence.

A.22. Definition. The sequence $\langle a \rangle$ is a **Cauchy sequence** if for every $k \in \mathbb{N}$ there is an N such that $n, m \geq N$ implies $|a_n - a_m| < 1/k$. The sequence $\langle a \rangle$ **converges** to $L \in \mathbb{Q}$ if for every $k \in \mathbb{N}$ there exists $N \in \mathbb{N}$ such that $n \geq N$ implies $|a_n - L| < 1/k$.

A.23. Lemma. If $a_n \to L$, then $\langle a \rangle$ is a Cauchy sequence.

Proof: Repeat the proof of Proposition 14.13, replacing ϵ with $1/k$. ∎

Let S denote the set of Cauchy sequences of rational numbers. We will partition S into equivalence classes in order to define $\mathbb{R}$. The following statements are easy to prove.

A.24. Proposition. The set S of Cauchy sequences of rational numbers is closed under addition, multiplication, and scalar multiplication:
   a) If $\langle a \rangle \in S$ and $\langle b \rangle \in S$, then $\langle a + b \rangle \in S$.
   b) If $\langle a \rangle \in S$ and $\langle b \rangle \in S$, then $\langle ab \rangle \in S$.
   c) If $\langle a \rangle \in S$ and $c \in \mathbb{Q}$, then $\langle ca \rangle \in S$.   ∎

A.25. Lemma. If a Cauchy sequence $\langle a \rangle \in S$ has a convergent subsequence, then $\langle a \rangle$ also converges and has the same limit.   ∎

We have defined convergence of sequences of rational numbers to a rational limit, in particular to zero. We write $\langle a \rangle \sim \langle b \rangle$ when $\langle a - b \rangle$ converges to the rational number 0. Let **0** denote the subset of S consisting of sequences converging to 0; we record some of its properties:

A.26. Lemma. Suppose that $\langle a \rangle$, $\langle b \rangle$, $\langle c \rangle$ are Cauchy sequences.
   a) If $\langle a \rangle$ and $\langle b \rangle$ converge to 0, then $\langle a + b \rangle$ converges to 0.
   b) If $\langle a \rangle$ converges to 0, then $\langle ca \rangle$ converges to 0.

Proof: These proofs are similar to the proofs in Theorem 14.5, after replacing ϵ with $1/k$.   ∎

Using the notation for **0**, these become the statements "$\langle a \rangle$, $\langle b \rangle \in \mathbf{0}$ imply $\langle a + b \rangle \in \mathbf{0}$" and "$\langle a \rangle \in \mathbf{0}$, $\langle c \rangle \in S$ imply $\langle ca \rangle \in \mathbf{0}$". In algebra, a **ring** is a set equipped with both addition and multiplication satisfying appropriate axioms. A subset of a ring that is closed under addition and under multiplication by elements of the full ring is an **ideal**. Thus **0** is an ideal in the ring S of Cauchy sequences of rational numbers.

A.27. Corollary. The relation $\sim$ is an equivalence relation on S.   ∎

A.28. Remark. Our approach here echoes our approach to modular arithmetic. We defined the integers modulo p by considering equivalence classes. Two integers are equivalent (modulo p) when their difference is a multiple of p. The set of multiples of p is an ideal in $\mathbb{Z}$, just as **0** is an ideal in S. In both cases we consider the equivalence classes modulo the ideal. Just as two integers are congruent modulo p when their difference is a multiple of p, so two Cauchy sequences represent the same real number when their difference converges to 0.   ∎

We define the set $\mathbb{R}$ of real numbers to be the set of equivalence classes of S under the relation $\sim$. This completes Step 1.

The class **0** consists of the elements of S converging to zero; this will be the additive identity **0** in $\mathbb{R}$. We let **1** denote the set of sequences converging to the rational number 1; this will be the multiplicative identity. This completes Step 2.

We next define the positive real numbers.

A.29. Definition. The real number α is **positive** if, for each sequence $\langle a \rangle \in \alpha$, there exist $k, N \in \mathbb{N}$ such that $n \geq N$ implies $a_n > 1/k$. The real number α is **negative** if $-\alpha$ is positive, where $-\alpha = \{\langle -a \rangle | a \in \alpha\}$.

In other words, if a sequence belongs to a positive equivalence class, then its terms are eventually positive and bounded away from zero. Proving the trichotomy property requires the following lemma.

A.30. Lemma. For any Cauchy sequence of rational numbers, precisely one of the following conditions holds:
a) The terms are eventually positive and bounded away from zero.
b) The terms are eventually negative and bounded away from zero.
c) The sequence converges to zero.

Proof: No two of the conditions can both hold for a given sequence. Hence it suffices to show that a sequence for which statements (a) and (b) are both false must converge to zero. To do so, find a subsequence converging to zero and use Lemma A.25. ∎

Consider a real number α; we emphasize that α is a set of sequences. We claim that all sequences in α satisfy the same property from Lemma A.30. If $\langle a \rangle$ and $\langle b \rangle$ are elements of α, then $\langle a - b \rangle$ converges to 0. If Lemma A.30a holds for $\langle a \rangle$, it therefore also holds for $\langle b \rangle$, because the terms of $\langle b \rangle$ are eventually arbitrarily close to those of $\langle a \rangle$. Similarly, if Lemma A.30b holds for $\langle a \rangle$, then it also holds for $\langle b \rangle$. Finally, if Lemma A.30c holds for $\langle a \rangle$, then by the definition of the equivalence relation this also holds for $\langle b \rangle$.

These remarks prove the claim. We define $\alpha > \mathbf{0}$, that is, α is *positive*, if Lemma A.30a holds for each $\langle a \rangle \in \alpha$. Similarly, $\alpha < \mathbf{0}$ if Lemma A.30b holds for each $\langle a \rangle \in \alpha$. If neither holds, then $\alpha = \mathbf{0}$. This proves the trichotomy property in Step 4.

Next we define the algebraic operations. When $\langle a \rangle$ is an element of S, we write $[\langle a \rangle]$ for the set of all elements equivalent to $\langle a \rangle$ (the class containing $\langle a \rangle$).

A.31. Definition. Let $\langle a \rangle$, $\langle b \rangle$ be sequences contained in the real numbers α, β, respectively. The **sum** and **product** of α and β are defined by

$$\alpha + \beta = [\langle a + b \rangle] \qquad \alpha \cdot \beta = [\langle ab \rangle].$$

For the definitions to be valid, we must show that the results do not depend upon which elements we choose from the classes.

A.32. Lemma. Addition and multiplication in $\mathbb{R}$ are well-defined.

Proof: Choose representatives $\langle a \rangle$, $\langle a' \rangle$ of α and $\langle b \rangle$, $\langle b' \rangle$ of β, and show that $\langle a + b \rangle - \langle a' + b' \rangle$ and $\langle ab \rangle - \langle a'b' \rangle$ converge to zero.   ∎

We have now defined *zero, one, positive, negative, sum,* and *product.* To define the additive inverse, we put $-\beta = [\langle -b \rangle]$, where $\langle b \rangle$ is any element of β. This definition is valid: if $\langle b \rangle$, $\langle b' \rangle \in \beta$, then $[\langle -b' \rangle] = [\langle -b \rangle]$, because $-\langle b \rangle - \langle -b' \rangle = \langle b' - b \rangle$ converges to 0.

To define the reciprocal of a nonzero real number, we need a preliminary observation. Let β be a nonzero real number. By Lemma A.30, we know that each sequence in β is eventually bounded away from zero. Therefore, omitting finitely many terms from some member of β yields a representative $\langle b \rangle$ all of whose terms are nonzero. For such a sequence $\langle b \rangle$, we define $\langle b^{-1} \rangle$ to be the sequence whose nth term is b_n^{-1}. Using such a representative $\langle b \rangle$ of β, we define $\beta^{-1} = [\langle b^{-1} \rangle] = \{c : \langle c \rangle \sim \langle b^{-1} \rangle\}$. Again the definition is valid.

We have now accomplished everything through Step 4. The laws in Steps 5–8 have similar proofs, which we leave to the reader.

A.33. Lemma. Addition in $\mathbb{R}$ is commutative.   ∎

A.34. Lemma. The distributive law holds in $\mathbb{R}$, and all other properties in Steps 5–9 hold.   ∎

We treat a rational number A as a real number α in the following way. Let $\alpha = [\langle a \rangle]$, where $\langle a \rangle$ is the constant sequence such that $a_n = A$ for all n. Thus it makes sense to write an inequality between a rational number and a real number. Furthermore, the additive identities and multiplicative identities of $\mathbb{Q}$ and $\mathbb{R}$ correspond.

Finally, we prove the Completeness Property. Recall that a nonempty subset $T \subset \mathbb{R}$ is bounded above if there exists a real number β such that $\alpha \in T$ implies $\alpha \leq \beta$. Adding a positive number to an upper bound yields another upper bound; hence every set having an upper bound has a rational upper bound. Similarly, for every nonempty set there are rational numbers that are not upper bounds.

A.35. Theorem. The real number system $\mathbb{R}$ satisfies the Completeness Property.

Proof: Suppose T is a nonempty set of real numbers having an upper bound. Let a_1 be a rational number that is not an upper bound for T, and let b_1 be a rational number that is an upper bound for T. Note that

$a_1 < b_1$. Put $c_1 = \frac{a_1+b_1}{2}$, and observe that c_1 is the average of the two numbers. Hence $a_1 < c_1 < b_1$, and c_1 is rational.

We define $\langle a \rangle$, $\langle b \rangle$, $\langle c \rangle$ inductively. Given a_n, b_n, we define c_n to be the average of a_n and b_n. If c_n is not an upper bound for S, then we put $a_{n+1} = c_n$ and $b_{n+1} = b_n$. If c_n is an upper bound for S, then we put $b_{n+1} = c_n$ and $a_{n+1} = a_n$. This defines three sequences of rational numbers, and we observe that $a_n < c_n < b_n$ holds for each n. The three sequences have the same limit, and hence all define the same real number $[\langle c \rangle]$, which is easily seen to be the least upper bound for T.   ∎

We have now constructed the real numbers and proved that they form a complete ordered field. There is only one complete ordered field, in the sense that we could label the elements of any complete ordered field $\mathbf{F}$ by the real numbers in such a way that $\mathbf{F}$ behaves just like $\mathbb{R}$.

A.36. Theorem. If $\mathbf{F}$ is a complete ordered field, then there is a unique bijection $f\colon \mathbb{R} \to \mathbf{F}$ that preserves addition, multiplication, and order.

Proof: Suppose $\mathbf{F}$ is an ordered field; to indicate the possibility that the elements of $\mathbf{F}$ differ from those of $\mathbb{R}$, we write the elements of $\mathbf{F}$ in bold type; thus $\mathbf{0}$ and $\mathbf{1}$ denote the additive and multiplicative identity elements in $\mathbf{F}$. We define a bijection $f\colon \mathbb{R} \to \mathbf{F}$ that preserves arithmetic and order. We define f in stages, first defining it on 0 and $\mathbb{N}$, then extending it to $\mathbb{Z}$ and $\mathbb{Q}$ before using the Completeness Axiom to extend it to $\mathbb{R}$.

Define $f(0) = \mathbf{0}$, $f(1) = \mathbf{1}$, and $f(n) = \mathbf{1} + \mathbf{1} + \cdots + \mathbf{1}$, meaning that $f(n)$ is the sum in $\mathbf{F}$ of n copies of $\mathbf{1}$. Using the existence of additive and multiplicative inverses in $\mathbf{F}$, we extend f to negative integers by defining $f(-n) = -f(n)$ and then to rational numbers by defining $f(m/n) = \frac{f(m)}{f(n)}$. (The division is taken in $\mathbf{F}$.) We then must show that f preserves the order relation on $\mathbb{Q}$.

Next we define f on the irrational numbers. Given $x \in \mathbb{R}$, let S denote the set of rational numbers less than x, and let $S' = \{f(y)\colon y \in S\}$. Since f preserves the order relation on $\mathbb{Q}$, the set S' is bounded above in $\mathbf{F}$ by the image of some rational number bigger than x. Since S' has an upper bound and $\mathbf{F}$ is complete, S' has a supremum in $\mathbf{F}$. Let $\mathbf{x}$ be the supremum of S' in $\mathbf{F}$; we define $f(x)$ to be $\mathbf{x}$. Now f is a bijection that preserves addition, multiplication, and positivity. We see that $\mathbf{F}$ behaves exactly like $\mathbb{R}$, with the role of $x \in \mathbb{R}$ played by its boldface counterpart $f(x) = \mathbf{x}$ in $\mathbf{F}$.   ∎

EXERCISES

A.1. Establish a bijection between $\mathbb{N} \cup (-\mathbb{N}) \cup \{0\}$ and the set of equivalence classes of $\mathbb{N} \times \mathbb{N}$ under the relation $\sim$ defined by putting $(a, b) \sim (c, d)$ if $a + d = b + c$.

A.2. Write an inductive definition of exponentiation by a natural number and prove that $x^{m+n} = x^m x^n$ when $x, m, n \in \mathbb{N}$.

A.3. Complete the proof of Theorem A.15 by verifying that multiplication in $\mathbb{Z}$ is well-defined, has identity element $[(n, n+1)]$, and is commutative and associative.

A.4. Use induction and the definition of multiplication to prove that the product of two nonzero integers is nonzero. Use this and the distributive law to prove that multiplication by a nonzero integer is an injective function from $\mathbb{Z}$ to $\mathbb{Z}$.

A.5. Prove that multiplication by a natural number is an order-preserving function from $\mathbb{Z}$ to $\mathbb{Z}$ ($x > y$ implies $f(x) > f(y)$), and use this to prove that multiplication by a nonzero integer is an injective function from $\mathbb{Z}$ to $\mathbb{Z}$.

A.6. Complete the proof that $\mathbb{Q}$ is an ordered field by verifying the field axioms for rational addition and multiplication. The desired statements should be reduced to statements about integers, and then properties of integer arithmetic should be used to prove them. Division is not allowed, but nonzero integers can be canceled from both sides of an equality.

A.7. Define $\langle a \rangle$ by $a_1 = 2$ and $a_{n+1} = \frac{1}{2}(a_n + \frac{2}{a_n})$ for $n \in \mathbb{N}$. Prove that $\langle a \rangle$ is a Cauchy sequence of rational numbers. Prove that $\langle a \rangle$ has no limit in $\mathbb{Q}$. What does this say about Lemma A.23?

A.8. Prove that the set S of Cauchy sequences of rational numbers is closed under addition, multiplication, and scalar multiplication.

A.9. Prove that if a Cauchy sequence of rational numbers has a convergent subsequence, then $\langle a \rangle$ also converges and has the same limit.

A.10. Prove that multiplication of real numbers is commutative and that addition and multiplication of real numbers are associative.

A.11. Prove that $\mathbf{0}$ is an identity element for addition and that $\mathbf{1}$ is an identity element for multiplication of real numbers. Given $\alpha \in \mathbb{R}$ with $\alpha \neq \mathbf{0}$, prove that $\alpha + (-\alpha) = \mathbf{0}$ and that $\alpha \cdot \alpha^{-1} = \mathbf{1}$. Prove that $\mathbf{0} < \mathbf{1}$.

A.12. Prove that the sum and the product of positive real numbers are positive.

A.13. Prove that the limit of any convergent sequence of upper bounds for S is an upper bound for S.

A.14. Suppose that $|a_{n+1} - a_n| \leq M/2^n$ for some constant $M > 0$. Prove that $\langle a \rangle$ is a Cauchy sequence. (Hint: Estimate $|a_m - a_n|$ by using a telescoping sum, and use the convergence of $\sum_{k=0}^{\infty} 1/2^k$.)

A.15. Prove that the function f constructed in Theorem A.36 preserves the order relation on $\mathbb{Q}$.

A.16. Prove that the function f constructed in Theorem A.36 is a bijection and preserves addition, multiplication, and positivity on $\mathbb{R}$.

A.17. Use the axioms of a complete order field (Definitions 1.39–1.41) to prove (some of) the properties that follow from them (Propositions 1.43–1.46).

Appendix B

Hints for Selected Exercises

1.8. The two sections need not have the same number of students.

1.18. The condition gives a quadratic equation. Solve it.

1.19. Relate this to the Babylonian Problem.

1.20. When x is a solution, $(x - r)(x - s) = 0$.

1.22. Think about the amount of liquid in each glass and the amount of liquid of each type.

1.24. Explain why there is no missing dollar!

1.25. Consider all the different ways to factor 36 into a product of three positive integers. The given story eliminates all but one possiblity.

1.26. Apply similar reasoning as in Exercise 1.25 for various possible ages of the mailman, until obtaining a case that fits the scenario.

1.27. Square both sides and simplify.

1.28. For part (b), make an appropriate substitution.

1.29. Start by grouping $x + y + z$ appropriately and expanding the square.

1.32–34. When is the inequality satisfied?

1.35. When x, y have the same sign, multiplying by xy gives an equivalent inequality.

1.37. If $a \neq 0$, then the graph is a parabola. Place the parabola at different heights, facing both up and down, to see the possibilities.

1.39. How many factors must be negative for x to satisfy the inequality?

1.42. Think about leap years.

1.45. Read the discussion about "well-defined".

1.49. For part (e), show that $f^2 + g^2$ is bounded.

1.52. Graph some level sets of $x + y$ and xy.

2.12. Consider the cost of each additional call.

2.13. Let x, y, z be the three ages. Rewrite each piece of information in terms of x, y, z and solve the resulting equations. Check your answer.

2.14. For part (b), complete the squares.

2.16. In part (a), why is there only one solution for $g(x)$ and $h(x)$?

2.18. How can $A^2 - B^2$ be computed?

2.19. There is more than one interpretation.

2.23. Pay careful attention to negation of quantifiers.

2.24–25. Consider Example 2.11.

2.26. In one case the same δ must work for every a; in the other case δ is allowed to depend on a.

2.28. Pay careful attention to quantifiers.

2.30. Vowel implies Odd is the same as Not Odd implies Not Vowel.

2.32. Try analysis by cases.

2.33. The front child must use the information that the other children were unable to decide instantly. Consider what the other children would see if the two front children had two red hats, two black hats, or one of each.

2.34. First simplify the equations.

2.38. Use the definitions of odd and even.

2.40. In both cases count the squares of each color in the defective checkerboard. Note that a T-shape may be centered on a black square or on a white square.

2.47. "x is odd" means we can write $x = 2n + 1$ for some n; "$x^2 - 1$ is divisible by 8" means we can write $x^2 - 1 = 8m$ for some m. Note that $n(n + 1)$ is always even. For part (b), use the contrapositive.

2.48. Pay attention to the scope of the quantifiers.

2.50. Convert intuition obtained from Venn diagrams into careful language about membership.

2.54. Is it possible to reach a configuration with an odd number of white tokens inside each circle?

3.6. Must $P(1)$ be true?

3.9. Notice the difference between this problem and Exercise 3.8.

3.10. Use induction and be precise.

3.14. Practice induction or relate the given sums to known sums.

3.15–17. In the algebraic computations for the induction step, extract the desired factors as early as possible.

3.19. Keep in mind what goes wrong when the exponent is even.

3.20. Shift the index in one sum to make it easy to combine terms.

3.26. Use induction and the formula for a_{n+1} in terms of a_n.

3.28. Use partial fractions to rewrite the fraction.

3.30. Try small values to guess the formula. Be sure to replace $2i - 1$ with $2(i + 1) - 1 = 2i + 1$ when replacing n with $n + 1$.

3.31–32. Try small values to guess the formula.

3.33. Relate this to a familiar sum.

3.34. Try weighing a different number of balls from each box.

3.36. Modify the Geometric Sum to obtain the desired formula.

3.37. Apply the Geometric Sum.

3.38. Determine who wins when the goal is a multiple of 4.

3.40. Use the approach in Solution 3.22.

3.43. In using induction on n, consider the case $u = 0$ separately.

3.47. Use induction on n and the inequality $1 < 5$.

3.48. Use the case $n = 1$ to obtain a necessary condition.

3.52. Find r and s, and then use the Method of Undetermined Coefficients.

3.54. Try the cases $n = 1, 2, 3$ to guess the formula.

3.55–57. The induction step uses two earlier instances of the claim.

3.58. Use induction on k.

3.59. Use induction. The case of the 5 by 9 rectangle must be treated separately.

3.60. Use induction on k. In part (a), one must find a good first place to look and show that it works. In part (b), one must show that every first place to look fails.

3.62. A person who can say "Nov. 30" will win. Work backwards from the end, determining the winning dates. Strong induction permits a formal proof.

4.3. Express the two alternatives in terms of n.

4.4. 50 is the average of 20 and 80.

4.5. When can two elements be interchanged?

4.6. List the values of f.

4.7. What do injective and surjective mean in terms of level sets?

4.8. See Example 4.29.

4.10. First find a formula for h.

4.12. Rely on the definitions.

4.13. Reduce to $a > c$. Compute the outcome in terms of a, b, c.

4.14. Use strong induction.

4.15. How can the induction hypothesis be used to obtain weighings for all the desired values?

4.16. For the necessity of the condition, show that when $w_j > 1 + 2\sum_{i=1}^{j-1} w_i$, the weight $(\sum_{i=1}^{k} w_i) - 1 - 2(\sum_{i=1}^{j-1} w_i)$ cannot be balanced.

4.17. Use strong induction. Prove that if the condition does not hold, then Player 1 can make a move so it will hold.

4.20. For part (c), think of the point $(p, q) \in \mathbb{R}^2$ as an arrow from $(0, 0)$ to (p, q). Then locate the tail of the arrow for $f(x, y)$ at the point (x, y).

4.21. Focus on one element of $[n]$. Use it to modify elements of A to obtain elements of B. Prove that the resulting function is a bijection.

4.24. What happens when $f = g$?

4.26. Start with $f(x) = f(y)$ and use the inequality.

4.28. The constant term is irrelevant. Replace x with $Ax + B$ for an appropriate choice of A and B to eliminate the quadratic term.

4.30. Consider two cases, depending on whether $ad - bc = 0$.

4.32. Show that f and g are both injective and surjective.

4.34. Consider the pictures accompanying Remark 1.22 and Definition 4.28.

4.35. In part (a), what happens if A has more elements than B?

4.36. The definitions of injection and surjection say what must be verified.

4.37. Suppose that $f(a) = f(b)$, and apply f again.

4.39–40. Use induction to be precise.

4.41. Use induction on n.

4.49. Imitate the proof that $\mathbb{N} \times \mathbb{N}$ is countable. Note that the sets need not be disjoint. Note also that proving countability of $A_1, \ldots, A_k \cup$ by induction on k *cannot* solve the problem.

4.51. Start by deciding which elements will be mapped to 0 and 1.

5.7. The number of choices for the second card may depend on the choice of the first card.

5.8. Don't multiply it out!

5.9. It may be necessary to consider cases. For part (b), there is more than one approach.

5.12. Count the ways of obtaining eleven.

5.14. This can be done by evaluating summations or by a more direct method; the desirable outcomes are ordered triples (x_1, x_2, x_3) that sum to n.

5.15. Use induction on k.

5.17. Assume without loss of generality that $n \le m < k$; cancel common factors.

5.19. After picking k digits to use, one must count the 6-tuples that can be formed using all k digits. This may involve cases.

5.20. Prove that this ratio counts some set. Induction also works, as does a discussion of divisibility.

5.21. What determines a rectangle within the grid?

5.22. It is possible to do this by counting the crossing pairs added with each additional point, applying Exercise 40 and Theorem 5.28. For a more direct proof, what determines a pair of crossing diagonals?

5.23. Use the rule of product. In parts (a) and (b), pick the ranks and then the cards within the ranks.

5.26. In summation notation, the proof requires shifting the index of summation to combine terms appropriately.

5.27. Show that counting the even subsets positively and the odd subsets negatively yields 0 when $n > 0$.

5.28–29. Consider Theorem 5.23.

5.30. Use Pascal's Formula.

5.32. There are n^2 dots in an n by n square of dots.

5.35. How many losers are there in each race?

5.37. Form committees with subcommittees.

5.38. Consider binary lists that are not all zero.

5.39. What does the formula on the right side count? Split this set into subsets corresponding to the terms of the left.

5.40. Split the pairs of elements in $[n]$ into groups. The ith group should have size $i - 1$.

5.41. Split the triples of elements in $[n]$ into groups. The ith group should have size $(i - 1)(n - i)$.

5.42. Split the elements of a set counted by the right side into groups. The kth group should be defined so its size is the term for k in the sum.

5.43. Group the selections of $r + s + 1$ positions from a row of $m + n + 1$ positions according to the position of the $r + 1$th chosen position.

5.44. Cut an appropriate set of selections with repetition into subsets counted by the terms on the left side.

5.46. Use induction or expand a cleverly chosen product of n factors.

5.48. Interpret the chains of subsets in terms of the elements of $[n]$.

5.50. The answer needs the information that all the labels are wrong.

5.51. Find a permutation so that different ways of reaching the end lead to different answers for the final drummer.

5.55. By the induction hypothesis, the first $n - 1$ entries of an element of B_n yield a permutation of $[n - 1]$. Use the last entry to incorporate the element n.

5.56. Find the first n that can be used as a basis step of an inductive proof; consider smaller values separately.

5.57. Guess a formula by considering small n. The combinatorial proof is similar to the combinatorial proof of the Summation Identity.

5.58. What condition on the functional digraph of a permutation corresponds to having no fixed points? Count the functional digraphs satisfying this.

5.59. Use induction on n.

5.61. Compute $f(f(f(x)))$ and set it equal to x. Be careful not to divide by zero, and be sure not to confuse fixed points with points in a 3-cycle.

5.62. For part (b), show that interchanging rows and columns of dots defines an appropriate bijection.

5.63. To define a bijection, think of a natural (and reversible!) way to transform a partition using parts of odd size into a partition using parts of distinct sizes.

6.9. Compare with Example 6.19.

6.11–12. Let x be the number of coins of each type. Determine the amount of money as a function of x in each case. Use the notion of "relatively prime".

6.13–14. Mimic the procedure in Exercise 6.11.

6.16. Use subtraction and induction.

6.17. Show that each pair has the same set of common divisors.

6.19. Apply Proposition 6.6.

6.20. The right side counts points in a rectangle. Split it into classes counted by terms on the left side.

6.21. Rely on the definitions of the floor and ceiling functions.

6.23. Consider divisibility by 3.

6.24. Use induction or case analysis.

6.25. Use induction, and be careful about the basis.

6.26. Combine induction and case analysis.

6.27. What happens after powers of 3 are removed?

6.28–29. Consider the prime factorization.

6.30. Use induction or cancel common factors or provide a combinatorial argument to show that this counts a set of odd size.

6.31. In part (a), what must hold for c when a, b are both odd? In part (b), consider cases.

6.32. Try small examples to guess the pattern; then prove it by induction. The proof is easy when the pattern is expressed using divisibility considerations.

6.33. Find a particular divisor of $abcabc$.

6.34. Letting S be the (finite) set of all primes, build a number that is not divisible by any member of S.

6.35. If $x + i - 2$ is divisible by i, then what else is divisible by i?

6.36. In part (a), count the contributions to the exponent of each prime p. In part (b), show that the contributions are at least this large for the product of any k consecutive number.

6.38. Provide an explicit factorization, using the factorization of $x^{2r+1} - y^{2r+1}$.

6.42. Observe that $f(n)$ depends only on the last digit of n.

6.43–44. Apply strong induction.

6.45. Consider diophantine equations. Remember that only 500 weights are available of each type.

6.47. Clear fractions and then express all solutions in terms of one solution.

6.50. For part (b), consider cases depending on the factors of p in k, m, n.

6.51. Let $x - 4$ be the original number of coconuts and follow the events to determine x.

6.52. Given that it is possible to post k, under what conditions is it also possible to post $k + 1$?

6.53. Consider that $\sum_{i=1}^{2n} i = n(2n + 1)$.

6.54. How can this expression for y in terms of x be used to reach a triple with a smaller minimum value?

6.56. Use polynomial long division.

6.61. Consider the set of polynomials p such that $p(0, 0) = 0$.

6.63. Consider the degrees of the polynomials and prove the contrapositive.

7.3. Consider remainders.

7.4. Write each number as a sum of powers of 10. What is the congruence class of 10^k modulo 9?

7.6–7. Operate with an appropriate representative of a congruence class.

7.8. Perform computations modulo 8.

7.9. What is the congruence class of 2^{12} mod 13?

7.10–15. Rely on the definitions.

7.16. Leap years can be considered separately. Work modulo 7, where Friday is congruent to 6 modulo 7. Consider the value of the 13th day in each month.

7.19. List the squares modulo 5.

7.20. After observing that $k \equiv 1 \mod (k-1)$, the proof is one line!

7.23. What is the congruence class of 10^k modulo 11?

7.24. When is x not congruent to $-x$?

7.25. Choose suitable members of congruence classes to simplify computations.

7.28–30. Use modular arithmetic.

7.31. First observe that (-1) is a square modulo $m^2 + 1$.

7.32. Consider remainders modulo d.

7.33–34. Use the Chinese Remainder Theorem.

7.36. Note that a, b, c may have common factors.

7.38–39. How big is each equivalence class?

7.40. The colorings group into equivalence classes of sizes 1 and 2.

7.41. Note that these operations permute the congruence classes.

7.42. The functional digraphs make it easy to see the partitions.

7.43. Multiply together all of $\{a, 2a, \ldots, (p-1)a\}$.

7.44. Since 341 is not prime, Fermat's Little Theorem cannot be used directly with the exponent 341. Note that $341 = 11 \cdot 31$.

7.45. Use Fermat's Little Theorem.

7.47. Use that $(p-1)! = (p-1)(p-2)(p-3)!$.

7.51. Suppose that $y \circ x = 1 = z \circ x$ and use the existence of an inverse.

7.53. Mimic the proofs of Lemmas 7.34–7.35.

8.2. Clear denominators in $f(x) = 0$.

8.8. Clear fractions, simplify, and compare with Example 4.27. Alternatively, compare $1/(x+y)$ and $1/x + 1/y$ to $1/x$, considering the sign of y.

8.10. Follow the definitions and prove that $an + bm$ and mn have a common prime factor if and only if m and n have a common prime factor.

8.11. Divide numerator and denominator by y. The result is very simple.

8.14. Again set $y = t(x+1)$. Pay attention to which values of t are valid.

8.16. Consider parity.

8.17. Use a bijection from $\mathbb{N}$ to $\mathbb{N} \times \mathbb{N}$.

8.20–21. Use the Rational Zeros Theorem.

8.22. Consider the equation $x^k - n = 0$ and imitate Example 8.15.

8.27. Compare with Exercise 1.30. There is also a geometric explanation using circles in the plane.

8.28. List the successful pairs.

8.31. Make a substitution so that Theorem 8.23 applies.

9.5–6. Use $P(B^c) = 1 - P(B)$ and the definition of independence.

9.7. Use the definition of independence, but be careful.

9.9. The answer is not $1/2$.

9.11. Compute the conditional probabilities for the prize to be behind each remaining door, given what has occurred.

9.12. Describe the portions of the probability space that correspond to the event that the length of the chord exceeds $\sqrt{3}$.

9.14. The switching argument of the Ballot Problem can be used.

9.15. For part (a), use induction on n. For part (b), establish a one-to-one correspondence between these arrangements (when $m = n + 1$) and ballot sequences.

9.17–18. Use Bayes' Formula.

9.23. Given that A does not win on the first flip, what is the probability that A wins later?

9.24–25. The expected payoff of switching can be computed by using conditional probability or by using a more direct ad hoc argument.

9.26. Show that each element of the probability space contributes the same amount to each side of the equation.

9.28. There are several ways to express the desired random variable as a sum of random variables that take only the values 0 and 1.

9.29. Use the linearity of expectation.

9.33. Since the monomials are equally likely, this does not involve multinomial coefficients. Monomials correspond to selections with repetition.

9.35. Don't multiply it out!

9.36. Consider the height of the path at each horizontal step.

9.37. Apply the solution to Bertrand's Ballot Problem.

9.39. Start at one point on the circle. Traverse the circle, recording 0 or 1 for each of the $2n$ points in a way that yields a ballet list from each noncrossing pairing. Prove that the function is a bijection.

10.2. How many committees can be formed from n professors?

10.4. Split $[2n]$ into n classes such that within each class the numbers are pairwise relatively prime.

10.5. Work modulo 10. If no two are congruent, then divide into classes by considering sums.

10.6. Consider the average sum of three consecutive numbers.

10.8. Partition the square into regions such that two points in the same region have the desired property.

10.10. Note that each 100-yard field contains its endpoints. What happens if no point in the first 100 yards or last 100 yards is used four times?

10.12. Use partial sums.

10.13. Show that the largest number in S is between 2 and $(k+1)/2$, and use this in a proof by induction.

10.14. Consider the acquaintances of one student.

10.19. Show that with 989 keys, some 90 members can't open all the rooms.

10.24–37. Define a universe and appropriate subsets $A_1, \ldots, A_m$ so that the desired set is the set of elements in the universe outside all of $A_1, \ldots, A_m$. The Inclusion-Exclusion Formula then yields the answer.

10.39–41. Invent a universe and sets so that the terms in the sum will be the terms computed in the Inclusion-Exclusion Formula for counting the elements outside the sets. The term for $k = 1$ tells what the size of each set should be.

11.5. When there are n vertices, can 0 and $n-1$ both occur as vertex degrees?

11.6. Consider an appropriate subgraph of a graph that models the schedule.

11.7. Consider the vertex degrees in $G - e$.

11.10. Constructing the d-dimensional cube from two copies of the $d-1$-dimensional cube facilitates inductive proofs about Q_d.

11.11. Start by showing that the vertices of a 4-cycle agree in all but two coordinates, and the vertices of a 6-cycle agree in all but three coordinates.

11.16. Obtain a formula for the number of edges of G in terms of the number of vertices.

11.17. The Petersen graph can be described as the graph whose vertices are the 2-element subsets of $\{1, 2, 3, 4, 5\}$, with two vertices adjacent if and only if the pairs are disjoint.

11.18. One can obtain a graph on the smaller vertex set by deleting v_n; how can this map be inverted?

11.19. A complete graph with n vertices has $\binom{n}{2}$ edges.

11.20. For part (a), consider the neighborhoods of x and y, and apply an appropriate identity about sets.

11.22. Define a graph to model the possible positions and moves. What is the condition for reaching the desired configuration?

11.24. When P and Q have no common vertex, obtain a longer path.

11.25. When the maximum degree is k, show that G has at most $k(n-k)$ edges.

11.26. Use induction on $n-k$, or use the properties of trees.

11.27. Use induction or apply Lemma 11.36 and Theorem 11.40 to appropriate subgraphs.

11.28. Show that if there are two x, y-paths, then their union contains a cycle.

11.30. Use properties of trees.

11.32. Show first that G contains the tree T' obtained by deleting a leaf of T.

11.33. For $n > 2^k$, observe that every partition of the vertices into two sets has one set with more than 2^{k-1} vertices. Use this to prove the upper bound by induction.

11.34. If an edge appears in two cycles, then the graph has three paths joining its endpoints.

11.35. Use induction or contradiction.

11.36. For part (a), count edges. For part (b), prove that Hall's Condition holds.

11.39. Prove the contrapositive.

11.40. For the first part, color the vertices in some order, always using the least-indexed color not yet appearing on a neighbor.

11.41. Color the vertices using a wise order determined geometrically.

11.42. How many times can each color be used?

11.43. How can we compute the sum of the coefficients of a polynomial? When we do that with a chromatic polynomial, what does it say about the graph?

11.45. Consider the number of edges.

11.46. The minimum vertex degree is at most the average vertex degree.

11.47. Use Euler's formula as in Theorem 11.65.

11.48. Show that the boundary of the unbounded face has length at least n.

11.49. Think of the "rooms" as alcoves off a corridor.

12.1. The answer is obvious. Prove it by induction.

12.2–5. Use the characteristic equation method.

12.8. Partition the pairings according to the partner of person X.

12.9. How many regions does the last circle add?

12.13. For part (a), relate the increase in the number of regions to the crossings involving chords from the new vertex.

12.15–18. Compare with Rabbits and Cadillacs.

12.20. Devise an encoding of $1, 2$-lists as $0, 1$-lists without consecutive 1s, and show that it defines a bijection.

12.21. Use the model of $1, 2$-lists summing to n.

12.22. Model the determination of a $1, 2$-list with sum n as a selection problem.

12.23. When a parking lot of length $m + n$ is filled, the first m spaces are filled, and so are the last n.

12.24. Use strong induction.

12.27. Suppose α is the name of the largest of the k cards that appear at the top. Show that the number of flips until α appears at the top is the same as the number of flips in some pile in which at most $k - 2$ cards appear at the top.

12.33. To solve the recurrence when n is a power of 2, let $b_k = a_{2^k}$, for $k \geq 0$.

12.35. Focus on the element $n \in [n]$.

12.36. What happens when we subtract 1 from the length of each side?

12.39. Partition the pairings according to the partner of person X.

12.40. Relate these arrangements to ballot lists, or group them appropriately so that the Catalan recurrence will apply.

12.42. How can we add the element n to a partition of $[n-1]$ to obtain a partition of $[n]$ with k blocks?

12.43. For part (a), partition the spanning trees into appropriate subsets. Part (b) can be solved using part (a) or by direct argument.

12.44. The chromatic polynomial counts something concerning G_n. Obtain a recurrence for this.

12.46. Treat each partition as an array of dots, with the number of dots in the ith row being the size of the ith largest part. Split these arrangements into two sets with the desired sizes.

12.48. To form a 1, 2-list, we pick some length and then decide for each item in the list of that length whether it is a 1 or a 2. Model this with a generating function so that the coefficient of x^n is the number of ways to do this and get a sum of n.

12.52. Express each sum as a coefficient in a product of generating functions.

12.53. The sum is a coefficient in the product of two generating functions.

12.55–56. Build generating functions using the allowed options for usage of parts of each size.

12.57. Given a partition of n into odd parts with repetitions, eliminate repetitions in a natural way to obtain a partition of n into distinct parts. Explain how to retrieve the original partition into odd parts from a partition into distinct parts.

13.6. Can this method ever list an irrational number?

13.9. Rewrite the condition as the inequality $f(x) \le 1$.

13.13. Compute $y^2 - 2$ in terms of x.

13.14. $\frac{1}{10} = 0\frac{1}{3} + 0\frac{1}{9} + 2\frac{1}{27} + \cdots$.

13.16. For the second part, first figure out what $.1111\cdots$ is.

13.21. Given a set S, consider the set $\{x: -x \in S\}$.

13.22. First give simpler descriptions of each set.

13.23. Caution: $\sup A$ need not be in A, and $\sup B$ need not be in B.

13.25. Show first that $\sqrt{1+n^{-1}} - 1 < n^{-1}$.

13.26. First simplify $(1+a_n)^{-1} - 1/2$.

13.28. Does $\lim y_n$ need to exist?

13.29. For the first part, verify the decreasing property directly. Boundedness below is easy.

13.30. Apply the Monotone Convergence Theorem.

13.38. Relate the binary expansion of a number in T to a subset of $\mathbb{N}$.

13.39. Use decimal or binary expansions.

13.40. Compare n with x^{-1}.

14.7. Try squaring binary numbers directly and comparing with 2. For example, in binary $(1.0)^2 = 1_{(2)} < 2_{(10)}$, while $(1.1)^2 = 10.01_{(2)} > 2_{(10)}$.

14.12. Try $a_n = b_n = 1/\sqrt{n}$.

14.13. Use the definitions of convergence and subsequence.

14.14. To show that $\frac{1}{b_n}$ converges to $\frac{1}{M}$, write $\left|\frac{1}{b_n} - \frac{1}{M}\right| = |M - b_n| \frac{1}{|M||b_n|}$, and bound the denominator by an appropriate constant for sufficiently large n.

14.15. Use contradiction or contrapositive.

14.18. Use induction.

14.19. This is similar to Example 14.9.

14.21. What happens if $\langle x \rangle$ is bounded?

14.24. If $y_n = x_n - 2$, then $y_n \to L$ implies $x_n \to L + 2$.

14.25. Translate to reduce to the case $f(x) = x^2 + c$. Graph this parabola and the line $y = x$ on the same graph. Iterate the function.

14.29. Build a sequence using each value in an infinite set infinitely often.

14.30. What would happen to $\sum x_j$ if the limit were not zero?

14.32. Consider how long the fly travels! (Avoid summing a series.)

14.33. Use partial sums and an $\epsilon/2$ argument.

14.39. Consider Exercise 14.38. Does the decimal expansion eventually repeat?

14.44. For the first part, write $\frac{1}{n(n+1)}$ as $\frac{1}{n} - \frac{1}{n+1}$ to obtain a telescoping series. Then use $\frac{1}{(n+1)^2} < \frac{1}{n(n+1)} < \frac{1}{n^2}$.

14.45. What is the difference between consecutive partial sums?

14.46. Compute the partial sums exactly.

14.50. Compare with $\sum \frac{1}{2n}$.

14.54. First add up enough positive terms to exceed L, then enough negative ones to be less than L, and continue this process.

14.55. The statement is false. Consider $a_k = \pm(1/k)$ for appropriate choices of the signs.

14.58. Read the Approaches at the end of Chapter 14.

14.59. Use the limit comparison test.

14.60. For part (a), use the limit comparison test with $a_n = n^{-2}$.

14.61. Use the limit comparison test with $a_n = n^{-p}$ for an appropriate p.

14.64. Factor b_1 from a partial sum. Write $\frac{b_{k+1}}{b_1}$ as a product of $\frac{b_{j+1}}{b_j}$. Use the hypothesis to obtain bounds in terms of corresponding expressions for $\{a_j\}$.

14.66. First reduce to the case of a small positive rational number x. Then study what happens to the numerator after subtracting the largest reciprocal of a integer that is less than x.

14.68–69. Compare (eventually) with appropriate geometric series.

15.4. Use the Intermediate Value Theorem.

15.6. Consider examples with $f = -g$.

15.8. What happens when $|f|$ is constant?

15.12. For the first part, let $a_n = 1/(2n + 1)$ and $b_n = 1/(2n)$ and draw an appropriate graph for f.

15.13. Deduce the inequality $||x| - |a|| \leq |x - a|$ by squaring both sides.

15.14. Draw a picture like that in Example 15.5 with $a = .5$ and $f(x) = 1/x$.

15.16. Let $\epsilon = 1/n$ in the definition of convergent sequence.

15.18. For part (a), show that the epsilon-delta definition fails with $\epsilon = 1$. Part (b) is similar when $a \neq 0$. For $a = 0$, use the epsilon-delta definition.

15.21. Apply the intermediate value theorem to the difference of the two sides.

15.22. Apply the Intermediate Value Theorem to the function $f - g$.

15.24. Study the behavior for x near both positive and negative infinity, and use the Intermediate Value Theorem.

15.25. Start with $0 \leq (ax - by)^2$ and choose a, b appropriately.

15.29. Consider separately the cases when $x \leq 1$ and $x > 1$. For the second part, reduce to the first part.

15.33. Write $f(1) - f(0)$ as a telescoping sum.

15.34. One method is to find $m(a, \epsilon)$ for the special case where $f(x) = 1/x$ and think about the answer.

15.35. First graph such a function for $k = 3$ to gain understanding.

16.1. Think about the definition of a derivative.

16.3. Think about the interpretation of the derivative as a linear approximation.

16.5. Make f^2 a constant.

16.7. Compute $f'(x)$, using linear approximations.

16.8. Compare with Exercise 16.30.

16.10. One approach is to study small values of n to guess the general formula and then prove it by induction on n.

16.11. Use an $\epsilon/2$ argument and the definition of derivative.

16.14. Multiply the numerator and denominator by an appropriate constant.

16.18. Use the difference quotient definition; prove that the derivative is 1.

16.19. Use the difference quotient definition; prove that the limits must be zero.

16.22. Think of V and r as functions of time and use the chain rule.

16.23. Maximize the function defined by $f(x) = x - x^2$.

16.26. For part (a), apply the induction hypothesis for each possible value of m_k, then choose the best m_k. For part (b), think of a set counted by $\sum_{i<j} m_i m_j$; how does a small change in $\{m_i\}$ without changing $\sum m_i$ affect the size of this set?

16.27. Differentiate the difference of the two functions.

16.28. Use $g(x) = x$.

16.30. Let $g_y(x) = f(x) - yx$. Prove that g_y has a minimum. What happens when $(g_y)'(x) = 0$?